OF THE ELEMENTS

			IIIA	IVA	VA	VIA	VIIA	0
								2 He 4.00260 2
			5 B 10.81 3	6 C 12.011 4	7 N 14.0067 5	8 O 15.9994 6	9 F 18.99840 7	10 Ne 20.179 8
	IB	IIB	13 Al 26.98154 3	14 Si 28.086 4	15 P 30.97376 5	16 S 32.06 6	17 Cl 35.453 7	18 Ar 39.948 8
28 Ni 58.71 16,2	29 Cu 63.546 18,1	30 Zn 65.38 18,2	31 Ga 69.72 18,3	32 Ge 72.59 18,4	33 As 74.9216 18,5	34 Se 78.96 18,6	35 Br 79.904 18,7	36 Kr 83.80 18,8
46 Pd 106.4 18	47 Ag 107.868 18,1	48 Cd 112.40 18,2	49 In 114.82 18,3	50 Sn 118.69 18,4	51 Sb 121.75 18,5	52 Te 127.60 18,6	53 I 126.9045 18,7	54 Xe 131.30 18,8
78 Pt 195.09 32,17,1	79 Au 196.9665 32,18,1	80 Hg 200.59 32,18,2	81 Tl 204.37 32,18,3	82 Pb 207.2 32,18,4	83 Bi 208.9804 32,18,5	84 Po (210)* 32,18,6	85 At (210)* 32,18,7	86 Rn (222)* 32,18,8

63 Eu 151.96 25,8,2	64 Gd 157.25 25,9,2	65 Tb 158.9254 27,8,2	66 Dy 162.50 28,8,2	67 Ho 164.9304 29,8,2	68 Er 167.26 30,8,2	69 Tm 168.9342 31,8,2	70 Yb 173.04 32,8,2	71 Lu 174.97 32,9,2
95 Am (243)* 25,8,2	96 Cm (247)* 25,9,2	97 Bk (249)* 26,9,2	98 Cf (251)* 28,8,2	99 Es (254)* 29,8,2	100 Fm (253)* 30,8,2	101 Md (256)* 31,8,2	102 No (254)* 32,8,2	103 Lr (257)* 32,9,2

Chemistry, Man And Society

THIRD EDITION

Mark M. Jones

Chairman, Department of Chemistry,
Vanderbilt University,
Nashville, Tennessee

John T. Netterville

Chairman, Department of Chemistry,
David Lipscomb College,
Nashville, Tennessee

David O. Johnston

Professor of Chemistry,
David Lipscomb College,
Nashville, Tennessee

James L. Wood

Professor of Chemistry,
David Lipscomb College,
Nashville, Tennessee

SAUNDERS GOLDEN SUNBURST SERIES

SAUNDERS COLLEGE   Philadelphia

Saunders College
West Washington Square
Philadelphia, PA 19105

Chemistry, Man and Society   ISBN   0-03-056766-1

0123   006   9 8 7 6 5 4 3

PREFACE

The third edition of *Chemistry, Man and Society* has been prepared for those students wishing a one-semester course in college chemistry. Extensive revision has reduced the amount and rigor of the theoretical material presented. However, the fundamental approach has been preserved—discoveries are presented along with the consequent thought processes that lead to new discoveries and the modern chemical world. The presentation of applied chemistry has been reorganized and modernized to emphasize the chemistry that is of the greatest interest and use for the college student.

Certainly, there is mystery in chemistry, but to leave the workings of the chemist as a mystery is inevitably to argue that the liberally educated person must be dependent upon the chemist for the chemical decisions that affect society as a whole. This text is based on the belief that the liberal arts student can see and appreciate the chain of events leading from chemical fact to chemical theory and the ingenious manipulation of materials based on the chemical theories. The thoughtful student will then see that the intellectual struggles in chemistry are closely akin to personal intellectual pursuits and will feel that each educated individual can be partner to the ongoing chemical manipulation of the human environment.

The entire presentation in this book is based on stated student needs. Over the years we, through innumerable discussions with our students, have observed an intense interest in the following areas:

1. There is considerable pleasure in understanding a sensible, theoretical explanation of an observed natural phenomenon. Such curiosity appears innate in every person.
2. There is a desire to make highly personal choices as rational as possible. The role of fluoride in tooth structure, for example, is appreciated when understood. A class always perks up when the role of aluminum chloride in a body deodorant is explained, realizing the personal choice that is to be made in the matter.
3. Chemical choices to be made in the immediate environment, in the home, in the automobile, and in the work area are of vital interest. For example, the choice of fuel additives in gasoline proves to be interesting to almost every student.
4. The great environmental questions of a chemical nature also have a compelling interest. The liberal arts student appreciates a sense of awareness in the dilemmas presented in gross pollution problems and in the depletion of natural resources.
5. Energy needs and resources for a growing world population may be the pivotal issue in modern politics. Realizing this, the typical college student is interested in knowing the facts relative to fuels and food.

Although an effort has been made to tailor the text to suit the student with a minimum of scientific training, the approach is sufficiently different to challenge and interest the student with a high school chemistry background. Most of the topics presented in applied chemistry are not covered in high school texts.

Numerous study aids for the benefit of the students have been included. Notes in the margins highlight various ideas presented in the text and add enrichment. Self-tests have been interspersed throughout the text to give the student an opportunity to see whether or not there is comprehension of the material presented. Matching sets to help build and maintain the chemical vocabulary and numerous questions have been included at the end of each chapter. Though some of these questions are based on simple recall, most of them are intended to provoke thought about significant chemical problems, and some of them are open-ended in the sense that they call for a highly critical evaluation of a chemical issue. Important words in the text are presented in boldface type to aid the student in developing the necessary vocabulary. Considerable attention has been given to enriching the illustrative material in order to help the student grasp as quickly as possible the ideas presented.

The authors wish to thank Professor Eugene G. Rochow, Professor David G. Hendricker, and Professor John Otto Olson for their extensive help in reviewing the manuscript.

As in previous texts, we dedicate this effort to our wives and gratefully acknowledge their support and understanding during the preparation of this manuscript.

MARK M. JONES
JOHN T. NETTERVILLE
DAVID O. JOHNSTON
JAMES L. WOOD

CONTENTS

CHAPTER 17

APPENDIX A

APPENDIX B

APPENDIX C

APPENDIX D

CHAPTER 1

THE CHEMICAL VIEW OF MATTER

Look around! Most of the things we use in our daily life are very different from the materials which are an obvious part of our natural surroundings. Practically everything we use has been transformed from a natural state of little or no utility to one of very different appearance and much greater utility. The processes by which the materials of nature can be transformed, and a detailed description of such changes, are highly intriguing. Wherever we look we see matter undergoing change. This is what *chemistry* is all about: *matter,* and the *changes* it undergoes.

An understanding of our environment is necessary if we are to know the consequences of our acts on our environment and, even more critical, if we are to undertake the successful repair of previous damage. Since life itself involves very intricate sequences of changes, each of us has a more personal reason for valuing knowledge of such processes—a desire to gain some insight into the manner in which specific kinds of matter in our food, air, and water can influence our individual health, happiness, and behavior.

Chemical change alters the kinds of matter present.

While all changes in matter are of concern, our attention here will be focused on a particular kind of change called *chemical change.* In any chemical change the starting material is changed into a different kind of matter. Matter may also undergo another kind of change which does not produce a new kind of matter, but simply results in a new form of the same material. This kind of change is a *physical change.*

Since the feature used to recognize a chemical change is the production of a different kind of matter, it is necessary to be able to classify matter into different kinds. In a natural source the kinds of matter are usually mixed together and the separation of such mixtures has to precede their systematic classification. After an examination of the methods of separating such mixtures into their components, we can appreciate some

1

of the problems involved in an accurate definition of the terms "kinds of matter" and "chemical change".

In spite of the difficulties experienced in establishing the exact bounds of chemistry, it will be helpful to think of *chemistry* as *the study of the kinds of matter and the changes that transform one kind of matter into another.*

Common examples of
A. Chemical Change: burning, rusting, souring
B. Physical Change: melting, boiling, reshaping

PURE SUBSTANCES

Most natural samples of matter are complex mixtures. Sometimes it is easy to see the various ingredients in a mixture, such as the bits of sand and clay in a sample of soil, or the glittering crystals (mica and quartz) and dark areas (feldspar and magnetite) in a piece of granite. More often, however, it is not apparent to the casual observer that a mixture is made up of distinctly separate ingredients. For example, it is only by an experiment that we can establish that the air we breathe is a mixture.

When a mixture is separated into its components, the components of the mixture are said to be purified. However, most efforts at separations are incomplete in a single operation or step, and repetition of the purification process results in a better separation. Ultimately in such a procedure the experimenter arrives at pure substances, samples of matter that cannot be further purified. For example, if sulfur and iron powder are ground together to form a mixture, the iron can be separated from the sulfur by repeated stirrings of the mixture with a magnet. When the mixture is stirred the first time and the magnet removed, much of the iron is removed with it, leaving the sulfur in a higher state of purity. However, after just one stirring the sulfur may still have a dirty appearance due to a small amount of iron that remains. Repeated stirring with the magnet, or perhaps the use of a very strong magnet, will finally leave a bright yellow sample of sulfur that apparently cannot be purified further by this technique. In this purification process a property of the mixture, its color, is a measure of the extent of purification. After the bright yellow color is obtained, it could be assumed that the sulfur has been purified. Drawing a conclusion based on one property of the mixture may be dangerous because other methods of purification might change some other properties of the sample. It is only safe to call the sulfur a pure substance when all possible methods of purification fail to change its properties. This assumes that all pure substances have a set of properties by which they can be recognized, just as a person can be recognized by a set of characteristics. *A pure substance, then, is a kind of matter with properties that cannot be changed by further purification.*

There are some naturally occurring pure substances. Rain is very nearly pure water, except for small amounts of dust and air. Gold, diamond, and sulfur are also found in very pure form. These substances are special cases. Man, a complex assemblage of mixtures, lives in a world of mixtures—eating them, wearing them, living in houses made of them, and making most of his tools out of them.

Although naturally occurring pure substances are not common, it is possible to produce many pure substances from natural mixtures. Relatively pure substances are now very common as a consequence of the development of modern purification techniques. Common examples are sugar, table salt (sodium chloride), copper, sodium bicarbonate, nitrogen,

Bits of iron and sulfur mixed

Iron — Sulfur

Purification separates the kinds of matter.

Most materials in nature are mixtures; a few are relatively pure substances.

dextrose, ammonia, uranium, and carbon dioxide—to mention just a few. In all, about four million pure substances have been identified and cataloged.

DEFINITIONS: OPERATIONAL AND THEORETICAL

Chemistry begins with observations and experiments.

The preceding definition given for a pure substance is an *operational* definition, or a definition in terms of specific experiments or operations. That is, if further purification efforts are unsuccessful in changing the properties of a substance, it is said to be a pure substance. It is evident then that operational definitions result from performing operations or tests on matter and summarizing the results in a statement. For example, pure sulfur is a yellow substance that boils at 833°F and has a density of 129 pounds per cubic foot. When all the properties of pure sulfur have been listed, we find that the pure substance has been characterized in a way that distinguishes it from any other pure substance.

A pure substance also can be defined in theoretical terms; that is, in terms of the molecules, atoms, and subatomic particles that compose it. Both types of definitions are important in the study of chemistry and both will be used in the presentation of this text. The *theoretical* definition will follow the development of the theory on which it is based.

---

1. Four common materials that could not be pure substances are: SELF-TEST 1-A*

   (a) _____   (c) _____

   (b) _____   (d) _____

2. Which must come first, the operational definition or the theoretical defini-

   tion? _____ .

3. Four common materials that are very nearly pure substances are:

   (a) _____   (c) _____

   (b) _____   (d) _____

4. Two different pure substances could have all properties identical. True (   )
   or False (   )

SEPARATION OF MIXTURES INTO PURE SUBSTANCES

The separation of mixtures is usually more difficult than the magnetic separation of iron and sulfur described previously. Most beginning chemistry students would find it a bewildering task to separate a piece of granite into pure substances. Indeed, a trained chemist would find this a difficult assignment. Since each of the pure substances in the granite has a set of properties unlike those of any other pure substance, it should be possible to use these properties to separate the pure substances, just as the attraction of iron to a magnet is used to separate it from sulfur.

Many different methods have been devised to separate the pure substances in a mixture. In each case, differing properties of the pure substances are exploited to effect the separation. Figure 1–1 illustrates four commonly used methods: chromatography, distillation, recrystallization and filtration.

ELEMENTS AND COMPOUNDS

Experimentally, pure substances can be classified into two categories: those that can be broken down by chemical change into simpler pure substances and those that cannot. Table sugar (sucrose), a pure substance, will decompose when heated in the oven, leaving carbon, another pure substance, and evolving water. No chemical operation has ever been devised that will decompose carbon into simpler pure substances. Obviously sucrose and carbon belong to two different categories of pure substances. Only 89 substances found in nature cannot be reduced chemically to simpler substances; 17 others are available artificially. These 106 substances are called **elements.** Pure substances that can be decomposed into two or more different pure substances are referred to as **compounds.** Even though there are presently only 106 known elements, there appears to be no practical limit to the number of compounds that can be made.

Elements are the basic building blocks of the universe and the world in which we live. Table 1–1 (p. 8) lists the properties of some of the common elements; a complete list of the elements is found inside the back cover of this text. A number of elements are found as the elementary substance in nature; examples include gold, silver, oxygen, nitrogen, carbon (graphite and diamond), copper, platinum, sulfur, and the nobel gases (helium, neon, argon, krypton, xenon, and radon). Many more elements, however, are found chemically combined with other elements in the form of compounds.

Elements in compounds no longer show all of their original, characteristic properties, such as color, hardness, and melting point. Consider ordinary sugar, which is properly called sucrose, as an example. It is made up of three elements: carbon (which is usually a black powder), hydrogen (the lightest gas known), and oxygen (a gas necessary for respiration). The compound sucrose is completely unlike any of the three elements; it is a white crystalline powder which, unlike solid carbon, is readily soluble in water.

A careful distinction should be made between a compound of two or more elements and a mixture of the same elements. The two gases, hydrogen and oxygen, can be mixed in all proportions. However, these two elements can and do react chemically to form the compound water.

Scanning electron micrograph of particles of asbestos filtered from a sample of air by a small pore filter. (Courtesy of Dr. Spurny, Institut für Aerobiologie.)

These are operational definitions of "element" and "compound."

Figure 1–3 lists the most common elements in the universe and on earth.

Compounds have a fixed composition in terms of the elements they contain.

Paper Chromatography

Water begins to move up paper

Water moves by ink spot, carrying different components of the ink at different rates

Separation of 3-component ink complete

Dry absorbent paper

Ink spot   Water

a   b   c

Distillation

Water out

Condenser

Cooling water in

Recrystallization

a. Dissolve solid in minimum quantity of hot solvent.

b. Cool solution (generally in ice + water). New crystals form.

Collect new crystals on filter

Solution of compound plus impurities

c. Collect new purified crystals on filter.
d. Repeat process if necessary.

Filtration

Solid + liquid

Solid is caught on filter

Liquid passes through and is collected

Figure 1—1 Four methods of purifying mixtures of elements and compounds. **Paper chromatography.** Owing to the absorbent character of paper, water moves against gravity and carries the ink dyes along its path. If the ink dyes move at different rates because of differing attraction to the paper, they will be separated in the developed chromatogram. **Distillation.** Sodium chloride dissolves in water to form a clear solution. When heated above the boiling point (indicated by thermometer), water will vaporize and pass into the condenser. Cool water injected into the glass jacket of the condenser circulates over the inner tube, causing the steam to liquefy and collect in the flask. In this simple example pure water collects in the receiving flask, while the salt remains in the boiling flask. **Recrystallization.** This can be used to separate some solid mixtures. **Filtration.** The separation of a solid from a liquid by filtration.

Figure 1—2 Distillation. Some of the most useful purification techniques copy processes in nature and date back to alchemical times. Distillation allows a more volatile substance to be separated from a less volatile one. In this case, alcohol is partially separated by evaporation from water and other ingredients. One distillation can produce a mixture that is 40 percent alcohol from one that is only 12 percent. Further distillations would produce an even better separation.

Not only does water exhibit properties peculiar to itself and different from those of hydrogen and oxygen, but it also has a definite percentage composition by weight (88.8 percent oxygen and 11.2 percent hydrogen). This is a second distinct difference between compounds and mixtures: *compounds have a definite percentage composition by weight of the combining elements.*

CHEMICAL CHANGES AND PHYSICAL CHANGES

All changes in matter can be distinguished as chemical or physical changes. A chemical change occurs when substances are used up in the production of different ones. For example, the burning of carbon (as charcoal) to form carbon dioxide is a chemical change. A physical change produces no new substances; it only alters the original state of the elements or compounds. An example is the boiling of water, a change of

Matter can exist in three states: solid, liquid, or gas, depending upon conditions.

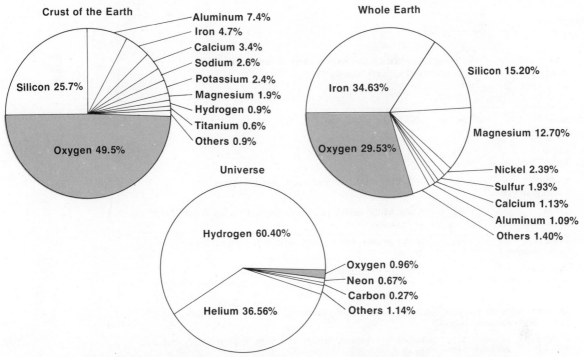

Figure 1–3 Relative abundance (by mass) of the most common elements in the earth's crust, the whole earth, and the universe. Note that the earth's crust differs significantly from the cosmic array of elements.

liquid water into gaseous water. Some additional examples of chemical and physical changes are the following:

Chemical Changes	*Physical Changes*
Rusting of iron	Evaporation of a liquid
Burning gasoline in an automobile engine	Melting of iron
	Drawing metal wire
Preparation of caramel by heating sugar	Freezing of water
Preparation of iron from its ores	Crystallizing salt from sea water
	Grinding or pulverizing of a solid
Solution of copper in nitric acid	Cutting and shaping wood

Nuclear changes, which are basic to radioactivity, atomic bombs, hydrogen bombs and solar energy, are a third type of change. Nuclear changes involve relatively large amounts of energy.

The properties of a new substance identify it. There are physical properties and there are chemical properties. A chemical property of an element or compound describes its ability to undergo chemical change. It follows, then, that to determine a chemical property we must attempt a chemical change. Physical properties, on the other hand, are characteristics of a substance which can be determined without chemically changing the element or compound into some other element or compound. Physical properties of a pure substance include such things as melting temperature, boiling temperature, density (weight of the substance in a given volume), color, physical state, crystalline form, and magnetic properties.

It should be emphasized that not all of the properties of a sample need to change in order to establish that a chemical change has occurred.

TABLE 1—1   Some Common Elements*

Metals

(A metal is a good conductor of electricity, can have a shiny or lustrous surface, and in the solid form usually can be deformed without breaking.)

Name	Symbol	Properties of Pure Element
Iron Latin, *ferrum*	Fe	strong, malleable, corrodes
Copper Latin, *cuprum*	Cu	soft, reddish-colored, ductile
Sodium Latin, *natrium*	Na	soft, light metal, very reactive, low melting point
Silver Latin, *argentum*	Ag	shiny white metal, relatively unreactive, good conductor of electricity
Gold Latin, *aurum*	Au	heavy yellow metal, very unreactive, ductile, good conductor
Chromium	Cr	resistant to corrosion, hard, bluish-gray

Nonmetals

(A nonmetal is often a poor conductor of electricity, normally lacks a shiny surface, and is brittle in crystal-solid form.)

Hydrogen	H	colorless, odorless, occurs as a very light gas (H_2), burns in air
Oxygen	O	colorless, odorless gas (O_2), reactive, constituent of air
Sulfur	S	odorless, yellow solid (S_8), low melting point, burns in air
Nitrogen	N	colorless, odorless gas (N_2), rather unreactive
Chlorine	Cl	greenish yellow gas (Cl_2), very sharp choking odor, poisonous
Iodine	I	dark purple solid (I_2), sublimes easily

*Chemists usually use the symbol rather than the name of the element. In addition to denoting the element, the chemical symbol has a very specialized meaning which is described later in this chapter. A complete list of the elements with the symbols can be found inside the back cover of this book.

For example, if table salt is treated with sulfuric acid and heated strongly, the solid product of this reaction is much like table salt in its appearance, but it has a different melting point and different chemical properties. A change in only one significant property usually indicates that a newly formed substance has resulted from a chemical change.

While most changes can easily be classified on the basis of whether a different substance is produced or not, some changes are difficult to put into such neat categories. Examples include those which accompany the dissolving of a solid in a liquid. We often apply a simple test to determine the type of change. If the solid can be recovered unchanged by the evaporation of the liquid, the solution process was a physical change. Thus, table salt or sugar dissolves in water, but either solid can be recovered by careful evaporation, so such solutions are said to be formed by physical changes although the interaction between constituents is considerable. An example of a chemical change resulting from solution is observed when the gas hydrogen chloride dissolves in water. In this case, the properties of the solution are very different from those of the constituents, and evaporation of the solution is not a feasible way to recover the starting substances completely. In general, however, solution processes are of such complexity that their detailed discussion is inappropriate at this point.

The classification of solution changes illustrates some difficulties about exact definitions as used in chemistry. Like the stamp collector arranging stamps and the cook arranging foods in the kitchen, we shall find our work easier if we group similar things and phenomena into categories. This makes it easier to organize knowledge. Of course, we try to define the limits of our categories so there is no question about what fits into each classification, but nature sometimes presents situations that are very hard to classify. There are borderline cases, and all efforts to be specific and limiting will usually meet with an item or idea that defies exact classification. However, a few exceptions do not stop us from originating simple categories for the things we observe.

WHY STUDY PURE SUBSTANCES AND THEIR CHEMICAL CHANGES?

A knowledge of chemical changes allows us to carry them out purposefully to achieve desired results.

Perhaps by now you are wondering why we should be interested in elements and compounds and their chemical properties. There are two basic reasons. The first is the belief that the knowledge of chemical substances and chemical changes will allow us to bring about desired changes in the nature of everyday life. Two hundred years ago most of the materials surrounding a normal person were only physically changed from the way the material occurred in nature. Only a few of his useful materials, such as iron, were the product of man's control over chemical change. By contrast, synthetic fibers, plastics, drugs, latex paints, detergents, new and better fuels, and photographic films are but a few of the materials produced by controlled chemical change. (We shall return to examine the chemistry of many of these later.) You will discover that it is difficult to find more than a few objects in your home that have not been altered by a desirable chemical change. Not only is it important to bring about desirable changes, but in the areas of toxicity and pollution it is important to avoid undesirable ones. The second reason for chemical studies is even more important—curiosity. Chemicals and chemical change are a part of nature that is open to investigation, and, like the mountain climber, we will find this task both interesting and challenging—simply because it is there. If we hope to understand matter, it is obviously necessary to discover the simplest forms of matter and to study their interactions.

STRUCTURE OF MATTER AND RESULTING PROPERTIES

For reasons which are partly theoretical and partly practical, we are deeply interested in the structure of matter—that is, the minute parts of matter and how these parts are fitted together to make larger units. Why does an element or compound have the properties it has? Why does one element or compound undergo a change that another element or compound will not undergo? Inanimate matter is the way it is because of the nature of its parts. A watch is what it is because of the nature of its individual parts. So is a car, a refrigerator, and the salt in your salt shaker. The individual parts (smaller than the whole) are what determine the nature (actions and properties) of the whole. The most basic parts of matter, as we shall see, are very, very small. If we even hope to understand the nature of matter, it is absolutely necessary that we have some understanding of these minute parts and how they are related to each other.

A very large portion of the research in chemistry being done today is aimed at sorting out and elucidating the structure of matter. Indeed, the basic theme of this text is the relationship between the structure of matter and its properties. This theme of structure and related properties is of great interest because if we know exactly how and with what strength the minute parts of matter are put together, we can discover exact relationships between structure and properties. Armed with this understanding, we can make changes that result in new substances and predict the properties of these substances. Such knowledge can save many months of trial and error which otherwise may be required to prepare a product with the desired qualities. While this day of predicting chemical changes based on structural characteristics has not completely arrived, such significant advances have been made that the practice of modern chemistry would not be possible without such knowledge.

Samples of matter large enough to be seen and felt and handled, and thus large enough for ordinary laboratory experiments, are called **macroscopic** samples, in contrast to **microscopic** samples, which are so small that they have to be viewed with the aid of a microscope. The structure of matter that really interests us, however, is at the **submicroscopic** level. Our senses have no direct access into this small world of structure, and any conclusions about it will have to be based on circumstantial evidence gathered in the macroscopic and microscopic worlds.

Submicroscopic structures help to explain chemistry.

We can now extend our concept of the science of chemistry. It is that science which investigates the properties and changes of pure substances. Chemistry is also deeply concerned with structure, both **macrostructure** and **submicrostructure,** in an effort to give plausible reasons for properties and change, with emphasis on chemical change.

Scope of chemistry.

FACTS, LAWS, AND THEORIES

A **scientific fact** is an observation about nature that can usually be reproduced at will. For example, carbon in some forms will readily burn in the presence of air. If you have any doubt about this fact it is easy enough to set up an experiment that will readily demonstrate the fact anew. You would only need some carbon, air, and a source of heat. The repeatability of a scientific fact distinguishes it from a historical fact, which obviously cannot be reproduced. Of course, some scientific facts are also historical facts—such as the movement of heavenly bodies—and are not repeatable at will.

A scientific fact can be verified independently of any particular observer.

Often a large number of related scientific facts can be summarized into broad, sweeping statements called natural **laws.** The law of gravity is a classic example of a natural law. This law, that all bodies in the universe have an attraction for all other bodies which is directly proportional to the product of their masses and inversely related to the square of their separation distance, summarizes in one sweeping statement an enormous number of facts. It implies that any object lifted a short distance from

A scientific law summarizes a large number of related facts.

the surface of the earth will fall back if released. Such a natural law can only be established in our minds by inductive reasoning; that is, you conclude that the law applies to all possible cases, since it applies in all of the cases studied or observed. A well-established law allows us to predict future events. When convinced of the generality of a scientific law, we may reason deductively, based on our belief that if the law holds for all situations, it will surely hold for the events in question.

A scientific law predicts what *will* happen. A governmental law describes what people *should do*.

The same procedure is used in the establishment of chemical laws, as can be seen from the following example. Suppose an experimenter carried out hundreds of different chemical changes in closed, leakproof containers, and suppose further that he weighed the containers and their contents before and after each of the chemical changes. Also, suppose that in every case he found that the container and its contents weighed exactly the same before and after the chemical change occurred. Finally, suppose that he repeated the same experiments over and over again, obtaining the same results each time, until he was absolutely sure that he was dealing in reproducible facts. It can be understood then that the experimenter would reasonably conclude: *"All chemical changes occur without any detectable loss or gain in weight."* This is indeed a basic chemical law and serves as one of the foundations of modern scientific theory.

Theories are ideas or models used to explain facts and laws.

After a natural law has been firmly established, its explanation must be sought. Chemists are not satisfied until they have explained chemical laws logically in terms of the submicroscopic structure of matter. This is indeed a difficult process, and the progress, until recent time, has been painfully slow because of our lack of direct access into the submicroscopic structure of matter with our physical senses. All we can do is to collect information in the macroscopic world in which we live, and then try, by circumstantial reasoning, to visualize what the submicroscopic world must be like in order to explain our macroscopic world. Such a visualization of the submicroscopic world is called a *theoretical* model. If the theoretical model is successful in explaining a number of chemical laws, a major scientific theory is built around it. The atomic theory and the electron theory of chemical bonding are two such major theories, and both will be discussed in relation to chemical laws in later chapters.

Chemical theories use the concepts of atoms to explain chemical observations.

Consider again the chemical law concerning the conservation of weight in chemical changes. What is a possible theoretical model that could explain this law? If we assume that matter is made up of atoms, and that these atoms are grouped in a particular way in a given pure substance, we can reason that a chemical change is simply the rearrangement of these atoms into new groupings, and, consequently, into new substances. If the same atoms are still there, they should have the same individual characteristic weight, and hence the law of conservation of weight is explained (Figure 1–4).

For a scientific theory to have much value, it must not only explain the pertinent facts and laws at hand, but it must also be able to explain new facts and laws that are obviously related. If the theory cannot consistently perform in this manner, it must be revised until it is consistent, or, if this is not possible, it must be completely discarded. You must not allow yourself to think that this process of trying to understand nature's secrets is nearing completion. The process is a continuing one.

The word theory is often used in a different sense from the one discussed above. If a student is absent from the chemistry class, his neighbor

Figure 1—4 Atomic explanation for weight conservation in chemical changes. If substances are composed of atoms, if chemical changes are atomic rearrangements in which atoms are not destroyed or changed into other atoms, and if atoms have fixed weights, the products in a reaction must weigh the same as the starting substances. Hence the law of conservation of weight is explained in terms of a theoretical model.

Figure 1—5 Direct observation stops at the microscopic level. Convinced of structure beyond the microscopic level, the chemist employs circumstantial evidence to construct the world of molecules, atoms, and subatomic parts in his mind's eye.

may say, "I do not know why he is absent, but my theory is that he is sick and unable to come to class." This use of the word theory is similar to the concept of the general scientific theory in that there is an element of speculation in both uses. But the speculative guess of the student about his absent friend is vastly different from the broad theoretical picture that is able to explain a number of laws. The reader should be alert for the considerable amount of confusion that has resulted from the diffferent meanings associated with this word. In this book the word *hypothesis* is used when speaking of a speculation about a particular event or set of data, reserving the word *theory* for the broad imaginative concepts that have gained wide acceptance.

THE LANGUAGE OF CHEMISTRY

"chemistry"

Chemical symbols are abbreviations for the different elements.

A *mole* contains 6.023 $\times$ 10^{23} particles.

Symbols, formulas, and equations are used in chemistry to convey ideas quickly and concisely. These shorthand notations are merely a convenience, and contain no mysterious concepts that cannot be expressed in words. Certain characters are used often, and a general familiarity with them is desirable.

A *chemical symbol* for an element is a one- or two-letter term, the first letter a capital and the second a lower case letter. The symbol is a sign for three concepts. First, the symbol stands for the element in general. H, O, N, Cl, Fe, Pt are shorthand notations for the elements hydrogen, oxygen, nitrogen, chlorine, iron, and platinum, respectively, and it is useful and timesaving to substitute these symbols for the words themselves in describing chemical changes. Some symbols originate from Latin words (such as Fe, from *ferrum,* the Latin word for iron); others came from English, French or German names. Second, the chemical symbol stands for a single atom of the element. The *atom* is the smallest particle of the element that can enter into chemical combinations. Third, the elemental symbol stands for a mole of the atoms of the element. The *mole* is a term in chemical usage (it is derived from the Latin for "a pile of" or "a quantity of") that means the quantity of a substance that contains 602 sextillion identical particles. Just as a dozen apples would be 12 apples, a mole of atoms would be 602,300,000,000,000,000,000,000 atoms or 6.023×10^{23} atoms. It turns out that a mole of atoms is usually a convenient amount for laboratory work. Thus, the symbol Ca can stand for the element calcium, or a single calcium atom, or a mole of calcium atoms. It will be evident from the context which of these meanings is implied.

Atoms can unite (bond together) to form molecules. A *molecule* is the smallest particle of an element or a compound that can have a stable existence in the close presence of like molecules. One or more of the same kind of atom can make a molecule of an element. For example, two atoms of hydrogen will bond together to form a molecule of ordinary hydrogen, and eight sulfur atoms will form a single molecule. Subscripts in a chemical formula show the number of atoms involved: H_2 means a hydrogen molecule is composed of two atoms, and S_8 means a sulfur molecule is composed of eight atoms. The noble gases, such as helium, He, have monatomic molecules (monatomic = one atom).

When unlike atoms combine, as in the case of water (H_2O) or sulfuric acid (H_2SO_4), the formulas tell what atoms and how many of each are present in the compound. For example, H_2SO_4 molecules are composed of two hydrogen atoms, one sulfur atom, and four oxygen atoms. A formula can stand not only for the substance itself, but also for one molecule of the substance or for a mole of such molecules, or for the substance in general, depending on the context.

H_2SO_4

4 atoms of oxygen

1 atom of sulfur

2 atoms of hydrogen

When elements or compounds undergo a chemical change, the formulas, arranged in the form of a **_chemical equation,_** can present the information in a very concise fashion. For example, carbon can react with oxygen to form carbon monoxide. Like most solid elements, carbon is written as though it had one atom per molecule; oxygen exists as diatomic (two-atom) molecules, and carbon monoxide molecules contain two atoms, one each of carbon and oxygen. Furthermore, one oxygen molecule will combine with two carbon atoms to form two carbon monoxide molecules. All of this information is contained in the equation

Chemical equations summarize information on chemical reactions in a concise fashion.

$$2C + O_2 \longrightarrow 2CO$$

The arrow is often read "yields"; the equation then states the following information:

(a) carbon plus oxygen yields carbon monoxide
(b) two atoms of carbon plus one diatomic molecule of oxygen yield two molecules of carbon monoxide
(c) two moles of carbon atoms plus one mole of diatomic oxygen molecules yield two moles of carbon monoxide

The number written before a formula, the **_coefficient,_** gives the amount of the substance involved, while the **_subscript_** is a part of the definition of the pure substance itself. Changing the coefficient only changes the amount of the element or compound involved, whereas changing the subscript would necessarily involve changing from one substance to another. For example, 2CO means either two molecules of carbon monoxide or two moles of these molecules.

MEASUREMENT

The heart and soul of reliable scientific facts, and, consequently, the basis for laws and theories, is the accurate measurement of weights, volumes, times, lengths, and other quantities. Perhaps more than any other person, Antoine Lavoisier, a French chemist who lived in the late 1700's, put chemistry on a quantitative basis by very accurately measuring the weights of materials involved in chemical reactions. Because of the significance of his work, Lavoisier has been called the Father of Modern Chemistry. The reproducible data which he obtained led directly to the formulation of several laws concerning chemical change.

Since 1960, a coherent system of units known as the International System of Units, abbreviated SI and bearing the authority of the International Bureau of Weights and Measures, has been in effect and is gaining acceptance among scientists. It is an extension of the metric system which began in 1790, and assigns each physical quantity an SI unit uniquely. We shall use the more familiar metric system in most of our discussions; many metric units are identical with SI units.

The metric system has one great advantage over older systems: it is simple. Its simplicity lies in its organized structure and its decimal

Figure 1—6 One meter = 39.4 inches.

counting, which require relatively few terms to be remembered. For example, units such as those of length are defined in such a way that a larger unit is ten (or some power of ten) times larger than a smaller unit. A *meter,* a unit of length equal to 39.4 inches, is 100 centimeters, and a centimeter is 10 millimeters. Making a conversion from one unit of length to another in the metric system merely involves shifting the decimal point. Compare this to the complexity of the conversion of miles to inches, wherein one would probably employ the factors 12 and 5280. Instead of having a number of different root words for length, as in the English system (league, mile, rod, yard, foot, inch, mil), the metric system has just one, the meter. With suitable prefixes, the meter can be expressive of a unit useful to the watchmaker (millimeter = 0.001 meter) or to the distance runner (kilometer = 1000 meters).

While the convenience of the metric system can scarcely be overstated, it should be emphasized that the science of chemistry need not rest on this system of measurement. Measurements made in the English system can be and are just as accurate. However, in the metric system only four basic units will be required in this course: a unit of length, the *meter;* a unit of volume, the *liter;* a unit of mass (weight), the *gram;*

Accuracy of English and metric systems of measurement is the same, but metric is easier to use.

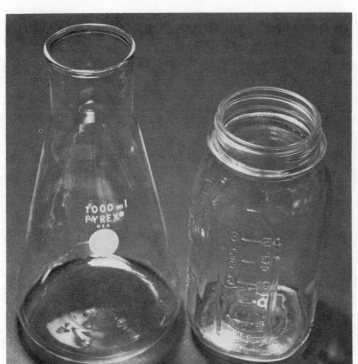

Figure 1—7 The volume of a liter is slightly more than that of a quart. One liter = 1.06 quarts.

Figure 1–8 Thermometer scales are defined in terms of the expansion of common materials such as mercury, and in terms of fixed reference points such as the changes of state of water and other common materials. It is only a matter of preference and convenience whether one scale or another is used.

and a unit of temperature, the *Celsius degree.* English equivalents of the first three are:

$$1 \text{ meter} = 39.4 \text{ inches (Figure 1-6)}$$
$$1 \text{ liter} = 1.06 \text{ quarts (Figure 1-7)}$$
$$1 \text{ gram} = 0.0352 \text{ ounce}$$

Temperature scales are defined in terms of the behavior of samples of matter. For example, the familiar Fahrenheit scale defines the temperature at which water freezes to be 32°F and that at which it boils to be 212°F. Thus there are 180 (212 − 32) Fahrenheit degrees between the freezing and boiling points of water. Most scientists prefer the Celsius scale, which defines the freezing point of water to be 0°C and the boiling point, 100°C. There are 100 Celsius degrees in this same temperature range (Figure 1–8).

For the interested reader and those who need to understand additional aspects of the metric system to support laboratory work, more information is presented in Appendices A, B, and C.

Mass is a measure of the amount of matter in a body, whereas weight is a measure of the attraction of the body for the earth. The weight of the body varies with its distance from the earth, whereas the mass remains constant. Unless otherwise stated, the term "weight" refers to the earth-surface weight.

Figure 1–9 Units of measure are selected for convenience; the millimeter for the small parts of this tiny radio (above), and the kilometer for measuring intercontinental distances. The metric system simply adds the ease of using multiples of 10.

1. The two large divisions into which elements can be divided are:

   (a) _____ and (b) _____

   A half dozen examples of each are:

   (a) _____   (a) _____

   (b) _____   (b) _____

   (c) _____   (c) _____

   (d) _____   (d) _____

   (e) _____   (e) _____

   (f) _____   (f) _____

2. How many elements are presently listed in the periodic table? _____

3. A compound has properties that are combinations of the elemental properties. True ( ) or False ( )

4. Four chemical changes not listed in the chapter are:

   (a) _____

   (b) _____

   (c) _____

   (d) _____

5. Four physical changes not listed in the chapter are:

   (a) _____

   (b) _____

   (c) _____

   (d) _____

6. A chemical change always produces a new _____.

7. _____ structures explain chemical properties.

8. Put in order of decreasing size: (1) microscopic, (2) molecular and (3) macroscopic.

   (a) _____, (b) _____, (c) _____

9. Arrange from most abstract to general to specific: laws, facts, and theories.

   (a) _____, (b) _____, (c) _____

10. Consider the equation: $2Na + 2HCl \longrightarrow H_2 + 2NaCl$. Explain what is meant by the symbols:

(a) Na _____

(b) 2Na _____

(c) HCl _____

(d) $\longrightarrow$ _____

(e) H_2 _____

(f) 2NaCl _____

MATCHING SET

_____	1. produces a new type of matter	(a)	filtration
_____	2. air	(b)	chemical change
_____	3. unchanged by further purification	(c)	element
_____	4. used to separate a solid from a liquid	(d)	properties of pure substance
_____	5. cannot be reduced to simpler substances	(e)	mixture
_____	6. scientific theory	(f)	used to explain facts and laws
_____	7. symbol for iron	(g)	O_3
_____	8. mole of atoms	(h)	CO
_____	9. molecule containing three oxygen atoms	(i)	Fe
_____	10. carbon monoxide	(j)	one meter
_____	11. 100 centimeters	(k)	6.023×10^{23} atoms

QUESTIONS

1. Name as many materials as you can that you have used during the past day that were not in any way the result of a man-made chemical change.

2. In pottery making, an object is shaped and then baked. Which part of this process is chemical and which part is physical?

3. Identify the following as physical or chemical changes. Justify in terms of the operational definitions for these types of changes.

   (a) formation of snow flakes
   (b) rusting of a piece of iron

(c) ripening of fruit
(d) fashioning a table leg from a piece of wood
(e) fermenting grapes

4. Would it be possible for two pure substances to have exactly the same set of properties? Give reasons for your answer.

5. Chemical changes can be both useful and destructive to man's purposes. Cite a few examples of each kind with which you have had personal experience. Also give observed evidence that each is indeed a chemical change and not a physical change.

6. Name two pure substances that are used at the dinner table. Identify each as an element or a compound.

7. Classify each of the following as a physical property or a chemical property. Justify in terms of operational definitions.

(a) density
(b) melting temperature
(c) substance decomposes into two elements upon heating
(d) electrical conductivity of a solid
(e) the substance does not react with sulfur
(f) ignition temperature of a piece of paper

8. Classify each of the following as element, compound, or mixture. Justify each answer.

(a) mercury   (e) ink
(b) milk   (f) iced tea
(c) pure water   (g) pure ice
(d) a tree   (h) carbon

9. Which of the materials listed in Question 8 can be pure substances?

10. Explain how the operational definition of a pure substance allows for the possibility that it is not actually pure.

11. Why do theoretical definitions come after operational definitions in a particular concept?

12. Suggest a method for purifying water slightly contaminated with a dissolved solid.

13. Distinguish between theory and law as used in chemistry.

(a) Which has a better chance of being true? Why?
(b) Which summarizes? Which explains?

14. What is the most abundant element in the universe?

15. Given the sentence below, express a chemical reaction in terms of symbols that the chemist uses to convey the same information. "One nitrogen molecule containing two nitrogen atoms per molecule reacts with three hydrogen molecules containing two hydrogen atoms per molecule to produce two ammonia molecules containing one nitrogen and three hydrogen atoms per molecule."

16. Aspirin is a pure substance, a compound of carbon, hydrogen, and oxygen. If two manufacturers produce equally pure aspirin samples, what can be said of the relative worth of the two products?

17. Is it possible to have a mixture of two elements and to also have a compound of the same two elements? Explain. Can you think of an example?

18. How is the salt content of the sea related to the purity of rain water? What method of purification does nature employ in the purification of rain water?

19. Give an example of a chemical fact.

20. Give an example of a chemical law.

21. Give an example of an assumption made to explain an observed fact.

22. How many times do you think a given experiment should give a result before a scientific fact is established? How many failures would you require before rejecting the "fact"?

23. Suppose a wife and children discover the family car missing on returning from an afternoon ball game, even though the father rode to work with a neighbor that morning. The mother says to the children, "Don't worry, Dad must have had an unexpected need for the car and got it after lunch." Would you call the statement made by the mother a theory or a hypothesis? Why?

24. From the molecular formulas given below, tell what kind of atoms and how many of each kind are present in a molecule. (You may look up the names of any elements whose symbols you do not know in the list printed inside the cover.)

$$SO_3, HCl, NH_3, H_2S$$

25. Describe in words the chemical process which is summarized in the following equation:

$$2Na + Cl_2 \longrightarrow 2NaCl$$

26. How many *atoms* are present in each of the following:
    (a) one mole of He
    (b) one mole of Cl_2
    (c) one mole of O_3

27. Would you think that tea in tea bags is a pure substance? Use the process of making tea to make an argument for your answer. How would your argument apply to instant tea?

28. Do elements sometimes retain their physical properties in the formation of compounds? Give two examples to support your argument.

29. What is the most fundamental assumption relative to structure and properties in chemical theory?

30. Consider the equation:

$$2C_7H_{14} \quad + \quad 21O_2 \quad \longrightarrow \quad 14CO_2 + 14H_2O$$

Heptane	Oxygen	Carbon	Water
(a component		dioxide	
of gasoline)			

Write out in words as much of the information presented in this equation as you can decipher.

31. How tall are you in meters?

32. What is your weight in kilograms?

33. Which is colder: 0°C or 0°F?

34. Convert the distance of your last auto trip into kilometers.

35. Library assignment. Look up other purification methods such as sublimation, extractions, and zone refining, and tell in each case how the method is used in the separation of pure substances.

ATOMS

Democritus

Since the writings of Leucippus and Democritus have been destroyed, we know about their ideas only by recorded opposition to atoms and by a lengthy poem (55 B.C.) by the Roman poet, Lucretius.

THE GREEK INFLUENCE

The ancient Greeks recorded the first theory of atoms. Leucippus and his student, Democritus (460–370 B.C.), argued for the concept of atoms. Democritus used the word *atom* to describe the ultimate particles of matter, particles that could not be divided further. He reasoned that in the division of a piece of matter, such as gold, into smaller and smaller pieces, one would ultimately arrive at a tiny particle of gold that could not be further divided and still retain the properties of gold. The atoms that Democritus envisioned representing different substances were all made of the same basic material. His atoms differed only in shape and size.

Democritus used his concept of atoms to explain the properties of substances. For example, the high density and softness of lead could be explained in terms of lead atoms that pack very close together like marbles in a box and move easily one over the other. Iron was known to be a less dense metal that is quite hard. Democritus argued that the properties of iron resulted from atoms shaped like cork screws, atoms that would entangle in a rigid but relatively lightweight structure. Although his concept of the atom was limited, Democritus did explain in simple terms some well-known phenomena such as the drying of clothes, how moisture appears on the outside of a cold glass of water, how an odor moves through a room, and how crystals grow from a solution. He imagined the scattering or collecting of atoms as needed to explain the events he saw. All atomic theory has been built on the assumption of Leucippus and Democritus: Atoms, which we cannot see, are the cause of the phenomena that we can see.

Plato (427–347 B.C.) and Aristotle (384–322 B.C.) led the arguments against the atom by asking to be shown atoms. They also argued that the idea of atoms was a challenge to God. If atoms could be used to explain nature, there would be no need for God. For centuries most of those in the mainstream of enlightened thought rejected or ignored the atoms of Democritus.

John Dalton (1766—1844), a self-taught English schoolteacher, moved to Manchester in 1793 and devoted the rest of his life to scientific investigations. His presentation of atomic theory in the early part of the 19th century served as the basis from which modern chemical theories have grown.

Ideas about atoms drifted in and out of philosophical discussions for about 2200 years without playing a major role. Galileo (1564–1642) reasoned that the appearance of a new substance through chemical change could be explained in terms of a rearrangement of parts too small to be seen. Francis Bacon (1561–1626) speculated that heat might be a form of motion by very small particles. Robert Boyle (1627–1691) and Isaac Newton (1642–1727) used atomic concepts to interpret physical phenomena.

It was John Dalton (1766–1844), an English schoolteacher, who forcefully revived the idea of the atom. By Dalton's time experimental results had gained a position of greater respect than authoritative opinions. More clearly than any before him, Dalton was able to explain general observations, experimental results, and laws relative to the composition of matter. Dalton was particularly influenced by the experiments of two Frenchmen, Antoine Lavoisier (1743–1794) and Joseph Louis Proust (1754–1826). We shall look at the major contributions of these two experimentalists before we examine Dalton's theory.

Although the idea was proposed three and a half centuries earlier by Roger Bacon, it was not until 1620 that Francis Bacon wrote his book, *New Organon,* which put experimental science in the most refined and scholarly terms and made it possible for other scholars to accept it.

ANTOINE LAURENT LAVOISIER: THE LAW OF CONSERVATION OF MATTER IN CHEMICAL CHANGE

There are many reasons why Antoine Laurent Lavoisier has been acclaimed the "Father of Chemistry." He clarified the confusion over the cause of burning. He wrote the first textbook of chemistry, *Elementary Treatise on Chemistry.* He was the first to use systematic names for the elements and a few of their compounds. While there were still other contributions, his most notable achievement was to show the importance of making very accurate weight measurements in describing chemical change. His work began the process of establishing chemistry as a quantitative science.

Lavoisier weighed the chemicals in such changes as the decomposition of mercury(II) oxide by heat into mercury and oxygen.

$$2HgO \longrightarrow 2Hg + O_2$$

MERCURY(II) OXIDE   MERCURY   OXYGEN

Figure 2–1 Mixing a solution of sodium chloride with a solution of silver nitrate produces a new substance, solid silver chloride, but the total weight of matter remains the same.

THE PERSONAL SIDE

With all of his success, Lavoisier had his problems and disappointments. His highest goal, that of discovering a new element, was never achieved. He lost some of the esteem of his colleagues when he was accused of saying the work of someone else was his own. In 1768, he invested half a million francs in a private firm retained by the French government to collect taxes. He used the earnings (about 100,000 francs a year) to support his research. Although Lavoisier was not actively engaged in tax-collecting, he was brought to trial as a "tax-farmer" during the French Revolution. Lavoisier, along with his father-in-law and other tax-farmers, was guillotined on May 8, 1794, just two months before the end of the revolution.

Very accurate measurements showed that the total weight of all the chemicals involved remained constant during the course of the chemical change. Similar measurements on many other chemical reactions led Lavoisier to the summarizing statement now known as the ***Law of Conservation of Matter: Matter is neither lost nor gained during a chemical reaction.*** In other words, if one weighed all of the products of a chemical reaction—solids, liquids, and gases—the total would be the same as the weight of the reactants. Substances can be destroyed or created in a chemical reaction, but matter cannot. As a further example of the law of conservation of matter, consider Figure 2–1.

JOSEPH LOUIS PROUST:
THE LAW OF CONSTANT COMPOSITION

Following the lead of Lavoisier, several chemists investigated the quantitative aspects of compound formation. One such study, made by Proust in 1799, involved copper carbonate. Proust discovered that, regardless of how copper carbonate was prepared in the laboratory or how it was isolated from nature, copper carbonate always contained five parts of copper, four parts of oxygen, and one part of carbon by weight. His careful analyses of this and other compounds led to the belief that a given compound has an unvarying composition. These and similar discoveries are summarized by the ***Law of Constant Composition: In a compound, the constituent elements are always present in a definite proportion by weight.***

THE PERSONAL SIDE

Proust's generalization has been verified many times for many compounds since its formulation, but its acceptance was delayed by controversy. Comte Claude Louis Berthollet (1748–1822), an eminent French chemist and medical doctor, believed and strongly argued that the nature of the final product was determined by the amount of reacting materials one had at the beginning of the reaction. The running controversy between Proust and Berthollet reached major proportions. But more careful measurements supported Proust. Proust showed that Berthollet had made inaccurate analyses and had purified his compounds insufficiently—two great errors in chemistry.

Unlike Lavoisier, Proust saved his head during the French Revolution. Proust fled to Spain, where he lived in Madrid and worked as a chemist under the sponsorship of Charles IV, King of Spain. When Napoleon's army ousted Charles IV, Proust's laboratory was looted and his work came to an end. Later, Proust returned to his homeland, where he lived out his life in retirement.

Pure water, a compound, is always made up of 11.2% hydrogen and 88.8% oxygen by weight. Pure table sugar, another compound, always contains 42.11% carbon by weight. Contrast these with 14-carat gold, a mixture, which should be at least 58% gold, from 14 to 28% copper, and 4 to 28% silver by weight. This mixture can vary in composition and still be properly called 14-carat gold, but a compound that is not 11.2% hydrogen and 88.8% oxygen is not water.

JOHN DALTON: THE LAW OF MULTIPLE PROPORTIONS

John Dalton made a quantitative study of different compounds that could be made from the same elements. Such compounds differed in composition from each other, but each obeyed the Law of Constant Composition. Examples of this concept are the compounds carbon monoxide, a poisonous gas, and carbon dioxide, a product of respiration. Both compounds contain only carbon and oxygen. Carbon monoxide is made up of carbon and oxygen in proportions by weight of 3 to 4. Carbon dioxide is made up of carbon and oxygen in proportions by weight of 3 to 8. Note that for equal amounts of carbon (three parts), the ratio of oxygen in the two compounds is a ratio of 8 to 4, or 2 to 1.

In 1803, after analyzing compounds of carbon and hydrogen such as methane (in which the ratio of carbon to hydrogen is 3 to 1 by weight) and ethylene (in which the ratio of carbon to hydrogen is 6 to 1 by weight), and compounds of nitrogen and oxygen, Dalton first clearly enunciated the *Law of Multiple Proportions:*

In the formation of two or more compounds from the same elements, the weights of one element that combine with a fixed weight of a second element are in a ratio of small whole numbers (integers) such as 2 to 1, 3 to 1, 3 to 2, or 4 to 3.

DALTON'S ATOMIC THEORY

Why do the Law of Conservation of Matter, the Law of Constant Composition, and the Law of Multiple Proportions exist? How can they be explained? John Dalton employed the idea of atoms and endowed them with properties that enabled him to explain these chemical laws.

THE PERSONAL SIDE

While Lavoisier is considered the father of chemical measurement, Dalton is considered the father of chemical theory. Dalton, a gentle man and a devout Quaker, gained acclaim because of his work. He made careful measurements, kept detailed records of his research, and expressed them convincingly in his writings. However, he was a very poor speaker and was not well received as a lecturer. When Dalton was 66 years old, some of his admirers sought to present him to King William IV. Dalton resisted because he would not wear the court dress. Since he had a doctor's degree from Oxford University, the scarlet robes of Oxford were deemed suitable, but a Quaker could not wear scarlet. Dalton, being color blind, saw scarlet as gray. So, he was presented in scarlet to the court but in gray to himself. This remarkable man was, in fact, the first to describe color blindness. He began teaching in a Quaker school when only 12 years old, discovered a basic law of physics, the Law of Partial Pressures of Gases, and helped found the British Association for the Advancement of Science. He kept over two hundred thousand notes on meteorology. In all of this, he shunned glory and maintained he could never find time for marriage.

The major points of Dalton's theory, presented in modernized statements, are:

1. Matter is composed of indestructible particles called atoms.

Figure 2—2   Features of Dalton's atomic theory.

Law	Statement of Law	Explanation of Law
Law of Conservation of Matter	Matter is neither lost nor gained in a chemical change.	A chemical change is the result of a new arrangement of the same atoms present initially; hence, the weight is the same before and after the change.

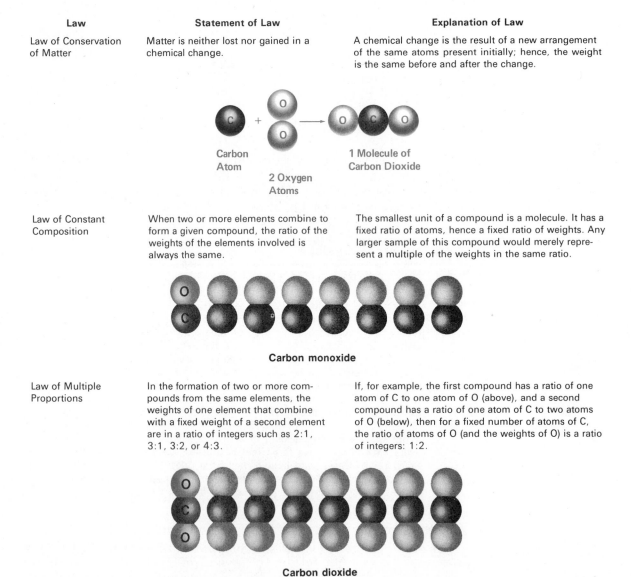

Law of Constant Composition	When two or more elements combine to form a given compound, the ratio of the weights of the elements involved is always the same.	The smallest unit of a compound is a molecule. It has a fixed ratio of atoms, hence a fixed ratio of weights. Any larger sample of this compound would merely represent a multiple of the weights in the same ratio.

Carbon monoxide

Law of Multiple Proportions	In the formation of two or more compounds from the same elements, the weights of one element that combine with a fixed weight of a second element are in a ratio of integers such as 2:1, 3:1, 3:2, or 4:3.	If, for example, the first compound has a ratio of one atom of C to one atom of O (above), and a second compound has a ratio of one atom of C to two atoms of O (below), then for a fixed number of atoms of C, the ratio of atoms of O (and the weights of O) is a ratio of integers: 1:2.

Carbon dioxide

Figure 2—3   John Dalton's explanation of three laws of chemistry in terms of atoms.

2. All atoms of a given element have the same properties such as size, shape, and weight, which differ from the properties of atoms of other elements.

3. Elements and compounds are composed of definite arrangements of atoms, and chemical change occurs when the atomic arrays are rearranged.

Dalton's theory was a monumental success in explaining the three laws of chemical composition and reaction. See Figure 2–3.

DALTON'S IDEA ABOUT ATOMIC WEIGHTS: THE FOLLOW-UP

John Dalton's idea about unique atomic weights for the atoms of the different elements generated interest in searching for the atomic weight characteristic of each element. Not until after 1860 did a consistent set of atomic weights appear, although several notable attempts were made before that, including the first table prepared by Dalton. In September 1860, many of the most brilliant minds in chemistry met in Karlsruhe, Germany, to discuss the inconsistencies in the atomic weights proposed at that time. There were differences of opinion on whether the formula of water was HO or H_2O, whether hydrogen gas was H_2 or H, and whether oxygen gas was O_2 or O. Water is 88.8 percent oxygen and 11.2 percent hydrogen by weight—a firmly established experimental fact by that time. If water is HO, as Dalton argued (based on his belief that the simplest formula is likely to be the correct one), then the weight of an oxygen atom should be about eight times that of a hydrogen atom:

$$\frac{\text{weight of an oxygen atom}}{\text{weight of a hydrogen atom}} = \frac{88.8}{11.2} = \frac{7.9}{1}$$

If the formula for water is H_2O, as one scientist, Amedeo Avogadro (1776–1856), had proposed in 1811, then one oxygen atom is 88.8% of the molecule but two hydrogen atoms are 11.2%. Each hydrogen atom would be $\frac{1}{2}(11.2\%)$ or 5.6%. With the formula H_2O, then, an oxygen atom would be about 16 times heavier than a hydrogen atom:

$$\frac{\text{weight of an oxygen atom}}{\text{weight of a hydrogen atom}} = \frac{88.8}{5.6} = \frac{15.9}{1}$$

Near the end of the meeting at Karlsruhe, Stanislao Cannizzaro (1826–1910) argued for the ideas of Avogadro. In spite of his arguments, which later proved to be correct, the confusion about atomic weights was not resolved during the conference. At the close of the meeting, however, a friend of Cannizzaro named Angelo Pavesi distributed copies of a paper written by Cannizzaro two years earlier. Several years later, the chemical community finally accepted H_2O as the formula of water, and a consistent set of atomic weights was generally agreed upon and used.

Just knowing that an oxygen atom is about 16 times heavier than a hydrogen atom does not tell us the weight of either atom. These are relative weights in the same way that a grapefruit may weigh twice as much as an orange. This information gives neither the weight of the grape-

Stanislao Cannizzaro (1826–1910), Italian chemist whose work resulted in the clarification of the atomic weight scale. (From Ihde, A. J. *The Development of Modern Chemistry,* Harper and Row, New York, 1964.)

fruit nor that of the orange. However, if a specific number is *assigned* as the weight of any particular atom, this fixes the numbers assigned to the weights of all other atoms. The standard for comparison of atomic weights was for many years the weight of the oxygen atom, which was taken as 16.0000 atomic weight units. This allowed the lightest atom, hydrogen, to have an atomic weight of 1.008, or approximately one.

The modern set of atomic weights (inside the back cover) is an outgrowth of the set of weights begun in the 1860's. The present atomic weight scale, adopted in 1961, is based on assigning the weight of a particular kind of carbon atom, the carbon-12 atom,* as exactly 12 atomic weight units. On this scale, an atom of Mg with an atomic weight of about 24 has twice the weight of a carbon-12 atom. An atom of titanium, Ti, with an atomic weight of 48 has four times the weight of a carbon-12 atom.

ATOMS ARE DIVISIBLE

Dalton's concept of the indivisibility of atoms was severely challenged by the subsequent discoveries of radioactivity and cathode rays, and was even in conflict with some previously known electrical phenomena such as static electrical charges.

Electrical charge was first observed and recorded by the ancient Egyptians, who noted that amber, when rubbed, attracted light objects. A bolt of lightning, a spark between a comb and hair in dry weather, and a shock on touching a doorknob are all results of the discharge of a buildup of electrical charge.

The two types of electrical charge had been discovered by the time of Benjamin Franklin (1706–1790). He named them positive (+) and negative (−) because they appear as opposites, in that they can neutralize each other. The existence and nature of the two kinds of charge, and their

---

*Contrary to Dalton's belief, all of the atoms of the same element do not have the same weight. The idea of isotopes is introduced later in this chapter.

Neutral

Rubber rod rubbed with silk or wool and touched to pith balls.

Rubber rod rubbed with silk and touched to pith balls repels balls that have the same charge as the rod.

Rubber rod rubbed with wool attracts pith balls that have a charge opposite to that on the rod.

Figure 2–4 Effects of charged matter on other charged matter. Like charges repel. Unlike charges attract.

effects on each other, can be shown with a simple electroscope (Figure 2–4). When a hard rubber rod is rubbed vigorously with silk and allowed to touch the lightweight balls, the balls spring apart immediately. The touching allowed the rod and the balls to share the same type of charge (positive). If the rod is then brought near one of the balls, the ball moves away from the rod. This movement indicates that *like charges repel*.

If the same rod is now rubbed vigorously with wool and brought near the charged balls, they move toward the rod. The opposite type of charge is now on the rod. The generalization is: *Unlike charges attract, and like charges repel*.

Radioactivity is a spontaneous process in which some natural materials give off very penetrating radiations. Henri Becquerel (1852–1908) discovered this property in natural uranium and radium ores in 1896. His student, Marie Curie (1867–1934), isolated the radioactive element radium and some of its pure compounds. It turns out that radioactivity is char-

Figure 2–5 Separation of alpha, beta, and gamma rays by electrostatic field.

acteristic of the elements, not the compounds, and that about 25 elements are naturally radioactive.

Radioactive elements commonly emit alpha, beta, and gamma rays, as shown in Figure 2–5. Alpha and beta rays are composed of particles, while gamma rays have no detectable mass. Alpha particles have a mass of 4 on the carbon-12 atomic weight scale, a positive charge, and low penetrating power (they will not penetrate skin, for example). In the arrangement shown in Figure 2–5, they are attracted toward the negatively charged plate. Beta particles have a mass of 0.000549 on the carbon-12 atomic weight scale, a negative charge (they are attracted toward the positive plate), and enough penetrating power to go through kitchen-strength aluminum foil. Gamma rays are a type of electromagnetic energy like light and X rays, but more penetrating. Gamma rays can penetrate a considerable thickness of aluminum, and even thin sheets of lead. They are not deflected at all by charged plates.

Cathode rays, as we shall see below, are similar to beta rays in that they are composed of negatively charged particles; their charge and mass are identical to those of beta particles.

The discoveries of natural radiation, cathode rays, and electrical charge are evidence that atoms can be divided and may even divide spontaneously. The smallest atom is 1836 times more massive than the beta or cathode-ray particle. Therefore, beta (and cathode ray) particles appear to be subatomic in origin.

SELF-TEST 2-A

1. Two Greek philosophers who were influential in advocating the concept of

   atoms were _____ and _____ .

2. The Greek approach to the "discovery" of atoms can best be described as:
   (a) experimentation.
   (b) philosophy (play on logic).
   (c) direct observation of atoms.
   (d) consistent explanation of well-known, established laws of nature.
   (e) deductive reasoning.

3. The Law of Conservation of Matter states that matter is neither lost nor

   _____ in a _____ reaction.

4. The Law of Multiple Proportions explains the existence of compounds like

   _____ and _____ .

5. (a) Assume that you are a chemist of many years ago. Your field of study is compounds composed of nitrogen and oxygen only. There are several that you know about. One contains 16 grams of oxygen for every 14 grams of nitrogen, while another contains 32 grams of oxygen for every 14 grams of nitrogen. Your assistant discovers what he claims is a new compound of nitrogen and oxygen. Upon analysis, the compound is found to contain 8 g of O for every 14 g of N. Has your assistant discov-

   ered a new compound or is it one of the others? _____

(b) What is the ratio by weight of oxygen in these compounds for a given

weight of N? _____

(c) What fundamental law of chemistry is illustrated by a comparison of

the compounds? _____

6. According to Dalton's atomic theory, what happens to atoms during a chemical change? Select one.

(a) Atoms are made into new and different kinds of atoms.
(b) Atoms are lost.
(c) Atoms are gained.
(d) Atoms are recombined into different arrangements.

7. According to Dalton's atomic theory, a compound has a definite percentage by weight of each element because

(a) all atoms of a given element weigh _____.
(b) all molecules of a given compound contain a definite number and kind

of _____.

8. Like charges _____; unlike charges _____.

9. The three types of radiation from a radioactive element such as radium are

_____, _____, and _____, of

which _____ pass through an electrostatic field without being deflected.

ELECTRONS

Cathode rays are streams of the negatively charged particles that are called electrons.

The first ideas about electrons came from experiments with cathode ray tubes. A forerunner of neon signs, fluorescent lights and TV picture tubes, a typical cathode ray tube is a partially evacuated glass tube with a piece of metal sealed in each end (Figure 2–6). The pieces of metal are called electrodes; the one given a negative charge is called the *cathode,* and the one given a positive charge is called the *anode.*

If a high electrical voltage is applied to the electrodes, an electrical discharge can be created between them. This discharge appears to be a stream of particles emanating from the cathode. This cathode ray will cause gases and fluorescent materials to glow, and will heat metal objects in its path to red heat. Cathode rays travel in straight lines and cast sharp shadows. Unlike light, however, cathode rays are attracted toward a positively charged plate. This led to the conclusion that cathode rays are negatively charged.

Careful microscopic study of a screen that emits light when struck by cathode rays shows that the light is emitted in tiny, random flashes. Thus, not only are cathode rays negatively charged, but they are composed of particles, each one of which produces a flash of light upon collision with the material of the screen. The cathode ray particles became known as *electrons.*

Figure 2—6 Deflection of a cathode ray by an electric field and by a magnetic field. When an external electric field is applied, the cathode ray is deflected toward the positive pole. When a magnetic field is applied, the cathode ray is deflected from its normal straight path into a curved path.

High voltage source

Cathode

Zinc sulfide screen

Anode

Magnet

Cathode rays

Figure 2—7 J. J. Thomson experiment. Electric field, applied by plates, and magnetic field, applied by magnet, cancel each other's effects to allow cathode ray (electron beam) to travel in straight line.

The charge and mass of an electron were determined by a combination of experiments by Sir Joseph John Thomson in 1897 and by Robert Andrews Millikan in 1911. Both scientists were awarded Nobel prizes, Thomson in 1906 and Millikan in 1923.

By using a specially designed cathode ray tube (Figure 2–7), Thomson applied electrical and magnetic fields to the rays. Using the basic laws of electricity and magnetism, he determined the charge-to-mass ratio of the electrons. He was able to measure neither the charge nor the mass of the electron, but he established the ratio between the two numbers and made it possible to calculate either one if the other could ever be measured. What Thomson did for the concept of the electron is like showing that a peach weighs 40 times more than its seed. What is the weight of the peach? What is the weight of the seed? Neither is known, but if it can be determined by other means that the peach weighs 120 grams, then the weight of the seed, by ratio, must be 3 grams.

An important part of Thomson's experimentation was his use of twenty different metals for cathodes and of several gases to conduct the discharge. Every combination of metals and gases yielded the same charge-to-mass ratio for the cathode rays. This led to the belief that electrons are common to all of the metals used in the experiments, and probably to all atoms in general.

Millikan, the first American to win a Nobel prize in physics, measured the fundamental charge of matter—the charge on an electron. A simplified drawing of his apparatus is shown in Figure 2–8. The experiment consisted of measuring the electrical charge carried by tiny drops of oil that are suspended in an electrical field. By means of an atomizer, oil droplets were sprayed into the test chamber. As the droplets settled slowly through the air, high energy X rays were passed through the chamber to charge the droplets negatively (the X rays caused air molecules to give up electrons to the oil). By using a beam of light and a small telescope, Millikan could study the motion of a single droplet. When the electrical charge on the plates was increased enough to balance the effect of gravity, a droplet could be suspended motionless. At this point, the gravitational force would equal the electrical force. Measurements made in the motion-

Thomson discovered the charge-to-mass ratio of the electron.

Electrons are present in all of the elements.

Millikan measured the charge on an electron.

Figure 2–8 Charged oil drops, suspended as a result of opposing gravitational and electrostatic forces, provided Millikan with the means of calculating the charge on the electron.

less state, when inserted into equations for the forces acting on the droplet, enabled Millikan to calculate the charge carried by the droplet.

Millikan found different amounts of negative charge on different drops, but the charge measured each time was always a whole-number multiple of a very small basic unit of charge. The *largest* common divisor of all charges measured by this experiment was 1.60×10^{-19} coulomb (the coulomb is a charge unit). Millikan assumed this to be the fundamental charge, which is the charge on the electron.

With a good estimate of the charge on an electron and the ratio of charge-to-mass as determined by Thomson, the very small mass of the electron could be calculated. The mass of an electron is 9.11×10^{-28} gram. On the carbon-12 relative scale, the electron would have a weight of 0.000549 atomic weight units. The negative charge on an electron of -1.60×10^{-19} coulomb is set as the standard charge of -1.

Only a whole number of electrons may be present in a sample of matter.

PROTONS

The first experimental evidence of a fundamental positive particle came from the study of canal rays. A special type of cathode ray tube produces canal rays (Figure 2–9). The cathode is perforated, and the tube contains a gas at very low presure. When high voltage is applied to the tube, cathode rays can be observed between the electrodes as in any cathode ray tube. On the other side of the perforated cathode, a different kind of a ray is observed. These rays are attracted to a negative plate brought alongside the rays. The rays therefore must be composed of positively charged particles. Each gas used in the tube gives a different charge-to-mass ratio for the positively charged particles. When hydrogen gas was used, the largest charge-to-mass ratio was obtained, indicating that hydrogen provides the positive particles with the smallest mass. This particle was considered to be the fundamental positively charged particle of atomic structure, and was called a ***proton*** (from Greek for "the primary one").

Experiments on canal rays were begun in 1886 by E. Goldstein and further work was done later by W. Wien. The production of canal rays is caused by high-energy electrons, moving from the negative cathode to the positive anode, hitting the molecules of gases occupying the tube. Electrons are knocked from some atoms by the high-energy electrons, leaving each molecule with a positive charge. The positively charged molecules are then attracted to the negative electrode. Since the electrode is perforated, some of the positive particles go through the holes or channels (hence the name canal rays).

The mass of the proton is 1.67261×10^{-24} gram, which is 1.007274 relative weight on the carbon-12 scale. The charge of $+1$ on the proton is equal in size but opposite in effect to the charge on the electron.

Canal rays are streams of positive ions derived from the gases present in the discharge tube.

An ion is an atom or a group of atoms carrying an unneutralized charge.

NEUTRONS

The existence of the neutron was proposed because more weight was needed to account for atomic weights than the number of protons and electrons in an atom could provide. A heavy neutral particle was proposed as a third constituent of atoms to account for the needed weight.

Positive electrode   **Negative electrode**

(a)

(b)

(c)

(d)

Faint glow

Figure 2–9 (*a*) Electrons rush from the negative electrode to the positive electrode due to high voltage. (*b*) Electrons collide with gas molecules to produce positive ions which are accelerated toward the negative electrode. (*c*) Some of the positive ions escape capture by the electrode and rush through the opening due to their kinetic energy. (*d*) Some of the positive ions in the positive ray collide with gas molecules to produce a characteristic glow, and strike the end of the glass tube to produce a luminous spot.

Direct evidence for the neutron came in 1932, when James Chadwick used a nuclear reaction to produce neutrons and detected them indirectly. A neutron has no electrical charge and has a mass of 1.67492×10^{-24} gram, which is a relative weight of 1.008665 on the carbon-12 scale.

THE NUCLEUS

Alpha particles are scattered by the nuclei of the gold atoms.

When Ernest Rutherford and his students directed alpha particles toward a very thin sheet of gold foil in 1909, they were amazed to find a totally unexpected result (Figure 2-10). As they had expected, the paths of most of the alpha particles were only slightly changed as they passed

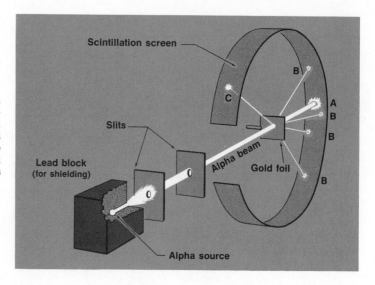

Figure 2–10 Rutherford's gold foil experiment. A circular scintillation screen is shown for simplicity; actually, a movable screen was employed. Most of the alpha particles pass straight through the foil to strike the screen at point A. Some alpha particles are deflected to points B, and some are even "bounced" backwards to points such as C.

through the gold foil. What was surprising was the extreme deflection of a few of the alpha particles. Some even "bounced" back toward the source. Rutherford expressed his astonishment by stating that he would have been no more surprised if someone had fired a 15-inch artillery shell into tissue paper and then found it in flight back toward the cannon.

What allowed most of the alpha particles to pass through the gold foil in a rather straight path? According to Rutherford's interpretation, the atom is mostly empty space and, therefore, offers little resistance to the alpha particles (Figure 2-11).

What caused a few alpha particles to be deflected? According to Rutherford's interpretation, concentrated at the center of the atom is a

Lord Rutherford (Ernest Rutherford, 1871–1937) was Professor of Physics at Manchester when he and his students discovered the scattering of α particles by matter. Such scattering led to the postulation of the nuclear atom. For this work he received the Nobel prize in 1908. In 1919, Rutherford discovered and characterized nuclear transformations.

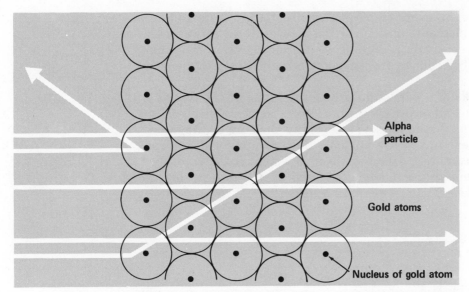

Figure 2—11 Rutherford's interpretation of how alpha particles interact with atoms in a thin gold foil. Actually, the gold foil was about 1000 atoms thick. For illustration purposes, points are used to represent the gold nuclei and the pathwidths of the alpha particles are drawn larger than scale.

Alpha particle scattering can be explained if the nucleus occupies a very small volume of the atom.

nucleus containing most of the mass of the atom and all of the positive charge. When an alpha particle passes near the nucleus, the positive charge of the nucleus repels the positive charge of the alpha particle; the path of the smaller alpha particle is deflected. The closer an alpha particle comes to a target nucleus, the more it is deflected. Those alpha particles that meet a nucleus head on are "bounced" back toward the source by the strong positive-positive repulsion, since the alpha particles do not have enough energy to penetrate the nucleus.

Based on Rutherford's calculations from electrostatics and the observed deflections, the nucleus is a very small part of an atom. An atom occupies about a million million times more space than does a nucleus; the radius of an atom is about 10,000 times greater than the radius of its nucleus. Thus, if a nucleus were the size of a baseball, then the edges of the atom would be about one third of a mile away. And most of the space in between would be absolutely empty.

Since the nucleus contains most of the mass and all of the positive charge of an atom, the nucleus must be composed of the most massive atomic particles, the protons and neutrons. The electrons are distributed in the near emptiness outside the nucleus.

Truly, Rutherford's model of the atom was one of the most dramatic interpretations of experimental evidence to come out of this period of significant discoveries.

ATOMIC NUMBER

The *atomic number* of an element indicates the number of protons in the nucleus of the atom, which is the same as the number of electrons outside the nucleus. The two types of particles must be present in equal numbers in order for the atom to be neutral in charge. Note that the periodic table of the elements, inside the back cover, is an arrangement of the elements consecutively according to atomic number. Beginning with the atomic number 1 for hydrogen, there is a different atomic number for each element.

The ***atomic mass*** of a particular atom is the sum of the masses of the protons, neutrons and electrons in that atom. Since an electron has such a small mass, the atomic mass is essentially the sum of the masses of the protons and neutrons in the nucleus. Both protons and neutrons have masses of approximately 1.0 on the atomic weight scale. Hydrogen, with an atomic weight of 1, must be composed of one proton (and no neutrons) in the nucleus, and one electron outside:

The lightest atom is the hydrogen atom.

Hydrogen atom

Helium has an atomic number of 2 and an atomic weight of 4. The atomic number of 2 indicates two protons and two electrons per atom of helium. The atomic weight of 4 means that, in addition to the two protons in the nucleus, there are also two neutrons.

A notation frequently used to show the atomic mass (also called *mass number*) and atomic number of an atom uses subscripts and superscripts to the left of the symbol:

For an atom of fluorine, $^{19}_{9}F$, the number of protons is 9, the number of electrons is also 9, and the number of neutrons is $19 - 9 = 10$.

ISOTOPES

Many of the elements, when analyzed by a special type of canal-ray tube called a mass spectrometer (Figure 2–12), are found to be composed of atoms of different masses (Figure 2–13). Atoms of the same element having different atomic masses are called ***isotopes*** of that element.

The element neon is a good example to consider. A natural sample of neon gas is found to be a mixture of three isotopes of neon:

$$^{20}_{10}Ne \qquad ^{21}_{10}Ne \qquad ^{22}_{10}Ne$$

The fundamental difference between isotopes is the different number of neutrons per atom. All atoms of neon have 10 electrons and 10 protons;

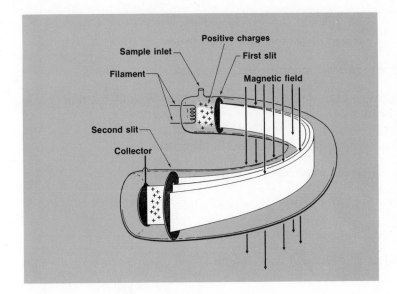

Figure 2—12 Mass spectrometer. Sample to be studied is injected near filament. Electrodes (not shown) subject sample to electron beam which ionizes a part of the sample. Electrodes are arranged to accelerate positive ions toward first slit. The positive ions that pass the first slit are immediately put into a magnetic field perpendicular to their path and follow a curved path determined by the charge-to-mass ratio of the ion. A collector plate, behind the second slit, detects charged particles passing through the second slit. The relative magnitudes of the electrical signals are a measure of the numbers of the different kinds of positive ions.

Isotopes are atoms of the same element having different numbers of neutrons.

about 90% of the atoms have 10 neutrons, some have 11 neutrons, and others have 12 neutrons. Because they have different numbers of neutrons, they must have different masses. Note that all the isotopes have the same atomic number, and indeed all behave the same chemically. They are all neon.

There are only 106 known elements, yet more than 1000 isotopes have been identified. Some elements have many isotopes; tin, for example, has ten natural isotopes. Hydrogen has three isotopes, and they are the only three which are generally referred to by different names: $_1^1H$ is called protium, $_1^2H$ is called deuterium, and $_1^3H$ is called tritium. Tritium is radioactive. The natural assortment of isotopes, each having its own distinctive atomic mass, results in fractional atomic weights for many elements. We need not concern ourselves with how these values are derived.

Figure 2—13 Mass spectrum of neon (+1 ions only). The principal peak corresponds to the most abundant isotope, neon-20. (From Masterton, W. L., and Slowinski, E. J., *Chemical Principles*, W. B. Saunders Co., Philadelphia, 1973.)

SELF-TEST 2-B

1. Isotopes of an element are atoms that have the same number of _____ _____ but different numbers of _____.

2. The nucleus of an atom occupies a relatively (large/small) fraction of the volume of the atom.

3. The positive charges in an atom are concentrated in its _____.

4. The negatively charged particles in an atom are _____; the positively charged particles are _____; and the neutral particles are _____.

5. In a neutral atom there are equal numbers of _____ and _____.

6. The number of protons per atom is called the _____ number of the element.

7. The mass of the proton is _____ times that of the electron.

8. An atom of arsenic, $_{33}^{75}$As, has _____ electrons, _____ protons, and _____ neutrons.

9. Positive (canal) rays obtained with different gases are (different/identical), while the cathode rays obtained using different cathodes are (different/identical).

10. Cathode rays are composed of a universal constituent of matter named _____.

11. The two fundamental particles revealed by studies using gas discharge (cathode ray) tubes are the _____ and _____.

12. All atoms of a given element are exactly alike (True/False).

ELECTRON THEORIES

Two major theories have been presented concerning the position, movement, and energy of electrons in an atom. The Bohr theory of the hydrogen atom was put forth in 1913 by Neils Bohr. This theory was extended and modified by Erwin Schrödinger, Werner Heisenberg, Louis de Broglie and others in 1926. The newer theory is referred to as the quantum mechanical theory, the wave mechanical theory, or Schrödinger's theory.

In Bohr's concept, electrons revolve around a nucleus in definite orbits, much as planets revolve around the sun. He equated classical

Niels Bohr (1885–1962) received his doctor's degree the same year that Rutherford announced his discovery of the atomic nucleus, 1911. After studying with Thomson and Rutherford in England, Bohr formulated his model of the atom. Bohr returned to the University of Copenhagen and, as Professor of Theoretical Physics, directed a program that produced a number of brilliant theoretical physicists. He received the Nobel Prize for physics in 1922.

Energy of matter in motion is kinetic energy.

Energy stored in matter is potential energy.

mathematical expressions for the force tending to keep the electron traveling in a straight line and the force tending to pull the electron inward (the positive-to-negative attraction between a proton and an electron). The total energy of the atom is the kinetic energy of the electrons plus the potential energy due to the electron's separation from the nucleus. In a revolutionary sort of way, Bohr suggested that electrons stay in rather stable orbits and can have only certain energies within a given atom. According to Bohr, an electron can travel in one orbit for a long period of time, or in another orbit some distance away for a long period of time, but cannot stay for any measurable time between the two orbits. A rough analogy is provided by considering books in a book case. Books may rest on one shelf or on another shelf for very long periods of time, but cannot rest between shelves. In moving a book from one shelf to another shelf, the potential energy of the book changes by a definite amount. When an electron moves from one orbit to another, its energy changes by a definite amount, called a *quantum* of energy.

Energy is required to separate objects attracted to each other. For example, energy is required to lift a rock from the earth, or to separate two magnets, or to pull a positive charge away from a negative charge. Bohr suggested that a very definite and characteristic amount of energy is required to move an electron from one energy level to another one. The energy added to move an electron further from its nucleus is stored in the system as potential energy (energy of position). Thus, the electron has more energy when it is in an orbit further from the nucleus, and less energy when it is in an orbit close to the nucleus. When the electron passes from an outer orbit to an inner orbit, energy is emitted from the atom, generally in the form of light energy.

Packets of light energy are called photons, or quanta.

Bohr used the idea of electrons moving up and down a "bookcase" of energy levels corresponding to orbits to explain the observable bright-line emission spectrum of hydrogen. A *spectrum* is the display produced by an instrument designed to separate or disperse light into its component colors. The spectrum of white light is the rainbow display of separated colors shown in Figure 2–14. An *emission* spectrum is produced when the light emitted by atoms energized by a flame or an electric arc is allowed to pass through a narrow vertical slit and then through a prism of glass or quartz. If sunlight or light from a white-hot solid is dispersed, all of the colors of the rainbow are seen. This is a *continuous emission spec-*

Dispersed light produces a spectrum.

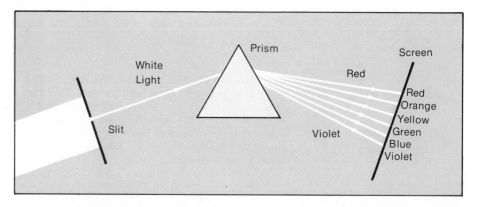

Figure 2–14 Spectrum from white light produced by refraction in a glass prism. The different colors blend into one another smoothly.

trum. However, if the light from an energized gaseous element is dispersed, a set of colored lines is produced, the lines being separated by black spaces. This is a ***bright-line emission spectrum.*** Each line is a pure color, and is really an image of the slit in that particular color. A quantum of light of any given color has a characteristic energy that is different from the energy of a quantum of any other color.

According to Bohr, the light forming the lines in the bright-line emission spectrum of hydrogen comes from electrons moving toward the nucleus after having first been energized and pushed to orbits further from the nucleus. A movement between two particular orbits involves a definite quantum of energy. Each time an electron moves from one orbit to another orbit closer to the nucleus, a quantum of light having a characteristic energy (and color) is emitted. Transitions toward the nucleus between two outer orbits emit quanta having smaller characteristic energies. Transitions from an outer orbit to orbits near the nucleus emit quanta having larger characteristic energies. Each line corresponds to its own particular energy of the quantum of light.

Not only could Bohr explain the cause of the lines in the bright-line emission spectrum of hydrogen, but he also calculated the expected wavelengths of the lines. He expressed the results of his calculations in the alternate view of the nature of light, its wave nature. The wave properties of light—wavelength, frequency, and speed—are considered in Figure 2–15 as they apply to any phenomenon possessing wave properties (water waves, sound waves from violin strings, radio waves, and light).

About thirty years before Bohr's work, Johann Balmer developed an equation which related the wavelengths of lines in that part of the hydrogen emission spectrum which was known to him. His expression was

$$\text{wavelength (in nanometers)} = 364.56\left(\frac{n^2}{n^2 - 4}\right)$$

Balmer arrived at this equation by measuring the wavelengths of the lines in the hydrogen emission spectrum, and then finding the algebraic relationship between the wavelengths. The values of n are integers beginning with 3 and progressing upward to 4, 5, 6, etc. The values of n appeared to have only certain quantities in a regular progression. Hence n is said to vary in quantum jumps, and n is a ***quantum number.*** Each value of n

Bohr assumed that an atom can exist only in certain energy states.

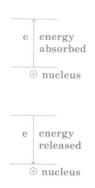

A nanometer is 10^{-9} meter.

Figure 2—15 The wave theory of light considers light to be waves of wavelength λ (lambda) vibrating at right angles to their path of motion. Red light, *A*, completes one vibration or wave in the same distance and time that involve two complete waves of ultraviolet light, *B*. Ultraviolet light is of shorter wavelength than violet light. A nanometer is 10^{-9} meter.

when substituted into the equation allows the calculation of a single wavelength. Each wavelength so calculated corresponds to a measured wavelength of a line in the hydrogen spectrum. Balmer had the equation, but he did not know what the numbers meant or why the relationship held. He had an ***empirical equation***—one obtained by arranging measured data into a form which included all the observations, and by which predicted values can be calculated. The usefulness of an empirical equation is that it does not require a theoretical explanation for its form.

TABLE 2—1   Agreement Between Bohr's Theory and the Lines of the Hydrogen Spectrum

Quantum Jumps (in Quantum Numbers)	Wavelength Predicted by Bohr's Theory (nm)	Wavelength Determined from Laboratory Measurement (nm)	Spectral Region
2 ⟶ 1	121.6	121.7	Ultraviolet
3 ⟶ 1	102.6	102.6	Ultraviolet
4 ⟶ 1	97.28	97.32	Ultraviolet
3 ⟶ 2	656.6	656.7	Visible red
4 ⟶ 2	486.5	486.5	Visible blue-green
5 ⟶ 2	434.3	434.4	Visible blue
4 ⟶ 3	1876	1876	Infrared

NOTE: These lines are typical; other lines could be cited as well, with equally good agreement between theory and experiment. The unit of wavelength is the nanometer (nm), 10^{-9} meter.

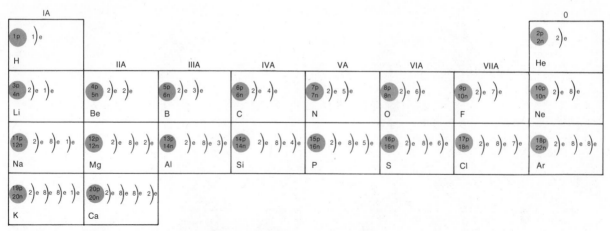

Figure 2—16 Electron arrangements of the first 20 elements. Above each column, or group, is the group number. The nuclear contents of a typical isotope are shown.

In contrast to the empirical approach of Balmer, Bohr started with the theoretical concepts of the electron-proton system for the hydrogen atom. He calculated the energies of the atomic system for different distances between the proton and electron and then took these to be the energy differences between the electron orbits. Since the energy differences corresponded to certain quanta or wavelengths of light, he was able to calculate a wavelength for each movement of an electron between all conceivable orbits of the atom. When Bohr assumed quantized orbits, he was able to calculate from his theory all the wavelengths of the hydrogen spectrum, even those which had not yet been observed. Table 2–1 shows the close agreement between the values calculated by the Bohr equation and the wavelengths measured to date.

The Bohr theory was accepted almost immediately after its presentation, and he was awarded the Nobel prize in physics in 1922 for his contribution to the understanding of the hydrogen atom.

THE PERSONAL SIDE
Bohr had a close call in his escape from Denmark when Hitler's forces ravished the country. Having done all he could to get Jewish physicists to safety, Bohr was still in Denmark when Hitler's army suddenly occupied the country in 1940. In 1943, to avoid imprisonment, he escaped to Sweden. There he helped to arrange the escape of nearly every Danish Jew from Hitler's gas chambers. He was later flown to England in a tiny plane, in which he passed into a coma and nearly died from lack of oxygen.

He went on to the United States, where until 1945 he worked with other physicists on the atomic bomb development at Los Alamos. His insistence upon sharing the secret of the atomic bomb with other allies, in order to have international control over nuclear energy, so angered Winston Churchill that he had to be restrained from ordering Bohr's arrest. Bohr worked hard and long on behalf of the development and use of atomic energy for peaceful purposes. For his efforts, he was awarded the first Atoms for Peace prize in 1957. He died in Copenhagen on November 18, 1962.

The Bohr theory introduced the idea of quantized electron orbits in the atom. With electrons appropriately placed in orbits, an enormous amount of chemical and physical information for all of the elements can be explained. The numbers of electrons in the orbits for the first twenty elements are given in Figure 2–16.

Bohr's orbits are often called "shells" as well as energy levels, and are given numbers beginning with 1 for the orbit closest to the nucleus. The *maximum* number of electrons that can circulate in each orbit is found to be $2n^2$, where n is the number of the orbit. The first orbit can have a maximum of 2 electrons; orbit number 2 can have a maximum of 8 electrons; orbit number 3 can have a maximum of 18 electrons, orbit number 4 can have up to 32 electrons, and so on. An examination of Figure 2–16 shows that the filling of the energy levels is very regular until element number 19, potassium, is reached.

The elements are arranged in Figure 2–16 in the form of a periodic table. When the elements are taken in the order of their atomic numbers, they show periodic recurrences of chemical properties. A **group** in the periodic table (a vertical column) is made up of elements with similar chemical properties. All elements in the same group tend to form the same types of compounds. For example, all of the alkali metals, lithium (Li), sodium (Na), potassium (K), rubidium (Rb), cesium (Cs), and francium (Fr), form oxides with the same general formula, M_2O, where M stands for the alkali metal atom.

Why do elements in a group exhibit similar chemical properties that are generally distinct from those of elements in other groups? Using Bohr's theory, we see that each element in a group has the same number of electrons in the outermost orbit. It is these outer electrons that are exposed to the immediate surroundings of the atom. Note in Figure 2 –16 that the alkali metal or group I elements all have one outer electron. On the other hand, the fact that oxygen and sulfur both have six outer electrons partially explains why the two elements are related chemically. For example, water, H_2O, has a formula like that of hydrogen sulfide, H_2S. Oxygen forms a compound with sodium, Na_2O, which has a formula similar to that of the compound of sulfur and sodium, Na_2S. Further study of members of chemical groups makes possible the understanding of other similarities in chemical formulas. The electron distribution in the major energy levels for all of the elements is given in the periodic table inside the back cover.

The understanding of the similarity of elements within a chemical group developed over a period of about fifty years (1820–1869) through the work of a number of chemists, and ultimately produced in the mind of a Russian chemist, Dmitri Mendeleev, the **periodic table.** Since all of the elements had not been discovered at the time, Mendeleev's periodic table contained gaps that provided a means of predicting the existence of unknown elements. Ten of these elements were discovered in the ten years after Mendeleev announced his periodic table in 1869.

Since Mendeleev had no concept of the modern idea of atomic number (which came from the atomic theory of the early 1900's), he arranged his table by inserting the elements in the order of a combination of atomic weights and chemical properties. Where the two conflicted, Mendeleev assumed that the atomic weight was in error. He was usually correct. One interesting case involves iodine (I) and tellurium (Te). The atomic weight of tellurium is 127.6, and that of iodine is 126.9. Had these two elements been inserted in order of atomic weight, iodine would precede tellurium. This would put tellurium in the group with fluorine, chlorine, and bromine (which tellurium does not resemble chemically); likewise, iodine would be out of place chemically in the group with oxygen, sulfur, and selenium. Mendeleev simply placed tellurium before iodine so that each would be in a group of similar elements.

Mendeleev's contribution to chemistry stood on experimental fact, not theory; and while scientists even today continue to attempt to understand atomic structure, his periodic table continues to stand unchanged.

WAVE MECHANICAL ATOM

The Bohr model failed when applied to elements other than hydrogen because it could not account exactly for the line spectra of atoms with more than one electron. It was also weak in explaining why the periods (the horizontal rows) of the periodic table vary considerably in length.

After Bohr's work, a more modern, mathematical theory of the atom was developed by Schrödinger, Heisenberg, Dirac and others. In this theory, electrons are treated as having both a particle and a wave nature. The locations of the electrons are treated as probabilities, without seeking

Electrons are described by both particle and wave theories.

Figure 2–17 Similar patterns are shown by light, X rays, and electrons as each is diffracted. Diffraction is the bending and spreading of wave motion around edges. The effect is prominent when the wavelength is large compared to the size of the obstacle and small when the wavelength is short compared to the size of the obstacle. Similar effects from light, X rays, and electrons indicate a property common to all: each has a wave nature.

to locate the exact spot for an electron at a given time. This approach suggested that the Bohr theory sought more precision in describing the atom than nature would allow.

A Frenchman, Louis de Broglie, was the first to suggest (in 1924) that electrons and other small particles should have wave properties. In this respect, he said, electrons should behave like light, a suggestion which scientists of the time found hard to accept. However, in a few years separate experiments by George Thomson (son of J. J. Thomson) in England and Clinton Davisson in the United States justified de Broglie's hypothesis (Figure 2–17). The electron microscopes found in many research laboratories today are built and operated on our understanding of the wave nature of the electron.

The wave theory of the atom was developed in the late 1920's principally by Erwin Schrödinger. The most fundamental aspects of the theory are the mathematical wave equations used to describe the electrons in atoms. Solutions to the equations are called wave functions or *orbitals*. Calculations involving the wave equations are complicated and time-consuming, but we do not need to do the elaborate calculations in order to use the results.

The principal result is a series of orbitals. Orbitals are different in their type, energy, and likely configuration in space (related to the probability of finding the electron there). The types of orbitals are distinguished by the letters s, p, d, and f. These letters were derived from terms in spectroscopy (sharp, principal, diffuse, and fundamental, respectively), and emphasize again that atomic theory developed very closely with atomic spectra.

The orbitals of the wave theory are simply subdivisions of the Bohr orbits. The first Bohr orbit is synonomous with the 1s orbital. (The 1 indicates the orbit; the s indicates the type of orbital.) According to the wave theory, an orbital may be occupied by a *maximum* of two electrons. Hence, the 1s orbital (and also the first orbit) can have a maximum of two electrons. When the electron of a hydrogen atom is as close to its nucleus as it can stably be (said to be in its *ground state*), the electron is in a 1s orbital. Likewise, the two electrons of a helium atom are in a 1s orbital when in their ground state.

The second Bohr orbit can have a maximum of eight electrons. Thus, four orbitals are required to accommodate the eight electrons. The four orbitals are a 2s orbital and a set of three 2p orbitals designated $2p_x$, $2p_y$, and $2p_z$. When the three electrons of a lithium atom are in their ground state, two electrons are in the 1s orbital and the other electron is in the 2s orbital. The five electrons of a boron atom are distributed with two in the 1s orbital, two in the 2s orbital, and one in a 2p orbital (either $2p_x$, $2p_y$, or $2p_z$, since these orbitals have the same energy unless the atom is in a strong magnetic field.)

Each orbit has an s orbital. Beginning with the second orbit, each orbit has also a set of three p orbitals. The third orbit and all orbits thereafter also have a set of five d orbitals. The fourth orbit and all orbits thereafter also have a set of seven f orbitals. We shall need only s and p orbitals for the explanations given in this book, since the electrons in the outer orbits of the various atoms are in the s and p orbitals of those atoms. These outer s and p electrons are the ones most involved with other atoms.

Instead of knowing precisely where an electron is located within an

Figure 2–18

An s orbital

A p orbital

atom, only the probability of finding an electron in a given volume of space around the nucleus can be calculated from the Schrödinger equation. In order to portray the probabilities of finding an electron, usually the surface of a region in space (similar to the surface of a balloon) is plotted that will enclose the volume in which the electron will be expected to be found 90 percent of the time.

As shown in Figures 2–18 and 2–19, the shape of an s orbital is always spherical. The shape of a p orbital is like that of a dumbbell. The d and f orbitals have more complicated shapes. The shapes of the orbitals are about the best that we can do to relate the wave theory of electron probabilities in a pictorial way, for we are trying to visualize where in space a given electron will be. The geometries associated with the orbitals help to explain the structures of molecules that result from the interaction and combinations of atoms.

What, then, is an atom really like? We have Dalton's concept of an atom as a hard sphere similar to a small billiard ball. We have Bohr's concept of the atom as a small three-dimensional solar system with a nucleus and electrons in paths called orbits. In the modern theory, we have more detail in that orbits now have sub-orbits called orbitals and we are given approximate spaces where electrons exert their greatest influence in an atom. Why present all three theories? First, an understanding of the simpler Dalton and Bohr theories helps us to understand the more complicated, more detailed modern theory of the atom. Second, all three theories help us to understand the phenomena that we observe. We simply use whatever detail is necessary to explain what we see. For example, the simpler Dalton concept adequately explains many properties of the gaseous, liquid and solid states. Most bonding between atoms of the light elements can be explained by application of the orbits of

Figure 2–19 Spatial orientations of s and p orbitals for hydrogen-like atoms. The x, y, and z subscripts denote p orbitals with different orientations in space with reference to the imaginary axes.

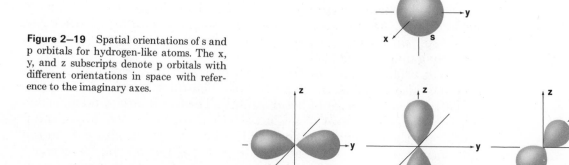

Bohr. The shapes of molecules and the arrangement of atoms with respect to each other can best be explained by the orbital representations of the modern theory. In the explanations given in this text, we shall follow the principle that simplest is best.

1. Under some conditions, light has properties of _____ and under other conditions exhibits the properties of _____ .

2. When light is dispersed into the different colors composing the light, a _____ is produced.

3. According to Bohr's theory, light of characteristic wavelength is produced as an electron passes from an orbit (closer to/farther from) the nucleus to an orbit (closer to/farther from) the nucleus.

4. Which of the following led to the modern theory of the atom and was not included in the Bohr theory?

    (a) concept of nucleus
    (b) quantum theory
    (c) particle nature of the electron
    (d) wave nature of the electron

5. According to de Broglie, every moving particle has not only a mass and velocity but also a characteristic _____ .

6. The maximum number of electrons in the $n = 3$ energy level is _____ , and the maximum number in any orbital is _____ .

7. In which group of the periodic table would elements with the following electron configuration be found?

    2 8 1 _____ 2 8 4 _____ 2 8 8 2 _____

8. Consider the meaning of the representations of the orbitals shown in Figure 2–18.

    (a) Are the representations those of the paths of electrons? _____

    (b) Are the representations the containers of electrons? _____
    (c) Do the representations show where an electron is most likely to be found? _____

_____ 1.	atomic mass	(a)	attract
_____ 2.	unlike electrical charges	(b)	equal to number of protons in nucleus
_____ 3.	$2n^2$	(c)	demonstrated wave nature of electron
_____ 4.	nucleus	(d)	cathode ray particle
_____ 5.	electron	(e)	neutrons plus protons
_____ 6.	^{22}Ne and ^{20}Ne	(f)	similar chemical properties
_____ 7.	atomic number	(g)	proton plus electron
_____ 8.	group of elements	(h)	uncharged elementary particle
_____ 9.	gamma ray	(i)	contains most of the mass in an atom
_____ 10.	particles in a H atom	(j)	maximum number of electrons in an orbit
_____ 11.	neutron	(k)	predicted wave nature of electron
_____ 12.	G. P. Thomson C. J. Davisson	(l)	probability for electrons in an atom
_____ 13.	de Broglie	(m)	a form of radiant energy
_____ 14.	orbital	(n)	isotopes

QUESTIONS

1. What kinds of evidence did Dalton have for atoms that the early Greeks (Democritus, Leucippus) did not have?

2. How does Dalton's atomic theory explain:

   (a) the law of conservation of matter?
   (b) the law of constant composition?
   (c) the law of multiple proportions?

3. The laws of chemical change presented in this chapter are often referred to as empirical laws. What does "empirical" mean? How does empirical differ from theoretical?

4. Although there may not be a very reliable way to check the conservation of matter in a large explosion of dynamite, what leads us to believe that the law of conservation of matter is obeyed?

5. Describe the potential energy and kinetic energy relationships as a rock tumbles off a cliff.

6. What experimental evidence indicates that

    (a) cathode rays have considerable energy?
    (b) cathode rays have mass?
    (c) cathode rays have charge?
    (d) cathode rays are a fundamental part of all matter?
    (e) two isotopes of neon exist?
    (f) atoms are destructible?

7. Describe in detail Rutherford's gold-foil experiment under the following headings:

    (a) experimental set-up.
    (b) observations.
    (c) interpretations.

8. Why was Thomson's charge-to-mass ratio determination for electrons very significant although he did not determine either the charge or the mass of the electron?

9. What part do electrons play in producing positive rays?

10. How do the following discoveries indicate that the Daltonian model of atoms is inadequate?

    (a) cathode rays.   (c) nucleus.   (e) isotopes.
    (b) positive rays.   (d) natural radioactivity.

11. Characterize the three types of emissions from naturally radioactive substances as to charge, relative mass, and relative penetrating power.

12. Use the information on the periodic chart to answer the following:

    (a) the nuclear charge on cadmium, Cd.
    (b) the atomic number of arsenic, As.
    (c) the atomic mass (or mass number) of an isotope of bromine, Br, having 46 neutrons.
    (d) the number of electrons in an atom of barium, Ba.
    (e) the number protons in an isotope of zinc, Zn.
    (f) the number of protons and neutrons in an isotope of strontium, Sr, atomic mass (or mass number) of 88.
    (g) an element forming compounds similar to those of gallium, Ga.

13. Explain what the following terms mean:

    (a) isotopes of an element.
    (b) atomic number.
    (c) an alpha emitter.
    (d) chemical group of elements.

14. If electrons are a part of all matter, why are we not electrically shocked continually by the abundance of electrons about and in us?

15. Sodium metal reacts violently with water and forms hydrogen gas in the process. Magnesium metal will react with water only when the water is very hot. Copper metal does not react with water. Suppose you find a bottle containing a lump of metal in a liquid and a label, "Cesium (Cs) Metal." Based on your knowledge of the periodic table, what danger is there, if any, of disposing of the metal by throwing it into a barrel of water?

16. There are more than 1000 kinds of atoms, each with different weight. Yet there are only 106 elements. How does one explain this in terms of subatomic particles?

17. What is a practical application of cathode ray tubes?

18. A common isotope of lithium (Li) has a mass of 7. The atomic number of lithium is 3. What are the constituent particles in its nucleus?

19. An element has 12 protons in its nucleus. How many electrons will the atoms of this element possess?

20. An isotope of atomic mass 60 had 33 neutrons in its nucleus. What is its atomic number, and what are the name and chemical symbol of the element?

21. An isotope of cerium (Ce) has 88 neutrons in its nucleus. How many protons plus neutrons does this nucleus contain?

22. The element iodine (I) occurs as a single isotope of mass 127; its atomic number is 53. How many protons and how many neutrons does it have in its nucleus?

23. An element with an atomic number of 8 is found to have three isotopes, with atomic masses of 16, 17, and 18. How many protons and neutrons are present in each nucleus? What is the element?

24. Suppose Millikan had determined the following charges on his oil drops:

    1.33×10^{-19} coulomb
    2.66×10^{-19} coulomb
    3.33×10^{-19} coulomb
    4.66×10^{-19} coulomb
    7.92×10^{-19} coulomb

    What do you think his value for the electron's charge would have been?

25. Suppose an isotope of aluminum has an atomic mass of 27.0. How many protons, neutrons, and electrons are in an atom of this isotope? What is the charge on the nucleus?

26. What is a quantum? What is a photon?

27. Discuss, in quantum terms, how a ladder works.

28. Distinguish between atomic number and atomic weight.

29. Distinguish between a continuous spectrum and a bright-line spectrum under the two headings:

    (a) general appearance.
    (b) source.

30. How does the Bohr theory explain the many lines in the spectrum of hydrogen although the hydrogen atom contains only one electron?

31. Helium, neon, argon, krypton, xenon and radon form a group of elements which are similar in that they form very, very few compounds. From their atomic structures, suggest a reason for this similarity in relative inactivity.

CHEMICAL BONDING

One very important aspect of atomic theory is its ability to explain the forces that hold atoms together in molecules, molecules together in molecular substances, and ions together in ionic structures. A link between submicroscopic particles, resulting from attractive forces between them, is called a ***chemical bond*** (Figure 3–1). Our understanding of chemical bonding has emerged through the interaction of two parallel lines of development: (1) the investigation of the properties of substances, and (2) the outgrowth of theories of chemical bonding based on concepts of atomic structure.

Chemical bonds serve to hold atoms, molecules, and ions together.

Studies of the properties of substances, such as chemical reactivity, volatility (ability to pass into gaseous state), melting point, electrical conductivity, and color can often give some indication as to how submicroscopic particles are bonded together. For example, since melting involves a situation in which atoms or molecules become less firmly bound to their neighbors, a high melting point implies that a solid is held together by very stable chemical bonds. As we shall see shortly, compounds

Figure 3–1 A water molecule is often represented by a ball-and-stick model. This model tells which atoms are bonded together and the angle involved, but gives no information as to why the atoms are bonded in a particular pattern, the relative sizes of the atoms, or the actual distances between them.

composed of a network of tightly bound ions or atoms tend to have relatively high melting points. The volatility of a substance also indicates how strongly molecules are attracted to each other. For example, in the case of carbon dioxide, CO_2, we must assume that the bonding between molecules (intermolecular bonding) is slight, since it takes relatively little energy to break up the solid CO_2.

There has been a continuing effort to develop theories which can explain chemical bonding. All of these theories deal with relationships between the submicroscopic structure of matter and the properties of matter. Theories have been used to establish detailed structure-property relationships for known substances and to predict properties for new substances.

CO_2 changes readily from a solid to a gas (sublimes).

Bonding theories must explain the observed behavior of chemicals.

TYPES OF CHEMICAL BONDING

Chemical bonds are of five major types, each having its own characteristics. These types of bonds, along with common materials in which they occur, are:

1.	Ionic bonding	Salts, such as table salt (sodium chloride), and metal oxides, such as lime, iron rust, ruby, and sapphire
2.	Covalent bonding	Molecular compounds, such as water, methane, and sucrose, and polymers, such as polyethylene
3.	Hydrogen bonding	Water, ammonia, and large molecules in living organisms
4.	Van der Waals' attractions	Liquid helium and solid CO_2
5.	Metallic bonding	Metals and alloys

We shall now look at these types of chemical bonds in more detail, with a major emphasis on describing various types of compounds and accounting for some of their properties in terms of the bonding that holds them together. We shall see that bonding is explained in terms of the outer electrons in the atoms. Such electrons are referred to as **bonding** or **valence** electrons.

IONIC BONDING

Ions and Ion Formation

Charged atomic or molecular-sized particles are called **ions.** They may be positively charged, as Na^+ (sodium ion), or negatively charged, as Cl^- (chloride ion). They may bear multiple charges, as in Ca^{2+} (calcium ion), Al^{3+} (aluminum ion), O^{2-} (oxide ion), and N^{3-} (nitride ion).

Since electrons constitute the outermost parts of the atom, it is reasonable to assume that only electrons are transferred between atoms to form ions. There are then two ways for ions to be formed: electron loss to form positive ions, and electron gain to form negative ions. Let us examine the formation of positive ions first.

Ions may bear one or more positive or negative charges, shown as a superscript to the right of the chemical symbol.

Figure 3–2 Probable distances from the nucleus for the electrons of a sodium atom. The farther the electron is from the nucleus, on the average, the easier it is to remove from the atom. Distances are in nanometers (nm).

Metals generally have electronic configurations in which there are three or fewer electrons in the outermost orbit. Since these are relatively far from the nucleus, they would be the easiest ones to remove from the metal atom (Figure 3–2). Metal atoms are neutral (uncharged), and the loss of a negatively charged electron results in a positive charge on the ion. We might guess that metals would form ions with charges of $+1$, $+2$, or $+3$ depending on how many of the outer electrons are lost in the ionization process. For example, sodium readily forms a $+1$ ion:

$$Na + \text{Ionization Energy} \longrightarrow Na^+ + 1e^-$$

SODIUM
ION

Metals may lose electrons to non-metals.

The energy required to remove each electron completely from an atom is called its ***ionization energy*** (for the first electron lost, the first ionization energy; for the second electron lost, the second ionization energy; and so on) (Table 3–1). Metals have low ionization energies; nonmetallic elements generally have higher ionization energies.

Ionization energies are measured experimentally and correspond to the reaction: Atom ⟶ Positive ion + electron.

As more and more electrons are removed from a single atom, it becomes progressively more difficult to remove the next one. The net positive charge builds up with the loss of each electron and this charge helps to hold the remaining electrons more securely. The difficulty of removing successive electrons from the same atom is emphasized in the ionization energies in Table 3–1.

The table of ionization energies also indicates how many electrons are relatively easy to remove from an atom. Note that Li, Na, and K have relatively low ionization energies for removal of the first electron, but require a big jump in energy for removal of the second. For Be, Mg, and Ca, the big jump is between the second and third ionization energies. This means that two electrons are relatively easy to remove, but the third, which would not be in the outer orbit of these atoms, requires considera-

TABLE 3–1 Ionization Energies of Selected Gaseous Atoms

An electron volt (ev) is the energy acquired by an electron when accelerated by a potential difference of 1 volt. For each element, electrons must be removed to the heavy vertical line in order to attain a noble gas electronic configuration.

Atomic Number	Atom	Ionization Energies (ev)							
		1st	2nd	3rd	4th	5th	6th	7th	8th
1	H	13.6							
2	He	24.6	54.4						
3	Li	5.4	75.6	122.4					
4	Be	9.3	18.2	153.9	217.7				
5	B	8.3	25.1	37.9	259.3	340.1			
6	C	11.3	24.4	47.9	64.5	392.0	489.8		
7	N	14.5	29.6	47.4	77.5	97.9	551.9	666.8	
8	O	13.6	35.1	54.9	77.4	113.9	138.1	739.1	871.1
9	F	17.4	35.0	62.6	87.2	114.2	157.1	185.1	953.6
10	Ne	21.6	41.1	64	97.2	126.4	157.9		
11	Na	5.1	47.3	71.7	98.9	138.6	172.4	208.4	264.2
12	Mg	7.6	15.0	80.1	109.3	141.2	186.9	225.3	266.0
13	Al	6.0	18.8	28.4	120.0	153.8	190.4	241.9	285.1
14	Si	8.1	16.3	33.5	45.1	166.7	205.1	264.4	303.9
15	P	10.6	19.7	30.2	51.4	65.0	220.4.	263.3	309.3
16	S	10.4	23.4	35.0	47.3	72.5	88.0	281.0	328.8
17	Cl	13.0	23.8	39.9	53.5	67.8	96.7	114.3	348.3
18	Ar	15.8	27.6	40.9	59.8	75.0	91.3	124.0	143.5
19	K	4.3	31.8	46	60.9	82.6	99.7	118	155
20	Ca	6.1	11.9	51.2	67	84.4	109	128	147

Table 3–1 shows the amounts of energy needed to remove one or more electrons from various atoms.

Figure 3–3 First ionization energies of the elements.

bly more energy to remove it. Figure 3–3 plots the first ionization energies for the elements. It can be seen that this pattern of ionization energies is periodic.

This periodic behavior of ionization energy is directly related to the electron configurations of the elements. For example, the elements known as the *noble gases* (helium, neon, argon, krypton, and xenon) are at the ends of the first five upward trends in Figure 3–3 (the element mercury, Hg, is a special case and will not be considered here). Each of these elements has the highest ionization energy in the period of elements to which it belongs, and they do not readily enter into chemical change. Indeed, they are called noble gases because of this indifference to chemical combination. It is reasonable to assume that both the high ionization energies and the relative chemical inertness are related to the electronic structure for these elements:

Of the elements in a given period of the periodic table, the noble gas found at the end of the period has the highest ionization energy.

He	2				
Ne	2	8			
Ar	2	8	8		
Kr	2	8	18	8	
Xe	2	8	18	18	8

The electrons in the outermost orbit of an atom are called its *valence electrons.* Note that except for helium (which has only two electrons), each of the noble gases has eight valence electrons. We must then also conclude that the electronic configurations of both metals and nonmetals represent arrangements that are more chemically reactive than those of the noble gases; that is, an atom that has fewer than eight valence electrons is more reactive than one that has exactly eight valence electrons.

Each of the Group IA metals (Li, Na, K, Rb, and Cs) has one more electron than the noble gas that precedes it in the periodic table:

Li	2	1				
Na	2	8	1			
K	2	8	8	1		
Rb	2	8	18	8	1	
Cs	2	8	18	18	8	1

The loss of one electron from any of these metals would produce an ion with a charge of $+1$, which would have an electronic configuration like that of the preceding noble gas. For example, the ion K^+ contains the same number of electrons in the same energy-level arrangement as does the argon atom. This further explains why the first ionization energies for these metals are relatively small and the second ionization energies are high: the removal of the second electron would have to disrupt the argon-like electron structure, which has a high ionization energy. We conclude, then, that the Group IA metals have a lesser degree of chemical stability (greater reactivity) than do noble gases, and readily undergo an ionization reaction to produce ions that resemble the noble gases in electronic structure.

For the Group IIA elements (Be, Mg, Ca, Sr, and Ba) to acquire a noble gas electronic structure, each would have to lose two electrons. For example:

$$Mg \longrightarrow \quad Mg^{2+} \quad + 2e^-$$

ELECTRONIC STRUCTURE
LIKE NEON

Again ionization energies are in agreement. For these elements, the first two ionization energies are low and the third one is relatively high.

It is not true that all stable metal ions have a noble gas electronic configuration. However, we can conclude that metal atoms can readily lose electrons to form positive ions and, in the process, some of them attain the noble gas electronic structures (Table 3–2).

The nonmetals have relatively high first ionization energies. The energies available in ordinary chemical change are simply not great enough to form positive ions from nonmetallic atoms. In order for a nonmetal to achieve a noble gas configuration, it *gains* one or more electrons. Thus, a nonmetal can add to its five, six, or seven valence electrons and, in so doing, can attain the configuration of a noble gas.

Nonmetal atoms gain electrons to form negative ions.

Consider an atom of fluorine (F, atomic number 9). Its electronic configuration is 2,7. If the fluorine atom gains one electron, a fluoride (F^-) ion is formed with the configuration of neon:

$$F + 1e^- \longrightarrow \quad F^-$$
$$2,7 \qquad\qquad\quad 2,8$$

(NEON CONFIGURATION)

The electronic configuration of oxygen is 2,6. With the gain of two electrons, an oxygen atom forms an oxide ion (O^{2-}) with the neon configuration.

Stable ions often have noble gas electronic configurations.

TABLE 3–2   Electronic Configuration
of the Noble Gases and Ions
with Identical Configurations

Species	Configuration
He, Li^+, Be^{2+}	2
Ne, Na^+, Mg^{2+}, F^-, O^{2-}	2,8
Ar, K^+, Ca^{2+}, Cl^-, S^{2-}	2,8,8
Kr, Rb^+, Sr^{2+}, Br^-, Se^{2-}	2,8,18,8
Xe, Cs^+, Ba^{2+}, I^-, Te^{2-}	2,8,18,18,8

$$O + 2e^- \longrightarrow \qquad O^{2-}$$
$$2,6 \qquad\qquad\qquad 2,8$$

(NEON CONFIGURATION)

On the basis of atomic structures, ionization energies, and the inert character of the noble gases, we have thus far reasoned that metals should form positive ions by losing electrons and nonmetals should form negative ions by gaining electrons. More specifically, we reasoned that Na, Ca, F, and O tend to form Na^+, Ca^{2+}, F^-, and O^{2-} ions, respectively. There is ample experimental evidence that this is actually the case. For example, consider the formulas of the compounds that can be formed from these elements and the equations for the reactions involved:

<table>
<tr><td>*Compound*</td><td>*Equation*</td></tr>
<tr><td>NaF</td><td>$2Na + F_2 \longrightarrow 2Na^+ + 2F^-$</td></tr>
<tr><td>CaF_2</td><td>$Ca + F_2 \longrightarrow Ca^{2+} + 2F^-$</td></tr>
<tr><td>Na_2O</td><td>$4Na + O_2 \longrightarrow 4Na^+ + 2O^{2-}$</td></tr>
<tr><td>CaO</td><td>$2Ca + O_2 \longrightarrow 2Ca^{2+} + 2O^{2-}$</td></tr>
</table>

Chemical formulas of ionic compounds result from the number of electrons lost or gained by the reacting atoms and the neutrality of the compound, which requires equal numbers of positive and negative charges.

Note that in each case the formula is compatible with the ideas developed previously, namely:

a. a sodium atom loses 1 electron to form Na^+,
b. a calcium atom loses 2 electrons to form Ca^{2+},
c. a fluorine atom gains 1 electron to form F^-,
d. an oxygen atom gains 2 electrons to form O^{2-}, and
e. the compound is neutral.

These four compounds are indeed ionic solids, as indicated by X ray studies as well as chemical properties. When molten, all four compounds conduct electricity; this behavior is characteristic of mobile ions (see Chapter 5). Mass spectrographic studies confirm that these ions have a real and stable existence and that they actually have the charges indicated.

There is extensive evidence for the existence of ions in solids and many solutions.

For ionic compounds of most of the light elements we can write:

1. There is a tendency for an atom with fewer than four electrons in its outer orbit to lose one or more electrons and form a positive ion with an electron configuration of a noble gas;
2. There is a tendency for an atom with 5, 6, or 7 electrons in its outer orbit to gain one or more electrons and form a negative ion with an electronic configuration of a noble gas;
3. When metal atoms react with nonmetal atoms, there is a tendency for the metal atoms to lose electrons to the nonmetal atoms and to form ions;
4. An atom with four electrons in its outer orbit has no pronounced tendency to lose or gain electrons or to form ions.

These generalizations should not be thought of as chemical laws, but merely as memory aids for the cases most frequently encountered.

Electron Dot Formulas

To keep track of the electrons involved in electron-transfer reactions such as those just discussed, we use a type of notation known as the *electron dot formula.* To write the electron dot formula of an atom or ion, the electrons in the outermost orbit are counted, and for each electron a dot is placed around the symbol for the element. For example, fluorine, with seven outer electrons, is $\cdot \ddot{F} :$. The reaction between sodium and fluorine is written with electron dot notation as follows

$$\text{Na} \cdot + \; : \ddot{F} \cdot \longrightarrow \text{Na}^+ + \; : \ddot{F} :^-$$

Note that Na⁺ has no dots around it, since it has none of its original valence electrons. The reaction between calcium and fluorine can be written using electron dot formulas as follows

$$\text{Ca} : + \; 2 : \ddot{F} \cdot \longrightarrow \; : \ddot{F} :^- + \; \text{Ca}^{2+} + \; : \ddot{F} :^-$$

Since the calcium atom has two outermost electrons and loses both of them, it must react with two fluorine atoms, each of which takes one electron.

Properties of Ions

Ion Sizes

The sizes (radii) of atoms of practically all the elements can be measured by several experimental methods. Let us consider briefly the sizes of the ions that are produced by either electron gain or electron loss. The sizes of the ions are important because the strength of the forces

METALS   NONMETALS

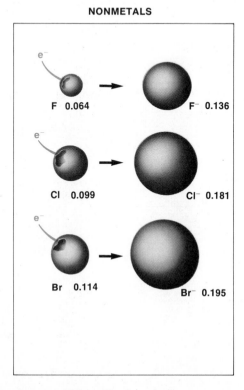

Figure 3—4 Relative sizes of selected atoms and ions. Numbers given are atomic or ionic radii in nanometers.

holding ions together in ionic compounds depends on the sizes of the ions involved.

One nanometer, nm, is 10^{-9} meter.

A sodium atom with its single outer electron has a radius of 0.186 nm. One would expect that when this electron is removed (forming the Na^+ ion) the resulting ion would be smaller. This decrease in size results because there are now only 10 electrons attracted to a charge of $+11$ on the nucleus, and these electrons are pulled closer to the nucleus by this charge imbalance. This same type of phenomenon is observed for all metal ions. Figure 3–4 compares atomic and ionic sizes for seven elements.

The metal ions with multiple charges (Al^{3+}, Fe^{2+}, and so on) are much smaller than the corresponding metal atom, because of still greater surplus positive charge on the nucleus.

Nonmetals gain electrons to form negative ions that are *larger* than the corresponding atoms. Consider the information in Figure 3–4. This phenomenon is due to the addition of electrons to the outer orbit of an atom without increasing the charge on the nucleus. The repulsion of the electrons causes the expansion.

Electrostatic Forces—The Ionic Bond

Ionic bonding is due to electrostatic interaction between ions of opposite charge. The geometrical array of ions is termed an ***ionic lattice,*** and the substance is an ionic solid (Figure 3–5, p. 62).

Ionic bonds are the attractions between ions of opposite charge.

Coulomb's law tells us that oppositely charged ions will attract each other with a force which depends on the product of the charges and inversely as the square of the distance between them. We can express this coulombic attraction as:

$$\text{Force} \left\{ \begin{array}{c} \text{is} \\ \text{proportional} \\ \text{to} \end{array} \right\} \frac{(\text{charge on positive ion}) \times (\text{charge on negative ion})}{(\text{distance of separation})^2}$$

Ionic attraction is determined by Coulomb's law.

This expression leads us to expect that strong ionic bonds are formed when the ions possess high charges and are separated by small distances (which will occur for the smaller ions). The effects of ionic charge and size can be seen rather easily if we compare the melting points of several salts, all with similar crystal geometries. Let us compare the melting points of NaCl and KCl (the K^+ ion is larger than the Na^+ ion) and also those of CaO and BaO (the Ba^{2+} ion is larger than the Ca^{2+} ion). All four salts have a sodium chloride (rock salt) crystalline structure (see Figure 3–5). Table 3–3 shows that in both cases the salts with smaller positive ions have the higher melting points. Those with doubly charged ions have dramatically increased melting points.

Ionic Lattices

There are numerous kinds of ionic lattices, depending on the charges and sizes of the ions involved. A common type is the sodium chloride lattice (Figure 3–5). Here, each Na^+ ion is surrounded by six Cl^- ions, and each Cl^- ion is in turn surrounded by six Na^+ ions. Each ion is attracted to all of the oppositely charged ions, and repelled by all of the like charged ions in the lattice. Thus, there are no sodium chloride diatomic molecules in this solid.

An ionic lattice is a regular geometrical array of ions.

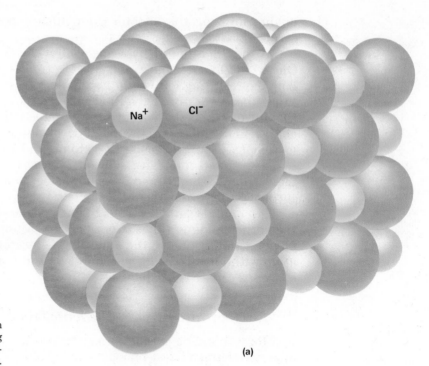

Figure 3—5 Structure of sodium chloride crystal. (a) Model showing relative sizes of the ions; (b) ball-and-stick model showing cubic geometry.

(a)

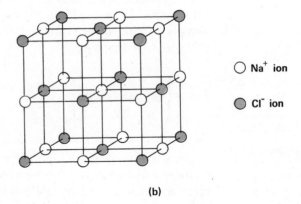

○ Na⁺ ion

⬤ Cl⁻ ion

(b)

TABLE 3—3   The Effects of Ionic Charge and Size on
Some Salts with Identical Structures

		M⁺, Radius, nm	Melting Point, °C
Salts with singly charged ions	NaCl	0.095	801
	KCl	0.133	776
		M²⁺, Radius, nm	Melting Point, °C
Salts with doubly charged ions	CaO	0.099	2580
	BaO	0.135	1923

One question arises at this point. If the oppositely charged ions attract one another, what keeps them from getting closer and closer to each other? This is prevented by the repulsion that arises when the electrons in the outer shell of one ion approach those in an adjacent ion. As a consequence, each ion has an ionic radius which is reasonably constant and which represents the closest distance of approach of another ion.

Other Properties of Ionic Compounds

The strong bonding between the ions in ionic compounds causes most of them to be rather hard and brittle crystalline solids with relatively high melting points. Since the electrons involved in bonding are localized on the individual ions and cannot move about in the lattice, solid ionic compounds are rather poor electrical conductors as compared to molten ionic compounds. Molten ionic compounds contain free ions, since the melting point partially breaks down the rigid structure that had existed in the lattice. These free ions can move in an electrical field and thus transport charge or an electric current from one place to another. Hence, molten salts are generally excellent conductors of electricity.

---

1. Charged atoms are called _____.

2. An array of attracting ions makes up an _____ _____.

3. A sodium atom loses _____ electron(s) in achieving a noble gas configuration.

4. Look at Figure 3–3. Which element requires the most energy for ionization?

   _____ The least? _____

5. What is the correct formula for calcium iodide? _____ The electron dot structures are Ca $:$ and $\cdot \ddot{I} :$.

6. Which ion gained an electron in its formation: Na^+ or Cl^-? _____

7. Electrons in the outer orbit may be called _____ electrons.

8. Positive ions are formed from neutral atoms by (losing/gaining) electrons.

9. Negative ions are formed from neutral atoms by (losing/gaining) electrons.

10. Predict the number of electrons lost or gained by the following atoms in forming ions. Indicate whether the electrons are gained or lost.

   Rb _____ ; S _____ ; Ca _____ ;

   Mg _____ ; Al _____ ; Br _____ .

COVALENT BONDING

A second type of chemical bonding, which is found in molecular compounds, nonmetallic elements, and polymers, is covalent bonding. This type of bonding is different from ionic bonding in that electrons are shared between atoms, rather than transferred from one atom to another.

The atoms that tend to bond covalently are those with relatively high ionization energies. These atoms do not readily lose electrons. Hence, the atoms must share electrons in their compounds if the atoms are to acquire a noble gas configuration. Some substances, such as carbon dioxide (CO_2), carbon tetrachloride (CCl_4), ammonia (NH_3), hydrogen (H_2), chlorine (Cl_2), and water (H_2O), are composed of molecules which are held together by covalent bonding.

A covalent bond results from sharing an electron pair.

Electron Sharing

The simplest example of covalent bonding is found in diatomic (two atoms) gaseous molecules such as H_2 (hydrogen), F_2 (fluorine), and Cl_2 (chlorine). A hydrogen atom has one bonding electron. Comparing this with the electronic configuration of the noble gas nearest to hydrogen in the periodic table, helium (two electrons), we see that, if the hydrogen atom could somehow obtain one more electron, a very stable configuration could be reached. For hydrogen, this configuration can be achieved by two hydrogen atoms sharing their single electrons. The electron dot formula for the resulting H_2 molecule is

Covalent bonding arises from the sharing of electrons between atoms.

$$2H \cdot \longrightarrow H : H$$

ATOMS   MOLECULE

Ionic bonding is not a reasonable possibility for the H—H bond since both hydrogen atoms have the same attraction for the two electrons, neither having the ability to remove the electron completely from the other.

For other atoms of the second period (elements 3 to 10) we may use the tendency to attain eight electrons as a guide in writing electron dot formulas. Two fluorine atoms bound together form a F_2 molecule. This bonding is due to each fluorine atom sharing one of its outer electrons.

$$2 : \ddot{F} \cdot \longrightarrow \quad : \ddot{F} : \ddot{F} :$$

ATOMS   FLUORINE MOLECULE

Orbital Overlap

Since hydrogen gas is molecular, and since the interaction between the two atoms must be other than ionic, the question arises as to how one hydrogen atom can be bonded to another. Recall that the electrons in an atom are subdivided into orbitals. Since an orbital can contain two electrons, it is reasonable to assume that the close approach of two orbitals containing one electron each on different atoms will result in the formation of a covalent bond in which both orbitals are made available to both of the electrons. Thus, the bond in the hydrogen molecule can be pictured as resulting from the overlap of two s orbitals, one from each atom, and

The theoretical model of sharing electrons involves overlap of atomic orbitals.

Figure 3–6 Overlap of s orbitals in H_2 molecule.

hydrogen
s orbital
hydrogen
s orbital
H_2 molecule

the pairing of the electrons involved (Figure 3–6). It is thought that the greater the overlap, the stronger the bond will be.

It is assumed that stability is achieved when two nuclei attract the same pair of electrons in the formation of a covalent bond. Figure 3–7 illustrates this for Cl atoms. The amount of energy released, ΔE, is approximately the *bond energy,* and is the same as the amount of energy necessary to break the Cl—Cl bond and give gaseous atoms. In general, energy is released in the formation of covalent bonds, and energy is required to break such bonds. Notice in Figure 3–7 that any attempt to force the two atoms closer together than their equilibrium bond length will require a large amount of energy, since there is a repulsion caused by the electrons in the filled atomic orbitals of the two atoms.

To break a bond requires energy; when bonds are formed, energy is released.

Single Bonds

A covalent bond formed by the sharing of two electrons (a single pair of electrons) between two atoms is called a ***single bond.*** Many molecules, both simple and complicated, contain one or more single covalent bonds.

Hydrogen fluoride (HF) contains a single covalent bond between the two atoms. Fluorine ($:\ddot{F}\cdot$) with seven outer electrons needs one more electron to attain a noble gas configuration, as does hydrogen ($H\cdot$). This could be accomplished by an overlap of the hydrogen orbital, a spherical s orbital, with a half-filled fluorine orbital. The fluorine orbital that is involved is a p orbital (Figure 3–8).

A single bond consists of one pair of electrons shared by two atoms.

Figure 3–7 The relative stabilities of Cl_2 molecules and Cl atoms. The bond energy of the Cl-Cl bond is approximately equal to ΔE.

Figure 3–8 Single bond formation in HF. (a) The electron dot representation; (b) the orbital overlap representation.

(b) s orbital in first energy level — p orbital in second energy level

An oxygen atom has six outer electrons and therefore needs two more electrons to attain a noble gas configuration. (We have already seen how oxygen atoms may gain two electrons and form O^{2-} ions.) If two hydrogen atoms are present, there can be sharing of electrons between them and the oxygen atom. The electron dot formula for the molecule that results, the water molecule, is given below. Oxygen is written as $:\ddot{O}\cdot$ because there are two orbitals that contain one electron each.

The choice of H:Ö: over H:Ö:H will be explained later in this chapter.

$$:\ddot{O}\cdot\ +\ 2H\cdot\ \longrightarrow\ :\ddot{O}\!:\!H$$
$$\overset{\displaystyle}{\ \ \ \ \ \ \ \ \ \ \ \ \ \ \ }\overset{H}{}$$

WATER

Consider next the nitrogen atom. Modern atomic theory predicts that three of its orbitals (the p orbitals) contain one electron each and one orbital (the s orbital in the second energy level) contains two electrons; thus, the nitrogen atom is written as $\cdot\ddot{N}\cdot$. The atom needs three more electrons to attain a noble gas configuration. One way this can be done is by forming three single bonds with some other atoms that can share one electron each. With hydrogen atoms, the result is ammonia (NH_3).

$$\cdot\ddot{N}\cdot\ +\ 3H\cdot\ \longrightarrow\ H\!:\!\ddot{N}\!:\!H$$
$$\overset{}{\ \ \ \ \ \ \ \ \ \ \ \ \ \ \ \ \ \ }\overset{H}{}$$

AMMONIA

Note that fluorine, oxygen, and nitrogen all end up with an octet (8) of electrons in HF, H_2O, and NH_3. This leads to the generalization known as the ***octet rule,*** which predicts that second period elements on the periodic table form covalent compounds in such a way as to complete their octet of electrons. Even though this rule can often be applied usefully to elements with heavier atoms, it does not always apply since some of these atoms can have more than eight electrons in their outer shells. Though the rule is generally true for second period elements, there are even a few exceptions among them.

The octet rule: eight electrons in the valence shell represent a stable state.

Multiple Bonding

When an atom has fewer than seven electrons in its outermost shell, it can complete its octet in two ways. In the first way, it simply shares a

Formula	Name	Electron Dot Structure	Line Structure

Double Bonds:

CO_2	Carbon dioxide	$:\ddot{O}::C::\ddot{O}:$	$O=C=O$
C_2H_4	Ethylene	(electron dot structure)	(line structure)
SO_3	Sulfur trioxide	(electron dot structure)	(line structure)

Figure 3–9 Electron dot structures of some molecules containing multiple bonds. Line structures are shown for comparison.

Triple Bonds:

N_2	Nitrogen	$:N:::N:$	$N\equiv N$
CO	Carbon monoxide	$:C:::O:$	$C\equiv O$
C_2H_2	Acetylene	$H:C:::C:H$	$H-C\equiv C-H$

One pair of electrons = one covalent bond.

single electron with each of several other atoms which can contribute a single electron each. This leads to *single* covalent bonds. But the atom can also share two (or three) pairs of electrons with a single other atom. In this case there will be two (or three) bonds between these two atoms.

A double bond consists of two electron pairs shared between two atoms.

When two shared pairs of electrons join together the same two atoms, we speak of a *double bond,* and when three shared pairs are involved, the bond is called a *triple bond.* Examples of these bonds are found in many compounds such as those shown in Figure 3–9.

As we can see from these structures, molecules may contain several types of bonds. Thus, ethylene (Figure 3–9) contains a double bond between the carbon atoms and single bonds between the hydrogen atoms and the carbon atoms. For convenience, an electron pair bond is often indicated by a dash as follows:

$$\begin{matrix} H & & H \\ & C=C & \\ H & & H \end{matrix}$$

A line between two atoms, as in H—H, represents a bonding pair of electrons.

The H_2 molecule with a single bond is shown as $H-H$; ethylene, with a double bond, is shown as $H_2C=CH_2$; and diatomic nitrogen, with a triple bond, is shown as $N\equiv N$. Note that in each of these cases the octet rule is obeyed if the shared electrons can be counted as belonging to both atoms.

Single, double, and triple bonds differ in length and strength. Triple bonds are shorter than double bonds, which in turn are shorter than sin-

TABLE 3—4   Some Bond Lengths and Bond Energies

Bond type	C—C	C=C	C≡C	N—N	N=N	N≡N
Bond length (nm)	0.154	0.134	0.120	0.140	0.124	0.109
Bond energy (kcal/mole)	83	146	200	40	100	225

kcal/mole = thousands of calories necessary to break 6.02×10^{23} bonds.

gle bonds. Bond energies normally increase with decreasing bond length owing to greater orbital overlaps. Bond lengths and energies in some typical cases are shown in Table 3–4.

Polar Bonds

In a molecule like H_2 or F_2, where both atoms are alike, there is equal sharing of the electron pair. Where two unlike atoms are bonded, however, the sharing of the electron pair is unequal, and results in a shift of electric charge toward one partner or the other. Such a bond is termed a **polar bond.** The negative portion is the region of high electron density, and the positive portion is the region of low electron density. In the molecule HF, the bonding pair of electrons is more under the control of the fluorine atom than of the hydrogen atom (Figure 3–10). The un-

In a polar bond there is an unequal sharing of the bonding electrons.

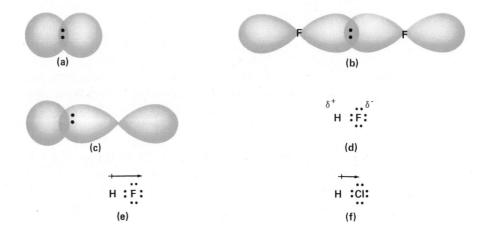

Figure 3—10 Polar bonds in HF and HCl. (a) Symmetrical distribution of electrons in H_2 results in the center of negative charge being identical with the center of positive charge. This is symbolized by the electron dots placed in the overlap area. (b) Overlap of p orbitals in F_2 also results in symmetrical distribution of charge. (c) In HF, the electron pair is displaced toward the fluorine nucleus since fluorine is more electronegative than hydrogen. Note the electron dots to the right of the overlap area which conveys the idea of polarity (separation of charge). (d) δ^+ (delta positive, meaning fractional positive charge) and δ^- (delta negative, meaning fractional negative charge) are used to indicate poles of charge. In (e) and (f) an arrow is used to indicate electron shift, the arrow having a "plus" tail to indicate partial positive charge on the hydrogen atom. Note that the longer arrow in the HF structure indicates a greater degree of polarity than in HCl. This should not be confused with the greater bond length in HCl.

equal sharing of electrons can also be expressed in terms of the centers of positive and negative charge. In the nonpolar hydrogen molecule, the center of negative charge is in the geometric center of the molecule, since there is a symmetrical distribution of negative charge (electrons) about that center. The center of positive charge is located there also.

In contrast to the nonpolar hydrogen molecule, the polar HF molecule has two distinct centers of charge—a positive center near the hydrogen atom, and a negative center near the fluorine atom. These are indicated by the symbols δ^+ and δ^-, or by the symbol $\longmapsto$ (Figure 3–10).

A measure of the net ability of an atom to attract electrons to it in a chemical bond is called its ***electronegativity.*** Since the fluorine atom attracts the shared electrons in the H—F bond more than does the hydrogen atom, we say that the electronegativity of the fluorine atom is greater. Electronegativity is a function of nuclear charge and atomic size. Small atoms with large nuclear charges are the most electronegative. The trend of electronegativities in the periodic table is periodic and follows an inverse trend in atomic size. Figure 3–11 shows these relationships.

Covalent bonds joining different atoms are generally polar in that one of the atoms has distorted the electron distribution toward itself due to its greater electronegativity. Thus, polar covalent bonds occur in practically every molecule that has different kinds of atoms covalently bonded together. Additional examples of molecules with polar covalent bonds include NO, SO_2, CO_2, CCl_4, and H_2S (Figures 3–12 and 3–13).

Unequal sharing of electrons results from differences in electron attraction by the bonding atoms.

The electronegativity of an atom is a measure of its ability to attract electrons to itself in a compound.

The most electronegative atom is fluorine.

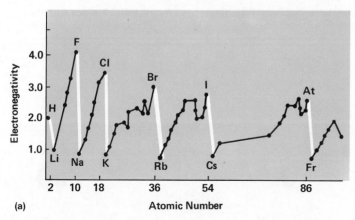

(a)

(b)

Figure 3–11 (a) Electronegativities of the elements. (b) Atomic radii of the elements.

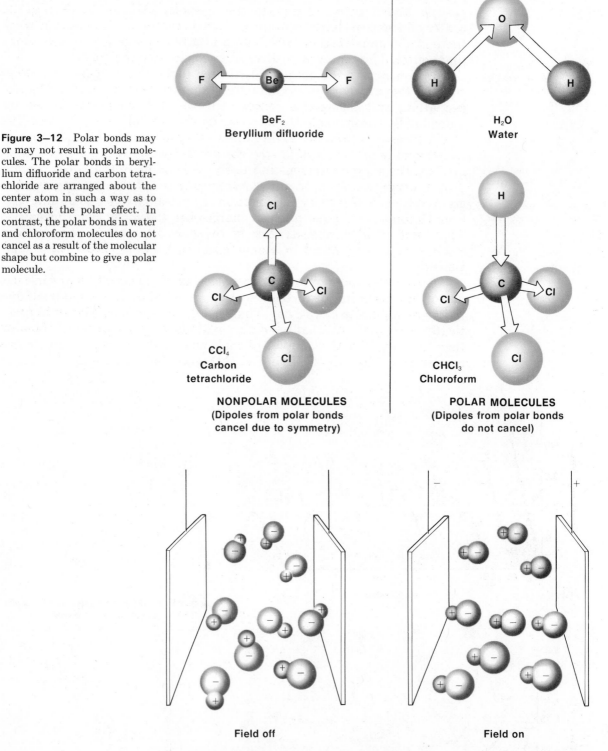

Figure 3–12 Polar bonds may or may not result in polar molecules. The polar bonds in beryllium difluoride and carbon tetrachloride are arranged about the center atom in such a way as to cancel out the polar effect. In contrast, the polar bonds in water and chloroform molecules do not cancel as a result of the molecular shape but combine to give a polar molecule.

BeF$_2$
Beryllium difluoride

H$_2$O
Water

CCl$_4$
Carbon tetrachloride

CHCl$_3$
Chloroform

NONPOLAR MOLECULES
(Dipoles from polar bonds cancel due to symmetry)

POLAR MOLECULES
(Dipoles from polar bonds do not cancel)

Field off

Field on

Figure 3–13 Physical evidence for both the existence of polar molecules and the degree of polarity is provided by a simple electrical capacitor. The capacitor is composed of two electrically conducting plates with nonconducting material (an electrical insulator) between. The storage of charge by the capacitor is least when there is a vacuum between the plates; charge storage is more effective when nonpolar substances are placed between the plates, and is most effective with polar substances between the plates. In the study of individual molecules, the material is placed between the plates in the gas phase.

A very common bond that we shall discuss frequently in the remainder of this text is the C—H bond. Since carbon has an electronegativity of 2.5 and hydrogen 2.1, the C—H bond is only slightly polar. The arrangement of C—H bonds around a carbon atom generally makes —CH$_2$— and —CH$_3$ groups essentially nonpolar.

HYDROGEN BONDING

For a series of molecular substances with similar structures, the boiling points ordinarily increase as the molecular weights increase. For example, the boiling points of fluorine (F$_2$), chlorine (Cl$_2$), bromine (Br$_2$), and iodine (I$_2$) increase rather regularly with increasing molecular weight (Figure 3–14).

The general relationship between boiling points and molecular weights also holds for hydrogen chloride (HCl), hydrogen bromide (HBr), and hydrogen iodide (HI). However, the boiling point of hydrogen fluoride (HF), the lightest member of this series of compounds, is abnormally high (Figure 3–15). Irregularities similar to this are also found in other compounds in which hydrogen is bonded to fluorine, oxygen, and nitrogen. The increased attraction between molecules containing H—F, H—O, or H—N bonds is termed *hydrogen bonding.*

Strongly electronegative atoms compete for a single hydrogen atom to give hydrogen bonds.

The explanation for hydrogen bonding is to be found in the extremely large electronegativity differences of the H—F, H—O, and H—N bonds, the polarity being due to the extreme electronegativity of fluorine, oxygen, and nitrogen. Consider the bonding in liquid HF. Since unlike-charged ends of these molecules should attract each other, we expect HF molecules to be associated with one another. This association is illustrated in Figure 3–16. The increased association between HF molecules compared to that found in HCl offers a ready explanation for the unusually high boiling point of HF.

For a substance to boil, its molecules must gain enough energy to break loose from each other.

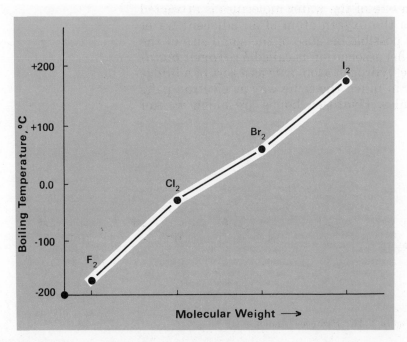

Figure 3–14 Boiling points of F$_2$, Cl$_2$, Br$_2$, and I$_2$, as a function of molecular weight.

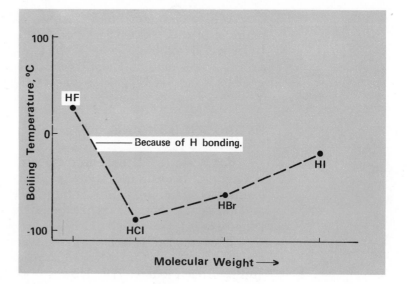

Figure 3–15 Boiling points of HF, HCl, HBr, and HI, plotted against molecular weights.

Water provides another good example of hydrogen bonding. Water molecules are not linear but rather are angular, with two nonbonding (unshared) electron pairs located toward one end of the molecule and the partially positive hydrogen atoms located toward the opposite end. The two polar bonds in this geometry result in a distinctly polar molecule.

$$\overset{\delta-}{\ddot{\text{O}}} \quad \text{H} \qquad \text{H} \quad {\scriptstyle \delta+}$$

In liquid and solid water, where the molecules are close enough to interact, the hydrogen atom on one of the water molecules is attracted to the nonbonding electrons on the oxygen atom of an adjacent water molecule (Figure 3–17). This is possible because of the small size of the hydrogen atom. The result of this association is called ***hydrogen bonding*** because the slightly positive hydrogen atom acts as a sort of a bridge to hold two molecules together in much the same way as electron pairs hold atoms together in molecules. Hydrogen bonds are much weaker than ordinary covalent bonds.

Figure 3–16 Hydrogen bonding in HF.

Figure 3—17   Hydrogen bonding in water.

The extent of hydrogen bonding in liquid water varies with temperature. In ice, hydrogen bonding is very extensive. This results in a very open structure for ice (Figure 3–18). Consequently, at ordinary pressures ice is *less* dense than water. As ice is melted and the molecules gain more energy and move about, this bonding begins to break down. Not all the hydrogen bonds are broken, however, and large aggregates of water mole-

Water has an abnormally high boiling point because of hydrogen bonding between the molecules.

Figure 3—18   Hydrogen bonding in the structure of ice. The hydrogen bonds are indicated by the dashed lines. In liquid water the hydrogen bonding is not as extensive as it is in ice.

cules exist in liquid water even near 100°C. As water is heated, thermal agitation disrupts the hydrogen bonding until, in water vapor, there is only a small fraction of the number of hydrogen bonds that are found in liquid or solid water.

Freezing water expands due to hydrogen bonding.

SELF-TEST 3-B

1. (a) An example of a molecule containing covalent bonding where the electrons are equally shared between the atoms is _____; (b) one where they are unequally shared is _____.

2. The number of electrons shared in a triple covalent bond is _____.

3. There are _____ covalent bonds in an ammonia (NH_3) molecule.

4. Which atom cannot form a double bond, fluorine or oxygen? _____

5. (a) How many valence electrons are thought to be involved in molecules containing covalently bound atoms of Period 2 and 3 elements? _____ (b) This is known as the _____ rule. (c) Is it true most of the time or all of the time? _____

6. Which is the most electronegative of all of the elements? _____

7. Hydrogen bonding very probably occurs when hydrogen is bound to atoms of _____, _____, or _____.

VAN DER WAALS ATTRACTIONS

The bonding between chemical species that cannot be explained by ionic, covalent, or hydrogen bonding is often due to a weak attraction known as van der Waals attraction. In helium, for example, the two electrons are nonbonding, and yet there is a slight attraction between two helium atoms. This slight attraction allows helium to become a solid at $-272°C$ under high pressure.

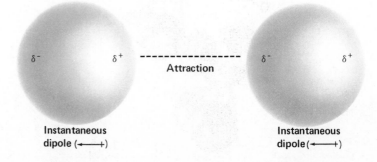

Figure 3–19 An illustration of van der Waals attraction. One instantaneous dipole interacts with another in a neighboring atom.

Van der Waals interactions can arise from a variety of causes. For instance, as two atoms or molecules approach each other, intermolecular interactions will cause a temporary shifting of their electron clouds. An uneven electron distribution in an atom makes the atom itself polar (a *dipole,* meaning "two poles"). These temporary poles on two adjacent atoms can interact with each other, resulting in the attractive force (Figure 3–19).

Van der Waals attractions explain the existence of solid helium.

This type of dipole-dipole interaction is called van der Waals attraction, after the Dutch physicist who suggested its possibility. The existence of the solid state of many nonpolar molecular substances (such as oxygen, nitrogen, and helium) can be explained only by invoking van der Waals attractions.

METALLIC BONDING

Metals have some properties totally unlike those of other substances. For example, most metals are very good electrical conductors; they are shiny solids and have relatively high melting points (with a few notable exceptions such as mercury). Any theory of the bonding of metal atoms must be consistent with these properties. Structural investigations of metals have led to the conclusion that metals are composed of very regular arrays (lattices) of metal ions in which the bonding electrons are very mobile (or delocalized). Figure 3–20 shows a diagram of a sodium lattice. The bonding electrons in sodium are third energy level electrons. These can be made to move rather easily through the lattice upon application of an electric field. In this way the metal acts as a conductor of electricity. As a consequence of this delocalization of bonding electrons, we cannot really write a satisfactory description of the bonding in a metal using an electron dot structure.

Delocalized electrons are found in metals. They can move freely and are not confined to the area between any particular pair of atoms.

MOLECULAR SHAPES

The modern theory of the atom is very effective in explaining bonding and periodicity of the elements. It is also capable of explaining related

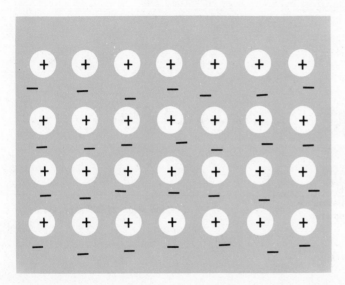

Figure 3–20 A cross-sectional view of the crystal lattice in metallic sodium. The bonding electrons are *delocalized,* that is, they are not fixed between any given atoms, but seem to be capable of movement throughout the metallic crystal.

chemical phenomena, such as molecular shapes. The shape of a molecule is determined by the arrangement of the atoms in the molecule with respect to each other, that is, by the angles between and the lengths of its covalent bonds. The angle formed by two intersecting lines drawn from the two nuclei of the attached atoms through the nucleus of the central atom is called the ***bond angle.*** Molecular shapes are established by many different experimental techniques. Here we shall see what molecular shapes are predicted by atomic theory and what modifications are needed for the theoretical and the experimental findings to agree.

Molecular Shapes Predicted by Atomic Orbitals

The starting point for a discussion of molecular shapes is the set of atomic orbitals, since these determine the preferred regions in space for electrons which will be involved in the bonding. Expected molecular geometries based on orbital overlap when s and p atomic orbitals are used to form bonds are shown in Figure 3–21.

Since an s orbital is spherically symmetrical, there is no preferred direction for the bonding with the other atom. There is only one s orbital per valence shell and only one bond can be formed from this kind of orbital. Hydrogen, with one electron in its outer s orbital, can form only one bond, and there is no specified direction from the hydrogen atom.

There are three p orbitals per valence shell. The electrons in a set of p orbitals concentrate charge along the x, y, and z axes, which are at 90° angles to each other. Any two atoms bonded to a third atom by overlap-

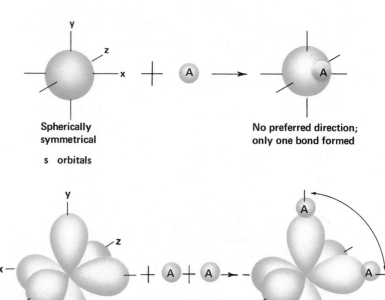

Figure 3–21 Predicted molecular shapes when s and p orbitals are used to form bonds. **A** represents any atom to be joined to the central atom.

ping of two p orbitals would be expected to form bonds enclosing an angle of 90°.

Is this predicted bond angle actually found in molecules? Some molecules, such as those of H_2S and H_2Se, come close. However, instead of the expected bond angle of 90°, when s orbitals of two hydrogens overlap two of the p orbitals of S or Se, the angles are 92° for H_2S and 91° for H_2Se. It is rare for the bond angles of molecules to agree this closely with the angles predicted by combinations of simple atomic orbitals. In H_2O, for example, the bond angle is 104.5°.

Before discarding the theory, we should look a little further. With a slight adjustment, the modern atomic theory can be brought into agreement with experimentally derived molecular shapes of relatively simple molecules. The theory is also available for use with complex molecules which contain large numbers of atoms.

Hybridization combines atomic orbitals to give new orbitals with different properties.

The method of adjusting the pure atomic orbitals is called ***hybridization.*** This process brings a modified theoretical model in agreement with experiment. Although hybridization is a mathematical process, we can gain an insight into its methods and its usefulness through a qualitative examination of its features. An alternative approach to this same problem is the ***valence-shell electron-pair repulsion theory.***

Both of these theories will be examined, along with some of their applications and some of the experimental observations used as a guide in their development.

Hybridization

The tetrahedral structure of species such as CH_4 and NH_4^+ requires an adjustment of atomic theory. The isolated carbon atom in its ground state has the electron configuration 2,4, with the four electrons in the second energy level arranged with two in the 2s orbital, one in the $2p_x$ orbital, one in the $2p_y$ orbital, and none in the $2p_z$ orbital, which we will write in a shorthand notation as 2s-2, $2p_x$-1, $2p_y$-1, $2p_z$-0. The letters x, y, and z are used to denote the orientation of the three p orbitals along lines that are perpendicular to each other. If the unpaired electrons in the 2p orbitals are used for bonding, this leads us to expect that carbon would form two bonds at an angle of 90°. Actually carbon commonly forms four single bonds, and each bond angle is 109.5°. This angle directs each bond from carbon toward a corner of a regular tetrahedron.

The ground state of an atom is its lowest possible energy state. Higher energy electron configurations result when energy is added, such as in an experiment in spectroscopy.

To explain these four equivalent bonds, we assume that prior to or during the formation of the bonds, four orbitals of the carbon atom are hybridized (or mixed) to form a new set of four equivalent orbitals. The carbon atom promotes one of the electrons from the 2s orbital to the empty $2p_z$ orbital to give the electronic arrangement: 2s-1, $2p_x$-1, $2p_y$-1, $2p_z$-1 for the outer electrons. These four orbitals are rearranged mathematically according to specific rules to obtain four new equivalent hybrid orbitals. These are designated sp^3 orbitals, since they are made from one s and three p orbitals. The new orbitals are directed toward the corners of a tetrahedron. The four valence electrons are distributed equally among the hybrid orbitals (one electron to each orbital). The hybrid orbitals are shown graphically in Figure 3–22. The energy required for this transition is less than that which is provided by the formation of additional, stronger bonds with the hybrid orbitals. Thus, we have a more

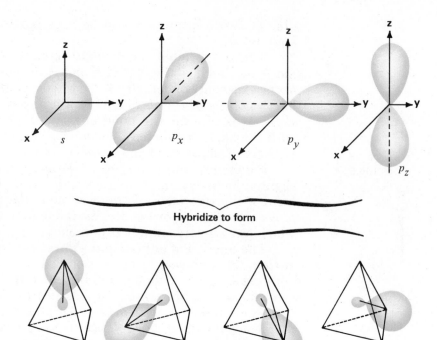

Figure 3–22 Hybridization of one s and three p orbitals into four sp³ orbitals results in new spatial representations of the electron density patterns.

favorable energy change than would result from bonding with p orbitals alone.

Other sets of bonding orbitals with geometries very closely in accord with established molecular geometries can be derived in a similar fashion. Some of these are listed in Figure 3–23. A hybridization analogy in orbitals, plants, and animals is illustrated in Figure 3–24.

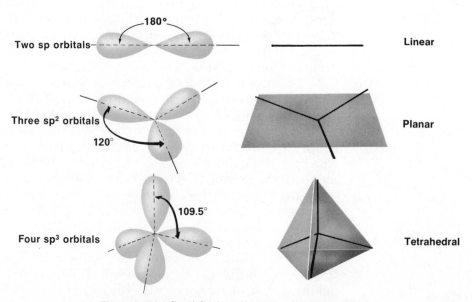

Figure 3–23 Spatial orientations of some hybrid orbitals.

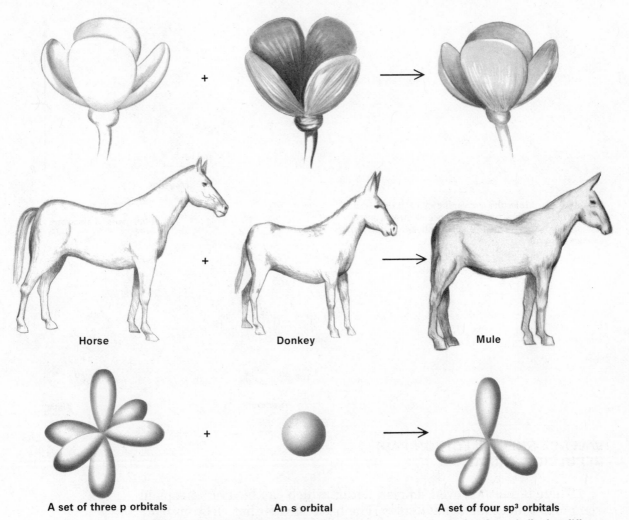

A set of three p orbitals   **An s orbital**   **A set of four sp³ orbitals**

Figure 3–24 Hybridization is a familiar concept in nature. A mixture of two similar but different entities produces an entity with intermediate characteristics. Biologic hybrids are analogous to hybrid orbitals in the sense of being the result of mixtures, yet hybrid orbitals are obtained much more exactly by mathematical calculations.

As examples of the use of hybrid orbitals, consider tetrahedral molecules such as CH_4, $SiCl_4$, and CCl_4. In all of these, the atomic orbitals of the central atom can be described as sp^3 hybrids. For molecules such as H_2O and NH_3, sp^3 hybrid orbitals can also be postulated if it is assumed that the ***nonbonding pairs*** of electrons (those on the O or N which are not involved in bonds to H) are also placed in hybridized orbitals. The experimentally determined bond angles in these molecules are not quite the same as normal tetrahedral bond angles (as shown in Figure 3–25); however, the agreement is relatively close.

NONBONDING PAIRS

Figure 3—25 Molecules containing sp³ orbitals. Methane has only bonding pairs of valence electrons; water and ammonia have bonding and nonbonding pairs of valence electrons.

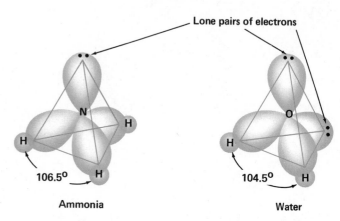

VALENCE-SHELL ELECTRON-PAIR REPULSION THEORY

There is another, even simpler, notion which can be used to explain bond angles in covalent molecules. The basic idea is that electron pairs repel each other. In fact, this theory assumes that electron pairs in the valence shell of a central atom behave like a group of electrically charged balloons that are connected to a central point by strings. If similarly charged, the balloons would tend to be as far apart from each other as possible. In a similar manner, electron-pair repulsion leads to the arrangements shown in Figure 3–26.

These ideas allow us to predict the shapes of molecules if the underlying electronic structure is known. It is particularly simple when every pair of electrons in the valence shell is a bonding pair. Examples can be seen in BCl_3, $SiCl_4$, and SF_6.

BCl_3

Atomic boron has three outer orbit electrons in its ground state electronic structure. When these are paired up with the unpaired electrons of three chlorine atoms, $:\ddot{C}l\cdot$, three electron-pair bonds are formed. This leads to the structure, shown below, which actually is the structure found experimentally.

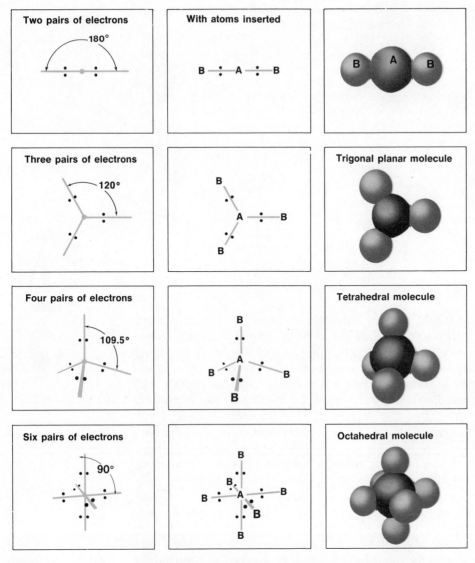

Figure 3—26 Arrangements of electron pairs according to the electron-pair repulsion theory.

The use of one s and two p orbitals to form sp² hybrid orbitals also leads to the same structure.

SiCl₄

Atomic silicon has four electrons in its outer orbit. When these are paired with the unpaired electrons of four chlorine atoms, $: \ddot{C}l \cdot$, four electron-pair bonds are formed. These four pairs of electrons in the valence shell will repel each other to form the predicted tetrahedral structure.

The combination of one s and three p orbitals to form sp³ hybrids leads to the same structure. The actual structure is, indeed, tetrahedral.

SF₆

The sulfur atom has six electrons in the outer orbit prior to formation of SF_6. Six electron pairs in the valence shell lead to the prediction of an octahedral structure for the molecule, which is the structure found.

90° angles

At this point you might ask how the presence of nonbonding pairs of electrons (in the outermost shell of the central atom) affects the disposition of the other electron pairs. In brief, a pair of electrons occupies somewhat more volume when it is not involved in bonding.

Nonbonding electron pairs help to explain the structures of NH_3 and H_2O molecules discussed earlier and shown in Figure 3–25. It can be postulated, for example, that the NH_3 and H_2O molecules have bond angles (106.5° in NH_3 and 104.5° in H_2O) slightly less than the tetrahedral angle of 109.5° because the lone, nonbonding pairs of electrons exert more repulsion than do the bonding pairs. This spreads the nonbonding pairs further apart and squeezes the bonding pairs closer together.

From the typical examples given here, it is obvious that both the electron-pair repulsion theory and the hybridization theory are consistent with experimental structures for molecules. The methods differ, however, in their approach. Hybridization utilizes the mathematical combination of atomic orbitals to form hybrid orbitals properly oriented to give the established bond angles. On the other hand, electron-pair repulsion theory assumes that electron pairs simply move as far away from each other as is possible while maintaining the proper bond length. The nonbonding pairs of electrons exert more repulsion than do bonding pairs.

Shapes of Molecules with Double Bonds and Triple Bonds

The electron dot structure for ethylene, C_2H_4, is:

Four electrons form a ***double bond*** between the two carbon atoms. Each carbon atom furnishes two unpaired electrons for the double bond.

Figure 3–27 Structure and bonding in ethylene, C_2H_4. (a) Electron dot structure; (b) sigma bond formation; (c) assembly of p orbitals to form a π bond; (d) representation of the sigma (σ) and pi (π) bonds.

This means two orbitals from each carbon atom are involved in the double bond.

A double bond is made up of one sigma bond and one pi bond.

Experimentally, it is observed that ethylene is a planar molecule (all six atoms are in the same plane) and that all bond angles are close to 120°. This kind of structure around a carbon atom can exist if the 2s and *two* of the 2p orbitals of carbon undergo hybridization to form three sp^2 hybrid orbitals. The hybridization can come about in the following way. Each carbon atom can be excited to the electronic arrangement, 2s-1, $2p_x$-1, $2p_y$-1, $2p_z$-1. When the 2s and two of the p orbitals are mixed, three equivalent orbitals are produced. The three sp^2 orbitals are all in the same plane and form angles of 120° with each other (Figure 3–23). As

Sigma (σ) bond: end-to-end overlap.
Pi (π) bond: side-by-side overlap.

shown in Figure 3–27, a **sigma (σ) bond** is formed between the two carbon atoms by overlapping the end of an sp^2 orbital of one carbon atom with the end of an sp^2 orbital of the other carbon atom. The second bond of the double bond between the two carbon atoms is formed by a lateral (or side-by-side) overlap of the two p orbitals not involved in the sp^2 hybridizations. The lateral overlap is called a **pi (π) bond.** It prevents rotation about the bond axis. The two CH_2 groups of ethylene cannot rotate freely with respect to each other because the rotation would reduce or eliminate the orbital overlap of the pi bond. As we shall see later, this rotational limitation about double bonds is quite important in the chemistry of certain compounds.

A pi bond is weaker than a sigma bond.

A pi bond, with less effective overlap of orbitals, is weaker than a sigma bond. However, a double bond, consisting of a sigma bond *and* a pi bond, is stronger than a single bond. A summary of covalent bond formation by orbital overlaps is shown in Figure 3–28.

Acetylene, C_2H_2, has a triple bond composed of one sigma and two pi bonds (Figure 3–29).

From two s orbitals

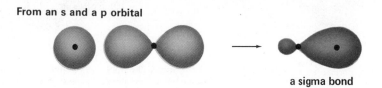

a sigma bond

From two p orbitals (end-to-end overlap)

a sigma bond

Figure 3—28 Ways of overlapping s and p orbitals to form sigma and pi bonds.

From an s and a p orbital

a sigma bond

From two p orbitals (lateral overlap)

a pi bond

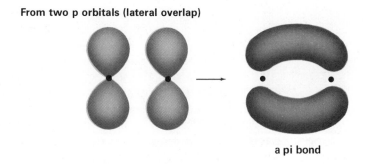

H : C ⦂⦂ C : H

(a)

sp sp   **(b)**

π

p   p   p   p

H σ C σ C σ H

p   p

π

p   p

(c)

H C   C H

(d)

Figure 3—29 Structure and bonding of acetylene, C_2H_2. (a) Electron dot structure; (b) sigma bond formation; (c) assembly of p orbitals to form two pi bonds; (d) representation of the electron density pattern, which is cylindrically symmetric in the molecule.

1. A double bond is composed of one _____ bond and one _____ bond.

2. A triple bond is composed of one _____ bond and two _____ bonds.

3. An sp³ hybrid orbital is one of a set of orbitals that can be formed by mixing one _____ orbital and three _____ orbitals.

4. (Bonding/nonbonding) pairs of electrons repel more than (bonding/nonbonding) pairs of electrons.

5. Around which kind of bond (single, double, triple) can atoms rotate on an axis drawn through the nuclei of the two atoms bonded? _____

6. When three atomic orbitals are combined in hybridization, how many new orbitals are produced? _____

7. Give the molecular shapes expected for the hypothetical molecules:

AX_2 _____, BZ_3 _____, DQ_6 _____. There are no nonbonding valence electrons in these structures, and all of the bonds are single bonds.

_____ 1. double covalent bond   (a) explanation for covalent bonding

_____ 2. ionic bonding   (b) sharing electrons

_____ 3. ionization energy   (c) requires O, F, or N

_____ 4. metallic bonding   (d) ions attracted to each other

_____ 5. van der Waals attractions   (e) ionic compound

_____ 6. noble gas   (f) electrons free to move

_____ 7. covalent bonding   (g) element with eight valence-shell electrons

_____ 8. NaCl   (h) covalent compound

_____ 9. metal ion   (i) attraction between nonpolar neutral particles

_____ 10. hydrogen bonding   (j) measures gaseous atom's hold on electron

_____ 11. NH_3 (k) smaller than parent atom

_____ 12. orbital overlap   (l) four electrons shared

_____ 13. single covalent bond   (m) two electrons shared

MATCHING SET II

_____ 1.	pi bond	(a)	an electrically neutral arrangement of covalently bonded atoms
_____ 2.	sigma bond	(b)	well-balanced arrangement of parts
_____ 3.	sp³ orbital	(c)	four-sided polygon
_____ 4.	hybridization	(d)	lateral overlap of p orbitals
_____ 5.	molecule	(e)	one pi bond and one sigma bond
_____ 6.	double bond	(f)	mixing of orbitals
_____ 7.	symmetry	(g)	end-to-end overlap of orbitals
_____ 8.	tetrahedral	(h)	hybrid orbital

MATCHING SET III

_____ 1.	$BeCl_2$ (no nonbonding valence electrons)	(a)	linear
_____ 2.	H_2O	(b)	trigonal planar
_____ 3.	$SiCl_4$	(c)	bent
_____ 4.	SF_6	(d)	tetrahedral
_____ 5.	BCl_3	(e)	octahedral

QUESTIONS

1. Diamond (a form of carbon) has a melting point of 3500°C, whereas carbon monoxide (CO) has a melting point of −207°C. What does this suggest about the kinds of bonding found in these two substances?

2. Write the electronic configuration for the element potassium (atomic number 19). What will be the electronic configuration when a K^+ ion is formed?

3. Is Ca^{3+} a possible ion under normal chemical conditions? Why?

4. Write the symbols for the six elements with the highest ionization potentials, selecting one from each period of the periodic table.

5. Match the electronic configurations that would be expected to lead to similar chemical behavior. The numbers denote the numbers of electrons in the orbits.

   (a) 2,2   (d) 2,8,2
   (b) 2,5   (e) 2,1
   (c) 2,8,8,1 (f) 2,8,5

6. Fluorine (atomic number 9) has an electronic configuration of 2,7. How many electrons will be involved in the formation of a single covalent bond?

7. Which compound would be expected to have the higher melting point, KCl or RbCl; MgO or BaO? Give a reason for your answer.

8. How many electrons would there be in an iodide ion, I^-?

9. Write the electron dot structures for the fluoride ion, F^-, the chloride ion, Cl^-, and the bromide ion, Br^-.

10. Draw the electron dot structure for water. Based on bonding theory, why is water's formula not H_3O?

11. Define the term *bond energy*.

12. Draw electron dot structures for the following molecules:

    (a) NF_3 (c) C_2Cl_2 (e) H_2S (g) N_2H_4 (i) Br_2
    (b) CCl_4 (d) OF_2 (f) CO (h) CH_3OH (j) HCl

13. The members of the nitrogen family N, P, As, and Sb, form compounds with hydrogen: NH_3, PH_3, AsH_3, and SbH_3. The boiling points of these compounds are given:

SbH_3	$-17°C$
AsH_3	$-55°C$
PH_3	$-87.4°C$
NH_3	$-33.4°C$

    Comment on why NH_3 doesn't follow the downward trend of boiling points.

14. Match the substances listed below with the type of bonding responsible for holding units in the solid together.

solid krypton (Kr)	ionic
ice	covalent
diamond	metallic
CaF_2	hydrogen bonding
iron	van der Waals

15. Predict the general kind of chemical behavior (that is, loss, gain, or sharing of electrons) you would expect from atoms with the following electron distributions:

    (a) 2,8,1
    (b) 2,7
    (c) 2,4

16. Show how two fluorine atoms can form a bond by the overlap of their half-filled p orbitals.

17. Select the *polar* molecules from the following list and explain why they are polar:
    N_2, HCl, CO, and NO

18. How many bonds join the two atoms in each of the following?
    CN^-, Cl_2, S_2

19. Boron trichloride has the electron dot formula: $\ddot{:}\overset{\displaystyle :\ddot{C}l:}{\underset{}{:\ddot{C}l:B:\ddot{C}l:}}$. What does this tell you about the octet rule even for Period 2 elements?

20. Use your chemical intuition and suggest a reaction that might occur between boron trichloride (problem 19 above) and ammonia, $H:\overset{\displaystyle \cdot}{\underset{\displaystyle H}{\ddot{N}}}:H$.

21. Sketch separately the spatial distribution of electronic charge associated with an s and a p atomic orbital.

22. What is the meaning of the notation sp^3?

23. What is the geometry of bonding around a central atom having the following hybridization?

    (a) sp (b) sp^2 (c) sp^3

24. What is the meaning of hybridization?

25. Account for the fact that carbon in methane, CH_4, forms four equivalent bonds.

26. What is a sigma bond; a pi bond?

27. Give the basic points of the valence-shell electron-pair repulsion theory.

28. Explain the 106.5° H—N—H bond angles of NH_3, the ammonia molecule.

29. How many atomic positions must be specified in order to define a bond angle?

30. $BeCl_2$ is a compound known to contain polar Be—Cl bonds, yet the $BeCl_2$ molecule is not polar. Explain.

31. What happens to the third p orbital in sp^2 hybridization?

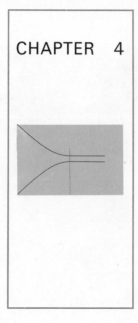

CHAPTER 4

SOME PRINCIPLES OF CHEMICAL REACTIVITY

There are 106 different known chemical elements, each with its own characteristics. In addition to the broad classifications of metals and nonmetals, smaller groups or families of elements can be recognized on the basis of similar chemical behavior. In this chapter we shall observe some general characteristics of chemical reactions and note some groupings of reactions that help to classify the elements.

REACTANTS BECOME PRODUCTS

In all chemical reactions some pure substances disappear and others appear. A familiar example is the change of shiny steel to a red-brown iron rust (Figure 4-1). This is a very important reaction since it has been estimated that the loss from corrosion of iron and steel in the United States is slightly over $60 per person per year.

Some other chemical changes are given in the following equations:

	REACTANTS		PRODUCTS	HEAT EFFECT*
+Heat means heat is liberated.	CaO Calcium Oxide (Quicklime)	$+ \quad H_2O$ Water	$\longrightarrow \quad Ca(OH)_2$ Calcium Hydroxide (Slaked Lime)	$+$ Heat [15.6 kcal per mole of $Ca(OH)_2$]
	$2Na$ Sodium	$+ Cl_2(gas)$ Chlorine	$\longrightarrow \quad 2NaCl$ Sodium Chloride (Table Salt)	$+$ Heat [98.2 kcal per mole of NaCl]
−Heat means heat is required.	$H_2(gas)$ Hydrogen	$+ \quad I_2(gas)$ Iodine	$\longrightarrow \quad 2HI(gas)$ Hydrogen Iodide	$-$ Heat [6.20 kcal per mole of HI]

*Heat energy can be measured in calories (cal). A kilocalorie (kcal) is 1000 calories. A calorie is the amount of heat required to raise the temperature of 1 gram of water 1 degree Celsius.

Figure 4–1 Chemical changes produce new substances, generally with properties very different from the starting substances. Bright, shiny nuts and bolts are changed to crumbly, dull rust by oxidation.

From these reactions we can note our first important point concerning chemical reactions:

> In chemical reactions, reactants become products; some substances are consumed and new ones appear.

In chemical reactions new substances are formed and old ones disappear.

QUANTITATIVE ENERGY CHANGES

Again consider the three reactions cited above. In the first reaction, calcium oxide (quicklime) reacts with water to give calcium hydroxide (slaked lime) with the evolution of heat. In the second reaction, metallic sodium reacts with the greenish yellow gas, chlorine, to give sodium chloride (table salt). If a piece of hot sodium is put into a flask containing chlorine, the sodium burns quickly, liberating a great deal of heat and light, to produce white crystals of sodium chloride. In the last reaction, gaseous hydrogen reacts with gaseous iodine to produce gaseous hydrogen iodide, with the absorption of heat. These facts, along with similar ones, lead to a second generalization about chemical reactions:

> A given amount of a particular chemical change corresponds to a proportional amount of energy change.

Energy changes in chemical reactions are proportional to the amount of reactant (or product).

For example, the preparation of one mole of $Ca(OH)_2$ from CaO and H_2O releases 15.6 kcal; for two moles of $Ca(OH)_2$, 2×15.6 or 31.2 kcal of heat is released.

Sometimes energy changes in reactions are difficult to observe because of the very slow rate of reaction. An example is the rusting of iron. Here the reaction involved is complicated, but we can represent it by the simplified equation:

$$4Fe + 3O_2 + 6H_2O \longrightarrow 4Fe(OH)_3 + 788 \text{ kcal}$$
IRON   MOIST AIR   IRON HYDROXIDE + HEAT
(RUST)

Ordinarily, the rusting of iron occurs so slowly that the liberation of heat is perceptible only with the aid of special instruments. The total amount of heat evolved in rusting is considerable, but it typically takes place over a long period of time.

RATE OF CHEMICAL REACTION

We noted earlier that sodium reacts more rapidly with chlorine than iron does with oxygen and water of the atmosphere. The whole notion of how fast or how slow chemical reactions proceed is one that can be put on a quantitative basis by the concept of **reaction rate**. The rate of a reaction is always defined in terms of the changes in the amounts of chemical substances present per unit of time. Thus, if we consider the burning of sulfur to produce sulfur dioxide,

$$S + O_2 \longrightarrow SO_2$$

we can discuss the rate of the reaction in terms of the amount of SO_2 formed per minute or of the amount of S or O_2 consumed per minute.

Effect of Temperature on Reaction Rate

It is possible to alter the rate of a chemical reaction by changing the temperature. If the temperature is raised, the rates of chemical reactions are increased; if the temperature is reduced, they are decreased. We make use of this principle in cooking foods (a roast will cook at a faster rate at a higher temperature) and in preserving foods (foods spoil less quickly if refrigerated). Figure 4–2 illustrates the effect of temperature

Raising the temperature speeds up chemical reactions.

Figure 4–2 The biochemical processes of decomposition occur more rapidly at higher temperatures. Half the peach shown in this photograph was refrigerated, while the other half was kept warm. The refrigerated half on the right shows little discoloration, while the other one shows the typical signs of decay.

on the reactions that take place in a slice of fruit exposed to air.

Another general characteristic of chemical reactions is:

> The rate of a typical chemical reaction is increased if the temperature is increased.

Effect of Concentration on Reaction Rate

It is also possible to alter the rate of a reaction by changing the concentrations of the reactants. For example, in the reaction of sulfur and oxygen given earlier, if air replaces oxygen, the reaction will proceed at a slower rate, since air is a mixture of about one part oxygen and four parts nitrogen. The rusting of iron can be retarded by painting or coating the surface of the metal to cut down on the concentration of the oxygen and moisture at the surface. A demonstration such as that shown in Figure 4–3 contrasts the oxidation of iron when oxygen is in relatively small and in relatively large concentrations. These facts illustrate still another characteristic of chemical reaction:

Increasing the concentration of reactants speeds up a reaction.

> The rate of a chemical reaction depends on the concentrations of the reactants.

A molecular explanation can be given for the faster reaction of sulfur in pure oxygen than of sulfur in air. The reaction proceeds more slowly in air because there are fewer oxygen molecules per volume of air to react with the sulfur. This example illustrates the useful technique of reducing the rate of a reaction by reducing the concentration of the reactants.

A theoretical (molecular) explanation for both the concentration effect and the temperature effect on the rate of chemical reactions is illustrated in Figure 4–4.

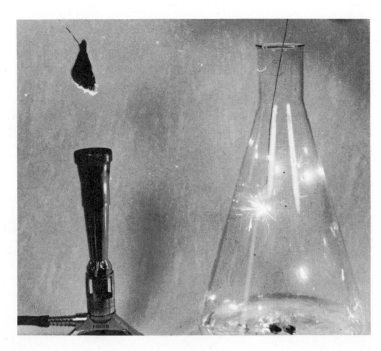

Figure 4–3 Effect of concentration on reaction rate. Steel wool held in the flame of a gas burner is oxidized rapidly. It is in contact with air, which is 20 percent oxygen. When the red hot metal is placed in pure oxygen in the flask, it oxidizes much more rapidly.

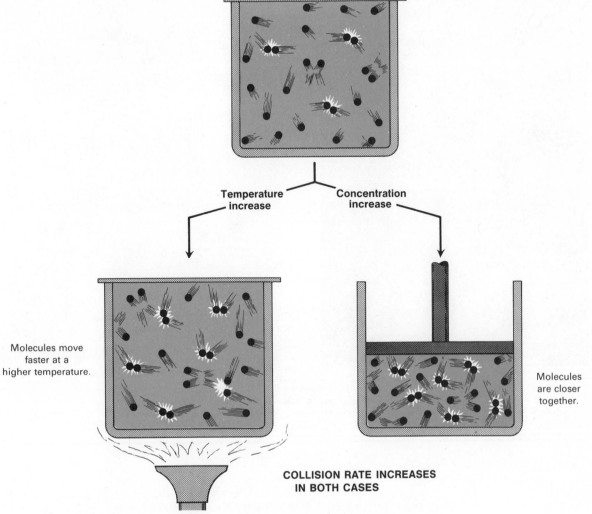

Temperature increase

Concentration increase

Molecules move faster at a higher temperature.

Molecules are closer together.

COLLISION RATE INCREASES IN BOTH CASES

Figure 4—4 Effects of temperature and concentration on rates of chemical reactions. At the higher temperature, more collisions occur between molecules, and a greater percentage of the collisions produce a chemical reaction. At the higher concentration (no temperature change), more collisions occur, but the percentage of effective collisions remains the same.

REVERSIBILITY OF CHEMICAL REACTIONS

Chemical reactions are capable of going forward or back- ward.

Most chemical processes are capable of being reversed under suitable conditions. When a chemical reaction is reversed, some of the products are converted back into reactants. For example, heating calcium hydroxide will drive off some of the water. This process is the reverse of adding water to quicklime.

$$Ca(OH)_2 + Heat \longrightarrow CaO + H_2O$$

Other methods can be used to reverse chemical reactions. Thus, if we put calcium hydroxide in a vacuum, there will soon be water vapor in the space around the solid. It is also possible to reverse chemical reactions when they are associated with small heat changes. For the compound

(a)

Porcelain dish

Water droplets

Small flame

Glass tube
with fine tip

Hydrogen gas

(b)

Figure 4–5 (a) Electrical energy is required to decompose water into hydrogen (right tube) and oxygen (left tube). Note that there are two volumes of hydrogen produced for each volume of oxygen. (b) Hydrogen and oxygen burning to produce water in the gaseous state. The water is condensed on the cooler porcelain dish.

hydrogen iodide, HI, it is possible to break the HI molecules apart into hydrogen and iodine,

$$2HI + Heat \longrightarrow H_2 + I_2$$

at slightly elevated temperatures. On the other hand, water is decomposed into hydrogen and oxygen only by the use of considerable amounts of electrical energy. The reaction that is the reverse of the electrolytic decomposition of water is also shown in Figure 4–5. Hydrogen when burned in air produces water vapor, which can be condensed rapidly.

The fact that many reactions can be approached from either direction leads to the conclusion:

Chemical reactions are generally reversible.

If atoms in one arrangement—as in the case of water—can be decomposed into another arrangement—in this case, hydrogen and oxygen—there is every reason to believe that the atoms could be rearranged into the original structure. After all, a set of building blocks can be used to construct a wall, then a bridge, and then recombined into the wall again.

There are many reversible reactions that are important to human life. One of these is involved in the transport of atmospheric oxygen from

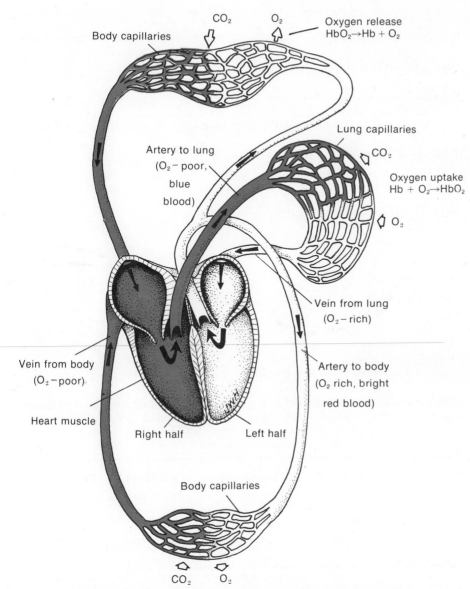

Figure 4–6 Simplified diagram of human circulation. The heart (shown in front view) is divided into two parallel halves. The right half pumps oxygen-poor blood to the lungs; the left half pumps oxygen-rich blood to the body. Hb = hemoglobin; HbO_2 = oxyhemoglobin. (Modified from Clark, M. E., *Contemporary Biology*, W. B. Saunders Co., Philadelphia, 1973.)

the lungs to the various parts of the body. This task is carried out by hemoglobin, a complex compound that is found in the blood. This substance takes up oxygen while in the lungs to form oxyhemoglobin.

$$\text{Hemoglobin} + O_2 \rightleftharpoons \text{Oxyhemoglobin}$$

The double arrows, $\rightleftharpoons$, indicate a reversible reaction.

The oxyhemoglobin is then carried by the bloodstream to the various parts of the body, where it releases the oxygen for use in metabolic processes, as shown in Figure 4–6.

SELF-TEST 4-A   1. Which factor affecting reaction rate—temperature or concentration—is most

closely related to freezing foods to prevent spoilage? _____

Why? _____

2. If water can be produced by burning hydrogen in oxygen, what are the

products of the decomposition of water? _____ and

_____

3. Name two reversible chemical changes.

_____

_____

4. The reaction of hydrogen with iodine (releases/absorbs) _____
energy.

5. Is it true that the faster a chemical reaction goes, the greater the amount of

energy released per mole of reactant? _____

6. If 68 kcal of energy is released in the formation of 18 g (one mole) of water,
how much energy would be released in the formation of 36 g of water?

_____

7. Which should burn faster: (1) a pound of flour in a sack, or (2) the same flour

in dust form in the air of a flour mill? _____

8. What chemical reaction can you name that is constantly reversed in your

bloodstream? _____

9. Name two chemical reactions that are not easily reversed.

_____ and _____

REACTIONS BY CHEMICAL GROUPS

When a survey is made of the known types of chemical reactions, a
diversity is found that is illustrated by the following examples:

$$C \; + \; 2F_2 \longrightarrow CF_4$$
Carbon   Fluorine   Carbon
tetrafluoride

$$PCl_3 \; + \; 3H_2O \longrightarrow H_3PO_3 \; + \; 3HCl$$
Phosphorus   Water   Phosphorous   Hydrogen
trichloride   acid   chloride

$$SO_3 \;+\; H_2O \;\longrightarrow\; H_2SO_4$$

Sulfur trioxide   Water   Sulfuric acid

$$C \;+\; 2S \;\xrightarrow{\text{Heat}}\; CS_2$$

Carbon   Sulfur   Carbon disulfide

$$Fe_2O_3 \;+\; 3CO \;\xrightarrow{\text{Heat}}\; 2Fe \;+\; 3CO_2$$

Iron(III) oxide   Carbon monoxide   Iron   Carbon dioxide

These reactions are not presented for you to learn at this point but only to show some of the wide variety of reactions and substances encountered in chemistry. Indeed, the number of known compounds runs into the millions and the number of reactions by which they are produced into the multimillions. However, becoming familiar with a large number of chemical reactions is not as hopeless as it may seem at first. There is much order to be found in the apparent chaos. For example, consider the following reactions:

Many chemical reactions follow similar patterns.

$$2Na \;+\; F_2 \;\longrightarrow\; 2NaF$$

Sodium   Fluorine   Sodium fluoride

$$2Na \;+\; Cl_2 \;\longrightarrow\; 2NaCl$$

Chlorine   Sodium chloride

$$2Na \;+\; Br_2 \;\longrightarrow\; 2NaBr$$

Bromine   Sodium bromide

$$2Na \;+\; I_2 \;\longrightarrow\; 2NaI$$

Iodine   Sodium iodide

Note that the group of elements, fluorine, chlorine, bromine, and iodine, all react with sodium in much the same way; that is, products with similar formulas are produced. Also note that this family of elements is arranged in a column of the periodic table, Group VIIA. See Figure 4–7. Once we have learned one reaction in such a group of elements, we have some idea about what is likely to happen to other members of the group.

There are elements that undergo similar chemical reactions. Elements showing such similarities are identified as a chemical group or family.

Another family includes the group of elements oxygen, sulfur, selenium, and tellurium, Group VIA. All react with calcium in much the same way.

$$2Ca \;+\; O_2 \;\longrightarrow\; 2CaO$$

Calcium   Oxygen   Calcium oxide

$$Ca \;+\; S \;\longrightarrow\; CaS$$

Sulfur   Calcium sulfide

Elements may be arranged in families; the members of a given family have similar chemical properties.

$$Ca \;+\; Se \;\longrightarrow\; CaSe$$

Selenium   Calcium selenide

$$Ca \;+\; Te \;\longrightarrow\; CaTe$$

Tellurium   Calcium telluride

Figure 4—7 The periodic table of the elements. The properties of the elements are periodic functions of the atomic numbers. Columns of elements form families (groups) with similar chemical properties.

Not only do elements react as members of groups, but groups of atoms may act as a single unit. Consider the following:

$$Mg + H_2SO_4 \longrightarrow MgSO_4 + H_2$$

Magnesium   Sulfuric acid   Magnesium sulfate   Hydrogen

$$Ca + H_2SO_4 \longrightarrow CaSO_4 + H_2$$

Calcium   Calcium sulfate

$$Sr + H_2SO_4 \longrightarrow SrSO_4 + H_2$$

Strontium   Strontium sulfate

$$Ba + H_2SO_4 \longrightarrow BaSO_4 + H_2$$

Barium   Barium sulfate

The sulfate group, SO_4, in this series of reactions acts as a unit and forms a doubly negative ion.

STOICHIOMETRY

The word **stoichiometry** comes from the Greek *stoicheion* (element) and literally means the measurement of amounts of elements. Stoichiometry deals with the weights of elements and compounds that are consumed or produced in a chemical reaction. For example, an important stoichiometric question is: How much aluminum can be produced from one ton of aluminum oxide? Once the atomic weight scale was established, such important calculations could readily be accomplished.

All calculations involving chemical equations are based on the Law of Conservation of Matter.

Since the law of conservation of matter states that matter is neither lost nor gained in a chemical reaction, the weight of the reactants in a chemical reaction must be the same as the weight of the products. For example, according to the law of conservation of matter, when hydrogen reacts with chlorine to form hydrogen chloride, the weight of the reactants, hydrogen and chlorine, used in the reaction must equal the weight of the product, hydrogen chloride. The reaction can be written more concisely by using the symbolism of a chemical equation.

$$H_2 + Cl_2 \longrightarrow 2HCl$$

Accounting for atoms requires balancing an equation. To balance a chemical equation:
1. *Place numbers (coefficients) only before formulae. ($2H_2$ means 2 molecules of H_2 and $2 \times 2 = 4$ atoms of H.)*
2. *Have same number of each kind of atom on each side of the arrow ($\longrightarrow$).*
3. *Do not change subscripts. Each symbol represents one atom. H_2 means a molecule composed of two atoms.*

Note that a coefficient, 2, has been placed in front of the hydrogen chloride so that 2 atoms of hydrogen and 2 atoms of chlorine are represented in both the reactants and the products; that is, none are gained or lost. What is the meaning of the symbolism of the equation? There are two alternative but equally meaningful ways to interpret the equation:

1 molecule of hydrogen reacts with 1 molecule of chlorine to form 2 molecules of hydrogen chloride;

or,

1 mole of hydrogen molecules reacts with 1 mole of chlorine molecules to form 2 moles of hydrogen chloride molecules.

Once the equation is balanced, the relative number of moles for each substance involved is given by the respective coefficients. Furthermore, since 1 mole weighs 1 gram molecular weight, a set of weights for all substances involved in the reaction can easily be obtained by adding up the atomic weights in each formula.

For example, the molecular weight of C_8H_{18}, a component of gasoline, is 114. This is calculated as follows:

	Atomic weight	$\times$	No. of atoms	=	Total weight
C	12.0	$\times$	8	=	96.0
H	1.0	$\times$	18	=	18.0
			Molecular weight	=	114.0

The molecular weight is simply the sum of the atomic weights. The gram molecular weight of C_8H_{18} is 114.0 *grams,* and this is the weight of 6.02×10^{23} molecules or one mole of C_8H_{18}. It follows that two moles of C_8H_{18} molecules weighs 2×114 grams, or 228 grams.

A mole of any item is 6.02×10^{23} of them.

These facts allow several types of calculations to be made for chemical reactions. The following examples will illustrate some of these.

EXAMPLE 1

How many moles of nitrogen (N_2) are required to react with 6 moles of hydrogen (H_2) in the formation of ammonia (NH_3)?

a. Write and balance the equation:

$$N_2 + 3H_2 \longrightarrow 2NH_3$$

Ammonia is used as a crop fertilizer.

b. Since 1 mole of N_2 reacts with 3 moles of H_2, how many moles of N_2 will react with 6 moles of H_2? The answer is 2 moles of N_2. If the number of moles of H_2 is doubled, then the number of moles of N_2 must be doubled to keep the same ratio of nitrogen and hydrogen that react with each other.

EXAMPLE 2

How many moles of nitrogen dioxide (NO_2) will be produced by 4 moles of oxygen reacting with sufficient nitric oxide (NO)?

a. Write and balance the equation:

$$2NO + O_2 \longrightarrow 2NO_2$$

Nitrogen dioxide is a major air pollutant.

b. From the balanced equation we see that the number of moles of NO_2 produced is twice that of the oxygen reacting. Therefore 4 moles of oxygen would produce 8 moles of NO_2.

EXAMPLE 3

How many grams of carbon dioxide (CO_2) can be produced by burning 2650 grams of gasoline (C_8H_{18})?

a. Write and balance the equation:

$$2C_8H_{18} + 25O_2 \longrightarrow 16CO_2 + 18H_2O$$

b. The balanced equation states that 2 moles of gasoline produce 16 moles of CO_2. Since the molecular weight of C_8H_{18} is 114, two moles would weigh 228 grams. The molecular weight of CO_2 is 44, i.e., $(1 \times 12) + (2 \times 16)$, so 16 moles would weigh $16 \times 44 = 704$ grams. Thus, 228 grams of gasoline would produce 704 grams of CO_2; that is, the weight of CO_2 produced is about three times the weight of gasoline burned. This means that our 2650 grams of gasoline should produce about 8000 grams of CO_2. To be more exact:

$$\text{grams of } CO_2 = 2650 \text{ g } C_8H_{18} \times \frac{704 \text{ g } CO_2}{228 \text{ g } C_8H_{18}}$$
$$= 8182 \text{ g } CO_2$$

1. Balance the following equations:

   (a) _____ Mg + _____ $O_2 \longrightarrow$ _____ MgO

   (b) _____ Si + _____ $Cl_2 \longrightarrow$ _____ $SiCl_4$

   (c) _____ Al + _____ $O_2 \longrightarrow$ _____ Al_2O_3

2. Selenium is directly below sulfur in the periodic table. Given the sulfur compounds below, write formulas that are expected for the analogous selenium compounds.

   (a) Oxides   (b) Fluorides   (c) Oxyacids

        Sulfur: SO_2 Sulfur: SF_4 Sulfur: H_2SO_3

              SO_3 SF_6 H_2SO_4

        Selenium: _____   Selenium: _____   Selenium: _____

        _____   _____   _____

3. An important source of hydrogen, used as a rocket fuel, is the decomposition of water by electrical energy. The reaction is:

$$H_2O \xrightarrow{\text{Electrical energy}} H_2 + O_2$$

   (a) Balance the equation.

   _____

   (b) What weight of water is necessary to produce 2.0 grams of hydrogen?

   _____

(c) How many grams of oxygen would be produced as a by-product?

_____

(d) How much water would be necessary to produce 2.0 tons of hydrogen?

_____

MATCHING SET I			
_____ 1.	stoichiometry	(a)	needed to express conservation of matter
_____ 2.	corrosion	(b)	F, Cl, Br, I, At
_____ 3.	refrigeration	(c)	react similarly with oxygen
_____ 4.	chemical family	(d)	atomic number ordering
_____ 5.	Mg and Ca	(e)	retards food spoilage
_____ 6.	periodic table	(f)	formation of rust
_____ 7.	balanced equation	(g)	would form an oxide similar to K_2O
_____ 8.	Li	(h)	chemical weight relationships

MATCHING SET II			
_____ 1.	6.02×10^{23}	(a)	18 grams
_____ 2.	gram molecular weight of P_4O_{10}	(b)	20
_____ 3.	weight of one mole of water molecules	(c)	one mole
_____ 4.	conserved during chemical reaction	(d)	carbon-12 isotope
_____ 5.	number of atoms in the formula $(NH_4)_3PO_4$	(e)	284 grams
_____ 6.	reference point for atomic weights	(f)	mass or weight

QUESTIONS

1. Based on information presented in this chapter on likenesses of elements in groups, predict the formulas of the products of the following reactions.

$Rb + Cl_2 \longrightarrow$ $K + Cl_2 \longrightarrow$
$Ba + O_2 \longrightarrow$ $Mg + Se \longrightarrow$
$Na + Br_2 \longrightarrow$ $Mg + S \longrightarrow$
$Sr + S \longrightarrow$ $Be + S \longrightarrow$

2. In the periodic table the elements are arranged in the order of their _____

_____ .

3. If you were arranging the elements A, B, C, D in a table and came upon an element, G, which had properties similar to those of A, where would you place it?

4. Why is it necessary to balance a chemical equation before it can serve as the basis of a stoichiometric calculation?

5. Identify four chemical reactions that we use in our daily lives in which energy plays an important role.

6. Give an example of a chemical reaction whose rate is fast and one whose rate is slow.

7. List three characteristics of all chemical reactions and illustrate each characteristic with an example.

8. Using the periodic table, select elements that can be expected to have chemical properties similar to:

   (a) Ca (calcium), atomic number 20
   (b) Fe (iron), atomic number 26
   (c) Sn (tin), atomic number 50
   (d) S (sulfur), atomic number 16

9. In 1968 an Apollo spacecraft cabin fire killed three astronauts. The fact that pure oxygen was used as the cabin atmosphere contributed to the severity of the fire. How?

10. The element helium is very unreactive chemically. What type of behavior would you expect for argon? Study their relationship in the periodic table.

11. Consider what would be the relative rates of rusting of iron in a dry climate as opposed to a damp climate. What principle(s) of reaction rate is (are) involved?

12. Copper forms a chloride salt, CuCl. Is this consistent with the group number into which copper is placed in the periodic table?

13. Fires have been started by water seeping into bags in which quicklime was stored. Why would this produce a fire?

14. How many moles of sulfur trioxide are needed to form 100 moles of sulfuric acid? The equation is: $SO_3 + H_2O \longrightarrow H_2SO_4$.

15. Hydrogen chloride (HCl) is produced by the action of sulfuric acid (H_2SO_4) on sodium chloride (NaCl).

$$NaCl + H_2SO_4 \longrightarrow HCl + Na_2SO_4$$

How many moles of HCl can be made from 50.0 moles of H_2SO_4? Be sure to balance the equation.

16. Chlorine can be made by the electrical decompositon of melted sodium chloride.

$$NaCl \xrightarrow{\text{Energy}} Na + Cl_2$$

How many moles of products can be made from one mole of sodium chloride? Be sure to include both products.

17. How many grams of hydrogen are liberated when 75.0 grams of sodium metal reacts with excess water?

$$2Na + 2H_2O \longrightarrow 2NaOH + H_2$$

18. Copper metal can be produced by heating copper sulfide with carbon and air.

$$CuS + O_2 + C \longrightarrow Cu + SO_2 + CO_2$$

   (a) Balance the equation.
   (b) How many moles of oxygen are required for each mole of CuS?
   (c) How many grams of copper can be produced from 100 grams of CuS?

19. How many grams of sulfur dioxide (SO_2) can be formed by burning 70.0 grams of sulfur?

$$S + O_2 \longrightarrow SO_2$$

20. Silver sulfide (Ag_2S) is the common tarnish on silver objects. What weight of silver sulfide can be made from 1.00 milligram of hydrogen sulfide (H_2S) obtained from a rotten egg?

$$4Ag + 2H_2S + O_2 \longrightarrow 2Ag_2S + 2H_2O$$

HYDROGEN ION TRANSFER (ACID-BASE) AND ELECTRON TRANSFER (OXIDATION-REDUCTION) REACTIONS

Two major classes of chemical reactions which have far-reaching applications in our lives are those of hydrogen ion transfer and electron transfer between molecular or ionic species. *Acids* such as vinegar or lemon juice, and *bases* such as lime or baking soda are commonly encountered. Reactions of acids with bases can be defined in terms of the transfer of hydrogen ions from an acid to a base. Chemical oxidation and reduction are equally common. The rusting of iron, the burning of wood, and the bleaching of hair or fabrics are examples of oxidation, while the transformation of iron ore in a blast furnace to iron metal is a classic example of chemical reduction. Such *redox* (short for *oxidation-reduction*) reactions can be described in terms of the transfer of electrons from one chemical species to another. If we can understand how hydro-

Redox is pronounced REE-DOX.

gen ions and electrons are transferred, we shall be better able to control many processes of importance.

Acid-base and redox reactions occur in the solid, liquid, or gaseous phase. However, the presentation will be made easier at this point if we begin our study with the liquid (solution) phase. This will require an understanding of some characteristics of the major types of solutions.

FORMATION OF SOLUTIONS

The process of **solution** is a familiar one. We can see sugar and salt dissolve in water, oil paints dissolve in turpentine, and grease dissolve in gasoline. Though they are less commonly observed, solid solutions (such as metal alloys) and gaseous solutions (like polluted air) are equally possible. In a solution the substance present in greater amount is defined as the **solvent** and the one in smaller amount the **solute.** For example, in a glass of tea, water is the solvent and sugar, lemon juice, and the tea itself are solutes.

Solution: homogeneous mixture of atoms, ions, or molecules.

In this presentation of acid-base and redox reactions, most of the chemistry studied will be in water or aqueous solutions, where water is the solvent. Generally, one of the species exchanging hydrogen ions or electrons is the solute. Complicating the study somewhat is the fact that solvent molecules themselves can enter into both acid-base and redox reactions. Water molecules can and do exchange hydrogen ions and electrons with solute particles under suitable conditions. In theoretical terms a solution of sugar in water can be thought of in terms of a collection of sugar molecules dispersed among the water molecules (Figure 5–1).

Aqueous solutions are water solutions.

IONIC SOLUTIONS (ELECTROLYTES) AND MOLECULAR SOLUTIONS (NONELECTROLYTES)

Solutes in aqueous solutions can be classified by their ability or inability to render the solution electrically conductive. When aqueous solutions are examined to see whether they conduct electricity, we find that solutions fall into one of two categories: **electrolytic** solutions, which conduct electricity, and **nonelectrolytic** solutions, which do not. A simple apparatus such as that shown in Figure 5–2 can be used to determine into which classification a given solution falls.

Solute particles may be ions or molecules.

The conductance of electrolytic solutions is readily explained, since the solute particles in such solutions are ions rather than molecules. Recall that sodium chloride crystals are composed of sodium ions, which

Figure 5–1 A schematic illustration at the molecular level of sugar solution in water. Large circles represent the sugar molecues and the small circles water. The size of the container and the size of the particles are not to scale.

● Sugar molecule

· Water molecule

(a) **Solution of table salt**
(an electrolytic solution)

(b) **Solution of table sugar**
(a nonelectrolytic solution)

(c) **Pure water**
(a nonelectrolyte)

Figure 5–2 A simple test for an electrolytic solution. In order for the light bulb to burn (*a*), electricity must flow from one pole of the battery and return to the battery via the other pole. To complete the circuit, the solution must conduct electricity. A solution of table salt, sodium chloride, results in a glowing light bulb. Hence, sodium chloride is an electrolyte. In (*b*), the light bulb does not glow. Hence, table sugar is a nonelectrolyte. In (*c*), it is evident that the solvent, water, does not qualify as an electrolyte since it does not conduct electricity in this test.

are positively charged, and chloride ions, which are negatively charged. When sodium chloride dissolves in water, ***ionic dissociation*** occurs. The resulting solution (Figure 5–3a) contains positive sodium ions and negative chloride ions dispersed in water. Of course, the solution as a whole is neutral since the total number of positive charges is equal to the total number of negative charges.

Ionic dissociation is the separation of ions of a solute when the substance is dissolved.

$$Na^+Cl^- \xrightarrow{\text{Water}} Na^+_{(aq)} + Cl^-_{(aq)}$$
$$\text{(SOLID)} \qquad \text{(AQUEOUS)} \quad \text{(AQUEOUS)}$$
$$\text{SODIUM} \qquad \text{CHLORIDE}$$
$$\text{ION} \qquad\qquad \text{ION}$$

The random motions of the sodium and chloride ions are not completely independent. The charges on the particles prevent all of the sodium ions from going spontaneously to one side of the container while all of the chloride ions are going to the other side. However, a net motion of ions occurs when charged electrodes are placed in an electrolytic solution (Figure 5–3b). If the negative ions give up electrons to one electrode while the positive ions receive electrons from the other electrode, a flow of electrons or electricity is maintained.

Ions migrate toward oppositely charged electrodes in an electric field.

Nonelectrolytic solutions composed of solute molecules dispersed throughout solvent molecules are insensitive to negatively and positively charged electrodes unless the voltage is so great that it breaks the molecules into ions.

Sometimes ionic solutions arise when a molecular substance dissolves in water. For example, hydrogen chloride, HCl, is a gas composed of di-

Figure 5–3 Conductance of electricity by ionic solution. (*a*) The hydrated ions are randomly distributed throughout the salt solution; the net charge is zero. (*b*) Negative electrode attracts positive ions; positive electrode attracts negative ions. If electrons are transferred from negative electrode to positive ions and from negative ions to positive electrode, the circuit is complete, and electricity will flow through the circuit.

atomic molecules, each having one hydrogen and one chlorine atom. When hydrogen chloride dissolves in water an *ionization* reaction occurs, producing ions from molecules. The resulting solution is composed of hydrogen ions and chloride ions dispersed among the water molecules; consequently it is a conducting solution, and hydrogen chloride in water is properly termed an electrolyte.

When a molecular solute dissolves in water to produce ions, the process is called ionization.

$$HCl \xrightarrow{\text{Water}} H^+_{(aq)} + Cl^-_{(aq)}$$
$$\text{MOLECULE} \qquad\qquad \text{IONS}$$

The hydrogen ion in aqueous systems is not free and unattached. Recall that water molecules are polar. A free proton (isolated hydrogen ion) could not exist in such a medium; it becomes attached to the negative end of one of the water dipoles. In fact, the attraction of water dipoles

The symbol $H^+_{(aq)}$ means a hydrogen ion dissolved in water, an aqueous system.

for the polar HCl molecule probably brings about its ionization in the first place.

$$H^+ + H\!:\!\overset{\displaystyle ..}{\underset{\displaystyle H}{O}}\!: \longrightarrow \left[H\!:\!\overset{\displaystyle ..}{\underset{\displaystyle H}{O}}\!:\!H\right]^+$$

HYDRONIUM
ION

H_3O^+ is the hydronium ion.

Thus, the hydrogen ion in water is *hydrated* and is often referred to as the *hydronium* ion, H_3O^+ or $H^+(H_2O)$. When one considers hydrogen bonding between water molecules, it is very likely that other water molecules are attached to the molecule to which the proton is attached. The best representation we can give for the hydrogen ion in water then is $H^+(H_2O)_n$, where n is a constantly changing number, perhaps averaging about 4 or 5 in dilute solutions at room temperature.

When sugar, sodium chloride, alcohol, or any other readily soluble material dissolves in water, we can have either a concentrated or a dilute solution. Such a qualitative description of concentration is much less satisfactory and useful than a quantitative description, which tells us just how much of a given substance is dissolved in a specified volume. In chemical work we often express *concentrations* in *moles* of solute *per liter* of solution.

Solutions of known concentration are prepared using volumetric flasks. These are glass vessels with the stems precisely marked to indicate specific volumes, such as 1.000 liter. The procedure involves the steps shown in Figure 5–4.

To show how concentrations are determined, let us consider a case where a 25 g sample of NaCl is carefully weighed, then transferred to a one-liter volumetric flask and dissolved in water. The next step is to add

1. Take a volumetric flask

2. Add carefully the weighed amount of solid

3. Add some water, shake, and dissolve solid

4. Fill flask to one liter mark and shake until homogeneous solution is obtained

Figure 5–4 Laboratory procedure for the preparation of a solution of known concentration.

water to the flask until the solution has a total volume of one liter. In order to determine the concentration of such a solution, we need to know the number of moles of NaCl present. The formula weight of NaCl is $23.0 + 35.5$ or 58.5. Thus, if we have 58.5 grams of NaCl, we have one mole. We have 25 grams, so the concentration is

$$\frac{25 \text{ grams}}{58.5 \text{ grams/mole}} = 0.43 \text{ mole of NaCl per liter of solution.}$$

We usually indicate this as 0.43 M NaCl, where M stands for moles of solute per liter of solution, and is read as "molar."

Suppose we have 86 grams of sucrose (table sugar: $C_{12}H_{22}O_{11}$, molecular weight 342) dissolved in a volume of 500 ml. What is the concentration of sugar in this solution? Since we have $^{86}/_{342}$ mole of sugar, or 0.25 mole, dissolved in $^{500}/_{1000}$ of a liter, the concentration of the sugar solution is

Molar concentration: Number of moles of a substance per liter of solution.

$$\frac{\dfrac{86 \text{ g}}{342 \text{ g/mole}}}{\dfrac{500 \text{ ml}}{1000 \text{ ml/liter}}} = 0.50 \text{ M sucrose.}$$

Molarity = M =

$$\frac{moles\ of\ solute}{liter\ of\ solution}$$

With this knowledge about solutions and their concentrations in mind, let us look at acid-base reactions occurring in solution.

ACIDS AND BASES

The terms "acid" and "base" have been used by chemists for several hundred years. These names were originally given to substances that showed certain properties. For example, acids have long been characterized as substances that are sour tasting, corrosive, and able to react with alkaline substances called bases. Acids turn blue litmus red. Bases, on the other hand, have a bitter taste, make the fingers or skin feel slippery on contact, and react with acids. Bases turn red litmus blue. As more and more information was collected on the properties of acids and bases, these simple definitions were refined to the point where current definitions of acids and bases are theoretical in nature.

Acids taste sour; bases taste bitter.

Litmus is an acid-base sensitive dye obtained from lichens.

Proton Transfer and Neutralization

Perhaps the most used definitions of acids and bases are those first given by J. N. Brønsted and T. M. Lowry in 1923.

Brønsted acid: A chemical species that can *donate* hydrogen ions (also called protons or H^+ ion) is an acid.

Brønsted base: A chemical species that can *accept* hydrogen ions is a base.

To illustrate these definitions we again consider the reaction between gaseous hydrogen chloride (HCl) and water:

$$HCl(gas) + H_2O \longrightarrow H_3O^+ + Cl^-$$

ACID   BASE   Hydronium Ion   Chloride Ion

HYDROCHLORIC ACID

Examination of the above reaction shows that the HCl molecule has donated a proton (H⁺ ion) to the water molecule. This behavior is understandable in terms of the electronic structures of the reacting molecules.

BASE   ACID

The partially positive (δ^+) hydrogen atom of the HCl molecule is attracted to the partially negative (δ^-) end of the water molecule, and the hydrogen ion is transferred. This reaction is essentially complete; that is, almost all the HCl is converted to H_3O^+ and Cl^-. A concentrated (about 12 M) solution of hydrogen chloride in water is mostly a solution of hydronium (H_3O^+) ions and chloride (Cl^-) ions.

If the ionic solid sodium oxide, Na_2O, is dissolved in water, a vigorous reaction produces a solution containing sodium ion (Na^+) and hydroxide ions (OH^-). In this process, the oxide ion (O^{2-}) reacts with water to form the hydroxide ion. In this, as in other such aqueous reactions, it is understood that the ions are hydrated (i.e., water molecules are bonded to the ions on a transitory basis).

Reaction is:
$Na_2O + H_2O \longrightarrow 2Na^+ + 2OH^-$

BASE   ACID   HYDROXIDE IONS

There are many other bases that take a proton from a water molecule in this way. For example:

$$S^{2-} + H_2O \longrightarrow HS^- + OH^-$$

SULFIDE ION   ACID   HYDROGEN   HYDROXIDE
BASE   SULFIDE ION   ION

$$CN^- + H_2O \longrightarrow HCN + OH^-$$

CYANIDE ION   ACID   HYDROGEN   HYDROXIDE
BASE   CYANIDE   ION

According to the Brønsted-Lowry definition, water acts as an acid in these reactions and donates a proton to the other molecule or ion, which acts as a base. A species such as water that can either donate or accept protons is called **amphiprotic.** The existence of amphiprotic species implies that acid-base reactions possess a reciprocal nature; an acid and a base react to form another acid (to which a proton has just been added) and another base (from which a proton has just been removed). Because water is the most commonly used solvent, it is also the most usual reference compound for acid-base reactions.

Also, one water molecule can transfer a proton to another water molecule.

$$2H_2O \rightleftharpoons H_3O^+ + OH^-$$

This reaction takes place to only a very small extent, as is indicated by arrows of unequal length. Since H_3O^+ and OH^- are produced in equal amounts when only water is present, pure water is neither acidic nor basic, but is described as **neutral.**

A chemical species in water solution is commonly spoken of as an acid if it donates protons to water and increases the concentration of H_3O^+ or $H_{(aq)}^+$. Similarly, a base in water solution is commonly described as a compound whose addition to water increases the concentration of OH^-. Since water is not the only possible solvent, these concepts are too narrow for general scientific use; they have been extended by the definitions given above, which focus on the essential feature of such acid-base behavior—that is, the donation or acceptance of a hydrogen ion (H^+) in a reaction.

Acids form H^+ ions in water; bases form OH^- ions in water.

When acids react with bases the properties of both species disappear. The process involved is called **neutralization.** To get a more precise picture of acid-base neutralization reactions, we will consider what happens when a solution of hydrochloric acid is mixed with a solution of sodium hydroxide. The hydrochloric acid contains H_3O^+ and Cl^- ions; the sodium hydroxide solution contains Na^+ and OH^- ions. When these two solutions are mixed, a reaction occurs between H_3O^+ and OH^-.

When an acid neutralizes a base, acid and base properties are suppressed.

$$\underset{\text{ACID}}{Na^+ + Cl^- + H_3O^+} + \underset{\text{BASE}}{OH^-} \longrightarrow H_2O + H_2O + Na^+ + Cl^-$$

If we have an equal number of H_3O^+ and OH^- ions, they will react to produce a neutral solution, with the hydronium ions (H_3O^+) donating their protons to the hydroxide ions (OH^-), forming molecules of water. Such reactions are called neutralization reactions because the acids and bases neutralize each other's properties. If we have more H_3O^+ ions than OH^- ions, the extra H_3O^+ will make the resulting solution **acidic.** If we have more OH^- ions than H_3O^+ ions, only a fraction of the OH^- ions will be neutralized, and the extra OH^- ions will make the resulting solution **basic.**

When an acid ionizes, it produces a hydronium ion plus a species which is called the **conjugate base** of that acid. For example:

Conjugate acids and bases differ by one hydrogen ion. The conjugate acid has one more hydrogen ion than its conjugate base.

$$\underset{\substack{\text{NITRIC}\\\text{ACID}}}{HNO_3} + \underset{\text{WATER}}{H_2O} \longrightarrow \underset{\substack{\text{HYDRONIUM}\\\text{ION}}}{H_3O^+} + \underset{\substack{\text{NITRATE ION, THE}\\\text{CONJUGATE BASE OF}\\\text{THE ACID } HNO_3}}{NO_3^-}$$

In the same manner we speak of nitric acid, HNO_3, as being the **conjugate acid** of the base NO_3^-.

1. When ammonia dissolves in water, the resulting solution conducts electricity. Ammonia in water is therefore a(n) _____.

2. A chemical species that can accept a hydrogen ion is a(n) _____.

   A chemical species that can donate a hydrogen ion is a(n) _____.

3. A compound HA is found to undergo a reaction forming a product H_2A^+. Therefore HA is a(n) (acid/base). If compound HA reacted to form A^-, then HA would be a(n) (acid/base).

4. If compound HA mentioned above undergoes both reactions described, then

   HA is termed _____.

5. A solution which contains equal concentrations of OH^- and H^+ ions is

   termed _____.

6. The word "aqueous" means _____.

7. In a neutralization reaction, a(n) _____ reacts with a(n)

   _____.

The Strengths of Acids and Bases

Because the water molecule is itself a weak base, the strongest acid (that is, the best proton donor) that can exist in water is the hydronium ion (H_3O^+). If strong acids such as sulfuric acid (H_2SO_4) or nitric acid (HNO_3) are added to water, they donate all of their protons to water molecules, and the resulting solutions contain H_3O^+ and SO_4^{2-}, and H_3O^+ and NO_3^-, respectively. All of these reactions are reversible, at least in principle, but when strong acids are dissolved in a large excess of water to form dilute solutions, virtually all of the strong acid is converted to products.

Reactants		Products	

$$HNO_3 + H_2O \longrightarrow H_3O^+ + NO_3^-$$
$$H_2SO_4 + H_2O \longrightarrow H_3O^+ + HSO_4^-$$

		CONJUGATE	
ACID	BASE	ACID	BASE

Strong electrolytes dissociate completely in water.

Since there are a large number of ions present, the solution conducts electricity well; HNO_3 and H_2SO_4 are termed ***strong electrolytes.***

Under equilibrium conditions the concentrations of the species in solution remain unchanged even though reactions in the forward and reverse directions continue.

Not all acids lose protons as readily to water as do nitric acid and sulfuric acid. Some anions are capable of competing with water for the proton being exchanged. The result of this competition is the establishment of an ***equilibrium*** (or balance) between neutral acid molecules and hydronium ions in water solution.

Acetic acid, found in vinegar, is a weak acid; that is, its conjugate base, acetate ion, competes well for a proton. The molecular structure of acetic acid ($HC_2H_3O_2$) is as follows:

$$
\begin{array}{c}
\text{H} \\
| \\
\text{H} - \text{C} - \text{C} \\
| \\
\text{H}
\end{array}
\quad
\begin{array}{c}
\text{O} \\
\diagup\diagdown \\
\quad \text{O} - \text{H} \quad \text{ACIDIC HYDROGEN ATOM}
\end{array}
$$

The hydrogen atom bonded to the oxygen in the molecule is the only one that is donated to a base in water solution; for that reason it is designated an acidic hydrogen. When acetic acid is dissolved in water, some ions are produced. However, most of the acetic acid molecules do not donate protons to water molecules. The result is a mixture containing a few H_3O^+ and $C_2H_3O_2^-$ ions and many $HC_2H_3O_2$ molecules. The reaction is:

$$
\underset{\text{ACETIC ACID}}{HC_2H_3O_2} + \underset{\text{WATER}}{H_2O} \rightleftharpoons \underset{\substack{\text{HYDRONIUM} \\ \text{ION}}}{H_3O^+} + \underset{\substack{\text{ACETATE} \\ \text{ION}}}{C_2H_3O_2^-}
$$

The relatively few ions in an acetic acid solution do not conduct electricity very effectively; consequently, acetic acid is a **_weak electrolyte._** (A dilute solution of acetic acid would barely conduct in the apparatus shown in Figure 5–2.)

Another way of looking at this reaction is to realize that in the mixture there are two bases: the water molecule H_2O, and the acetate ion $C_2H_3O_2^-$. Since the reverse reaction dominates, the acetate ion must be a stronger base than the water molecule.

The same kind of considerations can be made for other bases. Ammonia, dissolved in water, is a weak base. The resulting reaction produces relatively few ions; ammonia is mostly in the molecular form:

$$
\underset{\text{AMMONIA}}{NH_3} + \underset{\text{WATER}}{H_2O} \rightleftharpoons \underset{\substack{\text{AMMONIUM} \\ \text{ION}}}{NH_4^+} + \underset{\substack{\text{HYDROXIDE} \\ \text{ION}}}{OH^-}
$$

Consequently, the relatively few ions present do not conduct electricity well, and ammonia may thus be called a weak electrolyte.

Table 5–1 gives some common acids and bases ranked according to their relative strengths.

Ammonia is the number four commercial chemical in quantity produced, sodium hydroxide is number eight, nitric acid is number ten, and hydrochloric acid is number twenty-five. (See inside front cover.)

TABLE 5–1 Relative Strengths of Some Acids and Bases

	Acid					Conjugate Base		
perchloric acid	$HClO_4$	+	$H_2O \rightleftharpoons H_3O^+$	+	ClO_4^-	perchlorate ion		
hydrochloric acid	HCl	+	$H_2O \rightleftharpoons H_3O^+$	+	Cl^-	chloride ion		
nitric acid	HNO_3	+	$H_2O \rightleftharpoons H_3O^+$	+	NO_3^-	nitrate ion		
hydronium ion	H_3O^+	+	$H_2O \rightleftharpoons H_3O^+$	+	H_2O	water		
hydrofluoric acid	HF	+	$H_2O \rightleftharpoons H_3O^+$	+	F^-	fluoride ion		
acetic acid	$HC_2H_3O_2$	+	$H_2O \rightleftharpoons H_3O^+$	+	$C_2H_3O_2^-$	acetate ion		
water	H_2O	+	$H_2O \rightleftharpoons H_3O^+$	+	OH^-	hydroxide ion		
hydroxide ion	OH^-	+	$H_2O \rightleftharpoons H_3O^+$	+	O^{2-}	oxide ion		

Increasing Acid Strength (left margin, upward arrow)
Increasing Base Strength (right margin, downward arrow)

The pH Scale

Because water solutions of dilute acids and bases are used so extensively, it is convenient to have a simple way of designating the acidity or basicity of these solutions. The pH scale was devised for this purpose; it furnishes a number which describes the acidity of a solution. The **pH** is defined as the negative logarithm of the concentration of the hydrogen ion, $pH = -\log[H_3O^+]$. The brackets mean moles per liter of hydronium ions. The pH scale, for practical purposes, runs from 0 to 14. A pH of 7 indicates a neutral solution (such as pure water), a pH below 7 indicates an acidic solution, and a pH above 7 indicates a basic solution.

> At pH = 7, the number of H_3O^+ ions equals the number of OH^- ions.

The pH number is related to the concentration of the hydrogen ions in an aqueous solution expressed as the negative power of 10 (logarithm). A hydrogen ion concentration of 0.00001 or 1×10^{-5} M corresponds to a pH of 5. When the pH is 7, as in pure water, the hydrogen ion concentration would have to be 1×10^{-7} M. Note that for each unit decrease in the pH number, there is a ten-fold increase in the concentration of the hydrogen ion (that is, the acidity).

To understand why the pH of pure water is 7, we will need to re-examine the very slight acid-base reaction between water molecules:

$$H_2O + H_2O \rightleftharpoons H_3O^+ + OH^-$$

Figure 5—5   An illustration of what it may be like in an aqueous (water = ⊕) solution of a strong acid (HCl) and an aqueous solution of a weak acid (acetic acid, HAc). The strong acid is all ions; no molecules. The weak acid is mostly molecules with only a few ions.

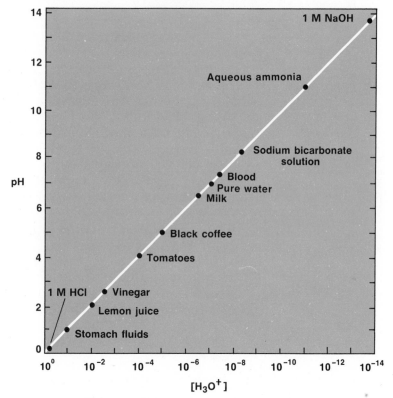

Figure 5–6 A plot of pH versus hydrogen ion concentration, $[H_3O^+]$. Note that the pH *increases* as the $[H_3O^+]$ decreases. The pH values of some common solutions are given for reference. (A solution in which $[H_3O^+] = 1$ M has a pH of 0 since $1 = 10^0$.)

Hydrogen or hydronium ion concentration in moles per liter

This ionization reaction produces ions, which should cause water to conduct electricity. But we have said that water is a nonelectrolyte. Actually, with very sensitive electrical equipment, pure water can be shown to conduct electricity *very slightly* because of the very small amount of ionization. Laboratory measurements reveal that 0.0000001 or 1×10^{-7} mole per liter of water is present as H_3O^+ and OH^- ions at 25°C. Pure water contains 55.6 moles of water per liter. Consequently, we can see that the actual amount of the water that ionizes is very small compared to the total amount of water present. Pure water, then, which is defined as being neutral, contains 1×10^{-7} mole of hydrogen ions (hydronium ions) per liter, and the pH is 7. In pure water the concentrations of hydronium ions (H_3O^+) and hydroxide ions (OH^-) must be equal, since each time a hydronium ion (H_3O^+) is produced in the ionization reaction a hydroxide ion (OH^-) is also produced. Therefore, in pure water the concentration of the hydroxide ion (OH^-) is also 1×10^{-7} mole per liter.

Figure 5–6 graphically displays the relationship between the pH number and the concentration of the hydrogen ion. It also gives the approximate pH values for common solutions.

A close examination of Figure 5–6 reveals that basic solutions such as ammonia water have hydrogen ion concentrations less than that of pure water. Note that the pH of a typical sample of ammonia water is 11. This is equivalent to a hydrogen ion concentration of 1×10^{-11}, or 0.00000000001 mole per liter.

One liter of water weighs 1000 grams and is 55.6 molar:

$$\frac{1000 \text{ g}}{\frac{18 \text{ g}}{\text{mole}}} = 55.6 \text{ moles}$$

SALTS

Preparation of Salts

Salts can be formed by acid-base neutralization.

The chemical compounds known as **salts** play a vital role in nature, in plant and animal growth and life, and in the manufacture of various chemicals for human use. They can be formed as the products of acid-base neutralizations, as in the following example:

$$(K^+ + OH^-) + (H_3O^+ + Cl^-) \longrightarrow \underline{K^+ + Cl^-} + 2H_2O$$

Potassium Hydroxide   Hydrochloric Acid
in Water   in Water

BASE   ACID

Crystallize the Salt by Removal of Solvent, Water

KCl (Solid)

SALT

Most salts contain ions held together by *ionic bonding* (see Chapter 3). Solid potassium chloride, for example, is composed of an equal number of K^+ ions and Cl^- ions arranged in definite positions with respect to one another in a lattice (see Figure 3–5). Since the salt crystal must be electrically neutral, it can have neither an excess nor a deficiency of positive or negative charge.

Let us imagine that we have at our disposal the ions listed below, and let us see what salts could result.

Ions		Possible Salts	Salt Name
Na^+	sodium	$NaCl$	sodium chloride
Ca^{2+}	calcium	$NaNO_3$	sodium nitrate
Cl^-	chloride	Na_2SO_4	sodium sulfate
NO_3^-	nitrate	$NaC_2H_3O_2$	sodium acetate
SO_4^{2-}	sulfate	$CaCl_2$	calcium chloride
$C_2H_3O_2^-$	acetate	$Ca(NO_3)_2$	calcium nitrate
		$Ca(C_2H_3O_2)_2$	calcium acetate
		$CaSO_4$	calcium sulfate

In the formula of a salt, the positive and negative charges are equal.

In the examples just given, notice that in order to attain an electrically neutral lattice, it is necessary to balance the charges of the ions. A sodium (Na^+) ion requires just one chloride ion (Cl^-), and the NaCl lattice contains an equal number of Na^+ and Cl^- ions. A sulfate ion (SO_4^{2-}) with two negative charges must have its negative charge balanced by two positive charges. This may be done by using two Na^+ ions:

$$2Na^+ + SO_4^{2-} \longrightarrow Na_2SO_4$$

or one Ca^{2+} ion:

$$Ca^{2+} + SO_4^{2-} \longrightarrow CaSO_4$$

It is possible to form many solid salts by mixing water solutions of different soluble salts with each other. For example, both lead acetate and sodium chloride are soluble in water. If we prepare solutions of these salts and then mix the solutions, we find the insoluble salt, lead(II) chloride, precipitates from the mixture and may be removed by filtration.

$$2Na^+ + 2Cl^- \; + \; Pb^{2+} + 2C_2H_3O_2^- \longrightarrow \; PbCl_2 \; + 2Na^+ + 2C_2H_3O_2^-$$

| SODIUM CHLORIDE IN SOLUTION | LEAD(II) ACETATE IN SOLUTION | SOLID LEAD(II) CHLORIDE | SODIUM ACETATE IN SOLUTION |

Sodium acetate may be recovered by evaporating the water.

This reaction illustrates an important principle based upon differences in solubility. If the component ions of a compound of low solubility are mixed in solution in great enough concentrations, the compound containing those ions will precipitate from solution.

Salts in Solution

An important property of many salts is their solubility in suitable solvents. The amount of a salt that will dissolve in a given quantity of solvent tells us the salt's solubility in that solvent. The preparation of lead(II) chloride shown above was made possible by the differences in solubilities of different salts in the same solvent. As a result of these differences in solubilities, we can make roads out of calcium carbonate (limestone), which is insoluble in water, but not calcium chloride, which is water soluble.

Consider what happens at the ionic level when a sodium chloride crystal is placed in contact with water. We know that sodium chloride is soluble in water. This means that most, if not all, of the attractive forces between the ions in the crystal lattice are somehow overcome in the solution process.

The surface of the salt crystal appears calm when the crystal is placed in water, but on the ionic level there is a great deal of agitation. Water molecules have sufficient polarity to interact strongly with the ions and bond with them. Once this takes place the ion is less strongly bound in the lattice, and so can be removed from the crystal. The crystal lattice now has a gap in it where the ion was removed. As stated earlier, ions that are bonded by solvent molecules are termed solvated ions, and the process of ion-solvent interaction is called **solvation.** But just how are these ions solvated? What causes this solvation?

The answers lie in the structure of the water molecule and its ability to interact with ions. As we saw in Chapter 3, water is a *polar* molecule. The negative end of the molecules is attracted to positive ions, and the positive end is attracted to negative ions. As a result, several water molecules will interact with each ion. Figure 5–7 shows several water molecules solvating a Na^+ ion. The positive end of the water molecule will tend to interact with a negative ion; this is shown for the Cl^- ion in Figure 5–7.

Every salt has what may be termed a "solubility limit" for a given solvent: at a given temperature, a certain number of grams of salt, and *no more,** will dissolve in a certain quantity of solvent. A solution that

While some salts are very soluble in water, others are quite insoluble. Salts are found with a wide range of water solubilities.

The two most abundant ions in ocean water are Cl^- and Na^+ ions, sufficient to recover about 27 grams of NaCl per kilogram of seawater. To put it another way, there are about 128 million tons of NaCl per cubic mile of seawater, and there are 328,000,000 cubic miles available!

Because of its polar nature, the water molecule is ideally suited to interact with ions.

Generally, for a given solvent, a salt will dissolve to a greater extent in the hot solvent than in the cold solvent.

---

*Supersaturated solutions can be formed under special conditions, but are not stable in the presence of the solid solute. The presence of the solid solute causes the excess dissolved solute to crystallize and a saturated solution is formed.

contains all the dissolved salt that it can hold is termed a "saturated solution." One might ask just why this type of solubility limit is found in nature. The reason for this can be understood if we remember that the oppositely charged ions of the salt in solution actually attract one another. If one crowds the solution with solvated ions to too great an extent and "ties up too many solvent molecules," the ions will begin

Figure 5–7   Dissolution of sodium chloride in water. (*a*) Geometry of the polar water molecule. (*b*) Solvation of sodium and chloride ions due to interaction (bonding) between these ions and water molecules. (*c*) Dissolution occurs as collisions between water molecules and crystal ions result in the removal of the crystal ion. In the process the ion becomes completely solvated.

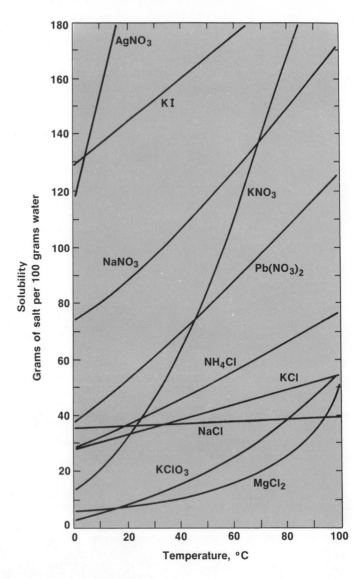

Figure 5—8 The effect of temperature on the solubility of some common salts in water.

re-forming the crystal lattice. This is *crystallization,* or solvation in reverse. When undissolved salt is in contact with a saturated solution of that salt, there is a dynamic equilibrium established with the salt crystal being broken down at one point while being formed at another. The effect of temperature on the solubility of some common salts is shown in Figure 5–8.

Uses of Salts

Salts are used for so many different applications that it would be impossible to list them all. We can, however, look at the uses of several typical salts and see just how versatile they are.

Sodium Chloride (Table Salt)

This is the best-known salt and perhaps the most important to us. Physiologically, mammals need this salt for an ionic balance in the blood-

Many body fluids contain potassium ions (K^+) in greater concentration than Na^+ ions.

stream and for the production of hydrochloric acid in the stomach to aid in digestion of food. Very early in our history, sodium chloride was used in barter, and wars have actually been fought for control of salt deposits. Commonly, sodium chloride is used for medicinal purposes, such as in water solution, to wash wounds and to treat burns.

Nitrate Salts

Ammonium nitrate is the number twelve commercial chemical. See inside the front cover.

Salts containing the nitrate ion (NO_3^-), such as sodium nitrate ($NaNO_3$), potassium nitrate (KNO_3), and ammonium nitrate (NH_4NO_3), find special uses in fertilizers and in the manufacture of explosives. Sodium nitrate and potassium nitrate occur in natural deposits, while ammonium nitrate can be prepared by a reaction involving ammonia (NH_3) and nitric acid.

$$NH_3 + HNO_3 \rightleftharpoons NH_4NO_3$$

$$\text{BASE} \quad \text{ACID} \qquad \text{SALT}$$
$$\text{(AMMONIUM NITRATE)}$$

Gunpowder contains potassium nitrate (KNO_3), sulfur, and charcoal.

Explosions are possible when nitrate salts and combustible materials are mixed. Gunpowder contains KNO_3. Ammonium nitrate will explode by itself under certain conditions, and a number of disasters have occurred when large amounts of this salt have exploded in warehouses and in the holds of ships in harbors. Under more controlled conditions, NH_4NO_3 decomposes to yield dinitrogen oxide (N_2O) and water. Dinitrogen oxide is called laughing gas and is used as an anesthetic in certain forms of surgery.

$$NH_4NO_3 \xrightarrow{\text{Heat}} N_2O + 2H_2O$$

$$\text{DINITROGEN}$$
$$\text{OXIDE}$$
$$\text{(LAUGHING GAS)}$$

This gas also has a limited solubility in milk and cream and finds a use as a propellant in canned whipped cream. When the valve is opened, the cream and gas escape together, and upon reaching the atmosphere the gas expands, thus "whipping" the cream.

Sodium Hydrogen Carbonate

Sodium hydrogen carbonate, also called sodium bicarbonate ($NaHCO_3$), has the common name, baking soda. The negative ion of the salt, the hydrogen carbonate ion, is a base and is capable of reacting with water in an acid-base reaction.

$$HCO_3^- + H_2O \rightleftharpoons H_2CO_3 + OH^-$$

$$\text{BASE} \quad \text{ACID} \qquad \text{ACID} \quad \text{BASE}$$

When this salt is placed in a mixture containing water, such as biscuit batter, this reaction takes place to a slight extent. The product of the reaction, carbonic acid (H_2CO_3), is quite unstable and decomposes under the influence of heat, yielding the products carbon dioxide (CO_2) and water. The decomposition of the H_2CO_3 causes the reaction of bicarbonate with water to take place to a greater extent. In effect, the equilibrium is shifted to the products side, and the overall reaction is essentially a complete conversion of the bicarbonate ion to carbon dioxide and water. The reaction is:

$$HCO_3^- + H_2O \longrightarrow H_2CO_3 + OH^-$$

$$H_2CO_3 \xrightarrow{\text{Heat}} H_2O + CO_2 \text{ (GAS)}$$

The CO_2 gas which is produced in the reaction causes the biscuit batter to rise. To hasten these reactions, pioneer women often added sour milk or buttermilk, which contain lactic acid, a weak acid. Baking powder contains baking soda along with a weak acid which shifts the equilibrium to the right. Self-rising flour contains a small amount of baking soda and a weak acid to produce a controlled amount of CO_2 gas as the mixture is heated in the oven.

The HCO_3^- ion can also act as a weak acid in the presence of a strong base:

$$\underset{\text{ACID}}{HCO_3^-} + \underset{\text{BASE}}{OH^-} \rightleftharpoons \underset{\text{ACID}}{HOH} + \underset{\text{BASE}}{CO_3^{2-}}$$

This explains its use as a neutralizer of spilled acids and bases in the laboratory.

Barium Sulfate

This salt may be prepared by taking advantage of its insolubility in water. If a solution of barium chloride ($BaCl_2$) is mixed with sodium sulfate solution (Na_2SO_4), insoluble $BaSO_4$ precipitates:

$$(Ba^{2+} + 2Cl^-) + (2Na^+ + SO_4^{2-}) \longrightarrow BaSO_4 \text{ (Solid)} + (2Na^+ + 2Cl^-)$$

In the diagnosis of stomach ulcers and intestinal disorders X rays are often used. Since these organs are normally transparent to X rays, a suitable material must be placed in the organ to render it partially opaque. Barium is a good element for this, since it is essentially opaque toward the type of X rays used in medicine (Figure 5–9). Barium ions, however, are quite poisonous and cannot be taken internally. The solution to this problem lies in the extreme insolubility of $BaSO_4$. A suspension of solid $BaSO_4$ is swallowed to introduce barium into the digestive system; since the barium is not in ionic form, it cannot be absorbed by the body.

Figure 5–9 $BaSO_4$ is used in intestinal X ray studies to minimize the danger of barium poisoning, since the very insoluble salt passes through the body unaltered.

1. Complete the matching set.

   pH
   10   (a)   acidic
   7   (b)   basic
   3   (c)   neutral

2. If the solubility of KI is 140 grams per 100 grams of water at 20°C, how much

   KI will dissolve in 1000 grams of water? _____

3. Write a chemical reaction showing water to be an acid. Use ammonia (NH_3)
   as the base.

$$H_2O + NH_3 \longrightarrow \text{_____} + \text{_____}$$

4. Write a chemical reaction showing water to be a base. Use HCl as the acid.

$$H_2O + HCl \longrightarrow \text{_____} + \text{_____}$$

5. Which is more acidic, a pH of 6 or a pH of 2? _____

6. Is a pH of 8 more basic than water? _____

7. The salt barium sulfate poses no threat of barium toxicity because it is

   _____ in the stomach and intestinal fluids.

8. When sodium bicarbonate reacts with an acid, the gas _____

   _____ is evolved.

9. Name two uses of nitrate salts. _____ and _____

ELECTRON TRANSFER (OXIDATION-REDUCTION)

A very common type of chemical reaction is one in which the control of loosely held electrons passes from one species to another. These reactions are called oxidation-reduction reactions. Examples include:

$$2Na + Cl_2 \longrightarrow 2NaCl \ (Na^+ \text{ and } Cl^-)$$
$$2H_2 + O_2 \longrightarrow 2H_2O$$
$$2Fe + 3Br_2 \longrightarrow 2FeBr_3 \ (Fe^{3+} \text{ and } Br^-)$$

The loss of electrons is called oxidation.

The gain of electrons is called reduction.

If the atom has either lost electrons or lost some control over the electrons, we say it has been **oxidized.** If it has gained electrons or gained a greater degree of control over the electrons, we say it has been **reduced.** In the reactions above, the sodium, hydrogen, and iron atoms have all lost electrons, so these elements have been oxidized. The species that have caused this change (Cl_2, O_2, and Br_2) are called **oxidizing agents.** Since these molecules have picked up electrons, they have been reduced, and the species that caused this change (Na, H_2, and Fe) are called **reducing agents.** In oxidation-reduction reactions, oxidizing agents become reduced and reducing agents become oxidized.

The term oxidation has its historical background in the fact that the first oxidizing agent whose chemical behavior was thoroughly studied was oxygen. The phenomenon was then named after the element. As the understanding of chemical reactions deepened, it became apparent that the reaction of a metal with oxygen was very similar to its reaction with fluorine, chlorine, or bromine. Thus, in each of the following reactions, the iron atom loses electrons to the other reactant:

> *Oxidizing agents gain electrons at the expense of reducing agents.*

$$4Fe + 3O_2 \longrightarrow 2Fe_2O_3 \ (Fe^{3+} \text{ and } O^{2-})$$
$$2Fe + 3F_2 \longrightarrow 2FeF_3 \ (Fe^{3+} \text{ and } F^-)$$
$$2Fe + 3Cl_2 \longrightarrow 2FeCl_3 \ (Fe^{3+} \text{ and } Cl^-)$$

After this similarity was noted, the concept of oxidation was generalized to cover all situations in which an atom, ion, or molecule lost electrons. In the same manner the concept of reduction is now used to cover all situations in which an atom, ion, or molecule gains electrons.

The formation of a salt by the direct combination of the elements is an example of an oxidation-reduction reaction. When sodium reacts with chlorine to form sodium chloride, the following reactions take place.

> *Here, chlorine is the oxidizing agent and gets reduced; sodium is the reducing agent and gets oxidized.*

$$Na \longrightarrow Na^+ + e^- \quad \text{(OXIDATION OF SODIUM)}$$
$$Cl_2 + 2e^- \longrightarrow 2Cl^- \quad \text{(REDUCTION OF CHLORINE)}$$

Since two electrons are gained per chlorine molecule, two sodium atoms must be oxidized for every Cl_2 molecule reduced; the overall chemical reaction is

$$2Na + Cl_2 \longrightarrow 2Na^+ + 2Cl^-$$

Reduction of Metals

One of the most practical applications of the oxidation-reduction principle is the separation of metals from their ores. The majority of metals are found in nature in compounds; that is, the metals are in an oxidized state. In order to obtain the metal, it must be reduced from the positive oxidation state to the neutral elemental state. This requires a gain of electrons. In addition to electricity, there are a variety of chemicals that can supply the electrons for the reduction process. Some of the ways that metals can be reduced from their ores are given in Table 5–2.

> *The reduction of metals for making tools is as old as recorded history.*

Electrolysis

Several metals either are separated from their ores or are purified afterward by electrolysis, as noted in Table 5–2. ***Electrolysis*** is a type of chemical reaction caused by the application of electrical energy.

The principal parts of an electrolysis apparatus are shown in Figure 5–10. Electrical contact between the external circuit and the solution is obtained by means of ***electrodes,*** which are often made of graphite or metal. The electrode at which electrons enter an electrolysis cell is termed the ***cathode,*** and this is the electrode at which reduction takes place. The electrode at which the electrons leave the cell is the ***anode.*** At the anode oxidation takes place. The battery or generator produces a current of electrons which flow toward one electrode (the cathode) and away

> *The suffix -lysis means splitting or decomposition; electrolysis is decomposition by electricity.*

TABLE 5–2 Some Ways Used to Separate Metals from Their Ores By Reduction

Metal	Occurrence	A Reduction Process	Uses of Metal
Cu	Cu_2S, chalcocite	Air blown through melted ore $Cu_2S + O_2 \longrightarrow 2Cu + SO_2\uparrow$ ($Cu^+ + e^- \longrightarrow Cu$)	Electrical wiring, boilers, pipes, brass (Cu 85%, Zn), bronze (Cu 90%, Sn, Zn), other alloys
Na	NaCl, rock salt	Electrolysis of fused chloride $2NaCl \longrightarrow 2Na + Cl_2$ ($Na^+ + e^- \longrightarrow Na$)	Coolant in nuclear reactors, orange street lights, making ethyl gasoline
Mg	Mg^{2+}, seawater	Electrolysis of fused chloride $MgCl_2 \longrightarrow Mg + Cl_2$ ($Mg^{2+} + 2e^- \longrightarrow Mg$)	Light alloys such as duralumin (0.5% Mg, Al), Dowmetal H (90.7% Mg, Al, Zn, Mn), flares, some flash bulbs
Ca	$CaCO_3$, chalk, limestone, marble	Thermal reduction of chloride obtained from carbonate $3CaCl_2 + 2Al \xrightarrow{heat} 3Ca + 2AlCl_3$ ($Ca^{2+} + 2e^- \longrightarrow Ca$)	Bearing metal alloys (0.7% Ca, Pb, Na, Li), storage battery electrodes
Al	$Al_2O_3 \cdot H_2O$, bauxite	Electrolysis in fused cryolite, Na_3AlF_6, at 800–900°C. $2Al_2O_3 \longrightarrow 4Al + 3O_2$ ($Al^{3+} + 3e^- \longrightarrow Al$)	Packaging, airplane and automobile parts, alloys, roofing, siding
Ag	Ag_2S, argentite	Reduction by a more active metal $2Ag(CN)_2^- + Zn \longrightarrow Zn(CN)_4^{2-} + 2Ag\downarrow$ ($Ag^+ + e^- \longrightarrow Ag$)	Jewelry, tableware, plating, electrical wiring, photography
Zn	ZnS, zinc blende	Roasting in air with carbon $ZnS + 2O_2 + C \longrightarrow Zn + SO_2\uparrow + CO_2\uparrow$ ($Zn^{2+} + 2e^- \longrightarrow Zn$)	Galvanizing iron and steel, die-cast auto parts
Hg	HgS, cinnabar	Roasting in air $HgS + O_2 \xrightarrow{heat} Hg + SO_2\uparrow$ ($Hg^{2+} + 2e^- \longrightarrow Hg$)	Some thermometers, electric switches, blue-green street lights, amalgams in dentistry, fluorescent lighting
Fe	Fe_2O_3, hematite	Reduction by carbon monoxide $Fe_2O_3 + 3CO \longrightarrow 3CO_2 + 2Fe$ ($Fe^{3+} + 3e^- \longrightarrow Fe$)	As cast and forged iron, alloyed with C (0.1 to 1.5%) to make steels: (stainless steel has 8% or more Cr)

Electron flow →

Copper anode (impure)   Cathode (pure)   ← Electron flow

Cu^{2+} →

← SO_4^{2-}

Copper sulfate solution

Generator

Anode reaction is an oxidation
$Cu \longrightarrow Cu^{2+} + 2e^-$

Cathode reaction is a reduction
$Cu^{2+} + 2e^- \longrightarrow Cu$

Figure 5–10 Electroplating from a copper sulfate solution.

from the other electrode (the anode). A salt solution can easily be observed to conduct electricity if a light bulb is part of the circuit (Figure 5–2). When the switch is closed, the positive ions in solution migrate toward the cathode. Soon, evidence of a chemical reaction can be seen at the electrodes. Depending on the substances present in the solution, gases may be evolved, metals deposited, or ionic species changed at the electrodes. The ions that migrate to the electrodes are not necessarily the species undergoing reaction at the electrodes, because sometimes the solvent undergoes reaction more easily. Whatever happens, the chemical reactions that are taking place at the cathode and anode are due to electrons going into and coming out of the solution. The chemical reaction at the cathode is one that furnishes electrons to solution species (reduction). At the anode, electrons are taken from species in solution, so the chemical reaction at the anode is one that gives up electrons (oxidation).

The electroplating of copper is illustrated in Figure 5–10. Such an electrolysis can be used either to plate an object with a layer of pure copper or to purify an impure sample of copper metal; copper is transferred from the positive electrode, into the solution, and eventually to the negative electrode. If the positive electrode is impure copper to be purified, electrolysis deposits the copper as very pure copper on the negative electrode.

Now let us examine how the electrolysis transfers the copper from the positive electrode to the negative electrode. Electrons flow out of the negative terminal of the generator through the wire, and are pumped into the negative electrode. Somehow this negative charge must be used up at the surface of the electrode.

Consider what happens when the electrons build up on the negative electrode. Since positive copper ions are nearby, they will be attracted to the surface and take electrons from it. Thus, the Cu^{2+} ions are reduced:

$$Cu^{2+} + 2e^- \longrightarrow Cu \qquad \text{(CATHODE REACTION)}$$

In a similar way the negative sulfate ions migrate to the positive electrode (anode). However, it turns out that it is easier to get electrons from the copper metal of the electrode than it is from the sulfate ions. As each copper atom gives up two electrons, it passes into solution:

$$Cu \longrightarrow Cu^{2+} + 2e^- \qquad \text{(ANODE REACTION)}$$

In any electrochemical cell, oxidation occurs at the anode; reduction occurs at the cathode.

In effect, then, the copper of the positive electrode (anode) passes into solution; the copper ions in solution migrate to the negative electrode (cathode) and plate out as copper metal. Large amounts of copper are purified in this way each year. Silver and gold purification can be done similarly.

If we desire to plate an object with copper, we have only to render the surface conducting and make it the negative electrode in a solution of copper sulfate. It will become coated with copper, with the copper coating growing thicker as the electrolysis is continued. If the object is a metal, it will conduct electricity by itself. If it is a nonmetal, its surface can be lightly dusted with graphite powder to render it conducting.

Copper can be plated onto an object by making that object the negative electrode in a cell containing dissolved copper salts.

A potentially very important electrolysis reaction is the electrolysis of water. When electricity is passed into graphite electrodes immersed in

Figure 5–11 Stamping masters for making high quality phonograph records are made by electroplating nickel onto a plastic record finely coated with silver. (Courtesy of GRT Record Pressing, Nashville.)

a dilute salt solution, water is reduced to hydrogen and hydroxide ions at the cathode:

$$2H_2O + 2e^- \longrightarrow H_2(g) + 2OH^- \qquad \text{(CATHODE REACTION)}$$

At the anode, water is oxidized to oxygen and hydrogen ions:

$$2H_2O \longrightarrow O_2 + 4H^+ + 4e^- \qquad \text{(ANODE REACTION)}$$

The OH^- and H^+ ions combine to re-form water. The overall, or net, cell reaction is:

$$2H_2O \xrightarrow{\text{Electricity}} 2H_2(g) + O_2(g)$$

The hydrogen produced by the reduction of water can be stored and used as a fuel—for example, to power rockets into space. Someday, if electricity becomes low enough in cost (see Chapter 17), water may be electrolyzed to produce hydrogen which can then be piped to the point of use, just as natural gas is today.

Relative Strengths of Oxidizing and Reducing Agents

When a piece of metallic zinc is placed in a solution containing hydrated copper ions (Cu^{2+}), an oxidation-reduction reaction occurs:

$$Zn + Cu^{2+} \longrightarrow Zn^{2+} + Cu$$

Evidence for this reaction is the deposit of copper on the zinc and the gradual decrease in the intensity of the blue color of the solution, indicating that the Cu^{2+} ions are being removed.

The oxidation of zinc by copper ions can be thought of as a competition between zinc ions (Zn^{2+}) and copper ions (Cu^{2+}) for the two electrons. Since the reaction proceeds almost to completion, the Cu^{2+} ions obviously win out in the competition. Other metals can compete similarly for electrons.

The *activity* of a metal is a measure of its tendency to lose electrons. Zinc is a more active metal than copper on the basis of the experiment just described. This means that given an equal opportunity, the first reaction will take place to a greater extent:

A copper atom has a greater attraction for the electrons than does the zinc atom.

$$Zn \longrightarrow Zn^{2+} + 2e^-$$
$$Cu \longrightarrow Cu^{2+} + 2e^-$$

Experiments of this type with various pairs of metals and other reducing agents yield an *activity series* of the elements which ranks each oxidizing and reducing agent according to its *strength* or *tendency* for the electron transfer to take place. An iron nail will be partly dissolved in a solution of a copper salt containing Cu^{2+} ions, with copper being deposited on the nail that remains. From this, it is determined that iron, like zinc, is more active than copper. The reaction that takes place is:

The active metals lose electrons more easily; hence, these free metals are not found in nature.

$$Fe + Cu^{2+} \longrightarrow Fe^{2+} + Cu$$

Now, which is more active, zinc or iron? This question can be answered by placing an iron nail in a solution containing Zn^{2+} ions and, in a separate container, a strip of zinc in a solution containing Fe^{2+} ions. It is found that the zinc strip is eaten away in the solution containing Fe^{2+} ions. The reaction is

$$Zn + Fe^{2+} \longrightarrow Fe + Zn^{2+}$$

Nothing happens to the iron nail in the solution of Zn^{2+} ions. We deduce that Zn loses electrons more readily than Fe.

Such an activity series can be extended to include other metals and even nonmetals as well. The concentrations of the ions in solution and other factors often must be considered for accurate work, but for our purposes these will be ignored. Table 5–3 is an activity series of some oxidizing and reducing agents.

Activity:
$Zn > Fe > Cu$

An application of the activity series can be seen in the use of *cathodic protection* to reduce corrosion (Figure 5–12). For this, the metal to be protected, say an iron pipeline, is connected to a more active metal, such as magnesium, by a heavy copper cable. In moist earth, an electrolytic cell is set up in which the magnesium is oxidized and transfers electrons to the iron. This keeps the iron from being oxidized by the air or any other reagent which might otherwise remove electrons from it.

The value of the activity series is that it allows us to predict the feasibility of reactions involving the species in it. Thus, a reducing agent in the table (right-hand column) is able to reduce the oxidized form of

TABLE 5–3 Relative Strengths of Some Oxidizing and Reducing Agents: The Activity Series

Oxidizing Agents		Reducing Agents
$Na^+ + e^-$	$\rightleftharpoons$	Na
$Ca^{2+} + 2e^-$	$\rightleftharpoons$	Ca
$Mg^{2+} + 2e^-$	$\rightleftharpoons$	Mg
$Zn^{2+} + 2e^-$	$\rightleftharpoons$	Zn
$Fe^{2+} + 2e^-$	$\rightleftharpoons$	Fe
$2H^+ + 2e^-$	$\rightleftharpoons$	H_2
$Cu^{2+} + 2e^-$	$\rightleftharpoons$	Cu
$Fe^{3+} + e^-$	$\rightleftharpoons$	Fe^{2+}
$Ag^+ + e^-$	$\rightleftharpoons$	Ag
$O_2 + 4H^+ + 4e^-$	$\rightleftharpoons$	$2H_2O$
$F_2 + 2e^-$	$\rightleftharpoons$	$2F^-$

(Increasing Strength of Oxidizing Agent — downward, left. Increasing Strength of Reducing Agent — upward, right.)

any species below it. For example, magnesium can reduce Cu^{2+} to Cu:

$$Mg + Cu^{2+} \longrightarrow Cu + Mg^{2+}$$

Magnesium can also reduce Ag^+ to Ag:

$$Mg + 2Ag^+ \longrightarrow Mg^{2+} + 2Ag$$

Figure 5–12 Cathodic protection. If magnesium is connected to the steel tank to be protected, the magnesium is more easily oxidized than the iron or copper connecting wire. The magnesium serves as a sacrificial anode. Hence, the cathode is protected with no points of oxidation occurring on its surface. The anode is the electrode where oxidation occurs; reduction occurs at the cathode. When the magnesium is used up, it is replaced by another block. The replacement is much easier and cheaper than replacing the tank.

Zinc can also reduce silver ions:

$$Zn + 2Ag^+ \longrightarrow 2Ag + Zn^{2+}$$

Zinc cannot reduce calcium ions, since calcium is above zinc in the series:

$$Zn + Ca^{2+} \longrightarrow \textit{No reaction}$$

The series also arranges oxidizing agents in the order of their effectiveness. Fluorine, F_2, can oxidize water, silver, iron(II) ion, or any species above it in the series; copper(II) ion, Cu^{2+}, can oxidize H_2, Fe, Zn, and the metals above Cu in the table, because it has a greater tendency to take on electrons than the ions that are formed. Thus:

The activity series is useful in predicting many chemical reactions.

$$Cu^{2+} + Fe \longrightarrow Fe^{2+} + Cu$$
$$Cu^{2+} + Mg \longrightarrow Mg^{2+} + Cu$$
$$Cu^{2+} + Ag \longrightarrow \textit{No reaction}$$

SELF-TEST 5-C

1. In a redox reaction, the oxidizing agent (gains/loses) electrons and gets (reduced/oxidized); the reducing agent (gains/loses) electrons and gets (reduced/oxidized).

2. Consider the reaction:

$$CH_4 + 2O_2 \longrightarrow CO_2 + 2H_2O$$

   Oxygen is the (oxidizing/reducing) agent. Methane (CH_4) is the (oxidizing/reducing) agent.

3. When silver-plated dinnerware is made, silver is deposited by the reaction $[Ag(CN)_2]^- + e^- \longrightarrow Ag + 2CN^-$. The silver is (oxidized/reduced). This reaction takes place at the (cathode/anode) of an electrolysis cell.

4. Consider the activity series in Table 5–3, write yes or no beside the following reactions, depending on whether or not they will proceed as written.

$$2Ag + Cu^{2+} \longrightarrow Cu + 2Ag^+ \qquad \underline{\hspace{2cm}}$$

$$F_2 + Zn \longrightarrow Zn^{2+} + 2F^- \qquad \underline{\hspace{2cm}}$$

$$Fe^{2+} + Ag^+ \longrightarrow Fe^{3+} + Ag \qquad \underline{\hspace{2cm}}$$

$$Mg^{2+} + 2Fe^{2+} \longrightarrow 2Fe^{3+} + Mg \qquad \underline{\hspace{2cm}}$$

5. When electrons are written on the left-hand side of an equation, as in $Cu^{2+} + 2e^- \longrightarrow Cu$, the reaction is a(n) (oxidation/reduction).

6. When electrons are written on the right-hand side of an equation, as in $Zn \longrightarrow Zn^{2+} + 2e^-$, the reaction is a(n) (oxidation/reduction).

Batteries

One of the most useful applications of oxidation-reduction reactions is in the production of electrical energy. A device that produces an electron flow (current) is called an ***electrochemical cell.*** Although a series of such cells is a ***battery,*** the term battery is commonly used even for single cells such as those we will describe.

Consider the reaction between zinc and copper ions that was previously discussed. If zinc is placed in a solution containing Cu^{2+} ions, the electron transfer takes place between the zinc metal and the copper ions, and the energy liberated simply causes a slight heating of the solution and the zinc strip. If the zinc could be separated from the copper solution, and the two connected in such a way to allow current flow, the reaction can proceed, but now the electrons are transferred through the connecting wires. Figure 5–13 shows a battery that can be constructed to make use of the energy evolved in the reaction of Zn with Cu^{2+}.

The anode reaction is the oxidation of zinc to Zn^{2+} ions.

$$Zn \longrightarrow Zn^{2+} + 2e^-$$

The electrons flow from the Zn electrode through the connecting wire, light the lamp in the circuit, and then flow into the copper cathode where reduction of Cu^{2+} ions takes place:

Historic battery used in the early telegraph.

$$Cu^{2+} + 2e^- \longrightarrow Cu$$

The copper is deposited on the copper cathode.

Figure 5–13 A simple battery involving the oxidation of zinc metal and the reduction of Cu^{2+} ions.

This flow of electrons (negative charge) from the anode to the cathode compartment in the battery must be neutralized electrically. In order to accomplish this, a "salt bridge" is provided which connects the two compartments. The salt bridge contains a solution of a salt such as K_2SO_4. The purpose of the salt bridge is to keep the two solutions neutral. Around the cathode the deposition of positive copper ions (Cu^{2+}) would tend to cause the solution to become negative owing to the presence of excess negative sulfate ions (SO_4^{2-}). Two actions can keep the solution around the cathode neutral: either positive potassium ions (K^+) pass into the solution, or negative sulfate ions pass out of the solution and into the salt bridge. Actually, both processes occur. Similarly, around the anode, the solution would tend to become positive because positive zinc ions (Zn^{2+}) are put into the solution. Two actions can keep the solution around the anode neutral: either negative sulfate ions pass into the solution or positive zinc ions pass out of the solution and into the salt bridge. In all of this exchange, it is necessary for the solution to maintain the same number of positive charges as negative charges. The reaction of zinc atoms and copper ions continues until one or the other is completely consumed.

In commercial batteries, the salt bridge is often replaced by a porous membrane.

Many different oxidation-reduction combinations are used in commercial batteries to produce a flow of electrons. A few of the more popular ones are listed in Table 5–4.

TABLE 5–4   Characteristics of Some Batteries

System	Anode (Oxidation)	Cathode (Reduction)	Electrolyte	Typical Operating Voltage Per Cell
Dry cell	Zn	MnO_2	NH_4Cl-$ZnCl_2$	0.9–1.4
Edison storage	Fe	Ni oxides	KOH	1.2–1.4
Nickel-Cadmium–NiCad	Cd	Ni oxides	KOH	1.1–1.3
Silver cell	Cd	Ag_2O	KOH	1.0–1.1
Lead storage	Pb	PbO_2	H_2SO_4	1.95–2.05
Mercury cell	Zn(Hg)	HgO	KOH–ZnO	1.30
Alkaline cell	Zn(Hg)	MnO_2	KOH	0.9–1.2

MATCHING SET

_____ 1. chemicals in automobile lead storage battery

(a) 7

_____ 2. most active metal in Table 5–3

(b) gain of electrons

_____ 3. ore of mercury

(c) proton donor

_____ 4. pH of pure water

(d) Na

_____ 5. solution

(e) proton acceptor

_____ 6. strong acid

(f) homogenous mixture of atoms, molecules, and/or ions

_____ 7. oxidation

(g) $HC_2H_3O_2$

_____ 8. reduction

_____ 9. acid definition

_____ 10. base definition

_____ 11. weak acid

_____ 12. electrolyte

(h) Pb, PbO_2, H_2SO_4

(i) causes solution to conduct

(j) loss of electrons

(k) cinnabar

(l) H_2SO_4

QUESTIONS

1. Define acid-base reactions in terms of protons, and oxidation-reduction reactions in terms of electrons.

2. Indicate the solute and solvent in (a) coffee, (b) a 5 percent alcohol in water solution, (c) a 5 percent water in alcohol solution, and (d) a solution of 50 percent alcohol and 50 percent water.

3. What is the one test that all aqueous electrolytes must pass?

4. Give an example of ionic dissociation. Give an example of ionization. What is the difference between the two?

5. Distinguish between the hydrogen ion and the hydronium ion.

6. Write the equation for a chemical reaction in which water acts as a Brønsted acid; as a Brønsted base.

7. A solution of hydrochloric acid is electrolyzed. The products are hydrogen at the cathode and chlorine at the anode. Write the reactions occurring at each electrode and tell which ions move toward the cathode and which toward the anode.

8. Classify each of the following as acids or bases, using the Brønsted-Lowry definitions: H_2SO_4, CO_3^{2-}, Cl^-, HCO_3^-, O^{2-}, H_2O.

9. Predict the formulas of salts formed with the following pairs of ions:

$$Na^+ \text{ and } SO_4^{2-}$$
$$Ca^{2+} \text{ and } I^-$$
$$Mg^{2+} \text{ and } NO_3^-$$
$$Ca^{2+} \text{ and } PO_4^{3-}$$
$$K^+ \text{ and } Br^-$$

10. Moist baking soda is often put on acid burns. Why? Write an equation for the reaction assuming the acid to be hydrochloric (HCl).

11. Using the table giving relative oxidizing and reducing strengths, predict whether or not the following reactions would be expected to proceed to the right.

$$Ag + Na^+ \rightleftharpoons Ag^+ + Na$$
$$H_2 + Cu^{2+} \rightleftharpoons Cu + 2H^+$$
$$F_2 + 2Fe^{2+} \rightleftharpoons 2Fe^{3+} + 2F^-$$
$$Mg + 2Ag^+ \rightleftharpoons 2Ag + Mg^{2+}$$

12. Describe what happens when an ionic solid dissolves in water.

13. What ions are present in water solutions of the following salts: Na_2SO_4, $CaBr_2$, $Mg(NO_3)_2$?

USEFUL MATERIALS FROM THE EARTH, SEA, AND AIR

Many of the things we use in everyday life result from chemical reactions that are carried out on an industrial scale. They include iron and steel products; fabricated items of aluminum, magnesium, and copper; glass, medicines, fertilizers, and numerous chemicals used to alter our surroundings. The purpose of this chapter is to discuss some of the chemistry involved in preparing these products from natural resources found in the earth, sea, and air.

Industrial chemistry has taken on some new demands. Not only must it produce a useful arrangement of atoms, but it must also provide for the return of unused atoms to the economy or to nature in a desirable form—that is, the eventual recycling of the atoms. In the process of producing chemicals commercially, energy must not be wasted and pollution must be minimized.

Recycling decreases our demands on many natural resources.

METALS AND THEIR PREPARATION

Metals occur mostly as compounds in the crust of the earth, though some of the less active metals such as copper, silver, and gold can be found also as free elements. Fortunately, the distribution of elements in the crust of the earth is not uniform. In order to obtain a metal from the crust, one must find an *ore,* that is, a *mineral* from which the metal can be extracted economically. Some common minerals are listed in Table 6–1.

A continual search is under way for new ore deposits.

Some elements that are not particularly abundant in the earth's crust are nevertheless very familiar to us because they tend to occur in very concentrated, localized deposits from which they can easily be extracted. Examples of these are lead, copper, and tin, none of which is

TABLE 6–1 Some Common Metals and Their Ores

Metal	Chemical Formula of Compound of the Element	Name of Ore
Aluminum	$Al_2O_3 \cdot xH_2O$	bauxite
Copper	Cu_2S	chalcocite
Zinc	ZnS	sphalerite
	$ZnCO_3$	smithsonite
Iron	Fe_2O_3	hematite
	Fe_3O_4	magnetite
Manganese	MnO_2	pyrolusite
Chromium	$FeO \cdot Cr_2O_3$	chromite
Calcium	$CaCO_3$	limestone
Lead	PbS	galena

among the more abundant elements in the crust of the earth (Figure 6–1). Other elements that actually form a much larger percentage of the crust of the earth are almost unknown to us because concentrated deposits of their ores are less commonly found. An example is titanium, the tenth most abundant element in the crust of the earth.

The preparation of metals from their ores involves chemical reduction (Chapter 5). Indeed, the concept of oxidation and reduction developed from metallurgical operations. Iron in the ore iron oxide (Fe_2O_3) is in the form of Fe^{3+}. If we *reduce* Fe^{3+} ions to Fe atoms, we must find a source of electrons. (Recall that oxidation is the loss of electrons and that reduction is the gain of electrons.) Sometimes the desired metal is in solution (e.g., magnesium in the sea), where it exists in the oxidized form (Mg^{2+} ions). To obtain the free metal magnesium we must add electrons to these ions (reduction) to produce neutral atoms.

Reduction of magnesium:
$$Mg^{2+} + 2e^- \longrightarrow Mg$$

Iron

The sources of most of the world's iron are large deposits of the iron oxides in Minnesota, Sweden, France, Venezuela, Russia, Australia, and England. In nature these oxides frequently are mixed with impurities, so the production of iron usually incorporates steps to remove such impurities. Iron ores are reduced to the metal by using carbon, in the form of coke, as the reducing agent.

Iron ores are iron compounds. To get iron from the ores, the iron in the compounds must be reduced.

The reduction of iron ore is carried out in a blast furnace (Figure 6–2). The solid material fed into the top of the blast furnace consists of a mixture of an oxide of iron(Fe_2O_3), coke (C), and limestone ($CaCO_3$). A blast of heated air is forced into the furnace near the bottom. The reactions that occur within the blast furnace are:

$$2C + O_2 \longrightarrow 2CO + heat$$
CARBON OXYGEN CARBON MONOXIDE

$$Fe_2O_3 + 3CO \longrightarrow 2Fe + 3CO_2 + heat$$
IRON OXIDE CARBON MONOXIDE IRON CARBON DIOXIDE

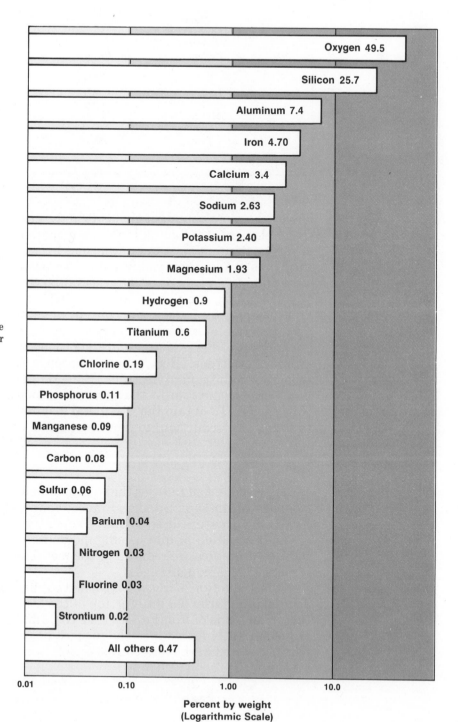

Figure 6–1 Abundance of the elements in the earth's crust (per cent by weight).

RAW MATERIAL
INPUT
Iron ore
Coke (from coal)
Limestone (flux)

CO, CO_2, NO_2

230°C

525°C

945°C

Hot air supply pipe

Hot air blast nozzle
(one of many)

1510°C

Slag

Molten iron

Adult human

Molten iron outlet

Figure 6–2 Diagram of a blast furnace.

Limestone is added to remove the silica (SiO_2) impurity.

$$CaCO_3 \xrightarrow{\text{heat}} CaO + CO_2$$

CALCIUM   CALCIUM   CARBON
CARBONATE   OXIDE   DIOXIDE

$$CaO + SiO_2 \longrightarrow CaSiO_3$$

CALCIUM   SILICON   CALCIUM
OXIDE   DIOXIDE   SILICATE

The calcium silicate, or *slag,* exists as a liquid in the furnace. Consequently, as the blast furnace operates, two molten layers collect in the bottom. The lower, denser layer is mostly liquid iron that contains a fair amount of dissolved carbon and often smaller amounts of other impurities. The upper, lighter layer is primarily molten calcium silicate with some impurities. From time to time the furnace is tapped at the bottom and the molten iron is drawn off. Another outlet somewhat higher in the blast furnace base can be opened to remove the liquid slag.

As it comes from the blast furnace, the iron contains too much carbon for most uses. If some of the carbon is removed, the mixture becomes structurally stronger and is known as *steel.* Steel is an alloy of iron with a relatively small amount of carbon (less than 1.5 percent); it may also contain other metals. In order to convert iron into steel, the excess carbon is burned out with oxygen.

There are several techniques for burning the excess carbon (see Figures 6–3 and 6–4). A recent development that has been very widely adopted is the basic oxygen process (Figure 6–4). In this process pure oxygen is blown into molten iron through a refractory tube (oxygen gun)

Alloy: a metal consisting of two or more elements.

Figure 6–3 The open hearth furnace for the conversion of iron to steel. Air and fuel are blown in first in one direction and then the other in order that heat, which is costly, might be stored in the brick checkerwork. (From Lee, Van Orden, and Ragsdale: *General and Organic Chemistry.* W. B. Saunders Co., Philadelphia, 1971.)

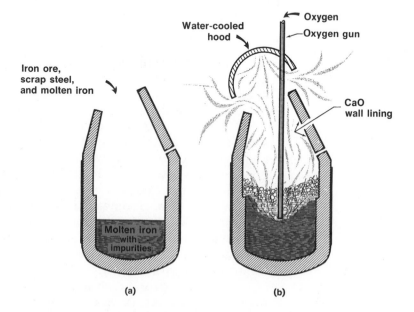

Figure 6–4 The basic oxygen process furnace. Much of the steel manufactured today is refined by blowing oxygen through a furnace charged with ore, scrap, and molten iron.

which is pushed below the surface of the iron. At elevated temperatures, the dissolved carbon reacts very rapidly with the oxygen to give gaseous carbon monoxide and carbon dioxide, which then escape.

After the carbon content has been reduced to a suitable level, the molten steel is formed into desired shapes. During the processing it is subjected to carefully controlled heat treatment to insure that the steel has the desired mechanical properties. This is necessary in order to control the crystalline form of the steel, which determines its pliability, toughness, and other properties.

Corrosion, Rusting, and Stainless Steel

Several billion dollars worth of valuable materials are destroyed each year by corrosion. A number of techniques are used to reduce this destruction.

Many metals undergo corrosion when exposed to moist air over a long period of time. The metal is transformed into its oxide or hydroxide. Corrosion of iron is called rusting and leads to the transformation of iron into rust ($Fe_2O_3 \cdot xH_2O$). The initial reaction is the oxidation of iron to iron(II) hydroxide:

$$2Fe + O_2 + 2H_2O \longrightarrow 2Fe(OH)_2$$
MOIST AIR IRON(II) HYDROXIDE

Iron(II) hydroxide is itself subject to further oxidation in moist air to give iron(III) hydroxide:

$$4Fe(OH)_2 + O_2 + 2H_2O \longrightarrow 4Fe(OH)_3 \qquad (or\ 2Fe_2O_3 \cdot 3H_2O)$$
MOIST AIR IRON(III) HYDROXIDE IRON(III) OXIDE
(RUST) TRIHYDRATE
(RUST)

The iron(III) hydroxide loses water readily to form iron(III) oxides with variable amounts of water. The rusting process changes iron back into a compound very similar in composition to its ore, and thus undoes all the effort expended in obtaining the metal.

Rusting also occurs when we make an iron object into an electrochemical cell or battery, often without realizing it. This happens when the iron is in contact with copper or nickel or any less active metal (Table 5–3), or when part of the iron surface becomes wet in contact with

Site of iron oxidation
$Fe \longrightarrow Fe^{2+} + 2e^-$

Moisture

Site of oxygen reduction
$O_2 + 2H_2O + 4e^- \longrightarrow 4OH^-$

electron flow

Iron

Figure 6–5 The site of iron oxidation may be different from the point of oxygen reduction owing to the ability of the electrons to flow through the iron. The point of oxygen reduction can be located with an acid-base indicator because of the OH^- ions produced.

air (Figure 6–5.) The oxygen in the air will dissolve in the water and remove electrons from the iron via the reactions:

$$2Fe \longrightarrow 2Fe^{2+} + 4e^-$$
$$O_2 + 2H_2O + 4e^- \longrightarrow 4OH^-$$

There are many ways to prevent or reduce corrosion of a metal object. Three of these are: (1) protective coatings, (2) cathodic protection, and (3) alloying (stainless steel is a corrosion-resistant alloy of iron).

Cathodic protection is discussed on page 128.

Protective coatings prevent access of water and atmospheric oxygen to the iron surface. Such coatings may be paint, enamel, grease, or another more resistant metal, such as chromium. This method is successful as long as cracks or holes do not develop in the coating. Galvanized iron has a coating of the more active metal zinc, which gives good cathodic protection (Table 5–3) and also forms an oxide that is hard and impervious to the air.

Stainless steels are alloys of iron and other metals such as nickel, chromium, or cobalt. They are made by melting iron and the alloying metals together in an electric furnace. The resulting alloys are resistant to corrosion, because in the presence of oxygen they form a very thin, tough, impervious, and adherent layer of metal oxide on their surfaces. The layer is so thin that it is essentially transparent (the luster of the metal is retained). This protects the underlying metal from further contact with the oxygen of the air and renders the objects "stainless," or very resistant to corrosion under normal circumstances.

A stainless steel that is widely used is the so-called 18-8: 18 percent chromium, 8 percent nickel, a trace of carbon, and the rest iron.

Stainless steels are used in household items; larger amounts are used in the construction of industrial plants that handle great quantities of hydrochloric acid and other highly corrosive materials.

Aluminum

Aluminum, in the form of Al^{3+} ions, constitutes 7.4 percent of the crust of the earth. However, because of the difficulty of reducing Al^{3+} to Al, only recently have we learned to isolate and use this abundant element. Aluminum metal is soft and has a low density. Many of its alloys, however, are quite strong. Hence, it is an excellent choice when a lightweight, strong metal is required. In structural aluminum, the high chemical reactivity of the element is offset by the fact that a transparent, hard film of aluminum oxide, Al_2O_3, forms over the surface, protecting it from further oxidation:

When aluminum was first made, it was very expensive and rare. A bar of aluminum was displayed next to the Crown Jewels at the Paris Exposition in 1855.

$$4Al + 3O_2 \longrightarrow 2Al_2O_3$$

The principal ore of aluminum is *bauxite,* a hydrated aluminum oxide, $Al_2O_3 \cdot xH_2O$. Because impurities in the ore have undesirable effects on the properties of aluminum, these must be removed, generally by the purification of the ore. This is accomplished with the Bayer process, which is based upon the fact that aluminum oxide and aluminum hydroxide react with strong bases. The principal impurities are iron oxides, which are not soluble in sodium hydroxide solution. In the Bayer process the mixture of oxides is treated with a sodium hydroxide solution, which dissolves the aluminum oxide and leaves the iron oxide:

$$Al_2O_3 \cdot xH_2O + Fe_2O_3 \xrightarrow[\text{solution}]{\text{NaOH}} Al(OH)_4^- + Na^+ + Fe_2O_3$$

(SOLID)   (SOLID)   (SOLUTION)   (SOLID)

The mixture is filtered; $Al(OH)_3$ is then carefully precipitated out of the clear solution by the addition of carbon dioxide (an acid), which makes the solution less basic:

$$CO_2 + Al(OH)_4^- \longrightarrow Al(OH)_3\downarrow + HCO_3^-$$

The aluminum hydroxide is heated to transform it into pure anhydrous aluminum oxide:

$$2Al(OH)_3 \xrightarrow{\text{Heat}} Al_2O_3 + 3H_2O$$

Aluminum metal is obtained from the purified oxide by electrolysis in molten cryolite (Figure 6–6 and Table 5–2). Cryolite, Na_3AlF_6, has a melting point of 1000°C; the molten compound dissolves considerable amounts of aluminum oxide, which lowers the melting point of the cryolite. This mixture of cryolite and aluminum oxide is electrolyzed in a cell with carbon anodes and a carbon cell lining that serves as the cathode on which aluminum is deposited. As the operation of the cell proceeds, the molten aluminum sinks to the bottom of the cell. From time to time the cell is tapped and the molten aluminum is run off into molds.

Figure 6–6 Schematic drawing of a furnace for producing aluminum by electrolysis of a melt of Al_2O_3 in Na_3AlF_6. The molten aluminum collects at the bottom of the carbon cathode container.

Aluminum is used both as a structural metal and as an electrical conductor in high-voltage transmission lines. It competes with copper as an electrical conductor because of the lower cost of aluminum. Larger diameter aluminum wires must be used to offset the lower electrical conductivity of aluminum compared to copper.

The top of the Washington monument is a casting of aluminum made in 1884.

Copper

Although copper metal occurs in the free state in some parts of the world, the supply available from such sources is quite insufficient for the world's need. The majority of the copper obtained today is from various copper sulfide ores, most of which must be concentrated prior to the chemical processes that produce the metal. These minerals include $CuFeS_2$ (chalcopyrite), Cu_2S (chalcocite), and CuS (covellite). Because the copper content of these ores is around 1 to 2 percent, the powdered ore is first concentrated by a *flotation process* (Figure 6–7).

In the flotation process, the powdered ore is mixed with water and a frothing agent such as pine oil. A stream of air is blown through to produce froth. The *gangue* in the ore, which is composed of sand, rock, and clay, is easily wet by the water and sinks to the bottom of the container. In contrast, a copper sulfide particle is wet by the oil, and so is hydrophobic—it is not wet by the water. The copper sulfide particle becomes coated with oil and is carried to the top of the container in the froth. The froth is removed continuously, and the copper sulfide minerals are recovered from it.

Gangue: commercially worthless material associated with valuable mineral in an ore.

The preparation of copper metal from a copper sulfide ore involves roasting in air to oxidize some of the copper sulfide and any iron sulfide present:

Sulfuric acid, H_2SO_4, is the number one commercial chemical. Some 40 million tons are produced annually in the U.S. Part of the H_2SO_4 comes from the conversion of waste SO_2 into SO_3 followed by dissolving in water.

$$2Cu_2S + 3O_2 \longrightarrow 2Cu_2O + 2SO_2\uparrow$$
$$2FeS + 3O_2 \longrightarrow 2FeO + 2SO_2\uparrow$$

Figure 6–7 Apparatus for flotation concentration.

Subsequently the mixture is heated to a higher temperature, and some copper is produced by the reaction:

$$Cu_2S + 2Cu_2O \longrightarrow 6Cu + SO_2 \uparrow$$

The iron oxides form a slag with silica added for that purpose. The product of this operation is a **matte**, a mixture of copper metal and sulfides of copper, iron, other ore constituents, and slag. The molten matte is heated in a converter with more silica. When air is blown through the molten material in the converter, two reaction sequences occur. First, the remaining iron is converted into a slag:

$$2FeS + 3O_2 \longrightarrow 2FeO + 2SO_2$$
$$FeO + SiO_2 \longrightarrow FeSiO_3$$
$$\text{(MOLTEN SLAG)}$$

The remaining copper sulfide is then converted to copper metal:

$$Cu_2S + 2Cu_2O \longrightarrow 6Cu + SO_2$$

The blister copper contains metals such as nickel, gold, and platinum!

The copper produced in this manner is crude or "blister" copper and is purified electrolytically.

In the electrolytic purification of copper, the crude copper is first cast into anodes; these are placed in a water solution of copper sulfate and sulfuric acid. The cathodes are made of pure copper. As electrolysis proceeds, copper is oxidized at the anode, moves through the solution as Cu^{2+} ions and is deposited on the cathode (Chapter 5). The voltage of the cell is regulated so that more active impurities (such as iron) are left in the solution and less active ones are not oxidized at all. These less active impurities include gold and silver, and they collect as "anode sludge," an insoluble residue. The anode sludge is subsequently worked up to recover these rarer metals.

The copper produced by the electrolytic cell is 99.95 percent pure and is suitable for use as an electrical conductor. Copper for this purpose must be pure because very small amounts of impurities, such as arsenic, considerably reduce the electrical conductivity of copper.

Magnesium

Magnesium, with a denstiy of 1.74 grams per cubic centimeter, is the lightest structural metal in common use. For this reason it is most often used in alloys designed for light weight and great strength. It is a relatively active metal chemically because it loses electrons easily. Magnesium "ores" include seawater, which has a magnesium concentration of 0.13 percent, and dolomite, a mineral with the composition $CaCO_3 \cdot MgCO_3$. Because there are six million tons of magnesium present as Mg^{2+} salts in every cubic mile of seawater, the sea can furnish an almost limitless amount of this element.

There are about 328 million cubic miles of sea water.

The recovery of magnesium from seawater (Figure 6–8) begins with the precipitation of magnesium hydroxide by the addition of lime to seawater:

$$CaO + H_2O \longrightarrow Ca^{2+} + 2OH^-$$
$$Mg^{2+} + 2OH^- \longrightarrow \underline{Mg(OH)_2}$$

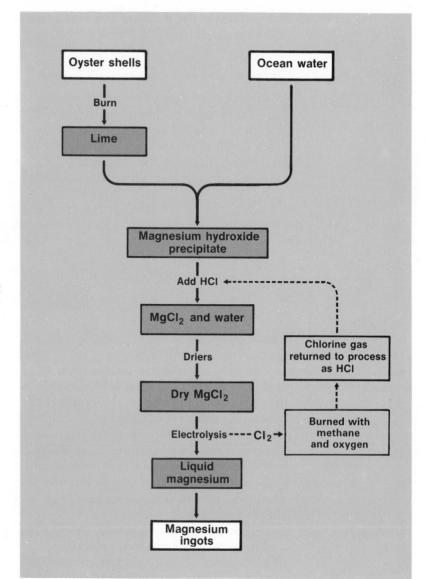

Figure 6–8 Flow diagram showing how magnesium metal is produced from seawater.

Figure 6–9 A cell for electrolyzing molten $MgCl_2$. The magnesium metal is formed on the steel cathode and rises to the top where it is dipped off periodically. Chlorine gas is formed on the graphite anode and is piped off.

The magnesium hydroxide is removed by filtration and then neutralized with hydrochloric acid to form the chloride:

$$\underline{Mg(OH)_2} + 2H^+ + 2Cl^- \rightleftharpoons Mg^{2+} + 2Cl^- + 2H_2O$$

The water is evaporated; this is followed by the electrolysis of molten magnesium chloride in a huge steel pot that serves as the cathode (Figure 6–9). Graphite bars serve as the anodes.

$$\underset{\text{(MELTED)}}{Mg^{2+} + 2Cl^-} \xrightarrow{\text{Electrolysis}} Mg + Cl_2\uparrow$$

AT THE CATHODE:

$$Mg^{2+} + 2e^- \longrightarrow Mg$$

AT THE ANODE:

$$2Cl^- \longrightarrow Cl_2\uparrow + 2e^-$$

To make magnesium one can use seawater, lime from oyster shells, methane from natural gas, and electricity.

As the melted magnesium forms, it floats to the surface and is removed periodically. The chlorine is recovered and reacted with air and natural gas (methane, CH_4) to form hydrochloric acid, which, in turn, is used in dissolving the magnesium hydroxide:

$$\underset{\text{METHANE}}{4Cl_2 + 2CH_4 + O_2} \longrightarrow 2CO + 8HCl$$

The lime used to precipitate the magnesium as the hydroxide is obtained by heating limestone or oyster shells:

$$CaCO_3 \xrightarrow{\text{Heat}} \underset{\text{LIME}}{CaO} + CO_2$$

The total world production of magnesium is only about 250,000 tons per year, although it is potentially available on a larger scale.

SELF-TEST 6-A

1. The most abundant element in the earth's crust is _____.

2. The most abundant metal in the earth's crust is _____.

3. In the United States, the largest iron deposits are found in the state of _____.

4. A natural material which is almost pure calcium carbonate is _____.

5. Which of the following metals may occur in the free or metallic state in mineral deposits? Iron, copper, aluminum, magnesium. _____.

6. Which of the metals listed are either produced or purified using electricity? Iron, copper, aluminum, magnesium. _____.

7. Another name for calcium silicate as it applies to production of iron is _____.

8. In order for most metals to be prepared from their ores, they must be (oxidized/reduced).

9. In an electrical refining process for metals, the purest metal will always be found at the (anode/cathode).

10. Which metal is sufficiently concentrated in the oceans to be extracted commercially? Magnesium, aluminum, copper, iron. _____.

11. Most metals are found in the earth as (neutral atoms/positive ions/negative ions).

GLASS

Silicon dioxide, SiO_2 (also called silica), occurs naturally in large amounts as sand or more rarely in much larger crystals (quartz). It has a melting point of 1710°C. If the melted material is cooled rapidly, a *glass* is obtained. Crystalline quartz consists of an extended structure in which each silicon atom is bonded tetrahedrally to four oxygen atoms (Figure 6–10a), and each oxygen atom is bonded to two silicon atoms. The bonding thus extends throughout the crystal (Figure 6–10b). When silica is melted, some of the bonds are broken and the units move with respect to each other. When the liquid is cooled, the re-formation of the original solid requires a reorganization that is hard to achieve because of the difficulty the groups experience in moving. The very viscous liquid structure is thus partially preserved on cooling to give the characteristic feature of a *glass,* which is an apparently solid material (pseudo-solid)

Glass will flow slowly like a liquid. Many colonial homes contain glass window panes thicker at the bottom than at the top.

Figure 6–10 (*a*) Tetrahedral structure of silicon and oxygen in silicates. (*b*) Chain of tetrahedra showing that an oxygen is common at each point of contact between tetrahedra.

Sodium carbonate (Na_2CO_3) is the number eleven commercial chemical. Calcium oxide is number two. See inside the front cover.

Viscosity is the resistance to flow.

Most glass consists of a mixture of the oxides of Si, Na, and Ca, which are melted together. Colored glass is produced by addition of other metal oxides.

with some of the randomness in structure characteristic of a liquid. This random structure accounts for one of the typical properties of a glass: it breaks irregularly rather than splitting along a plane like a crystal.

By the addition of metal oxides to silica, the melting temperature of the mixture can be reduced from 1710°C to about 700°C. The oxides most often added are sodium oxide (added as Na_2CO_3, soda ash) and calcium oxide (added as $CaCO_3$). The metal ions form ionic (nondirectional) bonds with oxygen atoms that previously had been bonded rigidly to specific Si atoms. As a result, the so-called soda-lime glass has a lower melting temperature and viscosity than pure SiO_2, and can be produced and fabricated more easily.

Soda-lime glass will be clear and colorless only if the purity of the ingredients has been controlled carefully. If, for example, too much iron oxide is present, the glass will be green. Other metal oxides produce other colors (see Table 6–2). To some extent, one color can counteract another.

TABLE 6–2 Substances Used in Colored Glasses

Substance	Color
Copper(I) oxide	red, green, or blue
Tin(IV) oxide	opaque
Calcium fluoride	milky white
Manganese(IV) oxide	violet
Cobalt(II) oxide	blue
Finely divided gold	red, purple, or blue
Uranium compounds	yellow, green
Iron(II) compounds	green
Iron(III) compounds	yellow

Figure 6—11 Craftsman working with molten glass. (Courtesy of the Corning Glass Company.)

The substances are melted together in a gas- or oil-fired furnace. As they react, bubbles of CO_2 gas are evolved.

$$CaCO_3 + SiO_2 \longrightarrow CaSiO_3 + CO_2$$

$$Na_2CO_3 + SiO_2 \longrightarrow Na_2SiO_3 + CO_2$$

TABLE 6—3   Special Glasses

	Special Addition or Composition	Desired Property
1.	Large amounts of PbO with SiO_2 and Na_2CO_3	brilliance, clarity, suitable for optical structures: crystal or flint glass
2.	SiO_2, B_2O_3, and small amounts of Al_2O_3	small coefficient of thermal expansion: borosilicate glass, "Pyrex," "Kimax," etc.
3.	One part SiO_2 and four parts PbO	ability to stop (absorb) large amounts of X rays and gamma rays: lead glass
4.	Large concentrations of CdO	ability to absorb neutrons
5.	Large concentrations of As_2O_3	transparency to infrared radiation
6.	Suspended Se particles	red color

The mixture is heated to about 1500°C to remove the bubbles of CO_2. At this temperature the viscosity is low, and the bubbles of entrapped gas easily escape. The mixture is cooled somewhat and then is blown into bottles by machines or is drawn into sheets, or molded into other forms (Figure 6–11).

It is possible to incorporate a wide variety of materials into glass for special purposes. Some examples are given in Table 6–3.

CHEMICAL FERTILIZERS—KEYS TO WORLD FOOD PROBLEMS

Regardless of the method used to predict future world populations, everyone agrees that food production for that population is a major problem that is sure to get worse. Indeed, adequate food production is a problem for most people on the earth today. Food production is complicated, involving the application of management, economics, mechanical and civil engineering, genetics, and chemistry in an area we usually call *agriculture.* In this section we shall deal with the chemical nutrients of plants as they relate to food production.

Crops were fertilized with manure, dead fish, or straw for centuries before it was understood that growing plants require 16 different nutrients (Table 6–4). Nine of these, the *macronutrients,* are needed in larger amounts, while seven, the *micronutrients,* are needed in smaller or trace amounts. Three of the macronutrients, carbon, hydrogen, and oxygen, come from the atmosphere and are available to all plants, while the

TABLE 6–4   Nutrients Needed by Plants

Element	Source Compound
Macronutrients	
C	CO_2, carbon dioxide
H	H_2O, water
O	H_2O, water
N	NH_3, ammonia; NH_4NO_3, ammonium nitrate; H_2NCONH_2, urea
P	$Ca(H_2PO_4)_2$, calcium dihydrogen phosphate
K	KCl, potassium chloride
Ca	$Ca(OH)_2$, slaked lime
Mg	$MgCO_3$, magnesium carbonate
S	Elemental sulfur
Micronutrients	
B	$Na_2B_4O_7 \cdot 10H_2O$, borax
Cu	$CuSO_4 \cdot 5H_2O$, copper(II) sulfate pentahydrate
Fe	$FeSO_4$, iron(II) sulfate
Mn	$MnSO_4$, manganese(II) sulfate
Zn	$ZnSO_4$, zinc sulfate
Mo	$(NH_4)_2MoO_4$, ammonium molybdate
Cl	KCl, potassium chloride

TABLE 6–5 Approximate Amounts of Nutrients
Required to Produce 150 Bushels of Corn

Nutrient	Approximate Pounds per Acre	Source
Oxygen	10,200	air
Carbon	7,800	air
Water	3,225 to 4,175 tons	29 to 36 inches of rain
Nitrogen	310	1,200 lbs. of high-grade fertilizer
Phosphorus	120 (as phosphate)	1,200 lbs. of high-grade fertilizer
Potassium	245 (as K_2O)	1,200 lbs. of high-grade fertilizer
Calcium	58	150 lbs. of agricultural limestone
Magnesium	50	275 lbs. of magnesium sulfate (epsom salt)
Sulfur	33	33 lbs. of powdered sulfur
Iron	3	15 lbs. of iron sulfate
Manganese	.45	1.3 lbs. of manganese sulfate
Boron	.05	1 lb. of borax
Zinc	trace	small amount of zinc sulfate
Copper	trace	small amount of copper sulfate
Molybdenum	trace	trace of ammonium molybdate

remainder of the nutrients must come from the soil. Growing plants deplete the available nutrients rather drastically (Table 6–5). Within the space of one or two growing cycles the soil nutrients must be replaced.

Soil nutrients must be replaced every one or two growing cycles.

With today's requirements on yields, the use of chemical fertilizers is the only* way to continue the high productivity of the world's farmland (Figure 6–12).

There are about 3.5 billion acres presently under cultivation worldwide. It has been calculated that the application of $35 worth of fertilizer per acre would increase crop production by 50 percent, equivalent to 1.7 billion more acres under cultivation. Of course this would be expensive, requiring about $40 per capita worldwide, and might be impossible due to energy limitations since present fertilizer manufacture uses a significant portion of our fossil fuels. Poorer countries such as India have almost been priced out of the world fertilizer markets by the recent steep increase in the prices of petroleum products.

There is a little less than one acre under cultivation for every individual on the earth.

The large amount of petroleum needed to make fertilizer causes fertilizer prices to rise with the price of petroleum.

Nitrogen Chemistry

Since the atmosphere is 78 percent nitrogen by volume, it might appear that this nutrient would be readily available to plants. However, because of the strong nitrogen-nitrogen triple bond in the N_2 molecule, gaseous nitrogen is rather unreactive and as such is not directly available to plants as a nutrient. To be useful, the nitrogen must be *fixed*—that is, oxidized or reduced to some species containing nitrogen such as nitrate ion, NO_3^- (oxidized form), or ammonia, NH_3 (reduced form), either of which is chemically more reactive than the nitrogen molecule. Only a few organisms, such as some bacteria and blue-green algae, are able to fix atmospheric nitrogen. Bacteria of the genus *Rhizobium* can fix nitro-

Fixed nitrogen refers to nitrogen present in chemical compounds, that is, in combinations other than N_2.

---

*Organic materials such as feedlot animal wastes, slaughter-house wastes, and plant wastes (stalks, leaves, etc.) could satisfy much of the soil nutrient demand, but social and sanitary restrictions have retarded their extensive application in this country.

(a)

(b)

Figure 6–12 Changes in fertilizer usage, crop acreage, and population in the United States in the last half century. Amounts of phosphorus, potassium, and especially nitrogen have increased markedly (*a*). As the population has increased (*b*), land use for agriculture has diminished sharply (*b*). (From Turk, A., J. Turk, J. T. Wittes, and R. Wittes, *Environmental Science,* W. B. Saunders Co., Philadelphia, 1974.)

Synergism is a cooperative action producing an effect greater than the sum of the individual effects.

gen when they grow in nodules in the roots of legumes such as peas, beans, and alfalfa. Generally more nitrogen is fixed than the plants themselves need; thus, the soil nitrogen is increased. In some parts of the world, crop rotation between legumes and nonlegumes works quite nicely for soil nitrogen requirements. Two of the most important cereal crops, wheat and corn, are not favored by nitrogen-fixing bacteria, and chemists and biologists are working at present on a newly discovered bacterium that grows in the root structures of certain South American grasses, in hopes of establishing a new **synergistic** relationship between nitrogen-fixing bacteria and plants.

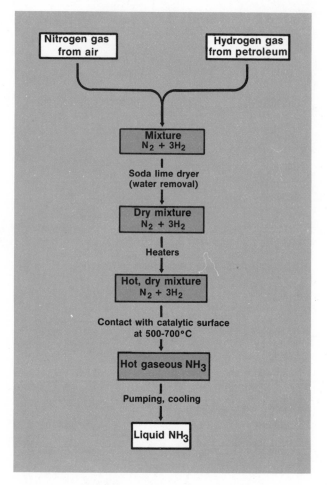

Figure 6-13 The Haber process for synthesizing ammonia. Ammonia is the number four commercial chemical, and nitrogen is number five. See inside the front cover.

Commercial fertilizers contain nitrogen fixed by the Haber process, the direct reaction of nitrogen with hydrogen to produce ammonia (Figure 6-13):

$$N_2 + 3H_2 \rightleftharpoons 2NH_3$$
AMMONIA

Pure nitrogen is obtained by distilling oxygen and other gases from liquid air. Hydrogen is more difficult to obtain. At present, petroleum products such as propane, $CH_3CH_2CH_3$, are made to react with steam in the presence of catalysts to produce hydrogen:

$$CH_3CH_2CH_3 + 6H_2O \xrightarrow{\text{Catalysts}} 3CO_2 + 10H_2$$
STEAM

This is one of the principal reasons why ammonia fertilizer costs are so closely tied to petroleum prices. Hydrogen can also be prepared by the electrolysis of water,

$$2H_2O \xrightarrow[\text{KOH}]{\text{Electricity}} 2H_2 + O_2$$

Figure 6–14 Liquid ammonia is being injected directly into the ground to provide nitrogen for the growing plants.

and by several other methods, but all of these require great quantities of energy. So, as energy costs continue to rise, food costs will necessarily rise due to the added costs of fertilizers.

Nitric acid is the number ten commercial chemical. See inside the front cover.

Nitrogen-containing fertilizers are all based upon ammonia. In the U.S., much of the ammonia is injected directly into the ground (Figure 6–14). Solid fertilizers can also be prepared from ammonia. ***Ammonium nitrate,*** NH_4NO_3, contains 35 percent nitrogen and is the salt of the reaction between ammonia and nitric acid:

$$NH_3 + HNO_3 \longrightarrow NH_4NO_3$$
AMMONIA   NITRIC ACID   AMMONIUM NITRATE

Ammonium nitrate is a solid fertilizer. It is often applied to crops as a solution.

Ammonium nitrate has an advantage over ammonia in ease of handling, although it is sensitive to violent shock and can explode. Nitric acid used to prepare ammonium nitrate is itself prepared from ammonia by oxidation to nitrogen dioxide followed by reaction with water:

Ammonium nitrate is the number twelve commercial chemical. See inside the front cover.

$$4NH_3 + 5O_2 \xrightarrow[\text{Pt catalyst}]{700°C} 4NO + 6H_2O$$

$$2NO + O_2 \longrightarrow 2NO_2$$

$$3NO_2 + H_2O \longrightarrow 2HNO_3 + NO$$
NITROGEN DIOXIDE   NITRIC ACID   NITRIC OXIDE (RECYCLED)

Urea (NH_2CONH_2) is probably one of the world's most important chemicals because of its wide use as a fertilizer and as a feed supplement for cattle. Ammonia and carbon dioxide react under high pressure near 200°C to produce first ammonium carbamate, which then decomposes into urea and water:

$$2NH_3 + CO_2 \longrightarrow \underset{\substack{\text{AMMONIUM} \\ \text{CARBAMATE}}}{H_2N-\overset{\displaystyle O}{\underset{\displaystyle ONH_4}{C}}} \longrightarrow \underset{\text{UREA}}{H_2N-\overset{\displaystyle O}{\overset{\|}{C}}-NH_2} + H_2O$$

Urea synthesis also produces the compound biuret as a side product. Biuret must be removed from fertilizer-grade urea because it retards seed germination.

$$\underset{\text{BIURET}}{H_2N-\overset{\displaystyle O}{\overset{\|}{C}}-NH-\overset{\displaystyle O}{\overset{\|}{C}}-NH_2}$$

A slurry of water, urea, and ammonium nitrate is often applied to crops under the name of "liquid nitrogen." Such a solution can contain up to 30 percent nitrogen and is very easy to store and apply.

Phosphate Rock and Potash

Phosphate rock and potash are two minerals that can be mined, pulverized, and dusted directly onto deficient soil. Often they are specially treated to produce desirable mixing properties. Phosphorus, for example, is found scattered throughout the world in deposits of *phosphate rock*, which when treated with sulfuric acid becomes more soluble and hence produces a product of greater phosphorus availability called "superphosphate."

Phosphoric acid, H_3PO_4, is the number nine commercial chemical. It is prepared from phosphate rock. See inside the front cover.

$$\underset{\substack{\text{PHOSPHATE} \\ \text{ROCK}}}{Ca_3(PO_4)_2} + 2H_2SO_4 + H_2O \longrightarrow \underset{\text{"SUPERPHOSPHATE"}}{\underbrace{Ca(H_2PO_4)_2 + 2CaSO_4}}$$

Phosphate rock itself is not useful to a growing plant because of its very low solubility. Recently, it has become evident that phosphate rock demand will eventually exceed supply unless large new deposits are discovered.

The world has limited deposits of the phosphate rock, which is essential to the manufacture of fertilizers.

Potassium in the form of *potash*, K_2CO_3, exists in enormous quantities throughout the world. A soluble form of potassium is its chloride (KCl), called muriate of potash, or simply potash. Because this compound often occurs with sodium chloride, which is toxic to plants, the potash ores must be treated by some process such as recrystallization to separate the two compounds.

Figure 6–15 Phosphate from agricultural uses is a principal cause of water pollution.

The Trace Nutrient Iron

Although they are needed in smaller amounts, the need for the micronutrients listed in Table 6–4 is as urgent as the need for macronutrients. For example, iron is an essential component of the catalyst involved in the formation of chlorophyll, the green plant pigment. When the soil is iron deficient, or when soil conditions exist such as too much "lime," $Ca(OH)_2$*, being present, iron availability will decrease. This condition is usually present when plant leaves lighten in color or even turn yellow. Often a gardener or lawn worker will apply phosphate and lime to adjust soil acidity, only to see his green plants turn yellow. What is happening is that both phosphate and the hydroxide from the lime tie up the iron and make it unavailable to the plants.

Lime, as CaO or $Ca(OH)_2$, is the number two commercial chemical. See inside the front cover.

The addition of lime, a basic substance, raises the pH of the soil.

$$Fe^{3+} + \underset{\text{PHOSPHATE}}{2PO_4^{3-}} \longrightarrow \underset{\text{TIGHTLY BOUND COMPLEX}}{Fe(PO_4)_2^{3-}}$$

$$Fe^{3+} + 3OH^- \longrightarrow \underset{\text{INSOLUBLE HYDROXIDE}}{Fe(OH)_3}$$

These problems can be overcome, since iron can be *chelated* with certain organic molecules which hold the iron tightly but will release the iron at the root structure. The word chelate comes from the Greek, *chela,* meaning claw.

Certain large negative ions, such as ethylenediaminetetraacetate (EDTA) will react with an iron ion and chelate it, holding the iron tightly.

---

*Actually, lime is calcium oxide, CaO, while hydrated lime, called slaked lime, is $Ca(OH)_2$. Slaked lime is sold as "lime."

EDTA

Since the resulting chelate is an ion, the iron can move through the soil to the root structures.

Chelate structures are important in many aspects of chemistry. Chlorophyll is a chelate of magnesium, while hemoglobin is an iron chelate. In Chapter 13, chelating agents such as EDTA will be discussed in relation to metal poisoning.

SELF-TEST 6-B

1. The principal element in glass is _____.

2. Flint or crystal glass contains a large amount of the metal _____.

3. The fertilizer produced by the Haber process is _____.

4. Three macronutrients that must be replaced in the soil and are the main ingredients in chemical fertilizers are _____, _____, and _____.

5. The principal source of hydrogen used to make ammonia is (water/petroleum hydrocarbons).

6. Superphosphate is manufactured from phosphate rock and (nitric/sulfuric) acid.

7. Potassium is usually added to the soil in the form of $(K_2O/KCl/KNO_3)$.

8. The green color of glass is usually due to the presence of _____.

MATCHING SET

_____ 1. copper

_____ 2. aluminum

_____ 3. milk glass

_____ 4. magnesium

_____ 5. oyster shells

_____ 6. atmosphere

(a) a nutrient necessary for the manufacture of chlorophyll

(b) calcium fluoride added to glass

(c) a nutrient in short supply

(d) source of carbon for plants

(e) made from nitrogen and hydrogen

(f) a limitless supply in seawater

_____ 7. ammonia (g) supply calcium hydroxide for magnesium production

_____ 8. iron (h) purified electrolytically

_____ 9. phosphate (i) the most abundant metal in the earth's crust

QUESTIONS

1. Name three metals that you would expect to find free in nature. Name three that you would not.

2. What is the primary reducing agent in the production of iron from its ore?

3. Why is CaO necessary for the production of iron in a blast furnace?

4. What is the chemical difference between iron and steel?

5. Both iron and magnesium will oxidize in the air. Why is the oxidation of iron a much greater problem than the oxidation of magnesium?

6. How is it possible that both oxidation and reduction can occur at different points on the same piece of iron?

7. Give examples of three ways in which metals can be protected from corrosion.

8. Describe the solution used in the commercial cell for the electrolytic reduction of aluminum.

9. Why is it so important to purify industrial quantities of copper electrolytically to a level above 99.9 percent pure?

10. What chemical is obtained from oyster shells in the production of magnesium from seawater? What is the role of this chemical in the process?

11. Explain how the structures of glass and a liquid are similar.

12. Give reactions involved in the preparation of:

$$Ca(OH)_2 \text{ from } CaCO_3$$
$$NH_3 \text{ from } N_2 \text{ and } H_2$$
$$HNO_3 \text{ from } NH_3$$

13. Explain what is meant by the term "flotation." Why is this an important process?

14. In the production of copper from its sulfide ores, large amounts of SO_2 gas are formed as a by-product. What useful product can be prepared from this substance? What are the consequences of allowing the gas to escape into the atmosphere?

15. Name four items in everyday use that contain (a) silicon, (b) aluminum, (c) calcium.

16. A typical soda-lime glass has a composition reported as 70 percent SiO_2, 15 percent Na_2O, and 10 percent CaO. What ratio of weights of sand (SiO_2), sodium carbonate (Na_2CO_3), and calcium carbonate ($CaCO_3$) must be melted together to make this glass? The carbonates are decomposed by heat to evolve carbon dioxide gas.

17. What is the maximum weight (in pounds) of magnesium that can be obtained from 1000 pounds of seawater? (See page 143.)

18. Explain why fertilizer costs are directly tied to petroleum costs.

19. Of the three major plant macronutrients, nitrogen, potassium, and phosphorus, which is in the most danger of running out? Explain.

THE UBIQUITOUS CARBON ATOM—AN INTRODUCTION TO ORGANIC CHEMISTRY

The importance of carbon compounds to life on earth cannot be overestimated. Consider what the world would be like if all the carbon and carbon compounds were suddenly removed. The result would be somewhat like the barren surface of the moon! Many of the little everyday things often taken for granted would be quite impossible without this versatile element. In an ordinary pencil, for example, the "lead" in the pencil (made from graphite, an elementary form of carbon), the wood, the rubber in the eraser, and the paint on the surface are all carbon or carbon compounds. The paper in this book, the cloth in its cover, and the glue holding it together are also made of carbon compounds. All of the clothes one wears, including the leather in shoes, would not exist without carbon. If carbon compounds were removed from the human body, there would be nothing left except water and a small residue of minerals, and the same is true of all forms of living matter. Fossil fuels, foods, and most drugs are essentially made of carbon compounds. In addition, many carbon compounds such as plastics and detergents, which are not directly connected with the life processes, play a vital role in our lives.

Several million different carbon compounds have been studied and described in the chemical literature, and thousands of new ones are

Carbon and its compounds are vital to life on this planet.

reported every year. Although there are 88 other naturally occurring elements, the number of known carbon compounds is many times greater than that of the known compounds which contain no carbon. The very large and important branch of chemistry devoted to the study of carbon compounds is ***organic chemistry.*** The name "organic" is actually a relic of the past, when chemical compounds produced from once-living matter were called "organic" and all other compounds were called "inorganic."

> Organic chemistry is the study of the compounds of carbon.

THE BASIS FOR THE LARGE NUMBER OF ORGANIC COMPOUNDS

The enormous number of organic compounds has intrigued chemists for over a hundred years. The atomic theory, as developed earlier for all atoms, describes a structure for the carbon atom which explains this multiplicity of carbon compounds. The peculiar structure of this atom allows it ***to form covalent bonds with other carbon atoms in a seemingly endless array of possible combinations.*** A simple organic molecule may contain a single carbon-carbon bond, whereas a complex one may contain literally thousands of such bonds. A few other elements are capable of forming stable bonds between like atoms. These include such elements as nitrogen, N_2, oxygen, O_2, and sulfur, S_8, to name a few. But only S, Sn, Si, and P can form long chain molecules.

> The large number of carbon compounds is due to:
> 1. stability of chains of carbon atoms;
> 2. occurrence of isomers;
> 3. reactivity of functional groups.

An additional factor in the large number of carbon compounds lies in the stability of carbon chains. The carbon chains are not normally subject to attack by water or, at ordinary temperatures, by oxygen. Chains formed by atoms of other elements undergo reaction with either water or oxygen, or both, much more easily than do carbon chains.

A further reason for the large number of organic compounds is the ability of a given number of atoms to combine in more than one molecular pattern and, hence, produce more than one compound. Such compounds, each of which has molecules containing the same number and kinds of atoms, but arranged differently relative to each other, are called ***isomers.*** For example, the molecular structure represented by A—B—C is different from the molecular structure A—C—B, as is C—A—B; these three species are isomers. If we consider the number of possible ways the digits one through nine can be ordered to make nine-digit numbers, we can begin to imagine how a single group of atoms could possibly form hundreds of different molecules. Carbon, with its ability to bond to other carbon atoms, is especially well suited to form isomers.

> Isomers are two or more different compounds with the same number of each kind of atom per molecule.

A final factor explaining the large number of organic compounds is the ability of the carbon atom to form strong covalent bonds with atoms of numerous other elements, such as nitrogen, oxygen, sulfur, chlorine, fluorine, bromine, iodine, silicon, boron, and even many metals. As a result there are large classes of organic compounds. A ***functional group,*** a particular combination of atoms, appears in each member of a class. For example, all organic acids have a carboxyl group attached to another carbon atom.

> A dash in a formula represents a single bond; two electrons are shared.

CARBOXYL GROUP, A FUNCTIONAL GROUP

CHAINS OF CARBON ATOMS—THE HYDROCARBONS

Carbon is intermediate among the elements in its ability to attract electrons to it in a covalent bond. Thus, a carbon atom is unable to remove electrons completely from metals, and even fluorine (the best attractor of electrons among the elements) is unable to remove an electron completely from a carbon atom. As a result, carbon atoms tend not to form ionic bonds but rather to share electrons in the formation of covalent bonds. In addition to the tendency for carbon to form covalent bonds with many other atoms, there is also a remarkable inclination for carbon atoms to form relatively strong covalent bonds with each other.

Only two elements, hydrogen and carbon, are required to explain the existence of literally thousands of compounds known as *hydrocarbons.* The simplest hydrocarbon is methane, CH_4. Methane is tetrahedral and has four C—H bonds, as shown in Figure 7–1.

A hydrocarbon with two carbon atoms is ethane, C_2H_6. Ethane has six C—H bonds and one C—C bond. The bonding in ethane is illustrated by the following formulas:

$$
\begin{array}{cc}
\text{H H} & \quad \text{H} \quad \text{H}\\
\text{H:C:C:H} \quad \text{or} & \text{H—C—C—H}\\
\text{H H} & \quad \text{H} \quad \text{H}
\end{array}
$$

Even though the picture suggests that the molecule is flat, in reality the bonds to each carbon atom are in a tetrahedral arrangement, as shown in Figure 7–2. Also shown in Figure 7–2 is the rotation that takes place around the carbon-carbon single bond. Rotation of this type is a common feature in many molecules.

By applying what we have learned, it is a simple matter to extend the concept of carbon-carbon bonding to a three-carbon molecule such as that of propane (C_3H_8).

$$
\begin{array}{cc}
\text{H H H} & \quad \text{H} \quad \text{H} \quad \text{H}\\
\text{H:C:C:C:H} \quad \text{or} & \text{H—C—C—C—H}\\
\text{H H H} & \quad \text{H} \quad \text{H} \quad \text{H}
\end{array}
$$

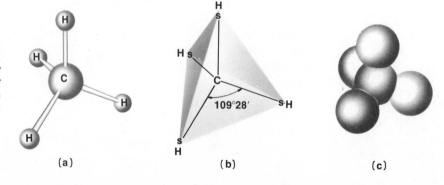

Figure 7–1 Methane. (*a*) Ball-and-stick model showing tetrahedral structure. (*b*) Geometry of regular tetrahedron. (*c*) Model of methane, CH_4, showing relative size of atoms in relationship to interatomic distances.

(a)   (b)   (c)

(a)

(b)

Figure 7—2 Two possible rotational forms of the ethane molecule. The hydrogen atoms in the methyl (CH$_3$) groups may be in an eclipsed position (*a*), staggered (*b*), or in any intermediate position. In ethane the two methyl groups can rotate easily about the carbon-carbon bond.

In Figure 7–3, note that the three carbon atoms in propane do not lie in a straight line because of the tetrahedral bonding about each carbon atom. Also, because of the rotation about the two C—C single bonds, the molecule is "flexible."

It is apparent that these bonding concepts can be extended to a four-carbon molecule and to a limitless number of larger hydrocarbon molecules. Actually, many such compounds are known; some, such as natural rubber, are known to contain over a thousand carbon atoms in a chain.

Figure 7–3 Ball-and-stick model of propane.

Propane

STRUCTURAL ISOMERS

When we try to write the structure for butane, C_4H_{10}, we soon discover that two structures are possible.

Structural isomers have the same molecular formulas but a different pattern of bonds.

	n-BUTANE	METHYLPROPANE (ISOBUTANE)
MELTING POINT	$-138.3°C$	$-160°C$
BOILING POINT	$-0.5°C$	$-12°C$
DENSITY (at $20°C$)	0.579 g/ml	0.557 g/ml

The two formulas represent two distinctly different compounds. Both are well known, each with its own particular set of properties. We must conclude, then, that molecular formulas such as C_4H_{10} are sometimes ambiguous, and that structural formulas are necessary. Since these are different compounds, they have different names, whose derivation will be explained later in this chapter. However, all hydrocarbons that have four carbon atoms can be generally referred to as *butanes*.

If no carbon atom is attached to more than two other carbon atoms, the carbon chain is said to be a **straight-chain** structure. Actually, as shown in Figure 7–4, the carbon chain is bent (109.5°) at each carbon atom, but it is called a straight chain because the carbon atoms are bonded together in succession one after the other. You might note that many molecular shapes are possible for n-butane because of the possible

(a)

Figure 7–4 The isomeric butanes, C_4H_{10}. (a) Normal butane, usually written n-butane. (b) Methylpropane (isobutane).

(b)

Rotational forms resulting from the twisting around C—C single bonds are called conformations; they are not isomers.

rotational motions about the single bonds. These arrangements (called **conformations**) do not constitute different molecules. Because of the ease of bond rotation, a sample of a single pure hydrocarbon contains all of its conformations, that is, molecules rotated into all of their possible shapes.

If one carbon atom is bonded to either three or four other carbon atoms in a molecule, the molecule is said to have a **branched chain**. Isobutane is an example of a branched-chain hydrocarbon (Figure 7–4). Isobutane and n-butane are called **structural isomers** because both molecules contain exactly the same number and kinds of atoms, C_4H_{10}, but the molecules are put together differently. Structural isomerism is somewhat like what you might expect from a child building many different structures with the same collection of building blocks, and using all of the blocks in each structure.

In all branched-chain structures, at least one carbon atom is bonded to three or four other carbon atoms.

The two butanes (and all hydrocarbon molecules) are essentially nonpolar, since the C—C bonds are nonpolar, and the slightly polar C—H bonds are symmetrically arranged to cancel each other out. The forces holding these molecules together in the liquid, therefore, are van der Waals attractions, which depend upon the surface area of a molecule and the closeness of approach of the molecules to each other. In general, a branched-chain isomer has a lower boiling point than a straight-chain isomer, since the branched-chain isomer does not permit intermolecular distances as short as those of a straight chain and has less surface area. Both of these mean less intermolecular attraction. Melting points of isomers generally do not follow the same pattern, since they also depend upon the ease with which the molecules fit into a crystalline array.

It is important to distinguish between different bond rotational arrangements (conformations) and structural isomers (configurations). To change from one rotational arrangement to another, only motion about a bond is required. However, to change from one structural isomer to another (for example, from isobutane to n-butane), it is necessary to break bonds and to form new ones. Structural isomers are "permanent" arrangements; conformations are transient.

Consider the isomeric pentanes, C_5H_{12}. There are three of these:

	n-PENTANE	2-METHYLBUTANE (ISOPENTANE)	2,2-DIMETHYLPROPANE (NEOPENTANE)
MELTING POINT	−130°C	−160°C	−17°C
BOILING POINT	36°C	28°C	9.5°C
DENSITY (at −20°C)	0.626 g/ml	0.62 g/ml	0.613 g/ml

According to the octet rule, each atom (except H) shares or controls eight valence shell electrons. Hydrogen shares only two electrons.

Note that the three isomers are predicted by bonding theory. Furthermore, the theory predicts no other possible isomers for C_5H_{12}, since there are no other ways to unite the 17 atoms and satisfy the octet rule for carbon. These three isomers of pentane are well known, and no others have ever been found.

Table 7–1 gives the number of isomers predicted for some larger molecular formulas, starting with C_6H_{14}. Every predicted isomer, *and no more,* has been isolated and identified for the C_6, C_7, and C_8 groups. However, not all of the C_{15}'s and C_{20}'s have been produced, but there is

TABLE 7–1 Structural
Isomers of Some
Hydrocarbons

Formula	Isomers Predicted	Found
C_6H_{14}	5	5
C_7H_{16}	9	9
C_8H_{18}	18	18
$C_{15}H_{32}$	4347	—
$C_{20}H_{42}$	366,319	—
$C_{30}H_{62}$	4,111,846,763	—

sufficient belief in the theory to presume that if enough time and effort were spent, all of the isomers could eventually be produced. Structural isomerism certainly helps to explain the vast number of carbon compounds.

Structural isomers can also exist in molecules with double and triple carbon-carbon bonds. For example, two of the six isomers of C_4H_8, 1-butene and *trans*-2-butene, have the following structures:

	1-BUTENE	TRANS-2-BUTENE
MELTING POINT	−185.4°C	−106.0°C
BOILING POINT	−6.3°C	1.0°C
DENSITY (at −20°C)	0.641 g/ml	0.649 g/ml

The number placed before the name butene indicates the position number of the double bond. *Trans*-2-butene is an example of a geometric isomer that results from double bonds within a molecule. Geometric isomers will be considered later. Note that the properties of 1-butene and 2-butene definitely indicate two *different* compounds. 1-Butyne and 2-butyne illustrate structural isomerism due to the positioning of triple bonds.

	1-BUTYNE	2-BUTYNE
MELTING POINT	−125.8°C	−32.2°C
BOILING POINT	8.1°C	27°C
DENSITY (AT −20°C)	0.65 g/ml	0.69 g/ml

Hydrocarbons containing a double bond have isomers with single bonded ring-type structures. For example, in addition to the isomers given before for C_4H_8, another isomer exists which has a cyclic structure with single bonds.

There are large numbers of single bonded ring structures which contain three or more carbon atoms in the rings.

$$
\begin{array}{c}
\text{H} \quad \text{H} \\
| \quad\quad | \\
\text{H}-\text{C}-\text{C}-\text{H} \\
| \quad\quad | \\
\text{H}-\text{C}-\text{C}-\text{H} \\
| \quad\quad | \\
\text{H} \quad \text{H}
\end{array}
$$

CYCLOBUTANE

MELTING POINT	−50°C
BOILING POINT	12°C
DENSITY (AT 0°C)	0.703 g/ml

In formulas such as CH_3CH_2I, the elements directly following a carbon are bonded to that carbon. CH_3CH_2I is the same as

$$
\begin{array}{c}
\text{H} \quad \text{H} \\
| \quad\quad | \\
\text{H}-\text{C}-\text{C}-\text{I}. \\
| \quad\quad | \\
\text{H} \quad \text{H}
\end{array}
$$

Compounds with double and triple bonds (called **alkenes** and **alkynes**, respectively) are much more reactive than those with *only* single bonds (called **alkanes**). For example, such compounds readily undergo **addition reactions** with halogens and compounds such as HI:

$$
\begin{aligned}
CH_2{=}CH_2 + Br_2 &\longrightarrow CH_2Br{-}CH_2Br \\
CH_2{=}CH_2 + HI &\longrightarrow CH_3CH_2I \\
HC{\equiv}CH + 2Br_2 &\longrightarrow CHBr_2{-}CHBr_2
\end{aligned}
$$

Unsaturated compounds contain double or triple bonds; saturated compounds do not contain such bonds.

The greater reactivity of alkenes and alkynes is one of the reasons why processes to transform alkanes (which occur naturally in petroleum) into alkenes and alkynes are so important. Because of their ability to add on other molecules, alkenes and alkynes are often called **unsaturated** compounds while alkanes are referred to as **saturated** compounds. Like the alkanes, the alkenes and alkynes undergo combustion reactions with oxygen and hence burn in air:

$$
\begin{aligned}
2CH_3{-}CH_3 + 7O_2 &\longrightarrow 4CO_2 + 6H_2O + \text{heat} \\
CH_2{=}CH_2 + 3O_2 &\longrightarrow 2CO_2 + 2H_2O + \text{heat} \\
2HC{\equiv}CH + 5O_2 &\longrightarrow 4CO_2 + 2H_2O + \text{heat}
\end{aligned}
$$

NOMENCLATURE

Alkanes: hydrocarbons with single bonds only; name ending: -ane.

Alkenes; hydrocarbons with one or more double bonds; name ending -ene.

Alkynes; hydrocarbons with one or more triple bonds; name ending -yne.

The -yl ending indicates an attached group such as $-CH_3$, methyl.

With so many organic compounds, a system of common names quickly fails owing to the shortage of unique names. It is evident that a system of nomenclature is needed that makes use of numbers as well as names. Much attention has been given to this problem, and several international conventions have been held to work out a satisfactory system that can be used throughout the world. The International Union of Pure and Applied Chemistry has given its approval to a very elaborate nomenclature system **(IUPAC System),** and this system is now in general use.

A few of these IUPAC names will be needed for our discussion, and an appreciation for the basic simplicity of the approach in naming organic compounds is desirable. Inscrutable names, such as some of those encountered on medicine bottles, are replaced by systematic names that are descriptive of the molecules involved. The names of a few of the hydrocarbons are presented in Table 7–2.

For branched-chain hydrocarbons it becomes necessary to name submolecular groups. The $-CH_3$ group is called the methyl group; this name is derived from methane by dropping the -ane and adding -yl. Any of the

TABLE 7–2   The First Ten Straight-Chain Saturated Hydrocarbons

Name	Formula	Structural Formula
Methane	CH_4	H—C—H with H above and below
Ethane	C_2H_6	$H-C-C-H$ with H above/below
Propane	C_3H_8	3-carbon chain
n-Butane	C_4H_{10}	4-carbon chain
n-Pentane	C_5H_{12}	5-carbon chain
n-Hexane	C_6H_{14}	6-carbon chain
n-Heptane	C_7H_{16}	7-carbon chain
n-Octane	C_8H_{18}	8-carbon chain
n-Nonane	C_9H_{20}	9-carbon chain
n-Decane	$C_{10}H_{22}$	10-carbon chain

An aliphatic compound is any hydrocarbon that has an open chain of single-bonded carbon atoms.

nine other hydrocarbons listed in Table 7–2 can give rise to a similar group. For example, the propyl group would be $-C_3H_7$. As an illustration of the use of the group names, consider this formula:

$$H-\overset{\overset{\textstyle H}{|}}{\underset{\underset{\textstyle H}{|}}{C}}-\overset{\overset{\textstyle H}{|}}{\underset{\underset{\textstyle H}{|}}{C}}-\overset{\overset{\textstyle H}{|}}{\underset{\underset{\textstyle H}{|}}{C}}-\overset{\overset{\textstyle H}{|}}{\underset{\underset{\textstyle H-\overset{\textstyle |}{C}-H}{|}}{C}}-\overset{\overset{\textstyle H}{|}}{\underset{\underset{\textstyle H}{|}}{C}}-H$$

position of group on chain

↓

2-*Methylpentane*

↑

group attached to chain · longest chain of C atoms

The *longest* carbon chain in the molecule is five carbon atoms long; hence, the root name is pentane. Furthermore, it is a methylpentane (written as one word) because a methyl group is attached to the pentane structure. In addition, a number is needed because the methyl group could be bonded to either the second or third carbon atom.

3-METHYLPENTANE   **2-METHYLPENTANE**

Positional numbers of groups should have the smallest possible sum.

Note that 2-methylpentane is the same as 4-methylpentane, since the latter would be the same molecule turned around; the accepted rule requires numbering from the end of the carbon chain that will result in the smallest numbers. Therefore, 2-methylpentane is the correct name.

Any number of substituted groups can be handled in this same fashion. Consider the name and the formula for 4,4-diethyl-3,5,6,6-tetramethyloctane.

If a double bond appears in a hydrocarbon, then the root name, which indicates the number of carbon atoms, must be modified to reflect the double bond structure and its position. Changing -ane to -ene indicates the presence of the double bond, and a number is used to indicate its position. For example:

No number is necessary for ethene or propene because there is only one possible position for the double bond.

$CH_2=CH_2$	ETHENE (COMMON NAME: ETHYLENE)
$CH_2=CHCH_3$	PROPENE
$CH_2=CHCH_2CH_3$	1-BUTENE
$CH_3CH=CHCH_3$	2-BUTENE
$CH_3-\overset{\overset{\textstyle CH_3}{|}}{C}=CHCH_2CH_3$	2-METHYL-2-PENTENE

If a triple bond is present, the -ane is changed to -yne. Examples:

$H-C\equiv C-H$ ETHYNE (COMMON NAME: ACETYLENE)

$CH_3CH_2C\equiv CH$ 1-BUTYNE

$CH_3-C\equiv C-CH_3$ 2-BUTYNE

When other groups are present in an organic compound, the names are developed on the same basis as those of the hydrocarbons, as can be seen from the following examples:

$CH_3CH_2CH_2Br$ 1-BROMOPROPANE

$CH_3CHBrCH_3$ 2-BROMOPROPANE

$BrC\equiv CH$ BROMOETHYNE

$$CH_3-\overset{\displaystyle CH_3}{\underset{\displaystyle CH_2Br}{C}}-H$$ 1-BROMO-2-METHYLPROPANE

$CH_3CHBrCH_2Br$ 1,2-DIBROMOPROPANE

SELF-TEST 7-A

1. The branch of chemistry that deals with compounds of carbon is known as

   _____ chemistry.

2. The structure of the CH_4 molecule is described as _____.

3. How many covalent bonds are in a molecule of butane? _____

   Does it make a difference which butane is considered? _____

4. A straight-chain hydrocarbon, such as pentane, actually has all of its carbon atoms in a straight line. (True/False)

5. How many different isomers of C_5H_{12} are shown below?

(a)
```
    H  H  H  H
    |  |  |  |
H - C- C- C- C- H
    |  |  |
    H  H  H
          |
       H- C- H
          |
          H
```

(b)
```
       H
       |
   H - C - H
       |
       H  H
       |  |
   H - C- C- C- H
       |  |
       H  H
             |
          H- C- H
             |
             H
```

(c)
```
    H  H  H  H
    |  |  |  |
H - C= C- C- C- H
       |
       H
       |
    H- C- H
       |
       H
```

(d)
```
    H       H     H
    |       |     |
H - C ----- C --- C - H
    |       |
 H- C- H  H- C- H
    |       |
    H       H
```

(e)
```
        H            H
        |            |
   H  H-C-H     H- C- H
   |    |          |
H- C ----- C ----------- C - H
   |    |          |
   H    H          H
```

(f)
```
    H  H  H  H
    |  |  |  |
H - C- C- C- C- H
    |  |     |
    H  H     H
          |
       H- C- H
          |
          H
```

6. When the name of a compound ends in -ene (for example, butene), what structural feature is indicated? _____

7. Name the compound shown on the right:

   _____

8. The formula for the methyl group is _____ .

```
                                H
                                |
                            H — C — H
                                |
        H   H   H               H   H
        |   |   |               |   |
    H — C — C — C — C — C — C — H
        |   |       |   |   |   |
        H   |       H   H   H   H
            |
        H — C — H
            |
            H
```

OPTICAL ISOMERS

Stereoisomers have the same atoms and the same bonds but the atoms are arranged in space differently. Optical isomers are one type of stereoisomers.

In the preceding section, it was pointed out that in order to change one *structural* isomer into another, at least two carbon atoms have to change the atoms to which they are bonded. However, it is possible for some sets of atoms to form two isomeric molecules, both of which have the same atoms bonded to each other. This special case of isomerism is called ***optical isomerism.***

Optical isomerism is possible when a molecular structure is asymmetric (without symmetry). One common example of an asymmetric molecule is one containing a tetrahedral carbon atom bonded to four *different* atoms or groups of atoms. Such a carbon atom is called an ***asymmetric*** carbon atom; an example is the carbon atom in the molecule $CBrClIH$.

Figure 7–5 shows the two ways to arrange four different atoms in the

Figure 7–5 Optical isomers. Four different atoms, or groups of atoms, are bonded to tetrahedral center atoms so that the upper isomeric form cannot be turned in any way and exactly match the lower structure. The upper structure and the lower structure are nonsuperimposable mirror images. See also Figure 7–6.

Figure 7–6 Optical isomers of the amino acid, alanine.

$$H_2N-\overset{\displaystyle COOH}{\underset{\displaystyle CH_3}{\overset{|}{\underset{|}{C}}}}-H$$

The D-form is the nonsuperimposable mirror image of the L-form (see Figure 7–5).

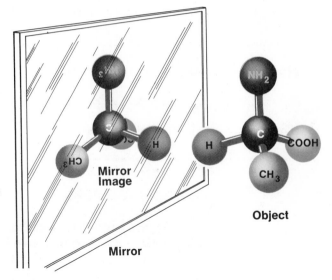

tetrahedral positions about the central carbon atom. These result in two nonsuperimposable, mirror-image molecules which are optical isomers.

There are many examples of nonsuperimposable mirror images in the macroscopic world. Consider right- and left-hand gloves, for instance. They are mirror images of one another and are nonsuperimposable. In Figure 7–6, this mirror image relationship is shown for isomeric forms of alanine, the molecules of which contain a tetrahedral carbon atom surrounded by an amino group ($-NH_2$), a methyl group ($-CH_3$), an acid group ($-COOH$), and a hydrogen atom. Note in Figure 7–6 that the carbon atoms in the methyl and acid groups are not asymmetric since these atoms are not bonded to four different groups.

The properties of optical isomers are almost identical. Different compounds whose molecules are mirror images of one another have the same melting point, the same boiling point, the same density, and many other identical physical and chemical properties. However, they always differ in one physical property: they rotate the plane of **polarized** light in opposite directions. According to the wave theory of light, a light wave traveling through space vibrates at right angles to its path (Figure 7–7). A group of such rays traveling together vibrate in random directions, all of which are at right angles to the path of travel. If such a group of waves is passed through a polarizing crystal, such as Iceland spar (a form of $CaCO_3$), or through a sheet of Polaroid material, the light is split into two rays and the waves emerging along the incoming axis will vibrate in only one plane perpendicular to the light path. Such light is said to be **plane polarized.** When plane-polarized light is passed through a solution of D-lactic acid, the light is still polarized but the plane of vibration is rotated somewhat in one direction. If the other lactic acid isomer is substituted (L-lactic acid), just the opposite rotation of the light is obtained.

Optical isomers can also differ in biological properties. An example is the hormone adrenalin (or epinephrine). Adrenalin is one of a pair of optical isomers, and its structure is shown on page 172. C* designates the asymmetric carbon atom. Only the isomer that rotates plane-polarized light to the left is effective in starting a heart that has stopped beating

D- and L- simply indicate that two structures are possible around an asymmetric C atom. The D- and L- notations do not indicate which way the substance will rotate the plane polarized light.

PUT MORE OXIDIZED GROUP SUCH AS ALDEHYDE (—CHO) AT TOP

Figure 7—7 Rotation of plane-polarized light by an optical iso-mer. (*a*) A sodium lamp provides a monochromatic yellow light. The original beam is nonpolarized; it vibrates in all directions at right angles to its path. After passing through a Polaroid filter, the light is vibrating in only one direction. This polarized light will pass through another Polaroid filter if the filter is lined up properly but will not pass through the third Polaroid filter if it is at right angles to the other two. The direction of the Polaroid filters deter-mines the direction of polarization. (*b*) The plane of polarized light is rotated by a solution of an optically active isomer. The analyzer can be a second Polaroid filter which can be rotated to find the angle for maximum transmission of light. If the solution rotates the plane of polarized light, the analyzer will not be at the same angle as the polarizer for maximum transmission.

momentarily, or in giving a person unusual strength during times of great emotional stress. The other isomer is inactive.

It is also interesting to note that during the contraction of muscles the body produces only the L-form of lactic acid and not the D-form. The concentration of this lactic acid in the blood is associated with the feeling of tiredness, and a period of rest is necessary to reduce the concentration of this chemical by oxidation.

Large organic molecules may have many asymmetric carbon atoms within the same molecule. At each such carbon atom there exists the possibility of *two* arrangements of the molecule. The total number of possible molecules, then, increases exponentially with the number of asymmetric centers. With two asymmetric carbon atoms there are 2^2 or four possible structures; for three, there are 2^3 or eight possible structures. It should be emphasized that each of the eight isomers can be made from the *same* set of atoms with the *same* set of chemical bonds. Glucose, a simple blood sugar also known as dextrose, contains four asymmetric carbon atoms per molecule. Thus, there are 2^4 or 16 isomers in the family of compounds to which glucose belongs. Obviously, then, the concept of optical isomerism helps to explain the vast number of carbon compounds.

ADRENALIN
(EPINEPHRINE)

GEOMETRIC ISOMERS

Where carbon-carbon double bonds exist in a molecule, geometric isomerism is possible. A double bond between carbon atoms does not allow free rotation, and this lack of rotation provides a structural basis for geometric isomerism.

Consider the compound ethene, C_2H_4. Its six atoms lie in the same plane, with bond angles of approximately 120°:

Geometric isomers are another type of stereoisomer.

If two chlorine atoms replace two hydrogen atoms, one on each carbon atom of ethene ($H_2C{=}CH_2$), the result is $CHCl{=}CHCl$. Experimental evidence confirms that there are two compounds with this general arrangement. If the two chlorine atoms are close together, this is one isomer

To have geometric isomers, each carbon connected by the double bond must have two unlike groups attached.

(the *cis* isomer), and if they are far apart, this forms another isomer (the *trans* isomer). Both compounds are called 1,2-dichloroethene (the 1 and 2 indicate that the two chlorine atoms are attached to different carbon atoms). They are distinguished from each other by the prefixes *cis* and *trans*. Note that the two isomeric compounds have significant differences in their properties.

$$
\begin{array}{cc}
Cl \qquad Cl \\
\diagdown \quad \diagup \\
C = C \\
\diagup \quad \diagdown \\
H \qquad H
\end{array}
\qquad
\begin{array}{cc}
H \qquad Cl \\
\diagdown \quad \diagup \\
C = C \\
\diagup \quad \diagdown \\
Cl \qquad H
\end{array}
$$

	CIS-1,2-DICHLOROETHENE	TRANS-1,2-DICHLOROETHENE
MELTING POINT	−80.5°C	−50°C
BOILING POINT	60.1°C	48.4°C
DENSITY (AT 15°C)	1.291 g/ml	1.265 g/ml

The third possible isomer, 1,1-dichloroethene (a *structural* isomer of the *cis* and *trans* isomers), does *not* have *cis* and *trans* structures.

As a general rule, *trans* isomers are higher melting than *cis* isomers because of the greater ease with which the *trans* molecules can fit into the lattice and form strong intermolecular bonds.

When there is a carbon-carbon double bond in an organic molecule, the possibility exists for *cis* and *trans* isomers. Sometimes a number of such bonds can be found in the same molecule, giving rise to numerous isomeric compounds.

FUNCTIONAL GROUPS

Carbon forms covalent bonds wth a number of elements besides hydrogen. As a result, certain groups of atoms called *functional groups*

TABLE 7–3   Alcohols Derived From Pentane (C_5H_{12})

Substitution of an —OH for an end-hydrogen **gives** 1-PENTANOL

Substitution of an —OH for a 2-carbon hydrogen **gives** 2-PENTANOL

Substitution of an —OH for a 3-carbon hydrogen **gives** 3-PENTANOL appear over and over again in different organic compounds. Consider, for example, the —OH group, called the hydroxyl group. A large number of molecules containing an —OH group have properties that are characteristic of a class of organic compounds called ***alcohols.***

The —OH group has the same combining power as an atom of hydrogen. This means that in any of the hydrocarbons considered thus far, any of the hydrogen atoms could be replaced by an —OH group to form an alcohol. A single hydrocarbon molecule can give rise to a number of alcohols if there are different isomeric positions for the —OH group. Three different alcohols result when a hydrogen atom is replaced by an —OH group in n-pentane, depending on which hydrogen atom is replaced (Table 7–3). When one or more functional groups appear in a molecule, the IUPAC name reveals the functional group name and position. For example, the name of an alcohol will use the root of the name of the hydrocarbon to which it corresponds to indicate the number of carbon atoms, and the suffix -ol to denote an alcohol. As before, a number is used to indicate the position of the alcohol group.

Alcohols contain the —OH functional group.

TABLE 7–4   Classes of Organic Compounds Based on Functional Groups*

General Formulas of Class Members	Class Name	Typical Compound	Compound Name	Common Use of Sample Compound
R—OH	Alcohol	$H-\overset{\displaystyle H}{\underset{\displaystyle H}{\overset{\displaystyle \vert}{\underset{\displaystyle \vert}{C}}}}-OH$	Methanol (Wood Alcohol)	Solvent
$R-\overset{\displaystyle O}{\overset{\Vert}{C}}-H$	Aldehyde	$H-\overset{\displaystyle O}{\overset{\Vert}{C}}-H$	Methanal (Formaldehyde)	Preservative
$R-\overset{\displaystyle O}{\overset{\Vert}{C}}-OH$	Carboxylic acid	$H-\overset{\displaystyle H}{\underset{\displaystyle H}{\overset{\vert}{\underset{\vert}{C}}}}-\overset{\displaystyle O}{\overset{\Vert}{C}}-OH$	Ethanoic acid (Acetic acid)	Vinegar
$R-\overset{\displaystyle O}{\overset{\Vert}{C}}-R'$	Ketone	$H-\overset{\displaystyle H}{\underset{\displaystyle H}{\overset{\vert}{\underset{\vert}{C}}}}-\overset{\displaystyle O}{\overset{\Vert}{C}}-\overset{\displaystyle H}{\underset{\displaystyle H}{\overset{\vert}{\underset{\vert}{C}}}}-H$	Propanone (Acetone)	Solvent
R—O—R′	Ether	$C_2H_5-O-C_2H_5$	Diethyl ether (Ethyl ether)	Anesthetic
$R-O-\overset{\displaystyle O}{\overset{\Vert}{C}}-R'$	Ester	$CH_3-CH_2-O-\overset{\displaystyle O}{\overset{\Vert}{C}}-CH_3$	Ethyl ethanoate (Ethyl acetate)	Solvent in fingernail polish
$R-N\overset{\displaystyle H}{\underset{\displaystyle H}{\big\langle}}$	Amine	$H-\overset{\displaystyle H}{\underset{\displaystyle H}{\overset{\vert}{\underset{\vert}{C}}}}-$	Methylamine	Tanning (Foul odor)
$R-\overset{\displaystyle O}{\overset{\Vert}{C}}-\overset{\displaystyle H}{\overset{\vert}{N}}-R'$	Amide	$CH_3-\overset{\displaystyle O}{\overset{\Vert}{C}}-N\overset{\displaystyle H}{\underset{\displaystyle H}{\big\langle}}$	Acetamide	Plasticizer

*R stands for an H or a hydrocarbon group such as CH_3-, C_2H_5-, etc.
R′ could be a different group from R

Some major functional groups that we shall consider in following chapters are shown in Table 7–4. The symbol —R stands for any hydrocarbon group such as methyl (—CH_3) or ethyl (—C_2H_5).

AROMATIC COMPOUNDS

Delocalized electrons can occupy orbitals on several nuclei.

All of the hydrocarbons we have discussed up to this point have localized electronic structures; that is, the bonding electrons are essentially fixed between two atomic centers as in C—C or C=C bonds. For a large group of organic compounds known as *aromatic* compounds, this type of complete electron localization is not found. Rather, these compounds have some delocalized electrons, which are spread over several atoms, or even the entire molecule, a feature that leads to some interesting chemical properties.

Benzene is the number fourteen commercial chemical.

The simplest aromatic compound is *benzene* (C_6H_6). The molecular structure of benzene is a ring of carbon atoms in a plane, with one hydrogen atom bonded to each carbon atom. The bonds between these carbon atoms are shorter than single bonds, but longer than double bonds. The measured bond angles are 120°.

One way to account for the bonding and structure of the benzene molecule is to use two of the three sp² hybrid orbitals of each carbon to form *sigma* bonds between the carbon atoms and to use the six remaining sp² orbitals to form six sigma bonds with six hydrogens.* The six p orbitals (one on each carbon) not involved in the formation of sp² orbitals can

_____

*At this point, you may wish to review sigma and pi bonds in Chapter 3.

The p orbitals on carbon are omitted

sigma skeleton only

Overlap of *sp²* orbitals to form sigma bonds

Figure 7–8 Bonding in an aromatic compound, benzene, C_6H_6.

Lateral overlap of six *p* orbitals to form pi bonds

overlap laterally to form *pi* bonds. These pi bonds are not exactly the same as those discussed earlier for double and triple bonds because there is a ring system of p orbitals in an aromatic molecule. Each p orbital overlaps with the p orbitals on both neighboring carbon atoms. This causes the pi bonds to be less well fixed (or less localized) between the three pairs of p orbitals. Figure 7–8 illustrates this type of pi bonding. The pi bonds are averaged evenly around the ring; we say that they are *delocalized.*

Various symbols have been used to represent benzene. The first two of the following structures suggest two different types of carbon-carbon bonding within the same ring, with three single bonds and three double bonds; however, there is only one type of bond. Hence, the structure on the right is preferred. The hydrogen atoms are not shown.

Aromatic compounds have electrons delocalized over the orbitals of carbon atoms which are arranged in a ring.

A large number of aromatic compounds can be derived from benzene by replacing one or more of the hydrogen atoms with other kinds of atoms of functional groups. These molecules exhibit a wide variety of properties and differ greatly in their chemical reactivity. Some interesting examples of isomerism are also possible. Consider the three different compounds with the formula C_8H_{10} found in some coal tars. There are several names given to these isomers:

Xylenes rank number twenty-three among commercial chemicals.

1,4-DIMETHYLBENZENE
(PARA-XYLENE)
m.p. 13.3°C

1,3-DIMETHYLBENZENE
(META-XYLENE)
m.p. −47.9°C

1,2-DIMETHYLBENZENE
(ORTHO-XYLENE)
m.p. −25°C

Each of these isomers has two methyl groups substituted for hydrogen atoms on the ring. The prefixes *para-, meta,* and *ortho-* are used if there are only two groups on the benzene ring.

If more than two groups occur, a number system is most useful. Consider the following compounds:

Nomenclature in aromatic compounds:

ortho—groups on adjacent carbon atoms (1,2 positions)

meta—groups have one carbon atom between them (1,3 positions)

para—groups have two carbon atoms between them (1,4 positions)

1,2,3-TRICHLOROBENZENE   1,2,4-TRICHLOROBENZENE   1,3,5-TRICHLOROBENZENE

There are no other ways of drawing these three isomers, and only three trichlorobenzenes have been isolated in the laboratory.

Some rings have nitrogen, oxygen, or sulfur atoms in place of a few carbon atoms in the ring. Only aromatic ring structures incorporating nitrogen will be encountered later in the text. Examples are pyridine, pyrimidine, and purine.

PYRIDINE   PYRIMIDINE   PURINE

Not all ring structures are aromatic. The bonds in structures such as cyclohexane and glucose are only localized single (or sigma) or double (sigma plus a pi) bonds.

CYCLOHEXANE   GLUCOSE

No wonder carbon is ubiquitous; there are over two million recorded compounds containing carbon. These millions of carbon compounds exist for the following reasons:

There are some specific reasons for the large number of carbon compounds.

1. The ability of carbon to form covalent bonds to other carbon atoms almost without limit.
2. The ability of a given number of carbon atoms to combine in more than one molecular pattern—isomers.
3. The ability of carbon to form stable covalent bonds to a large number of other atoms—functional groups.
4. The stability of carbon chains in the presence of substances such as oxygen and water which usually destroy chains of other atoms.

Structural differences explain how two compounds can have the same chemical composition by weight, and yet have different physical and chemical properties. Some order can be brought out of chaos with an understanding of isomerism and a systematic approach to the possible molecular structures.

SELF-TEST 7-B

1. In order to have optical isomers in carbon compounds, a carbon atom must have _____ different groups attached.

2. In what physical property do optical isomers that are mirror images differ? _____

3. How many optical isomers would be possible if five asymmetric carbon atoms are contained in a structure? _____

4. The benzene ring has both localized electrons (sigma bonds) and _____ _____ electrons (pi bonds).

5. How many atoms does the symbol, ⬡ , represent? _____

6. Name the following compound. _____

$$CH_3$$

$$CH_3$$

$$CH_3$$

7. All ring structures contain only carbon atoms in the ring and have delocalized electrons. (True/False)

MATCHING SET

_____ 1. organic chemistry

_____ 2. isomers

_____ 3. functional group

_____ 4. hydrocarbon

_____ 5. methyl group

_____ 6. asymmetric carbon atom

_____ 7. delocalized electrons

_____ 8. sigma bond

_____ 9. pi bond

(a) $-\overset{\overset{\textstyle O}{\|}}{C}-O-H$

(b) occurs when pi bonds overlap over three or more atomic centers

(c) compound such as C_3H_8

(d) lateral overlap of p orbitals

(e) chemistry of carbon compounds

(f) $-CH_3$

(g) same number and kinds of atoms differently arranged

(h) end-to-end overlap of p orbitals

(i) has four different groups attached

QUESTIONS

1. *Saturated hydrocarbons* are so named because they have the maximum amount of hydrogen present for a given amount of carbon. The saturated hydrocarbons have the general formula C_nH_{2n+2} where n is a whole number. What are the names and formulas of the first four members of this series of compounds?

2. Using diagrams and bonding theory, explain why nearly free rotation is allowed around a carbon-carbon single bond but not around a carbon-carbon double bond. Refer to Chapter 3 as needed.

3. Using the periodic table and electron-dot formulas, illustrate a sigma bond in a compound other than those given in this chapter.

4. Draw the structural formula for each of the five isomeric hexanes, C_6H_{14}.

5. Write the structural formulas for:

   (a) 2-methylbutane
   (b) ethylpentane
   (c) 4,4-dimethyl-5-ethyloctane
   (d) methylbutane
   (e) 2-methyl-2-hexene
   (f) 3-methyl-3-hexanol

6. Give the names for:

7. If cyclohexane has the formula C_6H_{12}, and if it has *no* double bonds, what would you suggest as its structural formula?

8. How can optical isomers be distinguished from each other experimentally?

9. (a) Which arrangement has a mirror image which is nonsuperimposable?
   (b) Explain the term: asymmetric carbon atom.

10. If a solution of the D optical isomer of a pair of optical isomers rotates plane polarized light clockwise, and the L isomer does the opposite, predict what a mixture containing equal amounts of both isomers would do.

11. How many optical isomers can there be for a molecular structure containing eight asymmetric carbon atoms?

12. Explain the statement: The formation of a pi bond between two carbon atoms does not release as much energy as the formation of a sigma bond, yet more energy is released in the formation of a carbon-carbon double bond than in the formation of a carbon-carbon single bond.

13. Which do you think would be better to use for medicinal purposes, pure adrenalin obtained from natural products (found in nature) or the pure compound as synthesized in the laboratory? Why?

14. Distinguish between the classical and modern use of the word "organic" in chemistry.

15. Look up the formula for ethyl alcohol (use the index), and explain how the formula for this compound indicates that it is a member of the alcohol family.

16. What structural feature characterizes aromatic compounds?

17. Indicate the functional groups present in the following molecules:

(a) $CH_3CH_2CH_2COOH$

(b) $CH_3CH_2NH_2$

(c) $CH_3CHCH_2CH_2COOH$
  |
  NH_2

(d) CH_3CHCH_2COOH
  |
  OH

(e) $CH_3CCH_2CH_2COOH$
  ‖
  O

(f) CH_3CHCH_2OH
  |
  NH_2

18. Name the following compounds: $CH_3CH_2CH_2COOH$, $CH_3CH_2CH_2C\equiv CH$, $CH_3CH_2CH_2OH$

19. Write structural formulas for butanoic acid, aminomethane, 2-butanol, and 3-aminopentane.

20. Given an example of:

(a) an alkane
(b) an amine
(c) a carboxylic acid
(d) an ether
(e) an ester
(f) an alkene
(g) an alkyne
(h) an alcohol
(i) a ketone

21. Write structural formulas for two compounds that can have each of the molecular formulas listed:

(a) $C_5H_{12}O$
(b) C_3H_6O
(c) $C_5H_{10}O_2$

22. Draw the *cis* and *trans* isomers for:

(a) 1,2-dibromoethene
(b) 1-bromo-2-chloroethene

CHAPTER 8

SOME APPLICA- TIONS OF ORGANIC CHEMISTRY

The preparation of new and different organic compounds through chemical reactions is called ***organic synthesis.*** Millions of organic compounds have been synthesized in the laboratories of the world during the past 150 years. Prior to 1828, it was widely believed that chemical compounds synthesized by living matter could not be made without living matter — a "vital force" was necessary for the synthesis. In 1828, a young German chemist, Friedrich Wöhler, destroyed the vital force myth and opened the door to modern organic syntheses. Wöhler heated a solution of silver cyanate and ammonium chloride, neither of which had been derived from any living substance. From these he prepared urea, a major animal waste product found in urine.

Organic compounds can be made from inorganic compounds.

$$AgOCN \quad + \quad NH_4Cl \quad \longrightarrow \quad AgCl \quad + \quad NH_4OCN$$

| SILVER CYANATE | AMMONIUM CHLORIDE | SILVER CHLORIDE (PRECIPITATE) | AMMONIUM CYANATE |

$$NH_4OCN \quad \xrightarrow{\text{Heat}} \quad H_2N\overset{\displaystyle O}{\overset{\|}{C}}NH_2$$

AMMONIUM CYANATE   UREA

The notion of a mysterious vital force declined as other chemists began to synthesize more and more organic chemicals without the aid of a living system. Soon it was shown that chemists could do more than imitate the products of living tissue; they could form unique materials of their own design.

Advances in understanding the structure of organic compounds gave organic synthesis a tremendous boost. Knowing the structure of com-

Friedrich Wöhler (1800–1882) was Professor of Chemistry at the University of Berlin and later at Göttingen. His preparation of the organic compound urea from the inorganic compound ammonium cyanate did much to overturn the theory that organic compounds must be prepared in living organisms. One of the first to study the properties of aluminum, he was the first to isolate the element beryllium and is known for many other outstanding contributions to chemistry.

pounds, the organic chemist could predict by analogy with simpler molecules what reactions might take place when organic reagents were used. Very elegant and reliable schemes of synthesis could now be constructed.

In this chapter emphasis will be given to some important organic compounds that not only are useful in themselves but are necessary for the synthesis of many other organic compounds. The dependence of synthesis upon a knowledge of structure will be pointed out from time to time.

SOME USES OF HYDROCARBONS

Complex mixtures of hydrocarbons, compounds containing only carbon and hydrogen, occur in enormous quantities in nature as petroleum and natural gas. These materials were formed from organisms that lived millions of years ago. After their death, they became covered with layers of sediment and ultimately were subjected to high temperatures and pressures in the depths of the earth's crust. In the absence of free oxygen, these conditions converted once living tissue into petroleum and coal. After petroleum and natural gas are brought to the surface of the earth, they can be separated into various fractions with different boiling points by the use of fractional distillation (Figure 8–1). Once distilled, the fractions are refined further and made to undergo various types of chemical change in order to make desirable substances.

Common sources of the hydrocarbons are the "fossil fuels" (coal, petroleum, and natural gas).

By heating coal or wood in the absence of air (destructive distillation), coal tar or wood tar can be separated from the coke or charcoal (carbon). A ton of a typical soft coal produces about 140 pounds of coal tar, which is about one-half pitch, the rest being composed of such chemicals as naphthalene, benzene, phenol, cresols, toluene, and xylenes.

Coal is one natural source of aromatic compounds.

The naturally occurring compounds not only are important in themselves, but they also serve as starting materials for making numerous organic compounds that do not occur in nature. From the simplest hydrocarbons come such diverse consumer products as plastic dishes, acrylic and polyester fibers for textiles, vinyl and latex paints, neoprene rubber, and such industrial products as Teflon and cattle feed.

Condenser

Gasoline vapors

Gas

Naphtha 40–200°C

Fractionating column

Kerosene 175–275°C

Heating oil 250–300°C

Lubricating oil
above 300°C

Vaporizer

Crude oil

Live steam

Wax distillate

Fuel oil

Figure 8–1 A diagram of a fractionating column for distilling petroleum. Notice that the higher boiling substances condense at the lower levels and the lower boiling substances do not condense until the higher, cooler levels. (From Routh, J. I., Eyman, D. P., and Burton, D. J.: *A Brief Introduction to General, Organic and Biochemistry,* W. B. Saunders Co., Philadelphia, 1971.)

Natural gas consists primarily of low molecular weight hydrocarbons. It is predominantly methane (CH_4), but ethane (C_2H_6), propane (C_3H_8), and butane (C_4H_{10}) are also present. This mixture is conveyed in long pipelines from the areas in which it occurs to cities where it is used as a fuel.

Energy is obtained by burning the constituents of natural gas in air:

$$CH_4 + 2O_2 \longrightarrow CO_2 + 2H_2O + 213 \text{ kcal}$$
$$C_2H_6 + 3\tfrac{1}{2}O_2 \longrightarrow 2CO_2 + 3H_2O + 372.8 \text{ kcal}$$

The energy released here can be used to heat homes, run electric power plants, or power special internal combustion engines.

Propane and butane, principal components of bottled gas, can be liquefied by the use of moderate pressure and cooling. Because propane is more volatile than butane, bottled gas is prepared so that it contains more propane in colder climates and more butane in warmer ones.

"Bottled gas" is actually stored in the liquid state under pressure. In order for bottled gas to be burned in stoves or furnaces, it must be vaporized.

The enormous increase in our use of organic compounds in the last century forces us to recognize that the sources of these materials are limited and subject to eventual exhaustion. Thus, petroleum sources, from which we draw so much of our energy and which serve as starting materials for the synthesis of industrial organic compounds, cannot be expected to last beyond 50 to 100 years at the most. The transformation of many of the natural sources of organic compounds into disposable convenience items which soon end up in a garbage dump represents an extremely short-sighted use of irreplaceable natural resources. Most of us realize that we must begin ultimately to recycle a much larger percentage of the material we use in our daily lives, but few seem ready to begin effective efforts in this direction now. It seems very probable that the next decade will force us to change many of our casual attitudes toward the utilization of our natural sources of organic compounds.

Our supplies of coal, and especially petroleum, are limited.

GASOLINE

From the time petroleum was first discovered in the United States in 1859 until 1900 when the automobile became popular, most oil was refined to yield kerosene, a mixture of hydrocarbons in the C_8 to C_{13} range. This liquid was used principally as a fuel for lamps.

The internal combustion engines used in early automobiles were designed to burn a more volatile mixture of hydrocarbons, the C_6 to C_{10} fraction, which became known as *gasoline.* These lower molecular weight hydrocarbons mix readily with air in simple carburetors and burn fairly completely.

Mixtures of hydrocarbons in the C_6 to C_{10} range are known as gasolines.

With the increasing popularity of the automobile, petroleum refiners had to shift the output of a barrel of crude oil from a reasonably large fraction of kerosene to almost no kerosene and a much greater fraction of gasoline (Table 8–1). This dramatic increase in the amount of gasoline from a barrel of crude oil was accomplished by the discovery of chemical processes which convert nongasoline molecules into ones that burn well in an automobile engine.

A barrel of crude oil is 42 gallons.

The first conversion discovered was ***thermal cracking,*** a process in which long-chain hydrocarbon molecules were heated under pressure in the absence of air. This heating strains the bonds of the molecule so that it may break or "crack," thus forming two smaller molecules, some of which will be in the gasoline range.

Cracking breaks larger molecules into smaller ones.

$$C_{16}H_{34} \xrightarrow[\text{Heat}]{\text{Pressure}} C_8H_{18} + C_8H_{16}$$

AN ALKANE   AN ALKANE   AN ALKENE

IN THE GASOLINE RANGE

Later, cracking was carried out in the presence of certain catalysts that allowed the processes to proceed at lower pressures and to produce even higher yields of gasoline. Today, these catalysts include specially processed clays. Each refiner of petroleum has his own special methods, which offer different advantages in cost and type of crude oil handled.

TABLE 8–1 Division of a Barrel of Crude Oil

	1920 %	1967 %
Gasoline	26.1	44.8
Kerosene	12.7	2.8
Jet Fuel	—	7.6
Heavy Distillates	48.6	22.2
Other (asphalt, etc.)	12.6	22.6
Total	100.0	100.0

The hydrocarbon molecules produced are much the same regardless of the methods used.

Cracking of petroleum fractions brought about not only an increase in the quantity of gasoline available from a barrel of crude oil, but also an increase in *quality*. That is, gasoline from a cracking process can be used at higher efficiency (in a high-compression engine) than can "straight run" gasoline, because the molecular structures of the hydrocarbons in the cracked gasoline allow them to oxidize more smoothly at high pressure. When the burning of the gasoline-air mixture is too rapid or irregular, ignition occurs in the combustion chamber too early, resulting in a small detonation which is heard as a "knock" in the engine. This knocking can spell trouble to the owner of an automobile because it will eventually lead to the breakdown of the internal parts of the automobile's engine.

An arbitrary scale for rating the relative knocking properties of gasolines has been developed. Normal heptane, typical of straight run gasoline, knocks considerably and is assigned an octane rating of 0,

Knocking in gasoline engines is a sign of improper combustion.

$$CH_3CH_2CH_2CH_2CH_2CH_2CH_3 \text{ (octane rating = 0)}$$
n-HEPTANE

while 2,2,4-trimethylpentane (isooctane) is far superior in this respect and is assigned an octane rating of 100.

$$CH_3-\overset{\overset{\displaystyle CH_3}{|}}{\underset{\underset{\displaystyle CH_3}{|}}{C}}-CH_2-\overset{\overset{\displaystyle CH_3}{|}}{\underset{\underset{\displaystyle H}{|}}{C}}-CH_3 \text{ (octane rating = 100)}$$
2,2,4-TRIMETHYLPENTANE

The octane scale measures the ability of a mixture to burn without knocking in a gasoline engine.

The octane rating of a gasoline is determined by first using the gasoline in a standard engine and recording its knocking properties. This is compared to the behavior of mixtures of n-heptane and isooctane, and the percentage of isooctane in the mixture with identical knocking properties is called the octane rating of the gasoline. Thus, if a gasoline has the same knocking characteristics as a mixture of 9 percent n-heptane and 91 percent isooctane, it is assigned an octane rating of 91. This corresponds to a regular grade of gasoline. Since the octane rating scale was established, fuels have been developed that are superior to isooctane, so the scale has been extended well above 100.

A catalyst increases the speed of a reaction without being consumed in the reaction.

A primary source of the present high-octane gasolines is a chemical process known as *catalytic reforming*. Under the influence of certain catalysts, such as finely divided platinum, straight-chain hydrocarbons with low octane numbers can be reformed into their branched-chain isomers, which have higher octane numbers.

$$CH_3CH_2CH_2CH_2CH_3 \xrightarrow[\text{Heat}]{\text{Platinum}} CH_3CH_2CHCH_3$$
$$\underset{\displaystyle CH_3}{|}$$
n-PENTANE 2-METHYLBUTANE

The knock caused by combustion of gasoline can be avoided by either of two procedures. The first way is to add "antiknock" agents to a motley mixture of hydrocarbons. The best-known antiknock additive is tetraethyllead, $(C_2H_5)_4Pb$. Tetraethyllead is extremely toxic and, as new car owners know, is being limited in use. The Environmental Protection Agency has approved a newer antiknock mixture, which is mostly 2-methyl-2-propanol (also called *tertiary* butyl alcohol). The second way is to use only hydrocarbons with structures known to burn satisfactorily in a high-compression engine. Branched-chain structures are particularly desirable in this respect, and gasolines with appreciable amounts of aromatic hydrocarbons (20 to 40 percent) are also less prone to knock. Some petroleum is just naturally better supplied with aromatic compounds. For example, California crude has a higher percentage of aromatic compounds than Pennsylvania crude. Regardless of the source, petroleum fractions such as naphtha, kerosene, and gas oil are mixtures of open-chain, branched-chain, and single-bonded-ring hydrocarbons. When these stocks, in the vapor phase, are passed over a copper catalyst at about 650°C, a high percentage of the original material is converted into a mixture of aromatic hydrocarbons from which benzene, toluene, xylenes, and similar compounds may be separated by fractional distillation. For example, n-hexane is converted into benzene,

$$CH_3CH_2CH_2CH_2CH_2CH_3 \longrightarrow \bigcirc$$

n-HEXANE   BENZENE

and n-heptane is changed into toluene.

$$CH_3CH_2CH_2CH_2CH_2CH_2CH_3 \longrightarrow \overset{CH_3}{\bigcirc}$$

n-HEPTANE   TOLUENE

SOME IMPORTANT COMPOUNDS OF CARBON, HYDROGEN, AND OXYGEN

Carbon, hydrogen, and oxygen can be combined to form an enormous number of compounds. As we have seen, considerable order is introduced into the study of these compounds when they are divided into classes on the basis of the functional groups they contain. The *alcohols,* the *organic acids* and their derivatives (compounds that can be made from them), and the *esters* and *soaps* are very important compounds of carbon, hydrogen, and oxygen since they find such wide application in our everyday lives.

Alcohols

When a hydroxyl (—OH) group is attached to a nonaromatic carbon skeleton of a hydrocarbon (an R-group), the resulting R—OH compound has properties common to a class of compounds called *alcohols.* Formulas and names for some important substances of this type are given in Table 8–2.

Alcohols are compounds with the structure R—O—H, where R is a hydrocarbon group.

TABLE 8–2   Some Important Alcohols

Formula	IUPAC Name	Common Name
H | H—C—OH | H	Methanol	Methyl alcohol (wood alcohol)
H H | | H—C—C—OH | | H H	Ethanol	Ethyl alcohol (grain alcohol)
H H H | | | H—C—C—C—OH | | | H H H	1-Propanol	n-Propyl alcohol

H H H
| | |
H—C—C—C—H
| | |
H O H
    |
    H
	2-Propanol	Isopropyl alcohol
		(rubbing alcohol)

H | H—C—OH | H—C—OH | H	1,2-Ethanediol	Ethylene glycol (permanent antifreeze)
H | H—C—OH | H—C—OH | H—C—OH | H	1,2,3-Propanetriol	Glycerol (glycerin)

Methanol (Methyl Alcohol)

Methanol is number 22 in the list of commercial chemicals.

Methanol was originally called wood alcohol since it was obtained by the destructive distillation of wood. It is the simplest of all alcohols and has the formula CH_3OH. In the older method for the production of wood alcohol, hardwoods such as beech, hickory, maple, or birch are heated in the absence of air in a retort (Figure 8–2). Methanol that is 92 to 95 percent pure can be obtained by fractional distillation of the resulting liquid.

In 1923, the price of wood alcohol in the United States was 88¢ per gallon. In that year German chemists discovered how to produce this useful compound synthetically. Methanol is formed when carbon monoxide, CO, and hydrogen are heated at a pressure of 200 to 300 atmospheres over a catalyst of mixed oxides (90% ZnO–10% Cr_2O_3).

$$CO + 2H_2 \xrightarrow[\substack{300°C \\ Pressure}]{ZnO-Cr_2O_3} CH_3OH$$

Figure 8–2 Destructive distillation of wood.

As a result of this synthetic process, German industrialists were able to sell pure methanol at 20¢ per gallon. Even a high tariff was not able to save the wood distillers in their outdated operations. The synthetic product soon dominated the market in the United States.

Industrial chemists are continually searching for less costly ways to prepare important chemicals such as methanol.

The production of synthetic methanol in the United States is over 4.0 billion pounds per year. About half of this is used in the production of formaldehyde (used in plastics, embalming fluid, germicides, and fungicides), 30 percent in the production of other chemicals, and smaller amounts for jet fuels, antifreeze mixtures, solvents, and as a denaturant (a poison added to ethanol to make it unfit for beverages). Methanol is a *deadly poison;* it causes blindness in less than lethal doses. Many deaths and injuries have resulted when this alcohol was mistakenly substituted for ethanol in beverages.

Ethanol (Ethyl Alcohol)

Ethanol (ethyl alcohol) is called grain alcohol because it can be fractionally distilled from the fermented mash made from corn, rice, barley, or other grains. Fermentation is a breakdown of starch in the grain by means of enzymes. Enzymes are catalysts that are complex organic molecules produced by living cells. If the enzyme diastase is mixed with ground grain and water, and the mixture is allowed to stand at 40°C for a period of time, the starch in the grain will be changed into maltose.

$$2(C_6H_{10}O_5)_n + nH_2O \xrightarrow{\text{diastase}} nC_{12}H_{22}O_{11}$$
STARCH   MALTOSE (A SUGAR)

The subscript n in the formula for starch indicates that starch is made up of many $C_6H_{10}O_5$ units.

Brewers call the resulting mixture of maltose and water the wort. The wort is diluted and mixed with yeast, and held at a temperature of 30°C for 40 to 60 hours. The living yeast cells secrete the enzymes maltase and zymase. The maltase causes the sugar, maltose, to hydrolyze into a simple sugar, glucose:

$$\underset{\text{MALTOSE}}{C_{12}H_{22}O_{11}} + H_2O \xrightarrow{\text{maltase}} \underset{\text{GLUCOSE}}{2C_6H_{12}O_6}$$

The glucose, in turn, is converted by zymase to ethanol and carbon dioxide:

$$\underset{\text{GLUCOSE}}{C_6H_{12}O_6} \xrightarrow{\text{zymase}} 2CO_2 + \underset{\text{ETHANOL}}{2C_2H_5OH}$$

Over 2 billion pounds of ethanol are produced synthetically in the U.S. each year.

A solution of 95 percent ethanol and 5 percent water can be recovered from the mash by fractional distillation.

Synthetic ethanol is produced on a large scale for industrial use. The direct chemical addition of water to ethylene accounts for more than 80 percent of all ethanol production. Under high pressure in the presence of a catalyst and a large excess of water vapor, ethylene produces ethanol:

Ethylene is number 6 in the list of commercial chemicals.

ETHYLENE   ETHANOL

Pure ethanol is 200 proof (exactly twice the percentage of alcohol). Apart from the alcoholic beverage industry, ethanol is used widely in solvents and in the preparation of chloroform, ether, and many other organic compounds.

Some of the most commonly encountered alcoholic beverages and their characteristics are presented in Table 8–3.

TABLE 8–3 Common Alcoholic Beverages

Name	Source of Fermented Carbohydrate	Amount of Ethyl Alcohol
Beer	Barley, wheat	3.2–9%
Wine	Grapes or other fruit	12% maximum, unless fortified*
Brandy	Distilled wine	40–45%
Whiskey	Barley, rye, corn, etc.	45–55%
Rum	Molasses	~45%

*The growth of yeast is inhibited at alcohol concentrations over 12% and fermentation comes to a stop. Beverages with a higher concentration are prepared either by distillation or by fortification with alcohol that has been obtained by the distillation of another fermentation product.

Propanols (Propyl Alcohols)

When one considers the possible structures for propyl alcohol, it is apparent that two isomers are possible.

$$\begin{array}{cccc} & H & H & H \\ & | & | & | \\ H- & C- & C- & C-OH \\ & | & | & | \\ & H & H & H \end{array} \qquad \begin{array}{cccc} & H & H & H \\ & | & | & | \\ H- & C- & C- & C-H \\ & | & | & | \\ & H & O & H \\ & & | & \\ & & H & \end{array}$$

1-PROPANOL
n-PROPYL ALCOHOL

2-PROPANOL
ISOPROPYL ALCOHOL

Of the two propanols, 1-propanol is the more expensive; it is prepared by the oxidation of simple hydrocarbons. It finds uses as a solvent and as a raw material in the manufacture of other organic compounds.

The hydration of propylene yields 2-propanol (isopropyl alcohol), which is rubbing alcohol. It has greater germicidal activity than the other simple alcohols and is used as an antiseptic.

Propylene is number 13 in the list of commercial chemicals.

Ethylene Glycol and Glycerol (Glycerin)

More than one alcohol group (—OH) can be present in a single molecule. Glycerol and ethylene glycol, the base of permanent anti-freeze, are examples of such compounds.

Permanent anti-freeze is ethylene glycol:

$$\begin{array}{cc} CH_2- & CH_2 \\ | & | \\ OH & OH \end{array}$$

$$\begin{array}{c} H \\ | \\ H-C-OH \\ | \\ H-C-OH \\ | \\ H \end{array} \qquad \begin{array}{c} H \\ | \\ H-C-OH \\ | \\ H-C-OH \\ | \\ H-C-OH \\ | \\ H \end{array}$$

1,2-ETHANEDIOL
ETHYLENE GLYCOL

1,2,3-PROPANETRIOL
GLYCEROL (GLYCERIN)

Glycerol has many uses in foods and tobacco as a digestible and nontoxic humectant (gathers and holds moisture), and in the manufacture of drugs and cosmetics. It is also used in the production of nitroglycerin and numerous other chemicals. Perhaps the most important compounds of glycerol are its natural esters (fats and oils), which we shall discuss later in this chapter.

Hydrogen Bonding in Alcohols

The physical properties of water, methanol, ethanol, the propanols, ethylene glycol, and glycerol offer another interesting example of the effects of **hydrogen bonding** between molecules in liquids. In Table 8–4 the boiling points for these compounds are listed.

Since boiling involves overcoming the attractions between liquid molecules as they pass into the gas phase, a higher boiling point indicates

TABLE 8–4   Boiling Points for Some —OH Compounds

Water	HOH	100°C
Methanol	CH_3OH	65.0°
Ethanol	CH_3CH_2OH	78.5°
1-Propanol	$CH_3CH_2CH_2OH$	97.4°
2-Propanol	$CH_3CHOHCH_3$	82.4°
Ethylene glycol	CH_2OHCH_2OH	198°
Glycerol	$CH_2OHCHOHCH_2OH$	290°

Note: The parent hydrocarbon of methanol is methane (bp −164°C); of ethanol and ethylene glycol, ethane (bp −88.6°C); and of the propanols and glycerol, propane (bp −42.1°C).

Hydrogen bonding is responsible for the fact that an alcohol has a higher boiling point than its parent hydrocarbon.

stronger intermolecular forces holding the molecules together. Another factor is also present: as the molecules become larger, higher boiling points result, since more energy is required to change the longer-chain molecules from the liquid to the gaseous phase, owing in part to the larger van der Waals attraction. A graph showing the boiling points of the normal alcohols (straight carbon chains with the —OH group on an end carbon) as a function of chain length is given in Figure 8–3.

Methanol, like water, has an —OH group, and some hydrogen bonding is to be expected, as shown in Figure 8–4. Hydrogen bonding explains why methanol (molecular weight = 32) is a liquid, whereas propane (C_3H_8, molecular weight = 44), an even heavier molecule, is a gas at room temperature. Methanol has only one hydrogen through which it can hydrogen bond, while water can hydrogen bond from either of its two hydrogen atoms. Thus, water, with more extensive intermolecular bonding, has the higher boiling point even though it is made up of lighter molecules.

Methanol and ethanol oxidize easily to acids, and so corrode the radiator.

Both methanol and ethanol can be used as antifreeze, but they tend to distill out of the coolant at the temperatures of a hot gasoline engine. Protection against freezing is then lost over a period of time. Ethylene glycol is equally effective (molecule for molecule) in lowering the freezing point of water, and its high boiling point (198°C) makes it a permanent

Figure 8–3   Boiling points of straight-chain alcohols (—OH group on an end carbon).

Figure 8—4   Hydrogen bonding in methanol.

type antifreeze. This property makes ethylene glycol more desirable even though it takes almost twice as much ethylene glycol by weight as methanol for the same amount of protection of a car's cooling system. Ethylene glycol with suitable additives to protect the radiator system is sold under a number of brand names. The higher boiling point of ethylene glycol is readily explained in terms of the two —OH groups per molecule and the enhanced possibility for hydrogen bonding. Glycerol, with three —OH groups per molecule, has an even higher boiling point, as well as a very high viscosity (resistance to flow).

Synthesis with Alcohols

Alcohols can serve as the starting substances for the synthesis of many other types of organic compounds. Oxidation of alcohols may yield aldehydes, ketones, or acids depending on the starting compound and the amount of oxygen added. For example, the oxidation of ethanol can be used to make acetaldehyde and acetic acid:

Alcohols can be used as the starting substances for the synthesis of other organic compounds.

$$\underset{\text{ETHANOL}}{CH_3\overset{\overset{\displaystyle H}{|}}{\underset{\underset{\displaystyle H}{|}}{C}}-OH} \xrightarrow{\text{Oxidation}} \underset{\text{ACETALDEHYDE}}{CH_3\overset{\overset{\displaystyle H}{|}}{C}=O} \xrightarrow{\text{Oxidation}} \underset{\text{ACETIC ACID}}{CH_3\overset{\overset{\displaystyle O}{\|}}{C}-OH}$$

The oxidation of 2-propanol provides the ketone, acetone:

$$\underset{\text{2-PROPANOL}}{CH_3\overset{\overset{\displaystyle H}{|}}{\underset{\underset{\displaystyle OH}{|}}{C}}CH_3} \xrightarrow{\text{Oxidation}} \underset{\text{ACETONE}}{CH_3\overset{\underset{\underset{\displaystyle O}{\|}}{}}{C}CH_3}$$

Substitution reactions produce alkyl halides from simple alcohols:

$$\underset{\text{2-PROPANOL}}{CH_3\overset{\overset{\displaystyle H}{|}}{\underset{\underset{\displaystyle OH}{|}}{C}}CH_3} + HBr \longrightarrow \underset{\text{2-BROMOPROPANE}}{CH_3\overset{\overset{\displaystyle H}{|}}{\underset{\underset{\displaystyle Br}{|}}{C}}CH_3} + H_2O$$

Organic Acids

Earlier an acid was defined as a species that has a tendency to donate hydrogen ions, or protons. We shall now consider the carboxylic acids, which contain the carboxyl group, $-C\overset{\displaystyle O}{\underset{\displaystyle OH}{\diagup}}$. The electronegative character of the $C=O$ group tends to drain electron density away from the region between the oxygen and hydrogen atoms. The partial positive charge assumed by the hydrogen then makes it possible for polar water molecules to remove the hydrogen ions from some of the carboxyl groups. The strength of an organic acid depends on the group that is attached to the carboxyl group. If the attached group has a tendency to pull electrons away from the carboxyl group, the acid is a stronger acid. For example, trichloroacetic acid is a much stronger acid than acetic acid:

Organic acids are compounds of the type $R-\overset{\displaystyle O}{\overset{\|}{C}}-OH$. They are generally weak acids.

TRICHLOROACETIC ACID   ACETIC ACID

The highly electronegative chlorine atoms withdraw electron density from the region of the carboxyl group and make the loss of the hydrogen ion even easier than in acetic acid. However, trichloroacetic acid is still weaker than a strong mineral acid such as sulfuric acid.

The ionization of carboxylic acids in water is as follows:

$$R-\overset{\displaystyle O}{\overset{\|}{C}}-OH + H_2O \rightleftharpoons R-\overset{\displaystyle O}{\overset{\|}{C}}-O^- + H_3O^+$$

They are *neutralized* by bases to form salts:

$$R-\overset{\displaystyle O}{\overset{\|}{C}}-OH + Na^+ + OH^- \longrightarrow R-\overset{\displaystyle O}{\overset{\|}{C}}-O^- Na^+ + H_2O$$

A SALT

Methanoic Acid (Formic Acid)

The simplest organic acid is methanoic acid, also called formic acid, in which the carboxyl group is attached directly to a hydrogen atom.

Weak bases such as ammonia, NH_3, which would neutralize formic acid, are used in the treatment of insect bites.

$$H-C\overset{\displaystyle O}{\underset{\displaystyle O-H}{\diagup}}$$

METHANOIC ACID
FORMIC ACID

This acid is found in ants and other insects and is part of the irritant that produces itching and swelling after a bite.

Formic acid may be prepared from its sodium salt, which is readily prepared by heating carbon monoxide (CO) with sodium hydroxide (NaOH):

$$CO + NaOH \xrightarrow[\text{6–10 min.}]{200°} HCOO^-Na^+$$

SODIUM FORMATE

If the resulting salt is mixed with a mineral acid, formic acid can be distilled from the mixture:

$$HCOO^-Na^+ + H_3O^+ + Cl^- \longrightarrow HCOOH + Na^+Cl^- + H_2O$$

| SODIUM FORMATE | HYDROCHLORIC ACID | FORMIC ACID | SODIUM CHLORIDE | |

Ethanoic Acid (Acetic Acid)

Ethanoic (acetic) acid is the most widely used of the organic acids. It is found in vinegar, an aqueous solution containing 4 to 5 percent acetic acid. Flavor and colors are imparted to vinegars by the constituents of the alcoholic solutions from which they are made. Ethanol in the presence of certain bacteria and air is oxidized to acetic acid:

Acetic acid can be made by oxidizing ethanol.

$$CH_3CH_2OH + O_2 \xrightarrow{\text{Bacteria}} CH_3COOH + H_2O$$

ETHANOL   OXYGEN   ETHANOIC ACID   WATER
(ACETIC ACID)

The bacteria, called mother of vinegar, form a slimy growth in a vinegar solution. The growth of bacteria can sometimes be observed in a bottle of commercially prepared vinegar after it has been opened to the air.

Acetic acid is an important starting substance for making textile fibers, vinyl plastics, and other chemicals, and is a convenient choice when a cheap organic acid is needed.

Fatty Acids

A fatty acid contains a carboxyl group attached to a long hydrocarbon chain. The chains often contain only single carbon-carbon bonds, but may contain carbon-carbon double bonds as well. Examples are stearic acid, palmitic acid, and oleic acid.

Fatty acids may be obtained from animal and vegetable fats or oils. The carbon chain in fatty acids is generally 8 to 18 carbon atoms in length.

$$CH_3CH_2CH_2CH_2CH_2CH_2CH_2CH_2CH_2CH_2CH_2CH_2CH_2CH_2CH_2CH_2CH_2C\overset{\displaystyle O}{\underset{\displaystyle OH}{}}$$

STEARIC ACID, $CH_3-(CH_2)_{16}-COOH$

$$CH_3CH_2CH_2CH_2CH_2CH_2CH_2CH_2CH_2CH_2CH_2CH_2CH_2CH_2CH_2C\overset{\displaystyle O}{\underset{\displaystyle OH}{}}$$

PALMITIC ACID, $CH_3-(CH_2)_{14}-COOH$

$$CH_3CH_2CH_2CH_2CH_2CH_2CH_2CH_2CH=CHCH_2CH_2CH_2CH_2CH_2CH_2CH_2C\overset{\displaystyle O}{\underset{\displaystyle OH}{}}$$

OLEIC ACID, $CH_3(CH_2)_7CH=CH(CH_2)_7COOH$

Stearic acid is obtained by the hydrolysis of animal fat, palmitic acid results from the hydrolysis of palm oil, and oleic acid is obtained from olive oil. These reactions are given in the next two sections. Stearic and palmitic acids are especially important in the manufacture of soaps.

1. Which of the following hydrocarbons would be expected to have the highest octane rating? **SELF-TEST 8-A**

$CH_3CH_2CH_2CH_2CH_2CH_2CH_3$
(a)

$$CH_3CH_2-\overset{\overset{\displaystyle CH_3}{|}}{CH}-CH_2CH_2CH_3$$
(b)

$$CH_3-\overset{\overset{\displaystyle CH_3}{|}}{\underset{\underset{\displaystyle CH_3}{|}}{C}}-\overset{\overset{\displaystyle CH_3}{|}}{\underset{\underset{\displaystyle H}{|}}{C}}-CH_3$$
(c)

2. Name the following compounds:

(a) CH_3OH _____

(b) CH_3CH_2OH _____

(c) $HCOOH$ _____

(d) $CH_3CH_2\overset{\underset{\underset{\displaystyle OH}{|}}{}}{C}HCH_3$ _____

(e) CH_3COOH _____

(f) $\overset{\displaystyle CH_2}{\underset{\underset{\displaystyle OH}{|}}{}}-\overset{\displaystyle CH_2}{\underset{\underset{\displaystyle OH}{|}}{}}$ _____

(g) CH_3CHO _____

(h) $CH_3\overset{\underset{\underset{\displaystyle Br}{|}}{}}{C}HCH_3$ _____

3. Two methods by which ethanol is made on a large scale are:

(a) _____

(b) _____

4. An example of a carboxylic acid that contains a long hydrocarbon chain is

_____ .

5. Identify the functional groups present in each of the following molecules:

(a) $R-OH$ _____

(b) $R-\overset{\overset{\displaystyle O}{\|}}{C}-OH$ _____

(c) $R-\overset{\overset{\displaystyle O}{\|}}{C}-H$ _____

(d) $R-\overset{\overset{\displaystyle O}{\|}}{C}-R'$ _____

TABLE 8–5 Some Alcohols, Acids, and Their Esters

Alcohol	Acid	Ester	Odor of the Ester
$CH_3CHCH_2CH_2OH$ | CH_3 **ISOPENTYL ALCOHOL**	CH_3COOH **ACETIC ACID**	$CH_3CHCH_2CH_2-O-CCH_3$ | | CH_3 O **ISOPENTYL ACETATE**	Banana
$CH_3CHCH_2CH_2OH$ | CH_3 **ISOPENTYL ALCOHOL**	$CH_3CH_2CH_2CH_2COOH$ **n-VALERIC ACID**	$CH_3CHCH_2CH_2-O-C-CH_2CH_2CH_3$ | | CH_3 O **ISOPENTYL n-VALERATE**	Apple
$CH_3CH_2CH_2CH_2OH$ **n-BUTYL ALCOHOL**	$CH_3CH_2CH_2COOH$ **n-BUTYRIC ACID**	$CH_3CH_2CH_2CH_2-O-C-CH_2CH_2CH_3$ | O **BUTYL n-BUTYRATE**	Pineapple
CH_3CHCH_2OH | CH_3 **ISOBUTYL ALCOHOL**	CH_3CH_2COOH **PROPIONIC ACID**	$CH_3CHCH_2-C-CH_2CH_3$ | | CH_3 O **ISOBUTYL PROPIONATE**	Rum

Esters

In the presence of strong mineral acids, organic acids react with alcohols to form compounds called *esters*. For example, when ethyl alcohol is mixed with acetic acid in the presence of sulfuric acid, ethyl acetate is formed. This reaction is a dehydration in which sulfuric acid acts as a catalyst and dehydrator.

Organic esters are compounds of the type

$$R-O-\underset{O}{\overset{||}{C}}-R'$$

formed by the reaction of organic acids and alcohols.

$$CH_3CH_2O\overline{[H + HO]}\underset{O}{\overset{||}{C}}CH_3 \xrightleftharpoons{H_2SO_4} CH_3CH_2O\underset{O}{\overset{||}{C}}CH_3 + H_2O$$

ETHYL ACETATE

Ethyl acetate is a common solvent for lacquers and plastics, and is often used as fingernail polish remover.

Some of the odors of common fruits are due to the presence of mixtures of volatile esters (Table 8–5). In contrast, higher molecular weight esters often have a distinctly unpleasant odor.

Fats, Oils, and Soaps

Fats and oils are esters of glycerol (glycerin) and a fatty acid. R, R′, and R″ stand for the hydrocarbon chains of the acids in the following equation.

Fats and oils are esters of fatty acids and glycerol. Fats are solids, and oils are liquids.

$$
\begin{array}{l}
CH_2-OH \\
| \\
CH-OH \\
| \\
CH_2-OH
\end{array}
+
\begin{array}{l}
\overset{O}{\overset{||}{HO-C-R}} \\[4pt]
\overset{O}{\overset{||}{HO-C-R'}} \\[4pt]
\overset{O}{\overset{||}{HO-C-R''}}
\end{array}
\rightleftharpoons
\begin{array}{l}
CH_2-O-\overset{O}{\overset{||}{C}}-R \\[4pt]
CH-O-\overset{O}{\overset{||}{C}}-R' \\[4pt]
CH_2-O-\overset{O}{\overset{||}{C}}-R''
\end{array}
+ \ 3H_2O
$$

GLYCEROL (1 MOLECULE)   FATTY ACID (3 MOLECULES WHICH MAY OR MAY NOT BE THE SAME)   FAT OR OIL (1 MOLECULE)   WATER (3 MOLECULES)

Lipids are soluble in fats and oils.

The term *fat* is usually reserved for solid glycerol esters (butter, lard, tallow) and *oil* for liquid esters (castor, olive, linseed, tung, and so forth). The term *lipid* includes fats, oils, and fat-soluble compounds.

Saturation (all single bonds with maximum hydrogen content) in the carbon chain of the fatty acids is usually found in solid or semi-solid fats, whereas unsaturated fatty acids (containing one or more double bonds) are usually found in oils. Hydrogen can be catalytically added to the double bonds of an oil to convert it into a semi-solid fat. For example, liquid soybean and other vegetable oils are hydrogenated to produce cooking fats and margarine.

Consumers in Europe and North America have historically valued butter as a source of fat. As the population increased, the advantages of a substitute for butter became apparent, and efforts to prepare such a product began about a hundred years ago. One problem that arose was the fact that common fats are almost all *animal* products with very pronounced tastes of their own. Analogous compounds from vegetable oils, which are bland or have mixed flavors, were generally *unsaturated* and consequently *oils*. A solid fat could be made from the much cheaper vegetable oils if an inexpensive way could be discovered to add hydrogen across the double bonds. After extensive experiments, many catalysts were found, of which finely divided nickel is among the most effective. The nature of the process can be illustrated by the reaction:

Catalytic hydrogenation can convert a liquid oil into a solid fat.

$$H_2C-O-\underset{\underset{O}{\parallel}}{C}-(CH_2)_7CH=CH(CH_2)_7CH_3 \qquad\qquad H_2C-O-\underset{\underset{O}{\parallel}}{C}-(CH_2)_7CH_2CH_2(CH_2)_7CH_3$$

$$HC-O-\underset{\underset{O}{\parallel}}{C}-(CH_2)_7CH=CH(CH_2)_7CH_3 \xrightarrow[\sim 200°C]{H_2,\ Ni} HC-O-\underset{\underset{O}{\parallel}}{C}-(CH_2)_7CH_2CH_2(CH_2)_7CH_3$$

$$H_2C-O-\underset{\underset{O}{\parallel}}{C}-(CH_2)_7CH=CH(CH_2)_7CH_3 \qquad\qquad H_2C-O-\underset{\underset{O}{\parallel}}{C}-(CH_2)_7CH_2CH_2(CH_2)_7CH_3$$

TRIOLEIN, A LIQUID OIL   TRISTEARIN, A SOLID FAT

Oils commonly subjected to this process include those from cottonseed, peanut, corn germ, soybean, coconut, and safflower seeds. In recent years, as it became apparent that saturated fats may encourage disease of the heart and arteries, soft margarines and cooking oils (which still contain some of the unhydrogenated fatty acid) have been placed on the market.

Soaps can be made by treating fats or oils with sodium hydroxide.

Naturally occurring fats and oils can be hydrolyzed in strongly basic solutions to form glycerol and salts of the fatty acids. Such hydrolysis reactions are called **saponification** reactions; the sodium or potassium salts of the fatty acids formed are **soaps**. Pioneers prepared their soap by boiling animal fat with an alkaline solution obtained from the ashes of hard wood. The resulting soap could be "salted out" by adding sodium chloride, making use of the fact that soap is less soluble in a salt solution than in water.

$$CH_3(CH_2)_{16}COO-CH_2$$
$$CH_3(CH_2)_{16}COO-CH + 3NaOH \longrightarrow 3CH_3(CH_2)_{16}COO^-Na^+ + HO-CH$$
$$CH_3(CH_2)_{16}COO-CH_2 \qquad\qquad\qquad\qquad HO-CH_2$$

TRISTEARIN   SODIUM STEARATE   GLYCEROL
(GLYCERYL TRISTEARATE)   (A SOAP)

The cleansing action of soap can be explained in terms of its molecular structure. Material that is water soluble can be readily removed from the skin or a surface by simply washing with an excess of water. To remove a sticky sugar syrup from one's hands, the sugar is dissolved in water and rinsed away. Many times the material to be removed is oily and water will merely run over the surface of the oil. Since the skin has natural oils, even substances such as ordinary dirt that are not oily themselves can cover the skin in a greasy layer. The cohesive forces (forces between like molecules tending to hold them together) within the water layer are too large to allow the oil and water to intermingle (Figure 8–5). When present in an oil-water system, soap molecules such as sodium stearate,

$$CH_3CH_2CH_2CH_2CH_2CH_2CH_2CH_2CH_2CH_2CH_2CH_2CH_2CH_2CH_2CH_2CH_2C \overset{\displaystyle O}{\underset{\displaystyle O^-Na^+}{\Bigg\langle}}$$

will move to the interface between the two liquids. The carbon chain, which is a nonpolar organic structure, will readily mix with the nonpolar grease molecules whereas the highly polar $-COO^-Na^+$ group enters the water layer because it becomes hydrated (Figure 8–5b). The soap molecule will then tend to lie across the oil-water interface. The grease is then

Figure 8–5 The cleansing action of soap. (a) A piece of glass coated with grease inserted in water gives evidence for the strong adhesion between water and glass at 1, 2 and 3. The water curves up against the pull of gravity to wet the glass. The relatively weak adhesion between oil and water is indicated at 4 by the curvature of the water away from the grease against the force tending to level the water. (b) A soap molecule, having oil-soluble and water-soluble ends, will become oriented at an oil-water interface such that the hydrocarbon chain is in the oil (with molecules that are electrically similar, nonpolar) and the COO^-Na^+ group is in the water (highly charged polar groups interacting electrically). (c) In an idealized molecular view, a grease particle, 1, is surrounded by soap molecules which in turn are strongly attracted to the water. At 2 another droplet is about to break away. At 3 the grease and clean glass interact before the water moves between them.

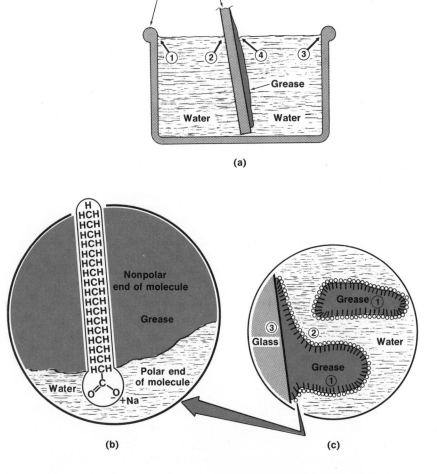

Soap, water, and oil together form an *emulsion*, with the soap acting as the emulsifying agent.

broken up into small droplets by agitation, each surrounded by hydrated soap molecules (Figure 8—5c). The surrounded oil droplets cannot come together again since the exterior of each is covered with $—COO^-Na^+$ groups which strongly interact with the surrounding water. If enough soap and water are available, the oil will be swept away, forming a clean and water-wet surface.

SYNTHETIC DETERGENTS (SYNDETS)

Ordinary soaps give precipitates with hard water ions such as Ca^{2+}, Mg^{2+} and Fe^{3+}.

If Ca^{2+} or Mg^{2+} ions are present in water, an ordinary soap will precipitate as an insoluble salt of these ions and the water is said to be ***hard.*** The ring on the bathtub and the scum in the washer are visible signs of this precipitate. Only after all these interfering ions have been precipitated can the added soap cause the water to form suds. Water softeners remove the hard water ions, replacing them with ions such as Na^+ which do not interfere with the soap's action, but this is often an expensive process. Synthetic detergents (syndets) are similar to soaps in that they are composed of molecules having water-soluble and oil-soluble ends, but syndets have an advantage over soaps in hard water in that they do not form precipitates with Ca^{2+} or Mg^{2+}. The synthesis of a typical syndet molecule with a water-soluble sodium sulfonate ($—SO_3^-Na^+$) group and an oil-soluble hydrocarbon group is shown below.

$$CH_3(CH_2)_{11}OH \xrightarrow{H_2SO_4} CH_3(CH_2)_{11}OSO_3H \xrightarrow{NaOH} CH_3(CH_2)_{11}OSO_3^-Na^+$$

LAURYL ALCOHOL   LAURYL HYDROGEN SULFATE   SODIUM LAURYL SULFATE
(AN INGREDIENT OF
SOME SHAMPOOS)

The most widely used syndets are sodium salts of substituted benzenesulfonic acids. Normally, a long hydrocarbon chain, designated $—R$, is attached to the benzene ring by a substitution reaction.

SO$_3^-$Na$^+$

R

SODIUM SALT OF A
BENZENESULFONIC ACID

Syndets are effective cleansing agents, but they pose problems in water treatment and purification. By careful choice of starting materials, chemists have been able to produce syndets which, once "used" can be broken down into simpler molecules by the action of bacteria. R must be a straight-chain hydrocarbon for proper breakdown. These biodegradable syndets have been used almost exclusively in the United States since about 1965.

Figure 8—6 Organic syntheses begin on a small scale in the laboratory (*top*), but graduate to a very large scale if they become commercially important. The top view shows a laboratory bench. The bottom view shows an automatic, modern chemical plant. In both cases the chemical reactions are the same; the scale is different. (*bottom*, courtesy of the Du Pont Company, Textile Fiber Department)

USEFUL PRODUCTS FROM AROMATIC ORGANIC REACTIONS

Many chemists are engaged in the synthesis of organic compounds. In educational and industrial laboratories throughout the world they prepare new and different compounds on a small scale. If the new compound has commercial value, the preparation is subsequently adapted for full scale plant operations (Figure 8–6).

The thousands of chemical changes required to synthesize the many known organic compounds have a few characteristics in common:

Organic syntheses are usually carried out so that only one part of the molecule changes in each reaction step.

1. Several chemical changes are usually required to synthesize a single organic compound. Each chemical change is called a step, and each step produces an intermediate compound that is used in the next step. The final step produces the desired end product.
2. Normally only one functional group undergoes change in each step. The rest of the molecule remains intact and unchanged. This is known as the ***principle of minimum structural change.***
3. From one principal starting substance can come many diverse products. The kind of product obtained depends upon the reactants and the conditions imposed.
4. The more steps in the synthesis, the lower the percentage yield of the final product. The starting substance and each intermediate are only partially converted to the next intermediate because of equilibrium considerations, side reactions, or both, which convert some of the starting substance into undesirable products. If an intermediate product must be removed and purified before proceeding to the next step, some additional material is lost. The principal purification methods are recrystallization and extraction for solids and distillation for liquids.

These principles will be illustrated by the preparation of chlorobenzene from a natural product, benzene, and then the use of chlorobenzene to prepare such useful products as aspirin and oil of wintergreen.

Chlorobenzene can be prepared from benzene and chlorine in the presence of a suitable catalyst, such as iron(III) chloride.

BENZENE   CHLOROBENZENE

Chlorobenzene is manufactured on an enormous scale and is widely used in industrial processes to prepare other organic compounds.

Aspirin and Oil of Wintergreen

The series of steps required to synthesize aspirin and oil of wintergreen from chlorobenzene is outlined in Figure 8–7. Each structure represents the beginning or end of a step in the synthesis. Only the principal organic substance is shown for each step. Other products such as NaCl (a coproduct with phenol) and water (a coproduct with sodium phenoxide) are sometimes important in the synthesis because they have to be removed to avoid interference with subsequent steps. However, in giving a broad outline of the synthetic process, the coproducts are generally omitted; only those products made in a previous step and required for subsequent steps are included. In Figure 8–7 the step-by-step structural changes can be followed by noting groups in color. Conditions and additional reactants for each step are written with the arrow. These conventions are generally used to summarize organic syntheses.

Some intermediates are useful compounds in their own right. Phenol, commonly called carbolic acid, is used to prepare plastics such as Bakelite, drugs, dyes, and other compounds. Phenol also has medical application as a topical anesthetic for some types of lesions and in the treatment of mange and colic in animals. Methyl salicylate, or oil of wintergreen, is used as a flavoring agent and as a component of rubbing alcohol for sore muscles.

The conversion of phenol into sodium phenoxide is an acid-base reaction. Phenol, with an acidic hydrogen in the hydroxyl group, reacts with a base, sodium hydroxide, to give a salt, sodium phenoxide, and water. The reaction of salicylic acid to form oil of wintergreen is an *esterification*. The organic acid reacts with an alcohol in the presence of strong mineral acid to produce an ester, methyl salicylate, and water.

Obviously, it is beyond the scope of this text to give an extensive overview of organic synthesis. You can be assured, however, that if the "miracle cancer drug" is found or whenever any other new and useful compounds are discovered, their production will involve step-by-step molecular modifications to produce the desired product. An exciting aspect of organic synthesis is the prediction of desired properties of new molecular arrangements and then the testing of the theoretical properties. With so many possibilities yet to be discovered, one can only guess at the potential power in organic synthesis.

Any given organic compound can usually serve as the starting substance for the synthesis of many other organic compounds.

Figure 8–7 Preparation of aspirin and oil of wintergreen. A discussion of the syntheses is given in the text.

1. Complete the following equation:

$$CH_3CH_2CH_2OH + CH_3CH_2COOH \xrightarrow{H_2SO_4}$$

2. Define the term soap. _____

3. (a) What is the difference between a fat and an oil when referring to edible

   lipids? _____

   (b) How can melting points of most edible oils be increased? _____

   _____

4. In a typical synthesis of organic compounds, the starting compounds must be stripped to their atoms before new compounds can be made. (True/False)

5. Aspirin and oil of wintergreen
   (a) are structurally similar.
   (b) are both acids.
   (c) can be made from chlorobenzene.

6. Phenol is the same as (carbonic/carbolic) acid.

_____ 1. Synthesized from NH_4OCN by Wöhler	(a) measures knocking behavior in engine	**MATCHING SET**
_____ 2. RCOOH	(b) breaks hydrocarbons into smaller molecules	
_____ 3. R—OH	(c) antiknocking agent	
_____ 4. R—O—C—R' ‖ O	(d) aldehyde	
_____ 5. octane rating	(e) carboxylic acid	
_____ 6. RCHO	(f) soap	
_____ 7. cracking	(g) urea	
_____ 8. $Pb(C_2H_5)_4$	(h) alcohol	
_____ 9. sodium stearate	(i) ester	

---

1. Would you expect $CH_3CH_2CH_2CH_2CH_2CH_2CH_2CH_2CH_3$ to be an important useful constituent of bottled gas? Explain your answer. Would it be useful in gasoline? Explain.   QUESTIONS

2. Indicate what products would be formed in the reaction of the following:

    (a) methanol and acetic acid
    (b) 1-propanol and stearic acid
    (c) ethylene glycol and acetic acid

3. Explain how hydrogen bonding could play a significant role in fixing the boiling point of acetic acid.

4. Write structural formulas for the four alcohols with the composition C_4H_9OH.

5. Draw a structural formula for each of the following:

    (a) an alcohol
    (b) an organic acid
    (c) an ester
    (d) glycerol

6. What is meant by each of the following terms?

    (a) proof rating of an alcohol
    (b) octane rating of a gasoline
    (c) denatured alcohol

7. Beginning with petroleum, outline the steps and write the chemical equations for the production of ethyl acetate.

8. How does a soap cleanse?

9. Discuss the fundamental characteristics of the following:

    (a) synthetic detergents
    (b) soaps

10. Which would you expect to boil at a higher temperature, $CH_3CH_2CH_2CH_3$ or CH_3CH_2OH? Why?

11. Why will the addition of strong alkali aid in unstopping a greasy sink drain?

12. Draw structural formulas for:

    (a) aspirin
    (b) oil of wintergreen
    (c) phenol

13. What chemical reactions can be used to distinguish between:

    (a) C_2H_5OH and CH_3COOH
    (b) $CH_3COOC_2H_5$ and CH_3COOH

14. Tell how you could prepare the following compounds:

    (a) chlorobenzene
    (b) urea
    (c) methanol

15. What hydrocarbons are gases in their natural state?

16. What is the use of tetraethyllead in gasoline?

17. Explain the common names for methanol and ethanol.

18. Is pure synthetic ethanol different from pure grain alcohol? Explain.

19. Which propanol is used as rubbing alcohol?

20. How many hydrogen bonds are possible per molecule of methanol? of ethylene glycol? of glycerol?

21. What is the primary chemical in permanent antifreeze?

22. What is the acid in vinegar?

MAN-MADE GIANT MOLECULES— THE SYNTHETIC POLYMERS

It is impossible for most Americans to get through a day without using a dozen or more materials based on synthetic *polymers.* Many of these materials are *plastics* of one sort or another. Examples of these include plastic dishes and cups, combs, automobile steering wheels and seat covers, telephones, pens, plastic bags for food and wastes, plastic pipes and fittings, plastic water-dispersed paints, false eyelashes and wigs, a wide range of synthetic fibers for clothing, synthetic glues, and flooring materials. In fact, these materials are so widely used they are usually taken for granted. All these materials are composed of *giant molecules.*

This "flood of plastic objects" did not arise out of whim or malice; it slowly became necessary over a period of 30 or 40 years because (1) natural resources dwindled and so many materials became scarce, (2) with rising labor costs, many items could be made less expensively by molding than by whittling, shaving, sawing, and gluing, and (3) the new materials were so superior in properties that they did the job better.

Some of our most useful polymer chemistry has been to copy giant molecules found in nature. Rayon is remanufactured cellulose. Synthetic rubber is copied from natural latex rubber. As useful as they may be, however, polymer chemistry is not restricted to nature's models. To name a few, nylon, Dacron, and polycarbonates do not have their exact duplicates in nature. We have gone to school on nature and extended our knowledge to new situations.

The purpose of this chapter is to examine some of the chemistry and technology that lie behind a few of these "indispensable" plastic objects.

A plastic is a substance that will flow under heat and pressure, and hence is capable of being molded into various shapes. All plastics are polymers, but not all polymers are plastic.

206

WHAT ARE GIANT MOLECULES?

Many chemists were reluctant to accept the concept of giant molecules, but in the 1920's a persistent German chemist, Hermann Staudinger (1881–1965; Nobel Prize, 1953), championed the idea and introduced a new term, **macromolecule**, for these giant molecules. Staudinger devised experiments that yielded accurate molecular weights, and, in addition, he synthesized "model compounds" to test his theory. One of his first model compounds was prepared from styrene.

A macromolecule is a molecule with a very high molecular weight.

$$H_2C = CH$$

STYRENE

Styrene is the number twenty commercial chemical.

Under the proper conditions, styrene molecules use the "extra" electrons of the double bond to undergo a **polymerization** reaction to yield polystyrene, a material composed of giant molecules. The word **polymer** means "many parts" (Greek: *poly* = many, *meros* = parts). The molecules of styrene are the **monomers** (Greek: *mono* = one); they provide the recurring units in the giant molecule analogous to identical railroad cars coupled together to make a long train.

A polymer has molecules composed of a large number of similar units.

The macromolecule polystyrene is represented as a long chain of monomer units bonded to each other. Each unit is bonded to the next by a strong covalent bond. The polymer chain is not an endless one; some polystyrenes made by Staudinger were found to have molecular weights of about 600,000, corresponding to a chain of about 5,700 styrene units. The polymer chain can be indicated as

$$R - CH_2 - CH - (CH_2 - CH - CH_2 - CH - R)_n$$

where R represents some terminal group, often an impurity, and n is a large number.

Polystyrene is a clear, hard, colorless solid at room temperature. Since it can be molded easily at 250°C, the term **plastic** has become associated with it and similar materials. Polystyrene has so many useful properties that its commercial production, which began in Germany in 1929, today exceeds 5 billion pounds per year. It is used to make combs, bowls, toys, electrical parts, and many other items such as tough plastics used for radio and TV cabinets and synthetic rubber for tires.

Synthetic polymers are commonly called *plastics* when in a solid form.

There are two broad categories of plastics. One, when heated repeatedly, will soften and flow; when it is cooled, it hardens. Materials which undergo such reversible changes when heated and cooled are called **thermoplastics;** polystyrene is one example. The other type is plastic when first heated, but when heated further it forms a set of interlocking bonds. When reheated, it cannot be softened and reformed without extensive degradation. These materials are called **thermosetting plastics**

Thermoplastic polymers can be repeatedly softened by merely heating.

and include Bakelite and rigid-foamed polyurethane, a polymer which is finding many new uses as a construction material.

In order to gain a better understanding of polymers it is necessary to look at representative examples of the different types of polymerization processes.

Thermosetting polymers form cross-linking bonds when heated and then become rigid.

ADDITION POLYMERS

In the previous section it was noted that some polymers, such as polystyrene, are made by adding monomer to monomer to form a polymer chain of great length. Perhaps the easiest addition reactions to understand chemically are those involving monomers containing double bonds. The simplest monomer of this group is ethylene, C_2H_4. When ethene (ethylene) is heated under pressure in the presence of oxygen, polymers with molecular weights of about 30,000 are formed. In order to enter into reaction, the double bond of ethylene must be broken. This forms *reactive sites* composed of unpaired electrons at either end of the molecule.

Ethene (ethylene) is the number six commercial chemical.

$$\underset{\substack{|\\H}}{\overset{\substack{H\\|}}{C}} = \underset{\substack{|\\H}}{\overset{\substack{H\\|}}{C}} \xrightarrow{\text{Energy}} \cdot \underset{\substack{|\\H}}{\overset{\substack{H\\|}}{C}} - \underset{\substack{|\\H}}{\overset{\substack{H\\|}}{C}} \cdot$$

REACTIVE SITE

The partial breaking of the double bond can be accomplished by physical means such as heat, ultraviolet light, X rays, and high-energy electrons. The *initiation* of the polymerization reaction can also be accomplished with chemicals such as organic peroxides. These initiators, which are very unstable, break apart into pieces with unpaired electrons. These fragments (called *free radicals*) are ravenous in trying to find a "buddy" for their unpaired electrons. They react readily with molecules containing carbon-carbon double bonds.

$$\overset{\frown}{CH_2} \overset{\bullet\bullet}{\frown} CH_2 \xrightarrow[\text{(peroxide)}]{\bullet OR} \dot{C}H_2 - CH_2OR \xrightarrow{nCH_2=CH_2} \left(CH_2 - CH_2\right)_{\overline{n+1}} OR$$

n is a very large number, 40,000 or more

The extension of the polyethylene chain summarized above comes from the unpaired electron bonding to an electron in an unreacted ethylene molecule. This leaves another unpaired electron to bond with yet another ethylene molecule. For example,

$$ROCH_2 - CH_2\cdot + \overset{\frown}{C}H_2 \overset{\bullet\bullet}{\frown} CH_2 \longrightarrow ROCH_2 - CH_2 - CH_2 - CH_2\cdot$$

Polyethylenes formed under various pressures and catalytic conditions have different molecular structures and hence different physical properties. For example, chromium oxide as a catalyst yields almost exclusively the linear polyethylene shown in the margin. Actually, a methyl group is attached to about every eighth or tenth carbon in the chain. If ethylene is heated to 230°C at a pressure of 200 atm, irregular branches result. Under these conditions, free radicals undoubtedly attack the chain at random positions, thus causing the irregular branching.

A portion of a polyethylene polymer. Ethylene is unsaturated; polyethylene is saturated.

$$-CH_2-CH_2-CH_2-CH_2- \longrightarrow -CH_2-CH-CH_2-CH_2- + H\cdot$$

$$RCH_2CH_2\cdot$$

$$
\begin{array}{c}
CH_2 \\
| \\
CH_2 \\
| \\
R
\end{array}
$$

BRANCHED POLYMER CHAINS

The molecules in linear polyethylene can line up with one another very easily, yielding a tough, high-density crystalline compound that is useful in making toys, bottles, and structural parts. The polyethylene with irregular branches is less dense, more flexible, and not nearly as tough as the linear polymer, since the molecules are generally farther apart and their arrangement is not as precisely ordered. This material is used for squeeze bottles and other similar applications.

There is a large group of derivatives of ethylene that undergo addition polymerization. Table 9–1 summarizes some information on these materials.

TABLE 9–1   Ethylene Derivatives Which Undergo Addition Polymerization

Formula	Monomer Name	Polymer Name	Uses
$CH_2{=}CH_2$	Ethylene	Polyethylene	coats, milk cartons, wire insulation, bread wrappers, toys, films
$HC{=}CH_2$ (phenyl)	Styrene	Polystyrene	synthetic rubber, combs, toys, bowls, packaging, appliance parts
$CH_2{=}CHCl$	Vinyl chloride (Number nineteen commercial chemical.)	Polyvinylchloride (PVC)	as a vinyl acetate copolymer in phonograph records, credit cards, rain wear, pipes, adhesives, films
$CH_2{=}CH$ $\mid$ $O-C-CH_3$ $\parallel$ O	Vinyl acetate	Polyvinylacetate	latex paint
$CH_2{=}CH$ $\mid$ CN	Acrylonitrile	Polyacrylonitrile (PAN)	rug fibers, high-impact plastics
$CH_2{=}CH-CH{=}CH_2$	Divinyl (1,3-Butadiene)	Buna rubbers	tires and hoses
$CH_2{=}C-C$ with CH_3 and O above, $O-CH_3$	Methyl methacrylate	Polymethyl methacrylate (Plexiglas, Lucite)	transparent objects, lightweight "pipes"
$CF_2{=}CF_2$	Tetrafluoroethylene	Polytetrafluoroethylene (TFE) (Teflon)	insulation, bearings, nonstick fry pan surfaces

Synthetic "Natural" Rubber—A Tailor-Made Addition Polymer

A very interesting application of stereochemical control over polymerization is the manufacture of synthetic rubber. When several structures are possible and only one is desired, stereochemical control must be exercised.

Natural rubber is a hydrocarbon with the composition $(C_5H_8)_n$, and when it is decomposed in the absence of oxygen it yields isoprene:

$$CH_2{=}\overset{\overset{\displaystyle CH_3}{|}}{C}{-}CH{=}CH_2$$
ISOPRENE

Natural rubber occurs as latex (an emulsion of rubber particles in water) that oozes from rubber trees when they are cut. Precipitation of the rubber particles yields a gummy mass that is not only elastic and water-repellent but also very sticky, especially when warm. In 1839, after 10 years' work on this material, Charles Goodyear (1800–1860) discovered that heating gum rubber with sulfur produced a material that was no longer sticky, but still elastic, water-repellent, and resilient.

Vulcanized rubber, as Goodyear called his product, contains short chains of sulfur atoms that bond together the polymer chains of the natural rubber and reduce its unsaturation. The sulfur chains help to align the polymer chains, so the material does not undergo a permanent change when stretched, but springs back to its original shape and size when the stress is removed (Figure 9–1).

In later years chemists searched for ways to make a synthetic rubber

Rubber is vulcanized by heating it with sulfur, which forms links between the polymer chains.

VULCANIZATION:

$$\begin{array}{c} \diagdown \\ C{=}C \\ \diagup \quad \diagdown \end{array}$$
+

Sulfur, or S_8 molecule

+

$$\begin{array}{c} \diagdown \\ C{=}C \\ \diagup \quad \diagdown \end{array}$$

↓

$$-\overset{|}{\underset{|}{C}}-\overset{|}{\underset{S}{C}}-$$

$$(S)_6$$

$$-\overset{|}{\underset{|}{C}}-\overset{|}{\underset{|}{C}}-$$

a. Before stretching

Figure 9–1 Stretched vulcanized rubber retains its elasticity

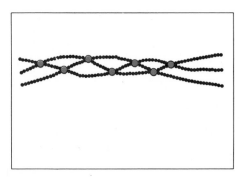

b. Stretched

$$CH_2=CH-CH=CH_2$$
Butadiene

so we would not be completely dependent on imported natural rubber during emergencies, such as during the first years of World War II. In the mid-1920's, German chemists polymerized butadiene (obtained from petroleum and structurally similar to isoprene, but without the methyl group side chain). The product was Buna rubber, so named because it was made from butadiene (Bu—) and catalyzed by sodium (—Na).

The behavior of natural rubber (polyisoprene), it was learned later, is due to the specific arrangement within the polymer chain. We can write the formula for polyisoprene with the CH_2 groups on opposite sides of the double bond (the *trans* arrangement),

POLY-*TRANS*-ISOPRENE (THE —CH_2—CH_2— GROUPS ARE *TRANS*)

or with the CH_2 groups on the same side of the double bond (the *cis* arrangement, from Latin meaning "on this side").

POLY-*CIS*-ISOPRENE (THE —CH_2—CH_2— GROUPS ARE *CIS*)

Natural rubber is poly-*cis*-isoprene.

Natural rubber is poly-*cis*-isoprene. However, the *trans* material also occurs in nature, in the leaves and bark of the sapotacea tree, and is known as *gutta-percha*. It is used as a thermoplastic for golf ball covers, electrical insulation and other such applications. Without an appropriate catalyst, polymerization of isoprene yields a solid that is like neither rubber nor gutta-percha. Neither the *trans* polymer nor the randomly arranged material is as good as natural rubber (*cis*) for making automobile tires.

Stereoregulating: controlling the arrangement of atoms in the polymer.

In 1955, chemists at the Goodyear and Firestone companies discovered, almost simultaneously, how to use stereoregulation catalysts to prepare synthetic poly-*cis*-isoprene. This material is, therefore, structurally identical to natural rubber. Today, synthetic poly-*cis*-isoprene can be manufactured cheaply and is used almost equally well (there is still an increased cost) when natural rubber is in short supply. More than 2.4 million tons of synthetic rubber are produced in the United States yearly.

Copolymers

After examining Table 9–1, one might well wonder what would happen if a mixture of two monomers was polymerized. If we polymerize pure monomer A, we get a *homopolymer,* poly A:

—AAAAAAAAAA—

Likewise, if pure monomer B is polymerized, we get a homopolymer, poly B:

—BBBBBBBBBB—

In contrast, if the monomers A and B are mixed and then polymerized, we get *copolymers* such as the following:

A copolymer is made by polymerizing two or more different monomers together.

$$-AABABAAABB-$$

$$-AABABABABB-$$

$$-BABABBAABA-$$

In such polymers the order of the units is often completely random, in which case the properties of the copolymer will be determined by the ratio of the amount of A to the amount of B.

A copolymer can have useful properties that are different from and often superior to those of the polymers of its pure constituents. As an example, consider synthetic rubbers again. During World War II it was apparent to our military planners that we would be hard-pressed if our rubber supplies from Asia were cut off by Japan. A crash program was begun to develop synthetic rubber that would be as good as natural rubber. The Germans had earlier polymerized styrene, but this is a hard thermoplastic with little elasticity. They had also polymerized butadiene to make the first synthetic rubber (Buna rubber), although it was not very serviceable. American chemists found, however, that a 1 to 3 copolymer of styrene and butadiene possessed properties closer to those of natural rubber.

$$CH_2=CH + CH_2=CH-CH=CH_2 \longrightarrow$$

STYRENE   BUTADIENE

$$-CH_2CH=CHCH_2CH_2CCH_2CH=CHCH_2CH_2CH=CHCH_2-$$

SBR COPOLYMER (STYRENE-BUTADIENE RUBBER)

The largest group of synthetic polymers is used to make rubber—about 10 billion pounds per year. About 40% is used for tires.

The double bonds remaining in the polymer chain allow them to undergo vulcanization like natural rubber polymer chains.

Copolymers are also used extensively in the plastics industry. For example, high-impact material for radio cabinets, tools, handles, and anything that might be dropped or struck is made of a copolymer of acrylate, butadiene, and styrene. Fibrous fillers and reinforcing agents add strength. A pure form of SBR has even replaced the latex in chewing gum.

Addition Polymers and Paints

All paints involve polymers in one form or another. Popular latex and acrylic paints contain addition polymers that serve as *binders*. A paint binder forms a molecular network to hold the *pigment* (coloring agent) in place and to hold the paint to the painted surface. In oil-based paints, a drying oil (such as linseed oil) or a resin is the binder. All paints also have

In 1970, nearly 830 million gallons of coatings were sold by U.S. companies. The $3 billion sales represented 5.5% of sales of all chemicals and allied products during 1970.

Figure 9–2 Photographers photographing a painter painting—in this picture the John Quincy Adams birthplace in Quincy, Massachusetts. The Sears Great American Home Series of advertisements illustrates the importance of surface coatings as preservatives. (Courtesy of Sears, Roebuck and Co., Chicago, Illinois.)

The first commercial water-based latex paint was Glidden's Spred Satin, introduced in 1948.

Water-based latex paints reduce fire hazards and air pollution associated with the handling and application of oil-based paints.

Monomer of polyvinylacetate:

$$\underset{H}{\overset{H}{}}\!\!C=C\underset{O-C-CH_3}{\overset{H}{}}$$

a volatile solvent or thinner; this is water in water-based paints and turpentine (or mineral spirits, or both) in oil-based paints.

Early latex paints were emulsions of partly polymerized styrene and butadiene in water (Figure 9–3). Some type of emulsifying agent (such as soap) was present to keep the small drops of nonpolar styrene and butadiene dispersed in the polar water.

Immediately after the application of a latex paint, the water begins to evaporate. When some of the water is gone, the emulsion breaks down, and the remaining water evaporates quickly, leaving the paint film. Further polymerization of the styrene and butadiene follows slowly, but the paint appears to be dry in a few minutes. The pigment is trapped in the network of the polymer. If the paint is white, the pigment is probably titanium dioxide, TiO_2, which has replaced the poisonous compound "white lead," $Pb(OH)_2 \cdot 2PbCO_3$, that was used in older paints.

The styrene-butadiene resin is the least expensive binder material used, but it has a relatively long curing period, relatively poor adhesion, and a tendency to yellow with age. Polyvinylacetate is only a little more

Oil-in-water emulsion   Water-in-oil emulsion

Figure 9–3 Two kinds of emulsions. An emulsion is composed of two immiscible liquids, one dispersed as tiny droplets in the other. An emulsifying agent is required to stabilize an emulsion.

expensive and is an improvement over the styrene-butadiene resin. It quickly captured 50 percent of the latex market for interior paints. Another type with rapidly growing popularity, though about one third more expensive, includes the acrylic resins and the "acrylic latex" paints. These are more washable and much more resistant to light damage. They are especially useful as exterior paints.

The fluoropolymers, similar to Teflon, are especially promising as surface coatings because of their great stability. Fluorine atoms are substituted for hydrogen atoms in the organic structure. Metals covered with polyvinylidene fluoride carry up to a 20-year guarantee against failure from exposure.

In the past few years, paint manufacturers have begun to blend linseed oil emulsions with latex emulsions in order to take advantage of the penetrating ability of the triglyceride molecules in the linseed oil. Presently, some "latex" paints contain as much as 75 percent linseed oil emulsion, but still have the desirable characteristic of latex. Table 9–2 gives a summary of various additives to emulsion paints and the rationale for their use.

The drying of modern oil-based paints involves much more than the evaporation of the mineral spirits or turpentine solvent. The chemical reaction between a drying oil and oxygen from the air completes the drying process. Common drying oils are soybean, castor, coconut, and linseed oils; the most widely used is linseed, which comes from the seed of the flax plant. All of these oils are glyceryl esters of fatty acids, as discussed in Chapter 7. Hydrolysis of a typical linseed oil yields the following assortment of fatty acids:

| 4–7% | Palmitic acid (16 C atoms) | (Saturated) |
| 2–5% | Stearic acid (18 C atoms) | (Saturated) |

Acrylic polymers have a sheen that allows latex paint to compete in the exterior gloss market traditionally monopolized by oil-based coatings. Acrylics adhere well and control corrosion. Acrylics are polymers of acrylonitrile,

$$\underset{H}{\overset{H}{C}}=\underset{H}{\overset{CN}{C}}$$

See Tables 9–1 and 9–3.

Mineral spirits are petroleum fractions of moderate volatility.

TABLE 9–2   Additives Used in Emulsion Paints

(1)	Dispersing agents for pigments	Example: tetrasodium pyrophosphate ($Na_4P_2O_7$). The principle of like-charged particles repelling.
(2)	Protective colloids and thickeners	A thicker paint is slower to settle and drips and runs less. A protective colloid tends to stabilize the organic-water interface in the emulsion. Examples: sodium polyacrylates, carboxymethylcellulose, clays, gums. (Same mechanism as soap dispersing oil in water, Chapter 8.)
(3)	Defoamers	Foaming presents a serious problem if not corrected. The surface tension of the water must be decreased. Chemicals used: tri-n-butylphosphate, n-octyl alcohol and other higher alcohols, silicone oil.
(4)	Coalescing agents	As the water evaporates and the paint dries, an agent is needed to stick the pigment particles together. As the resin film forms, the agent evaporates. Coalescing agents must volatilize very slowly. Examples: hexylene glycol and ethylene glycol.
(5)	Freeze-thaw additives	Freezing will destroy the emulsion. Antifreezes such as ethylene glycol are used.
(6)	pH Controllers	The effectiveness of the ionic or molecular form of the emulsifier depends on the acid or alkaline conditions (pH). The wrong pH will break down the emulsion. Most paints tend to be too acidic. Ammonia, NH_3, is added to neutralize the acid.

9–38%   Oleic acid (18 C atoms)   (Unsaturated)
3–43%   Linoleic acid (18 C atoms)   (Unsaturated)
25–58%   Linolenic acid (18 C atoms) (Unsaturated)

The chemical action of oxygen on a drying oil is to replace a hydrogen atom on a carbon atom next to a C=C double bond in an unsaturated fatty acid chain. When oxygen reacts with two fatty acids on two oil molecules, the result is cross-linking between the two molecules.

Part of one molecule

$$-CH_2-CH_2-CH=CH-CH_2- \qquad -CH_2-CH-CH=CH-CH_2-$$
$$+ O_2 \longrightarrow \qquad\qquad O \text{ } ether linkage \qquad + H_2O$$
$$-CH_2-CH_2-CH=CH-CH_2- \qquad -CH_2-CH-CH=CH-CH_2-$$

Part of another molecule

The polymeric network produced by the cross-linking hardens the paint, traps the pigment, and secures the paint in the crevices of the painted surface.

Metal ions such as Zn^{2+}, Co^{2+}, Fe^{3+}, Mn^{2+}, and Ca^{2+} are added to oil-based paints to catalyze the drying process. These ions decompose peroxides (compounds containing the $-O-O-$ group) formed during the cross-linking process and precipitate free fatty acids as salts of the ions.

CONDENSATION POLYMERS

Polyesters

A chemical reaction in which two molecules react by splitting out or eliminating a small molecule is called a ***condensation reaction.*** For example, acetic acid and ethyl alcohol will react, splitting out a water molecule, to form ethyl acetate, an *ester*.

$$\underset{\text{ACETIC ACID}}{CH_3C\overset{O}{-}OH} + \underset{\text{ETHANOL}}{HOCH_2CH_3} \xrightarrow[\text{Catalyst}]{H^+} \underset{\text{ETHYL ACETATE (AN ESTER)}}{CH_3C\overset{O}{-}OCH_2CH_3} + H_2O$$

This important type of chemical reaction does not depend upon the presence of a double bond in the reacting molecules. Rather, it requires the presence of two kinds of functional groups on two different molecules. If each reacting molecule has *two* functional groups, both of which can react, it is then possible for condensation reactions to lead to a long-chain polymer. If we take a molecule with two carboxyl groups, such as terephthalic acid, and another molecule with two alcohol groups, such as ethylene glycol, each molecule can react at each end. The reaction of one acid group of terephthalic acid with one alcohol group of ethylene glycol

● Carbon atom
○ Oxygen atom

Typical molecule in linseed oil.

In a condensation polymerization, molecules are linked when they react to split out a small molecule such as water. The backbone of the polymer contains functional groups.

initially produces an ester molecule with an acid group left over on one end and an alcohol group left over on the other:

TEREPHTHALIC ACID   ETHYLENE GLYCOL

The esterification of a dialcohol and a diacid involves two positions on each molecule.

(AN ESTER)

Subsequently, the remaining acid group can react with another alcohol group, and the alcohol group can react with another acid molecule. The process continues until an extremely large polymer molecule, known as a *polyester,* is produced with a molecular weight in the range of 10,000–20,000.

Terephthalic acid is the number twenty-four commercial chemical.

POLY(ETHYLENE GLYCOL TEREPHTHALATE)

Poly(ethylene glycol terephthalate) is used in making polyester textile fibers marketed under such names as "Dacron" and "Terylene," and films such as "Mylar." The film material has unusual strength and can be rolled into sheets one-thirtieth the thickness of a human hair. In film form this polyester is often used as a base for magnetic recording tape and for packaging frozen food. Dacron (and Teflon) tubes substitute for human blood vessels in heart bypass operations. The inert, nontoxic, non-allergenic, noninflammatory, non-blood-clotting natures of these polymers make them excellent substitutes.

A typical polyester is produced from a dialcohol and a diacid.

Condensation Polymers and Baked-On Paints

If you have ever had a car repainted, perhaps you have seen the baking oven in which the paint is dried (Figure 9–4). Automobile finishes, and those on major appliances (such as refrigerators, washing machines, and stoves), require very tough, adherent paints in order to withstand abuse. The tough coating is produced by extensively cross-linked condensation polymers.

A popular type of baked-on paint is the *alkyd* variety. The term comes from a combination of the words *alc*ohol and ac*id*. Alkyds, then,

When General Motors lacquered the 1923 Oakland with a nitrocellulose lacquer, the protective coatings industry first began its expansion into the use of a wide variety of materials instead of a few naturally occurring oils and minerals.

Figure 9—4 The high temperatures in a drying oven cause numerous cross-linking reactions to take place, which increase the surface strength of an alkyd paint.

are *polyesters* with extensive cross-linking. One of the simpler alkyds is formed from the diacid, phthalic acid, and the *tri*alcohol, glycerol.

The —OH and —COOH groups continue to react with more and more reactant molecules until extensive cross-linking occurs. Heating to about 130°C for about one hour causes maximum cross-linking. A portion of the resin's structure is shown below.

Polyamides (Nylons)

Another useful condensation reaction is that occurring between an acid and an amine to split out a water molecule and form an *amide.* Reactions of this type yield a group of polymers that perhaps have had a greater impact upon society than any other type. These are the *polyamides,* or nylons.

In 1928, the Du Pont Company embarked upon a program of basic research headed by Dr. Wallace Carothers (1896–1937), who came to Du Pont from the Harvard University faculty. His research interests were high molecular weight compounds, such as rubber, proteins, and resins, and the reaction mechanisms that produced these compounds. In February, 1935, his research produced a product known as nylon 66, prepared from adipic acid (a diacid) and hexamethylenediamine (a diamine):

$$\underset{\text{ADIPIC ACID}}{\overset{O}{\underset{HO}{\diagdown}}C-(CH_2)_4-\overset{O}{\underset{OH}{\diagup}}C} + \underset{\text{HEXAMETHYLENEDIAMINE}}{H_2N-(CH_2)_6-NH_2} \longrightarrow$$

$$-\overset{O}{\overset{\|}{C}}-(CH_2)_4-\boxed{\overset{O}{\overset{\|}{C}}-\underset{H}{\overset{|}{N}}}-(CH_2)_6-\boxed{\underset{H}{\overset{|}{N}}-\overset{O}{\overset{\|}{C}}}-(CH_2)_4-\boxed{\overset{O}{\overset{\|}{C}}-\underset{H}{\overset{|}{N}}}-(CH_2)_6- \quad + \quad xH_2O$$

NYLON 66

(The amide groups are outlined for emphasis.)

Common nylon can be made by the reaction of adipic acid and hexamethylenediamine.

This material could easily be extruded into fibers that were stronger than natural fibers and chemically more inert. The discovery of nylon jolted the American textile industry at almost precisely the right time. Natural fibers were not meeting the needs of twentieth-century Americans. Silk was not durable and was very expensive, wool was scratchy, linen crushed easily, and cotton did not lend itself to high fashion. All four had to be pressed after cleaning. As women's hemlines rose in the mid-1930's silk stockings were in great demand, but they were very expensive and short-lived. Nylon changed all that almost overnight. It could be knitted into the sheer hosiery women wanted, and it was much more durable than silk. The first public sale of nylon hose took place in Wilmington, Delaware (the hometown of Du Pont's main office), on October 24, 1939. They were so popular they had to be rationed. World War II caused all commercial use of nylon to be abandoned until 1945, as the industry turned to making parachutes and other war materials. Not until 1952 was the nylon industry able to meet the demands of the hosiery industry and to release nylon for other uses as a fiber and as a thermoplastic.

Many kinds of nylon have been prepared and tried on the consumer market, but two, nylon 66 and nylon 6, have been most successful. Nylon 6 is prepared from caprolactam, which comes from aminocaproic acid. Notice how aminocaproic acid contains an amine group on one end of the molecule and an acid group on the other end.

Figure 9–5 Structure and hydrogen bonding in nylon 6.

AMINOCAPROIC ACID   PORTION OF NYLON 6

$$H_2N-(CH_2)_5-C\underset{OH}{\overset{O}{\big<}} \xrightarrow[\text{Polymerization}]{-H_2O} -N-(CH_2)_5-C-N-(CH_2)_5-C-$$

Figure 9–5 illustrates another facet of the structure of nylon—***hydrogen bonding.*** This type of bonding explains why the nylons make such good fibers. In order to have good tensile strength, the chains of atoms in a polymer should be able to attract one another, but not so strongly that the plastic cannot be initially extended to form the fibers. Ordinary covalent chemical bonds linking the chains together would be too strong. Hydrogen bonds, with a strength about one tenth that of an ordinary covalent bond, link the chains in the desired manner. We shall see later that this type of bonding is also of great importance in protein structures.

Hydrogen bonding is important in determining the properties of nylon fibers.

Polycarbonates

The tough, clear polycarbonates constitute another important group of condensation plastics. One type of polycarbonate, commonly called Lexan or Merlon, was first made in Germany in 1953. It is as "clear as glass" and nearly as tough as steel. A one-inch sheet can stop a .38-caliber bullet fired from 12 feet away. Such unusual properties have resulted in Lexan's use in "burglar-proof" windows and as visors in astronauts' space helmets. More than 100 million pounds of polycarbonates were produced in the U.S. in 1976.

The polycarbonates are formed by condensing phosgene with a substance containing two phenol structures. A molecule of HCl is condensed out in the formation of a C—O bond to complete the ester ($-\overset{\displaystyle O}{\overset{\|}{C}}-O-$) linkage. Other chlorines then react with other alcohol groups to give a polymer chain containing $-O-\overset{\displaystyle O}{\overset{\|}{C}}-O-$ functional groups in the backbone of the chain. A representative portion of Lexan is made as follows.

PHENOL

The name polycarbonate comes from the linkage's similarity to an inorganic carbonate ion, CO_3^{2-}.

SELF-TEST 9-A
1. The individual molecules from which plastics are made are called _____ _____.

2. Write the formulas of the monomers used to prepare the polymers listed below. For example, $CH_2{=}CH_2$ is used to prepare polyethylene.

   (a) polypropylene

   _____

   (b) polystyrene

   _____

   (c) Teflon

   _____

3. Many molecules of a carboxylic diacid reacting with many molecules of a dialcohol produce a _____.

4. Natural rubber is a polymer of _____.

5. When styrene and butadiene are polymerized together the product is a

   type of _____.

6. Nylon is an example of a _____ polymer.

7. Polyamides are formed when _____ is split out from the reaction of many organic acid groups and many amine groups.

8. When an acid such as terephthalic acid

   reacts with ethylene glycol, $HOCH_2CH_2OH$, the structure of the resulting polymer is:

   _____

9. Polyesters are formed by (addition/condensation) reactions.

10. In paints, the components are dispersed in a liquid called the _____

    and the color is supplied by the _____.

11. The drying oil used most often in paints is _____ oil.

12. Water-based or "latex-type" paints are _____

    because two liquids that are ordinarily incompatible are stabilized.

13. Baked-on paints are usually alkyds. This means that they are usually made

    from polyfunctional _____ and _____.

SILICONES

Silane, SiH_4, is structurally like methane, CH_4, in that both are tetrahedral.

The element silicon, in the same chemical family as carbon, also forms many compounds with numerous Si—Si and Si—H bonds, analogous to C—C and C—H bonds. However, the Si—Si bonds and the Si—H bonds are reactive toward both oxygen and water; hence, there are no useful silicon counterparts to most hydrocarbons. However, silicon does form stable bonds with carbon, and especially oxygen, and this fact gives rise to an interesting group of condensation polymers containing silicon, oxygen, carbon, and hydrogen (bonded to carbon).

In 1945, E. G. Rochow, at the General Electric Research Laboratory, discovered that a silicon-copper alloy will react with organic chlorides to produce a whole class of reactive compounds, the *organo-silanes.*

$$2CH_3Cl + Si(Cu) \longrightarrow (CH_3)_2SiCl_2 + Cu$$

METHYL CHLORIDE   SILICON-COPPER ALLOY   DIMETHYLDICHLOROSILANE

The chlorosilanes readily react with water and replace the chlorine atoms with hydroxyl (—OH) groups. The resulting molecule is similar to a dialcohol.

$$(CH_3)_2SiCl_2 + 2H_2O \longrightarrow (CH_3)_2Si(OH)_2 + 2HCl$$

Two dihydroxysilane molecules undergo a condensation reaction in which a water molecule is split out. The resulting Si—O—Si linkage is very strong; the same linkage holds together all the natural silicate rocks and minerals. Continuation of this condensation process results in polymer molecules with molecular weights in the millions:

Silicones are polymers held together by a series of covalent Si—O bonds.

By using different starting silanes, polymers with very different properties result. For example, two methyl groups on each silicon atom results in *silicone oils,* which are more stable at high temperatures than hydrocarbon oils and also have less tendency to thicken at low temperatures.

Silicone rubbers are very high molecular weight chains cross-linked by Si—O—Si bonds. Room-temperature-vulcanizing (RTV) silicone rubbers are commercially available; they contain groups that readily cross-link in the presence of atmospheric moisture. The —OH groups are first produced, and then they condense in a cross-linking "cure" similar to the vulcanization of organic rubbers.

Over 3,000,000 pounds of silicone rubber are produced each year in the United States. The uses include window gaskets, o-rings, insulation, sealants for buildings, space ships, and jet planes and even some wearing apparel. The first footprints on the moon were made with silicone rubber boots, which readily withstood the extreme surface temperatures (Figure 9–6).

Silicone oils and rubbers find many medical uses.

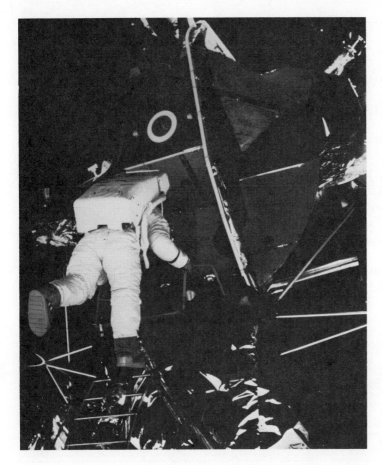

Figure 9–6 Examples of the use of silicone in the space program. Soles of lunar boots worn by the Apollo astronauts were made of high-strength silicone rubber. A silicone compound was also used for the air-tight seal of the lunar module hatch from which Astronaut Edwin E. Aldrin, Jr., has just emerged in this photo of the first manned landing on the moon on July 20, 1969.

"Silly Putty," a silicone widely distributed as a toy, is intermediate between silicone oils and silicone rubber. It is an interesting material with elastic properties on sudden deformation, but its elasticity is quickly overcome by its ability to flow like a liquid when allowed to stand.

REARRANGEMENT POLYMERS

Some molecules polymerize by rearrangement reactions to yield very useful products. Molecules containing the isocyanate group ($-$NCO), for example, will react with almost any other molecule containing an active hydrogen atom (such as in an $-$OH or $-$NH$_2$ group) in a rearrangement process. An example is the reaction of hexamethylene diisocyanate and butanediol. The urethane linkage $\left(\begin{array}{c} -\text{N}-\text{C}-\text{O}- \\ | \quad \parallel \\ \text{H} \quad \text{O} \end{array} \right)$ is produced by a shift of a hydrogen atom, moving it from the alcohol (butanediol) to a nitrogen atom on the isocyanate group; the linkage ($-$N$-$C$-$O$-$) is similar to, but not the same as, the amide bond ($-$N$-$C$-$C$-$) in nylons.

$$\text{OCN(CH}_2)_6\text{NCO} + \text{HO(CH}_2)_4\text{OH} \longrightarrow \text{OCN(CH}_2)_6\text{—}\overset{\text{H}}{\underset{}{\text{N}}}\text{—}\overset{\text{O}}{\underset{}{\text{C}}}\text{—O—(CH}_2)_4\text{OH}$$

HEXAMETHYLENE   1,4-BUTANEDIOL   PRODUCT MOLECULE (A URETHANE)
DIISOCYANATE

The continued reaction of the other groups gives rise to a polymer chain—a polyurethane.

Polyurethanes are structurally similar to many polyamides.

$$-\overset{\overset{\textstyle O}{\|}}{C}-\overset{\overset{\textstyle H}{|}}{N}-(CH_2)_6-\overset{\overset{\textstyle H}{|}}{N}-\overset{\overset{\textstyle O}{\|}}{C}-O-(CH_2)_4-O-\overset{\overset{\textstyle O}{\|}}{C}-\overset{\overset{\textstyle H}{|}}{N}-(CH_2)_6-\overset{\overset{\textstyle H}{|}}{N}-\overset{\overset{\textstyle O}{\|}}{C}-O-(CH_2)_4-O-$$

A PORTION OF POLYURETHANE

A polyurethane is structurally similar to a polyamide (nylon). In Europe polyurethanes have applications similar to those of nylon in this country. Polyurethanes have viscosities and melting points that make them useful for foam applications. Foamed polyurethanes are known as "foam rubber" and "foamed hard plastics," depending on the degree of cross-linking.

Table 9–3 lists the trade names and compositions of some of the addition, condensation, and rearrangement polymers on the market today.

TABLE 9–3   Composition of Some Trade-Name Polymers*

Trade Name	Composition
Acrilan	85% acrylonitrile plus vinyl acetate or vinyl pyridine
Acrylic	At least 85% acrylonitrile
Arnel	Cellulose triacetate
Bakelite	Phenol-formaldehyde condensation product
Caprolan	Nylon 6
Cellophane	Regenerated cellulose
Celluloid	Cellulose nitrate
Creslan	Copolymer of acrylonitrile and acrylamide
Dacron	Ethylene glycol-terephthalic acid condensation product
Delrin	Polyacetal
Carvan	Vinylidene dinitrile plus vinyl acetate
Dynel	60% vinyl chloride plus 40% acrylonitrile
Epoxy	Phenol-acetone-epichlorohydrin condensation product
Formica	Phenol-formaldehyde condensation product
Fortrel	Polyester similar to Dacron

$-CH_2C(OC_2H_5)_2-$
Acetal

$-CH_2CH(CONH_2)-$
Acrylamide

$HOCH_2-\overset{\overset{\displaystyle CH_2-CH_2}{\diagup\diagdown}}{\underset{\underset{\displaystyle CH_2-CH_2}{\diagdown\diagup}}{C}}-H \quad H-\overset{\overset{\displaystyle }{}}{C}-CH_2OH$
1,4-Cyclohexane Dimethanol

$H_2C\overset{\overset{\textstyle O}{\diagup\diagdown}}{-\!\!-\!\!-}CHCH_2Cl$
Epichlorohydrin

*Structures are given for those monomers not found elsewhere in the text. Consult the index for formulas not given.

Trade Name	Composition
Herculon	Polypropylene
Kodel	Polyester; terephthalic acid plus 1,4-cyclohexane dimethanol
Lucite	Methyl methacrylate
Melinex	Polyethylene terephthalate
Melmac	Melamine-formaldehyde condensation product
Meraklon	Polypropylene
Mylar	Polyethylene terephthalate
Neoprene	2-Chlorobutadiene
Nylon 501	Nylon 66
Nytril	At least 85% vinylidene dinitrile
Orlon	Originally pure acrylonitrile; now up to 14% of another monomer
Plexiglas	Methyl methacrylate
Polythene	Polyethylene
Polyzote	Polystyrene
Rayon	Regenerated cellulose
Saran	Vinylidene chloride-vinyl chloride addition product
Spandelle	Polyurethane; ethylene glycol-diisocyanate condensation product
Spandex	Polyurethane; ethylene glycol-diisocyanate condensation product
Teflon	Polytetrafluoroethylene
Terylene	Polyester similar to Dacron
Vectra	Polypropylene
Vectra nylon	Nylon 6
Velon	Vinylidene chloride-vinyl chloride addition product; see Saran
Vinyon	Vinyl chloride or copolymer with vinyl acetate
Zantrel	Rayon fiber

$$-CH_2CH(O\overset{\displaystyle O}{\overset{\|}{C}}CH_3)-$$
Vinyl Acetate

$$-CH_2C(CN)_2-$$
Vinylidene Dinitrile

Melamine

$$-CH_2CH\!\!\left(\!\!\bigcirc\!\!N\!\!\right)\!\!-$$
Vinyl Pyridine

POLYMER ADDITIVES—TOWARD AN END USE

Few plastics produced today find end uses without some kind of modification. Polyurethanes are a good example. In order for polyurethane to be used as insulation in refrigerators, refrigerated trucks and railroad cars, and as construction insulation, it is foamed.

Foaming Agents

Two methods of producing foamed plastics are commonly used. Hydrocarbons such as pentane or fluorocarbons like trichlorofluoromethane are **physical foaming agents** because they can foam a plastic simply by boiling to produce bubbles in the plastic. Polyurethane can be foamed by dissolving pentane in the liquid under pressure. When this mixture is extruded from an outlet into the atmosphere, the pentane vaporizes, leaving the polyurethane full of small holes. The polyurethane is quickly solidified by cooling. Planks, boards, or logs of foamed polystyrene (Styrofoam) can also be made in this way.

Foaming agents are used to make low-density plastics that are full of gas bubbles.

Chemical foaming agents produce a gas by chemical reaction, as their name implies. For example, when polyurethanes are formed, it is sometimes possible to foam the plastic by using the gaseous carbon dioxide generated when an isocyanate group reacts with a water molecule.

$$R-NCO \quad + H_2O \longrightarrow R-\underset{\underset{H}{|}}{N}-H + CO_2$$

AN ISOCYANATE   AN AMINE

Plasticizers

Many times a plastic such as polystyrene, polyethylene, or polypropylene is too stiff for its intended application. The structure of a polymer largely determines its flexibility. The polymer chains are intertwined and somewhat aligned in the solid state. This is enough to prevent adequate flexing of the solid in many instances. By the addition of *plasticizers,* polymers are made more flexible.

Plasticizers such as dioctyl phthalate (DOP) and dioctyl adipate (DOA) exert a partial dissolving effect by fitting in between the polymer chains and weakening the attractions between chains, thereby increasing the flexibility (Figure 9–7). For example, polyvinyl chloride (PVC) without plasticizers is hard and brittle, but with proper plasticizers it becomes soft and pliable. The modified PVC is used to make garden hose, auto seat covers, and raincoats. Plasticizers tend to evaporate in heat or sunlight, leaving the PVC hard and brittle again.

Plasticizers are added to plastics to make them more flexible.

The use of plastic food wraps would be much more limited without DOA, which has Food and Drug Administration approval. DOP and

DOP

DOA

Figure 9—7 Schematic representation of how a plasticizer works.

similar compounds are more questionable; they tend to migrate from discarded plastics and so are found in many areas of our environment. Recently, DOP and similar plasticizers were found in the heart muscle of cattle, dogs, rabbits, and rats and in the bloodstreams of a group of workers at one laboratory. Possible adverse effects of their presence are currently being studied. Very high dosages can induce mutations and birth defects in laboratory animals. What implications, if any, such plasticizers have for human health is not known at this time.

One form of the synthetic styrene-butadiene rubber (SBR, discussed earlier in this chapter) requires only ordinary hydrocarbon oil as a plasticizer.

Stabilizers

Most plastics used outdoors in such things as signs, tarpaulins, artificial turf, indoor-outdoor carpeting, auto seat covers, and toys, must be protected from sunlight, as anyone who has repeatedly parked his plastic-upholstered convertible in the outdoors knows.

Stabilizers are added to plastics to reduce the rate of damage from exposure to sunlight.

Photons of the 290 to 400 nanometer spectral region have sufficient energy to break some chemical bonds found in most organic molecules. However, there are other molecules that absorb ultraviolet light and liberate the energy as heat, thus preventing broken bonds in the polymer structure. These ultraviolet absorbers may be added directly to the plastic (about 0.1 percent by weight) that is to be subjected to sunlight. Unfortunately, many plastic formulations intended for outdoor use do not contain sufficient ultraviolet stabilizer to prevent decay.

Fillers and Reinforcing Agents

Very few polymers are used in the pure state, because they can always be strengthened or improved by incorporating solid particles or fibers. Tires would be no good without carbon black, which greatly increases strength and wear resistance. Glass fibers strengthen the polyester and polyurethane plastics used to make boats or fishing rods—without the fibers, the resin would be quite useless. Silicone rubber would have almost no strength at all without finely-divided silica as filler or reinforcing agent. Even if the dispersed solid only colors the plastic, or shades it from sunlight, it performs a necessary function. Such *compos-*

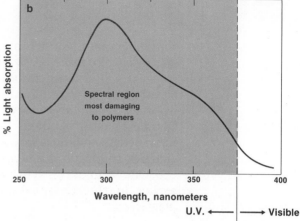

Figure 9–8 (*a*) A possible mechanism by which a molecule can absorb ultraviolet radiation and thus protect a plastic. This particular molecule is almost 100 per cent efficient in the process and finds wide use in plastics. (*b*) Typical ultraviolet absorption by a 2-hydroxybenzophenone-type compound used to protect plastics from sunlight.

Figure 9–9 Sunlight (mostly ultraviolet) damages most plastics such as polypropylene webbing. The results are shorter product life and higher costs to the consumer. Plastics containing ultraviolet absorbing chemicals have a longer outdoor life.

ites have elevated polymers to the stage of dependable plastics. "Fiberglass" boats are composites, not made just of glass, and almost all "plastics" in common use are heterogeneous composites.

THE FUTURE OF POLYMERS

As we have seen in this chapter, the development and use of synthetic polymers is quite recent. Polyethylene, for example, was not discovered until 1933, yet by 1978, its production in the United States amounted to eleven billion pounds. Chemists are constantly synthesizing new polymers and finding applications for them. The space age has brought with it the need for new polymers, especially in electronics and as special coatings that can withstand high temperatures without breaking down. Among the newcomers are the polyimides, prepared from the polycondensation of a diacid anhydride and a diamine. Some of these polymers have very high service temperatures (Figure 9–10).

Plastic materials are being improved constantly. Some have been made with the strength and rigidity of steel while having only 15 to 20% of the density of steel. The structural strength of such plastics offers the possibility of self-supporting domes for buildings and automobiles that

Anhydride: an HOH has been removed from the structure.

$$\underset{\substack{\text{PYROMELLITIC}\\\text{ANHYDRIDE}}}{} + \underset{\substack{\text{1,2-DIAMINOETHANE}}}{H_2N-CH_2-CH_2-NH_2} \xrightarrow{-H_2O} \underset{\substack{\text{A POLYIMIDE}}}{}$$

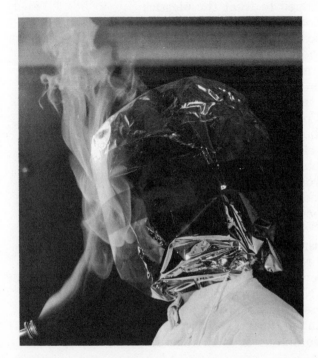

Figure 9–10 Preparation of a polyimide (*above*) and an example of how a polyimide film can withstand, for a short period, the flame of a blowtorch.

Figure 9—11 A method of separation of plastics from other wastes prior to recycling the plastics for reuse.

contain more plastic than metal. New low-temperature polymerizations without the use of a solvent are being developed. An application is to make "spray-on" clothes. Simply spray the monomers onto a mannequin—a little more here—a little less there. Then cut along desired lines, add buttons and zippers, and wear. Some plastics are being developed to replace wood fiber in paper for the printed page. These papers offer smooth surfaces, without the graininess of paper, an improvement especially in the quality of microfilming.

Because polymers are used so extensively throughout the world today, the problem of waste disposal is inevitable. Plants are in operation in which solid wastes undergo first a magnetic separation to remove iron and steel objects, then a ballistic separation (Figure 9–11) based on density, since glass and aluminum objects are denser than plastics. Another method is to have consumers separate waste into garbage, plastics, metals, paper, wood, and glass. Plans are being developed to treat the plastics thus separated in two ways. If suitable separation methods could be developed, thermoplastics could be reprocessed into new items (e.g., if all the nylon could be separated from polystyrene). Thermosetting plastics could not be treated this way, however, because breaking the cross-linking would cause complete molecular degradation. If separation and reuse were not feasible, asphalt could be made from the mixed plastics. In their original composition, combustion units built near cities could actually use plastics as fuels, since they are mostly carbon and hydrogen. There is danger, though, in that some plastics contain elements that could create massive pollution if released into the atmosphere. An example is polyvinyl chloride, which on burning yields hydrogen chloride, a very corrosive gas. However, some of the products of incomplete combustion (such as benzene, styrene, and acetylene) could be recycled as raw materials for other chemical syntheses.

The long-range future of plastics looks dismal unless we curtail our ravenous burning of petroleum. Most of the raw materials for plastics come from petroleum, and less from coal. Petroleum and coal are the principal sources of energy in this country. Not only are plastics and energy linked through the raw materials—fossil fuels relationship, but considerable energy is required to purify starting materials and to change them into the desired plastic in the preferred shape. It is the age-old

In the U.S., 95% of our petroleum is used as fuel; only 5% is used to make products such as medicines, textiles and plastics.

principle that we cannot continue to eat our "cake" (burn petroleum and coal) and have it, too (use petroleum and coal to make plastics, fibers, and medicines).

Hopefully, these and similar problems will be solved as we begin to understand more fully how to use what we have on this planet, and how to live in greater harmony with nature.

---

1. When $(CH_3)_2SiCl_2$ reacts with water, a representative portion of the structure of the polymer obtained is

2. Stabilizers protect plastics against the action of _____.

3. A plastic which is too stiff can be rendered more flexible by the addition of a

_____.

4. A silicone polymer contains Si— _____ bonds.

5. The process of burning plastics always produces CO_2 and H_2O (True/False).

6. (Ultraviolet/visible) light is more destructive to plastics.

7. Plastics become brittle usually because the _____ evaporate.

_____ 1. nylon

_____ 2. monomer

_____ 3. thermoplastic

_____ 4. thermosetting plastic

_____ 5. polymer with a memory

_____ 6. vulcanize

_____ 7. stereochemical control

_____ 8. styrene-butadiene copolymer

_____ 9. poly-*cis*-isoprene

_____ 10. polyester

(a) plastic which forms interlocking bonds when heated

(b) rubber

(c) hardening via reaction with sulfur

(d) forms polymers of desired structure

(e) natural rubber

(f) a synthetic rubber

(g) building unit for a polymer

(h) plastic softened by heat

(i) a polyamide

(j) formed from a dialcohol and a diacid

1. Explain how polymers could be prepared from each of the following compounds. **QUESTIONS** (Other substances may be used.)

   (a) $CH_3-\overset{\displaystyle H}{\underset{\displaystyle |}{C}}=\overset{\displaystyle H}{\underset{\displaystyle |}{C}}-CH_3$ (b) $HO-\overset{\displaystyle O}{\overset{\displaystyle ||}{C}}-CH_2-CH_2-\overset{\displaystyle O}{\overset{\displaystyle ||}{C}}-OH$

   (c) $\underset{\displaystyle OH}{\overset{\displaystyle |}{CH_2}}-\underset{\displaystyle OH}{\overset{\displaystyle |}{CH}}-\underset{\displaystyle OH}{\overset{\displaystyle |}{CH_2}}$ (d) $H_2N-CH_2-\langle\bigcirc\rangle-CH_2-NH_2$

2. Do modern latex paints always contain a rubber latex? Explain.

3. Draw the monomer used to prepare:

   (a) polyethylene $\Big($ for example: $\overset{\displaystyle H}{\underset{\displaystyle H}{}}\!C=C\!\overset{\displaystyle H}{\underset{\displaystyle H}{}}$ $\Big)$
   (b) Neoprene
   (c) polystyrene
   (d) Teflon

4. What are the monomers used to prepare the following polymers?

   (a) $-CH_2CH_2CH_2CH_2CH_2CH_2CH_2CH_2CH_2-$

   (b) $-\overset{\displaystyle CH_3}{\underset{\displaystyle |}{C}}HCH_2\overset{\displaystyle CH_3}{\underset{\displaystyle |}{C}}HCH_2\overset{\displaystyle CH_3}{\underset{\displaystyle |}{C}}HCH_2-$

   (c) $-CH_2\overset{\displaystyle H}{\underset{\displaystyle |}{C}}CH_2-\overset{\displaystyle H}{\underset{\displaystyle |}{C}}CH_2-\overset{\displaystyle H}{\underset{\displaystyle |}{C}}CH_2-\overset{\displaystyle H}{\underset{\displaystyle |}{C}}-$

   (with benzene rings below each $\overset{H}{C}$)

5. Write equations showing the formation of polymers by the reaction of the following pairs of molecules:

   (a) $\underset{COOH}{\overset{COOH}{\langle\bigcirc\rangle}}$ and $HOCH_2CH_2OH$

   (b) $HOOCCH_2CH_2COOH$ and $H_2NCH_2CH_2NH_2$

   (c) $\underset{\displaystyle CH_2OH}{\overset{\displaystyle CH_2OH}{\underset{\displaystyle |}{HCOH}}}$ and $\langle\bigcirc\rangle\overset{\displaystyle \overset{O}{\diagup}C-OH}{\underset{\displaystyle C-OH\,\underset{O}{}}{}}$

6. You are given two specimens of plastic, A and B, to identify. One is known to be nylon, and the other poly(methylmethacrylate), but we do not know which is which. Analysis of A shows it to contain C, H, and O, while B contains C, H, O, and N. What are A and B?

7. What structural features are necessary in a molecule for it to undergo addition polymerization?

8. What is meant by the term macromolecule?

9. Orlon has a polymeric chain structure of

$$-CH_2-CH-CH_2-CH-CH_2-CH-$$
$$\quad\quad\quad | \quad\quad\quad\quad | \quad\quad\quad\quad |$$
$$\quad\quad\quad CN \quad\quad\quad\quad CN \quad\quad\quad\quad CN$$

What is the monomer from which this can be made?

10. Which white pigment is banned in interior paints? Explain.

11. What is a copolymer? Give an example of one.

12. What feature do all condensation polymerization reactions have in common?

13. Give an example of the possibilities that exist if a trifunctional acid reacts with a difunctional alcohol.

14. What are the starting materials for nylon 66?

15. Suggest a major difference in the bonding of thermosetting and thermoplastic polymers. Which is more likely to have an interlacing (cross-linking) of covalent bonds throughout the structure? Which is more likely to have weak bonds between large molecules?

16. Explain how a plasticizer can make a polymer more flexible.

17. Could an oil-based paint "dry" in a vacuum? Explain.

18. Explain how a latex paint dries.

19. Acrylics are a popular type of water-based paint. The monomer of polyacrylates is

$$CH_2=CH$$
$$\quad\quad | $$
$$O=C-OR$$

where R is CH_3 or C_2H_5. Draw a representative portion of a polyacrylate molecule.

20. Draw representative portions of Acrilan, Delrin, Saran, and Plexiglas. Refer to Table 9–3.

21. (a) Should we stop the burning of petroleum? What are the problems involved?
    (b) Should we start now to develop research on how to change wood and straw into plastics? What are a few of the problems involved?

22. Would isoprene make a good motor fuel? Explain.

BIO-CHEMISTRY—BASIC STRUCTURES

The goal of biochemistry is to develop a chemically based understanding of living cells of all types. This includes the determination of the kinds of atoms present, the investigation of how they are joined together to form the larger structural units present in cells, and the study of the chemical reactions by which living cells obtain the energy required for the life processes of growth, movement, and reproduction.

In this chapter emphasis will be placed on fundamental biochemical structures and their preparation. These structures are (1) *carbohydrates;* (2) *fats;* (3) *amino acids, proteins,* and *enzymes;* and (4) *nucleic acids.*

CARBOHYDRATES

Carbohydrates are composed of the three elements carbon, hydrogen, and oxygen. Three structural groups are prevalent in carbohydrates: alcohol ($-OH$), aldehyde $\left(\begin{smallmatrix} & O \\ & \| \\ -C & H \end{smallmatrix}\right)$, and ketone $\left(\begin{smallmatrix} & O \\ & \| \\ -C & - \end{smallmatrix}\right)$. The carbohydrates can be classified into three main groups: ***monosaccharides*** (Latin, *saccharum,* sugar), ***oligosaccharides,*** and ***polysaccharides.*** Monosaccharides are simple sugars that cannot be broken down into smaller units by mild acid hydrolysis. *Hydrolysis is a water-splitting reaction in which water is split into $H\cdot$ and $\cdot OH$ units.* Hydrolysis of a molecule of an oligosaccharide yields two to six molecules of a simple sugar; complete hydrolysis of a polysaccharide produces many monosaccharide units.

Carbohydrates are synthesized by plants from water and atmospheric carbon dioxide. The process is called ***photosynthesis*** since it is a synthetic reaction that occurs when energized by photons (that is, quanta) of light energy. It requires an input of energy; consequently, carbohydrates are energy-rich compounds. These compounds serve as

Carbohydrates contain the elements carbon, hydrogen, and oxygen, with hydrogen atoms and oxygen atoms generally in the ratio of 2 to 1.

mono—one
oligo—few
poly—many

Carbohydrates are synthesized by plants from CO_2 and H_2O, using energy from the sun.

important sources of energy for the metabolic processes of plants and animals. Glucose, $C_6H_{12}O_6$, along with some of the other simple sugars, are quick energy sources for the cell. Large amounts of energy are stored in polysaccharides, such as starch. The stored energy is usable by living cells only if polysaccharides are broken down into monosaccharides.

Some complex carbohydrates are also used by cells for structural purposes. Cellulose, for example, partially accounts for the structural properties of wood.

Monosaccharides

Approximately 70 monosaccharides are known; 20 of these simple sugars occur naturally. Unlike many organic compounds, these sugars are very soluble in water owing to the numerous —OH groups present, which can form hydrogen bonds with water.

D-glucose, the most important monosaccharide, is found in fruits, blood, and living cells.

The most common simple sugar is ***D-glucose.*** This monosaccharide requires three structures for its adequate representation (Figure 10–1). Structure (*a*), in which the carbon atoms are numbered for later reference, depicts the "straight-chain" structure with the aldehyde group (—CHO) in position 1. The properties of a water solution of D-glucose cannot be explained by this structure alone. At any given time, most of the molecules exist in the ring form, structures (*b*) and (*c*), which results from a molecular rearrangement in which carbon 1 bonds to carbon 5 through an oxygen atom. Both ring structures are possible since the OH group on carbon 1 may form in such a way to point either along the plane of the molecule or out of the plane. It should be emphasized that a solution of D-glucose contains a mixture of three forms in a dynamic state of change from one form to another. There is more of the ring forms and much less of the straight-chain form.

(*a*) D-*Glucose*

(*b*) α-D-*Glucose*

(*c*) β-D-*Glucose*

(*d*) α-D-*Glucose*

(*e*) β-D-*Glucose*

Figure 10–1 The structures of D-glucose; *d* and *e* are two-dimensional representations of *b* and *c*, respectively. Note the difference in the positions of the —OH groups (color) in the α and β forms of glucose: the —OH groups on the 1 and 4 carbons are *trans* when the structure is beta (β), and the —OH groups are *cis* when the structure is alpha (α). In both alpha and beta glucose, the —OH group on the number 4 carbon atom must be in the same position.

$$CH_2OH$$
$$C=O$$
$$HO-C-H$$
$$H-C-OH$$
$$H-C-OH$$
$$CH_2OH$$

(a) *Ketone structure*

(b) *β-Ring structure (Pyranose structure: 6-membered ring with an oxygen atom in the ring)*

(c) *β-Ring structure (Furanose structure: 5-membered ring with an oxygen atom in the ring)*

Figure 10–2 The structures of D-fructose. The α-ring structure (not shown) differs from the β-ring structure in that the CH_2OH and OH groups are in reversed positions on carbon 2.

Because it is sweet, D-glucose is used in the manufacture of candy and in commercial baking. This simple sugar, also called *dextrose, grape sugar,* and *blood sugar,* is prevalent in fruits, vegetables, blood, and tissue fluids. A solution of D-glucose is fed intravenously when a readily available source of energy is needed to sustain life. As will be discussed later, many polysaccharides, including starch, are composed of glucose units and serve as a source of this important chemical upon hydrolysis of the complex structures.

Another very important monosaccharide is **D-fructose.** Its structure, which has a ketone group, is given in Figure 10–2.

Glucose has a relative sweetness of 74.3, compared to sucrose having an assigned value of 100.0. The value for fructose is 173.3.

Oligosaccharides

The most important oligosaccharides are the disaccharides (two simple sugar units per molecule). Examples include these widely used disaccharide sugars:

sucrose (from sugar cane or sugar beets), which consists of a glucose unit and a fructose unit,
maltose (from starch), which consists of two glucose units, and
lactose (from milk), which consists of a glucose unit and a galactose (an optical isomer of glucose) unit.

The formula for these disaccharides, $C_{12}H_{22}O_{11}$, is not simply the sum of two monosaccharides, $C_6H_{12}O_6 + C_6H_{12}O_6$. A water molecule must be eliminated as two monosaccharides are united to form the disaccharide. The structures for sucrose, maltose, and lactose, along with their hydrolysis reactions, are given in Figure 10–3.

The disaccharides are important as foods. Sucrose is produced in a very high state of purity on an enormous scale: the annual production amounts to over 80 million tons per year. Originally produced in India and Persia, sucrose is now used universally as a sweetener. About 40 percent of the world sucrose production comes from sugar beets and 60 percent from sugar cane. Sugar provides a high caloric value (1794 kcal per pound); it is also used as a preservative in jams, jellies, and candied fruit.

Disaccharides are molecules containing two simple sugars bound together, such as in sucrose, which contains a glucose and a fructose unit in each molecule. A water molecule is eliminated when the bond forms between the two simple sugars.

Figure 10—3 Hydrolysis of disaccharides (sucrose, maltose, and lactose).

Polysaccharides

There is an almost limitless number of possible structures in which monosaccharide units (monosaccharide molecules minus one water molecule at each bond between units) can be combined. Molecular weights are known to go above 1,000,000 amu. Apparently nature has been very selective in that only a few of the many possible monosaccharide units are found in polysaccharides.

Starches and Glycogen

Starch molecules consist of many glucose units bonded together.

Starch is found in plants in protein-covered granules. These granules are disrupted by heat, and part of the starch content is soluble in hot water. Soluble starch is *amylose* and constitutes 22 to 26 percent of most natural starches; the remainder is *amylopectin.* Amylose gives the

Figure 10–4 Amylose structure. From 60 to 300 α-D-glucose units are bonded together by **alpha** linkages to form amylose molecules.

familiar blue-black starch test with iodine solutions; amylopectin turns red on contact with iodine.

Structurally amylose is a straight-chain polymer of α-D-glucose units, each one bonded to the next, just as the two units are bonded in maltose (Figure 10–3). Molecular weight studies on amylose indicate the average chain contains about 200 units. The structure of amylose is illustrated in Figure 10–4.

Amylopectin is made up of branched chains of α-D-glucose units (Figure 10–5). Its molecular weight generally corresponds to about 1000 glucose units. Partial hydrolysis of amylopectin yields mixtures called **_dextrins._** Complete hydrolysis, of course, yields D-glucose. Dextrins are used as food additives, mucilages, paste, and in finishes for paper and fabrics.

Glycogen serves as an energy reservoir in animals as does starch in plants. Glycogen has a structure similar to that of amylopectin (branched

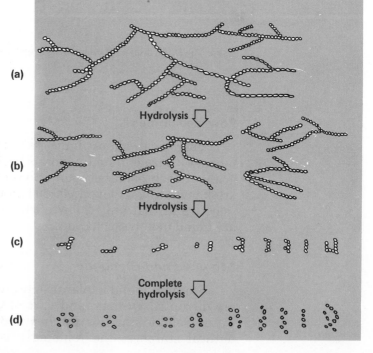

Figure 10–5 (_a_) Partial schematic amylopectin structure. (_b_) Dextrins from incomplete hydrolysis of _a_. (_c_) Oligosaccharides from hydrolysis of dextrins. (_d_) Final hydrolysis product: D-glucose. Each circle represents a glucose unit.

Figure 10–6 Cellulose structure. About 2800 β-D-glucose units are bonded together by **beta** linkages to form a cellulose molecule.

chains of glucose units), except that glycogen is even more highly branched.

Cellulose

Cellulose is the most abundant polysaccharide in nature. Like amylose, it is composed of D-glucose units. The difference between the structure of cellulose and that of amylose lies in the bonding between the D-glucose units. In cellulose, all of the glucose units are in the β-ring form in contrast to the α-ring form in amylose. Review the ring forms in Figure 10–1 and compare the structures in Figures 10–4 and 10–6.

The different structures of starch and cellulose account for their difference in digestibility. Human beings and carnivorous animals do not have the necessary enzymes (biochemical catalysts) to break down the cellulose structure as do numerous microorganisms. Cellulose is readily hydrolyzed to D-glucose in the laboratory by heating a suspension of the polysaccharide in the presence of a strong acid. Unfortunately, at present

Humans do not have an enzyme to split cellulose into its glucose units.

Figure 10–7 The properties of cotton, about 98 percent cellulose, can be explained in terms of this submicroscopic structure. A small group of cellulose molecules, each with 2000 to 9000 units of D-glucose, are held together in an approximately parallel fashion by hydrogen bonding (- - - -). When several of these *chain bundles* cling together in a relatively vast network of hydrogen bonds, a *microfibril* results; the microfibril is the smallest microscopic unit that can be seen. The macroscopic *fibril* is a collection of numerous microfibrils. The absorbent nature of cotton is readily explained in terms of the smaller water molecules being held by hydrogen bonding in the numerous capillaries that exist between the cellulose chains.

this is not an economically feasible solution to the world's growing need for an adequate food supply.

Paper, rayon, cellophane, and cotton are principally cellulose. A representative portion of the structure of cotton is shown in Figure 10–7. Note the hydrogen bonding between cellulose chains.

DIETARY FATS AND ESSENTIAL FATTY ACIDS

Most diets in the United States furnish 40 to 50% of their calories in the form of fats or oils. This is rather high when compared to diets in most other parts of the world. Natural fats and oils are generally mixtures of various esters of glycerol with more than one kind of fatty acid. In our diets most of the fatty acids are *saturated* fatty acids (Table 10–1). Such fatty acids can be (1) used as a source of energy if the body burns them to CO_2 and H_2O, (2) stored for possible future use in fat cells, or (3) used as starting materials for the synthesis of other compounds needed by the body. Fats are the most concentrated source of food energy in our diets, as they furnish about 9000 calories/g when burned for energy. The human body can make some fats from carbohydrates and carries out such processes to store the excess energy furnished in the diet.

A high intake of dietary fat has been implicated as one of the factors that can give rise to *atherosclerosis,* a complex process in which the walls of the arteries suffer damage and ultimately develop scar tissue and fatty deposits. Atherosclerosis is generally considered to be a precursor to certain types of heart disease and strokes. Atherosclerosis may be also related to the amount of cholesterol in the diet, but the relationship of both dietary fat and cholesterol intake to atherosclerosis does not appear at this time to be a simple one.

It has been known for about 50 years that the human body has a small requirement for certain types of fatty acids (called *essential fatty*

Refer to Chapter 8 for introductory material on fats and fatty acids.

TABLE 10–1 Ratio of Saturated and Unsaturated Fatty Acids from Common Fats and Oils*

Oil or Fat	Percent of Total Fatty Acids by Weight		
	Saturated	Monounsaturated	Polyunsaturated
Coconut oil	93	6	1
Corn oil	14	29	57
Cottonseed oil	26	22	52
Lard	44	46	10
Olive oil	15	73	12
Palm oil	57	36	7
Peanut oil	21	49	30
Safflower oil	10	14	76
Soybean oil	14	24	62
Sunflower oil	11	19	70

*Saturated means full complement of hydrogen (no C=C double bonds); monounsaturated means one C=C double bond per fatty acid molecule; polyunsaturated means two or more C=C double bonds per molecule of fatty acid. The chief unsaturated fatty acid is linoleic acid. Although derived from vegetable rather than animal fats, both coconut oil and peanut oil have been associated recently with hardening of the arteries when combined with a high cholesterol intake.

acids) and in recent years the basis for the need for these essential fatty acids has been determined.

The essential fatty acids are *linoleic, linolenic,* and *arachidonic* acids.

$$CH_3\underset{18}{C}H_2\underset{17}{C}H_2\underset{16}{C}H_2\underset{15}{C}H_2\underset{14}{C}H_2\underset{13}{C}H=\underset{12}{C}H\underset{11}{C}H_2\underset{10}{C}H=\underset{9}{C}H\underset{8}{C}H_2\underset{7}{C}H_2\underset{6}{C}H_2\underset{5}{C}H_2\underset{4}{C}H_2\underset{3}{C}H_2\underset{2}{C}\underset{1}{C}\overset{O}{\underset{OH}{}}$$

LINOLEIC ACID (C$_{18}\Delta_{9,12}$)

$$CH_3\underset{18}{C}H_2\underset{17}{C}H=\underset{16}{C}H\underset{15}{C}H_2\underset{14}{C}H=\underset{13}{C}H\underset{12}{C}H_2\underset{11}{C}H=\underset{10}{C}H\underset{9}{C}H_2\underset{8}{C}H_2\underset{7}{C}H_2\underset{6}{C}H_2\underset{5}{C}H_2\underset{4}{C}H_2\underset{3}{C}H_2\underset{2}{C}\underset{1}{C}\overset{O}{\underset{OH}{}}$$

LINOLENIC ACID (C$_{18}\Delta_{9,12,15}$)

$$CH_3CH_2CH_2CH_2CH_2CH=CHCH_2CH=CHCH_2CH=CHCH_2CH=CHCH_2CH_2CH_2C\overset{O}{\underset{OH}{}}$$

ARACHIDONIC ACID (C$_{20}\Delta_{5,8,11,14}$)

The presence of any one of these in the diet permits the body to synthesize a very important group of compounds, the prostaglandins. The key compound here is linoleic acid, which the body cannot make from more saturated fatty acids. If linoleic acid is available, the body can make arachidonic acid and linolenic acid.

Prostaglandins are a group of more than a dozen related compounds with very potent effects on physiologic activity such as blood pressure, relaxation and contraction of smooth muscle, gastric acid secretion, body temperature, food intake, and blood platelet aggregation. Their potential use as drugs is currently under widespread investigation. Two of the prostaglandins that have been characterized are Prostaglandin E$_1$ (used to induce labor to terminate pregnancy) and Prostaglandin E$_2$.

Prostaglandins are synthesized from the essential fatty acids. Even in very small amounts, prostaglandins have very powerful effects on the human body.

PROSTAGLANDIN E$_1$ (C$_{20}$H$_{34}$O$_5$)

PROSTAGLANDIN E$_2$ (C$_{20}$H$_{32}$O$_5$)

Note that both of these prostaglandins contain exactly the same number of carbon atoms as arachidonic acid.

SELF-TEST 10-A

1. Carbohydrates contain the elements _____ , _____ , and _____ .

2. The complete hydrolysis of a polysaccharide yields a _____ .

3. When a molecule of sucrose is hydrolyzed, one obtains one molecule of each of the monosaccharides _____ and _____ .

4. The sugar referred to as blood sugar, grape sugar, or dextrose is actually the compound _____ .

5. Starch is a polymer built up out of _____ units.

6. Essential fatty acids are needed by the body to synthesize _____ .

PROTEINS, AMINO ACIDS, AND THE PEPTIDE BOND

Proteins occur in all the major regions of living cells. These compounds serve a wide variety of functions, including motion of the organism, defense mechanism against foreign substances, metabolic regulation of cellular processes, and cell structure. The close relationship between proteins and living organisms was first noted by the German chemist G. T. Mulder in 1835. He named these compounds proteins from the Greek *proteios,* meaning first, indicating this to be the starting point in the chemical understanding of life.

Proteins are high molecular weight compounds made up of amino acid units.

Proteins are macromolecules with molecular weights ranging from 5000 to several million amu. Like the polysaccharides, these macrostructures are composed of recurring units of molecular structure. The fundamental units in the case of proteins are *amino acids.* Proteins and amino acids are made primarily from four elements: carbon, oxygen, hydrogen, and nitrogen. Other elements occur in trace amounts; the one most often encountered is sulfur.

Amino Acids

The complete hydrolysis of a typical protein yields a mixture of about 20 different amino acids. Some proteins lack one or more of these acids, others have small amounts of other amino acids characteristic of a given protein, but the 20 given in Table 10–2 are predominant. In a few instances, one amino acid will constitute a major fraction of a protein (the protein in silk, for example, is 44 percent glycine), but this is not common.

As the name suggests, amino acids contain an amino group ($-NH_2$) and an acid (carboxyl) group ($-COOH$). In all of the amino acids listed in

Amino acids are compounds that generally have the structure

$$R-\overset{\overset{\displaystyle H}{|}}{\underset{\underset{\displaystyle NH_2}{|}}{C}}-C\overset{\displaystyle O}{\underset{\displaystyle OH}{{\Large\diagup}\kern-0.3em{\Large\diagdown}}}$$

TABLE 10–2 Common Amino Acids

All of the amino acids except proline and hydroxyproline have the general formula:

$$R-\underset{\underset{NH_2}{|}}{\overset{\overset{H}{|}}{C}}-C\overset{O}{\underset{OH}{\big<}}$$

in which R is the characteristic group for each acid. The R groups are as follows.

1. Glycine —H
2. Alanine —CH_3
3. Serine —CH_2OH
4. Cysteine —CH_2SH
5. Cystine —CH_2—S—S—CH_2—

*6. Threonine —CH—CH_3
         |
        OH

*7. Valine CH_3—CH—CH_3
                    |

*8. Leucine —CH_2—CH—CH_3
                        |
                       CH_3

*9. Isoleucine —CH
         CH_3 /
         \ CH_2—CH_3

*10. Methionine —CH_2—CH_2—S—CH_3
11. Aspartic acid —CH_2CO_2H
12. Glutamic acid —CH_2—CH_2—CO_2H
*13. Lysine —CH_2—CH_2—CH_2—CH_2—NH_2

*14. Arginine —CH_2—CH_2—CH_2—NHCNH$_2$, with NH double bond

*15. Phenylalanine —CH_2—⬡

16. Tyrosine —CH_2—⬡—OH

*17. Tryptophan —CH_2—(indole)

*18. Histidine —CH_2—(imidazole N—H)

The structures for the other two are:

19. Proline H_2C——CH_2 / H_2C, N, CHCO$_2$H, H

20. Hydroxyproline HOHC——CH_2 / H_2C, N, CHCO$_2$H, H

*Essential amino acids; arginine and histidine are essential for children but not essential for adults.

Table 10–2, the amine group and the acid group are bonded to the same carbon atom. Of these acids, 18 have the general formula

There are about 20 common amino acids.

$$R-\underset{\underset{\displaystyle NH_2}{|}}{\overset{\overset{\displaystyle H}{|}}{C}}-C\overset{\displaystyle O}{\underset{\displaystyle OH}{}}$$

where R is a characteristic group for each amino acid. The simplest amino acid is *glycine,* in which R is a hydrogen atom:

$$H-\underset{\underset{\displaystyle NH_2}{|}}{\overset{\overset{\displaystyle H}{|}}{C}}-C\overset{\displaystyle O}{\underset{\displaystyle OH}{}}$$

GLYCINE

The human body is capable of synthesizing some amino acids needed for protein structures, but it is unable to provide others necessary for normal growth and development. The latter are designated *essential amino acids* and must be ingested in the food supply. The *nonessential amino acids* are just as necessary as the essential amino acids for life but can be made by the body from other compounds. The essential amino acids are indicated in Table 10–2 by asterisks.

Essential amino acids are amino acids that the body needs but cannot make.

For good nutrition we require *all* of the essential amino acids in our daily diet, but the amount required does not exceed 1.5 g per day for any of them.

The Peptide Bond

Amino acid units are linked together in protein structures by peptide bonds. This same linkage is found in nylon 66, in which a carboxylic acid and an amine are condensed to form the polymer and the amide bond. As it applies to proteins, the peptide bond can be understood in terms of the reaction between two glycine molecules.

If the acid group of one glycine molecule reacts with the basic amine group of another, the two are joined through the peptide linkage, and one molecule of water is eliminated for each bond formed.

The peptide bond,

$$-\overset{\overset{\displaystyle O}{\|}}{C}-\underset{\underset{\displaystyle H}{|}}{N}-,$$

serves to bind amino acid units together in proteins.

GLYCINE   GLYCINE

PEPTIDE BOND

GLYCYLGLYCINE

$+ \; HOH$

If this hypothetical reaction is carried out with two different amino acids, glycine and alanine, two different *dipeptides* are possible.

PEPTIDE BONDS

GLYCYLALANINE

ALANYLGLYCINE

Twenty-four *tetra*peptides are possible if four amino acids (for example, glycine, Gly; alanine, Ala; serine, Ser; and cystine, Cy) are linked in all possible combinations. They are:

Gly-Ala-Ser-Cy	Ala-Gly-Ser-Cy	Ser-Ala-Gly-Cy	Cy-Ala-Gly-Ser
Gly-Ala-Cy-Ser	Ala-Gly-Cy-Ser	Ser-Ala-Cy-Gly	Cy-Ala-Ser-Gly
Gly-Ser-Ala-Cy	Ala-Ser-Gly-Cy	Ser-Gly-Ala-Cy	Cy-Gly-Ala-Ser
Gly-Ser-Cy-Ala	Ala-Ser-Cy-Gly	Ser-Gly-Cy-Ala	Cy-Gly-Ser-Ala
Gly-Cy-Ser-Ala	Ala-Cy-Gly-Ser	Ser-Cy-Ala-Gly	Cy-Ser-Ala-Gly
Gly-Cy-Ala-Ser	Ala-Cy-Ser-Gly	Ser-Cy-Gly-Ala	Cy-Ser-Gly-Ala

A very large number of different proteins can be prepared from a small number of different amino acids.

If 17 different amino acids are used, the sequences alone would make 3.56×10^{14} uniquely different 17-unit molecules.* Although there are numerous protein structures in nature, these represent an extremely small fraction of the possible structures. Of all the many different proteins that could possibly be made from a set of amino acids, a living cell will make only a relatively small, select number.

Protein Structures

The **primary structure** of a protein indicates only the sequence of amino acid units in the polypeptide chain. Since the single bonds in the chain allow free rotation around the bond, it is reasonable to assume that there is an almost infinite number of possible conformations. Because of interactions, such as hydrogen bonding, between atoms in the same chain, certain conformations, called **secondary structures,** are favored. Linus Pauling, along with R. B. Corey, suggested the two secondary structures for polypeptides discussed below, the sheet structure and the helical structure.

Polyglycine is a synthetic protein made entirely of the amino acid glycine. In polyglycine the hydrogen attached to the nitrogen atom and the oxygen bonded to the carbon are both well suited to engage in hydrogen bonding. In the two stable conformations of polyglycine, maximum advantage is taken of the hydrogen bonding available. In the sheet structure, the hydrogen bonding is between adjacent chains of the polypeptide; in the helical structure, hydrogen bonding occurs between atoms within the same chain.

The amino acids in a protein chain interact with each other via hydrogen bonding.

Figure 10–8 illustrates a sheetlike structure in which several chains of the polypeptide are joined by hydrogen bonding. Note that all the oxygen and nitrogen atoms are involved in hydrogen bonding. Most of the prop-

---

*If the amino acids are all different, the number of arrangements is n! (read n factorial). For five different amino acids, the number of different arrangements is 5! (or $5 \times 4 \times 3 \times 2 \times 1 = 120$).

Linus Pauling (1901–   ): A scientist of great versatility and accomplishment. His interests have included the determination of the molecular structures of crystals by x-ray diffraction and theories of the chemical bond. His work led to the Nobel Prize in 1954. For his fight against the nuclear danger confronting the world, he was awarded the 1963 Nobel Peace Prize.

erties of silk can be explained in terms of this type of structure for fibroin, the protein of silk.

Hydrogen bonding is possible within a single polypeptide chain if the secondary structure is helical (Figure 10–9). Bond angles and bond lengths

(a)

Figure 10–8 Sheet structure for polypeptide. In (*a*) the two-dimensional drawing emphasizes that all of the oxygen and nitrogen atoms are involved in hydrogen bonding for the most stable structure. (*b*) Illustrates the bonding in perspective showing that the sheet is not flat; rather, it is sometimes called a pleated sheet structure.

(b)

Figure 10–9 Helix structure for a polypeptide in which each oxygen atom can be hydrogen bonded to a nitrogen atom in the third amino acid unit down the chain.

Figure 10–10 α-Helix structure of proteins. The sketch represents the actual position of the atoms and shows where intra-chain hydrogen bonding occurs.

Figure 10—11 The imaginary structure of collagen.

are such that the nitrogen atom forms hydrogen bonds with the oxygen atom in the third amino acid unit down the chain (Figure 10–10).

A coiled spring is helical in structure.

Collagen is the principal fibrous protein in mammalian tissue. It has remarkable tensile strength, which makes it important in the structure of bones, tendons, teeth, and cartilage. Three polypeptide chains, each of which is twisted into a left-handed helix, are twisted into a right-handed super-helix to form an extremely strong fibril, as shown in Figure 10–11. A bundle of such fibrils forms the macroscopic protein.

The structure of collagen illustrates a third level of protein structure, **tertiary structure.** The primary structure is the sequence of amino acids in the protein, the secondary structure is the helical form of the protein chain, and the tertiary structure is the twisted or folded form of the helix. Another tertiary structure is found in globular proteins. In these structures, the helix chain (secondary structure) is folded and twisted into a definite geometric pattern. This pattern may be held in place by one or more of several different kinds of chemical bonds, such as —S—S— bonds, depending on the particular functional groups in the amino acids involved (Table 10–2). Figure 10–12 illustrates the folded structure of a typical globular protein.

Figure 10–12 Imaginary folded structure of the helix in a globular protein.

Figure 10–13 (a) The structure of heme. (b) Two views of a model of the hemoglobin structure. Two light-colored chains of protein, two dark-colored chains, and two heme "disks" in proper arrangement compose the quaternary structure of hemoglobin. M. F. Perutz received a Nobel Prize for determining this structure. (Courtesy of M. F. Perutz and *Science*, 140:863, 1963.)

(a)

(b)

The ***quaternary structure*** of proteins refers to the degree of aggregation of protein units. Native hemoglobin (molecular weight of 68,000) must have its four polypeptide chains properly aggregated in order to form active hemoglobin. Insulin is also composed of subunits, properly arranged into its quaternary structure.

If hemoglobin, a globular protein, has an abnormal primary, secondary, tertiary, or quaternary structure because of a wrong amino acid in a given position, it may be unable to transfer oxygen in the blood. In sickle cell anemia, only one specific amino acid of the 146 in one of the hemoglobin chains is altered with respect to normal hemoglobin. Models of hemoglobin are shown in Figure 10–13.

1. The fundamental building units in proteins are the _____. SELF-TEST 10-B

2. Amino acids that the body cannot synthesize from other molecules are called

_____.

3. The peptide linkage that bonds amino acids together in protein chains has the structure

4. The basic structure present in almost all of the amino acids can be represented as

5. The formula for glycylglycine is:

6. (a) The primary structure of a protein refers to its _____ ;

   (b) the secondary structure refers to its _____ ;

   (c) its tertiary structure refers to _____ ;

   (d) and its quaternary structure refers to _____ .

7. (a) If we have three different amino acids, we can make a total of _____ different tripeptides from them if we can use an amino acid up to three times in any given tripeptide.

   (b) If we can use each amino acid only once, there are still _____ possible different tripeptides.

ENZYMES

An important group of globular proteins is the *enzymes,* molecules that function as catalysts for reactions in living systems. Like other catalysts, a given enzyme increases the rate of a reaction without requiring an increase in temperature. As an example of a simple type of catalysis, consider the oxidation of glucose, which burns in air with some difficulty and is hard to light with a match. If cigarette ashes, or other catalysts, are placed on its surface, combustion can be initiated easily with a match. When the glucose burns, it liberates a large amount of energy, 688 kilocalories, or 688,000 calories per mole.

Enzymes are protein molecules that speed up chemical reactions.

$$C_6H_{12}O_6 + 6O_2 \longrightarrow 6CO_2 + 6H_2O + 688 \text{ kcal}$$

GLUCOSE   OXYGEN   CARBON   WATER
                           DIOXIDE

The energy required to get the reaction started is the ***activation energy;*** catalysts, in general, work by lowering the activation energy. If an enzyme can lower the activation energy to a point where the average kinetic energy of the molecules in a living cell (or in a laboratory system) is sufficient for reaction, then the reaction can proceed rapidly. Glucose is oxidized rapidly and efficiently at ordinary temperatures in the presence of the proper enzymes. To be sure, the oxidation of glucose in a living cell requires many enzymes and many steps, but enzymatic catalysis produces the same final result as combustion at elevated temperature, namely carbon dioxide, water, and 688 kilocalories of usable energy per mole of sugar. Figure 10–14 graphically illustrates the concepts of activation energy, the energy available from an energy-producing reaction, and the reduction of the activation energy by an enzyme.

Enzymes are remarkable catalysts in that they are highly specific for a given reaction. Maltase is an enzyme that catalyzes the hydrolysis of maltose into two molecules of D-glucose. This is the only known function of maltase, and no other enzyme can substitute for it. The explanation for the specific activity of enzymes can be found in their molecular structures.

Enzymes are globular proteins with very definite tertiary structures. The highly specific action of maltase can be explained if its globular structure accurately accommodates a maltose molecule at the point where the reaction occurs, the reactive site. When the two units come together, strain is placed on the bonds holding the two simple sugar units together.

Enzymes are very specific catalytic molecules with a specific catalytic task.

(a) No enzyme present   (b) Enzyme present

Figure 10–14 Enzyme effect on activation energy. The vertical coordinate represents increasing energy, and the horizontal one represents time. For energy-producing reactions the reactant molecules are at a higher energy than the product molecules, as illustrated in (*a*). The difference between these energies is the net useful energy provided by the chemical reaction. The useful energy is tabulated as the free energy change of the reaction (ΔG). However, it is necessary for the reactant molecules to "get over" the energy barrier (acquire the activation energy, E) in going from reactants to products. Note that the activation energy is given back along with the free energy. The enzyme lowers the activation energy to E′, as illustrated in (*b*), while the free energy change remains the same. The net effect is to obtain the free energy of the reaction with a smaller expenditure of activation energy. Free energy is discussed in the next chapter.

As a result, water is allowed to enter and hydrolysis occurs. Sucrose cannot be hydrolyzed by maltase because of the different structure involved. Another enzyme, sucrase, hydrolyzes sucrose effectively. Some enzymes, however, are less specific. The digestive enzyme, trypsin, for example, acts predominantly on peptide bonds in proteins, but it will also catalyze the hydrolysis of some esters because of somewhat similar structure and polarity at the active site.

How can an enzyme lower the activation energy and be so specific for a given reaction? Just as a key can separate a padlock into two parts and subsequently remain unchanged, ready to unlock other identical locks, so the enzyme makes possible a molecular change (Figure 10–15). With enough energy the lock could be separated without the key, and with enough energy the molecular alteration could occur without the enzyme. An enzyme cannot make an unnatural chemical reaction occur.

The tremendous speed of enzyme-catalyzed reactions requires more than just random collisions to fit "the key in the lock." For example, a molecule of β-amylase catalyzes the breaking of four thousand bonds between the α-glucose units in amylose per second. Speed like this requires something to attract the key into the lock, such as electrically polar regions, partially charged groupings, or ionic sections on the enzyme and the **substrate** (the reactant molecule). These regions attract as well as guide the substrate to the proper position on the enzyme and thereby speed up the reaction. The electrically charged portions of the enzyme are believed to be the chemically **active sites** in the enzyme.

Enzyme structure is the key to specific catalytic activity.

Sometimes an enzyme is more than a globular protein; in such cases, the protein is not a catalyst by itself. In addition to the protein part of the enzyme there is also another chemical species called a **coenzyme.** The coenzyme, required for catalytic activity, may be an ion (e.g., Co^{3+}, Fe^{3+}, Mg^{2+}, or another of the essential minerals) or may be in part derived from a vitamin (as we shall see later in this section). The protein part of such an enzyme is called the **apoenzyme.** The coenzyme alone does not have enzymatic activity; neither does the apoenzyme. Before the enzyme becomes active, the apoenzyme and the coenzyme must combine like the two keys required to open a bank lock-box. Neither your key nor the bank's key will open the box when used alone, but both together will (Figure 10–16).

Some enzymes require coenzymes: metal ions or vitamins.

Key
(enzyme) **Padlock**
(molecule) **Key in Padlock**
(enzyme-molecule
complex) **Key**
(enzyme) **Opened Padlock**
(products)

Active site

Figure 10–15 Lock-and-key theory for enzymatic catalysis. While it is generally agreed that this analogy is an oversimplification, it does make one very important point: the enzyme makes a difficult job easy by reducing the energy required to get the job started. It also suggests that the enzyme has a particular structure at an active site which will allow it to work only for certain molecules, similar to a key that fits the shape of a particular keyhole and a particular sequence of tumblers.

Figure 10—16 A reaction utilizing an enzyme which requires a coenzyme. Here the essential feature is that the combination of enzyme plus coenzyme allows the reaction to proceed with a lower activation energy. Apoenzyme is the name given to the enzyme structure that combines with the coenzyme.

Let's use a simple chemical example to demonstrate the relationship between coenzyme, substrate, and apoenzyme. The peptide glycylglycine is hydrolyzed very slowly by water to give two molecules of glycine.

$$HC-C-N-CH_2-C + H_2O \longrightarrow 2CH_2-C$$

GLYCYLGLYCINE   GLYCINE

The glycylglycine substrate forms an intermediate compound with an enzyme, and as a result the substrate is activated for further reaction. Activation can result from extensive hydrogen bonding, interaction with a metal ion in the enzyme, or a number of other processes. In this case the coordination of the glycylglycine to the positive charge of a cobalt ion (Co^{2+}), the coenzyme, makes it more susceptible to attack by the negative end of a water molecule (Figure 10—17).

Most of the names of enzymes end in *ase.* There are a few exceptions, such as pepsin and trypsin, which are both digestive enzymes. Hydrolases promote the breakdown of foodstuffs and other substances by hydrolysis. Carbohydrases (such as maltase, lactase, sucrase, ptyalin and amylase) help to effect hydrolysis of carbohydrates. Proteases (such as pepsin, trypsin, and chymotrypsin) hydrolyze large protein molecules into groups of smaller proteins. Lipases hydrolyze esters such as fats and oils. Examples of the oxidizing enzymes, called oxidases, are catalase, which speeds up the conversion of hydrogen peroxide to water and oxygen, and dehydrogenases, which assist in the removal of hydrogen from molecules.

The names of most enzymes end in *ase.*

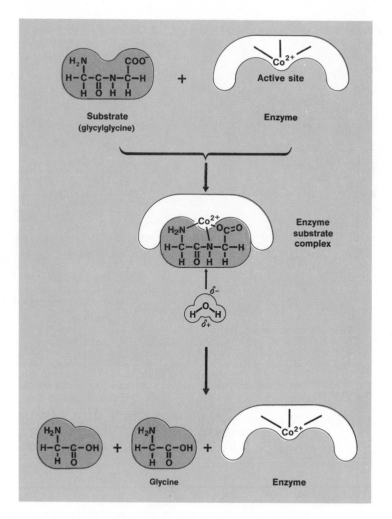

Figure 10–17 Action of an enzyme. The substrate molecule is chemically bonded to the enzyme (glycylglycine dipeptidase). The negative oxygen and the nitrogen atoms of the substrate bond to the positive cobalt ion in the enzyme. The bonding of the substrate makes it more susceptible to attack by water. Hydrolysis occurs and the glycine molecules are released by the enzyme, which is then ready to play its catalytic role again.

There are other categories of enzymes, but these illustrate the wide variety of biochemical catalysts.

Besides being biochemical middlemen, speeding up and directing all the chemical reactions that go into the continuous breakdown and buildup of our cells (three million red blood cells are renewed in the human body every second), enzymes may be the answer to future food problems. Scientists have already developed a way to produce sugar (on a limited experimental basis) by bubbling carbon dioxide into water containing enzymes. Trash fish can be converted into palatable animal feed by using enzymes. Work is underway to convert oil spills into edible products for sea organisms. A little later in this text we shall encounter the use of enzymes as meat tenderizers.

Enzymes are also important in commercial products.

Many **vitamins** function as coenzymes. They not only trigger specific enzymes, they pitch in and help out. The B vitamins are found in every cell as coenzymes in various oxidative processes. For example, niacin (vitamin B_5) becomes part of an enzyme that prevents pellagra, at one time a common vitamin-deficiency disease in the United States. Doctors and biochemists now know that the body suffers from pellagra when it lacks sufficient niacin. Some foods such as yeast, liver, meats, fish, eggs, whole wheat, brown rice, and peanuts contain niacin. The body needs only 0.06 gram of niacin each day in order to prevent pellagra. This

Niacin
(Nicotinic acid)

isn't much but it is vital. The coenzyme of which niacin is a part is necessary to the energy production in the body. If energy is not provided, the whole process of renewing cells and building needed compounds slows down and eventually stops. The coenzyme involved is a large molecule, ***nicotinamide adenine dinucleotide (NAD$^+$).***

Some of the bonds in this structure are shown abnormally long for clarity.

NAD$^+$, R = H
NADP$^+$, R = PO(OH)$_2$

Niacin is also an integral part of another oxidative enzyme, ***nicotinamide adenine dinucleotide phosphate*** (NADP$^+$). In NAD$^+$, R = H; in NADP$^+$, R = H$_2$PO$_3$. The part of the structure printed in color above comes from niacin and is the active part in oxidation-reduction. It accepts a hydrogen atom during biological oxidation processes.

Riboflavin (vitamin B$_2$) is a necessary part of another important hydrogen acceptor coenzyme, ***flavin adenine dinucleotide*** (FAD).

FLAVIN ADENINE DINUCLEOTIDE (FAD)
(The portion in color is riboflavin.)

FAD accepts hydrogen atoms in the following manner:

Glucose and glycogen are the principal sources of energy in the body. How NAD^+, $NADP^+$, and FAD fit into the oxidation of glucose and glycogen will be considered in the next chapter.

NUCLEIC ACIDS

Like the polysaccharides and the polypeptides, the *nucleic acids* are polymeric substances, with molecular weights up to several million. Nucleic acids are found in all living cells, with the exception of the red blood cells of mammals. The almost infinite variety of possible structures for nucleic acids allows information in coded form to be recorded in molecular structures in a somewhat similar fashion to the way a few language

The coded information of nucleic acids tells cells which molecules to synthesize.

Figure 10–18 The structure of α-D-ribose and α-2-deoxy-D-ribose. In the IUPAC names given, α indicates the one of two ring-forms possible; D distinguishes the isomers that rotate plane polarized light in opposite directions, and the 2 indicates the carbon to which no oxygen is attached in the second sugar.

α-D-*ribose*

α-2-*deoxy-*
D-*ribose*

Uracil

Figure 10—19 Some nitrogenous bases obtained from the hydrolysis of nucleic acids.

symbols can be used to convey the many ideas in this book. Such stored information is believed to control the inherited characteristics of the next generation as well as many of the ongoing life processes of the organism.

Hydrolysis of nucleic acids yields one of two simple sugars, phosphoric acid (H_3PO_4), and a group of nitrogen compounds that have basic

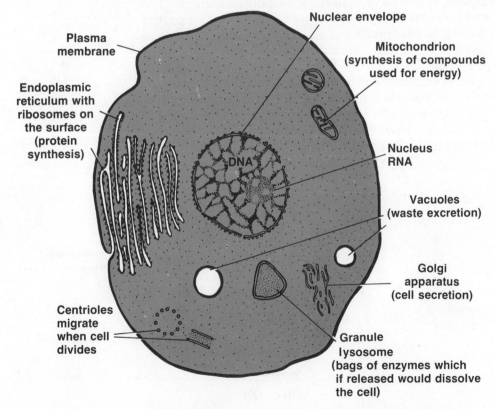

Figure 10—20 Diagrammatic generalized cell to show the relationships between the various components of the cell. Cytoplasm is the material of the cell, exclusive of the nucleus. Many of the components shown are not visible through an ordinary optical microscope.

(alkaline) properties. The structures of the two sugars in nucleic acids are shown in Figure 10–18. The names and formulas of the basic nitrogen compounds are given in Figure 10–19.

The nucleic acids can be classified as either those containing the sugar *α-2-deoxy-D-ribose,* or those containing *α-D-ribose.* The former are called *deoxyribonucleic acids* (DNA) and the latter *ribonucleic acids* (RNA). DNA is found primarily in the nucleus of the cell, whereas RNA is found mainly in the cytoplasm outside the nucleus (Figure 10–20).

Nucleotides

The repeating units of DNA or RNA are *nucleotides.* These substances contain a simple sugar unit, one of the nitrogenous bases, and one or two units of phosphoric acid. An example of a nucleotide structure is illustrated in Figure 10–21.

Polynucleotides

In addition to the mononucleotides, partial hydrolysis of DNA or RNA yields oligonucleotides that have a few nucleotide units bonded as shown in Figure 10–22. Obviously, a large number of oligonucleotides is possible when one considers the choice of base structures and the different sequence possibilities for the chain of nucleotides.

DNA and RNA are polynucleotides. The number of possible structures for these molecules, which have molecular weights as high as a few million amu, appears to be almost limitless. Since DNA is a major part of the chromosome material in the nucleus of a cell, it seems reasonable to assume that the organism's characteristics are coded in the DNA structure. Even with over two million different species of organisms and if each individual of each species requires a different DNA structure, there are ample combinations of nucleotides for each individual to be unique. It is now believed that some kinds of RNA transfer the information coded in the DNA structure to the cytoplasmic region of the cell, where they control the thousands of reactions that occur.

One nucleotide is joined to another by an ester-forming reaction.

$$-\overset{|}{\underset{\underset{O}{\|}}{P}}-OH + HO-\overset{|}{\underset{|}{C}}-$$

$$\longrightarrow \overset{|}{\underset{\underset{O}{\|}}{P}}-O-\overset{|}{\underset{|}{C}}- + H_2O$$

The heredity of an organism is contained in DNA molecules.

(*phosphoric acid unit*)

(*adenine unit— a nitrogenous base*)

(*ribose unit—a simple sugar*)

Figure 10–21 A nucleotide. If other bases are substituted for adenine, a number of nucleotides are possible for each of the two sugars shown in Figure 10–18. There is ample evidence that the nucleotides found in both DNA and RNA have the general structure indicated for this nucleotide.

Figure 10–22 Bonding structure of a trinucleotide. Bases 1, 2, and 3 represent any of the nitrogenous bases obtained in the hydrolysis of DNA and RNA. The primary structure of both DNA and RNA is an extension of this structure to produce molecular weights as high as a few million amu.

Three major types of RNA have been identified. They are messenger RNA (mRNA), transfer RNA (tRNA), and ribosomal RNA (rRNA). Each has a characteristic molecular weight and base composition. Messenger RNA's are generally the largest, with molecular weights between 25,000 and one million amu. They contain from 75 to 3000 mononucleotide units. Transfer RNA's have molecular weights in the range of 23,000 to 30,000 amu and contain 75 to 90 mononucleotide units. Ribosomal RNA's have molecular weights between those of mRNA's and tRNA's and make up as much as 80 percent of the total cell RNA.

Most RNA is found in the cytoplasm and ribosomes of the cell (Figure 10–20), but in liver cells as much as 11 percent (largely mRNA) of the total cell RNA is found in the nucleus. Besides having different molecular weights, the three types of RNA appear to differ in function. One difference in function is described in the next chapter in the discussion of natural protein synthesis.

Secondary Structure of DNA and RNA

In 1953, James D. Watson and Francis H. C. Crick (Figure 10–23) proposed a secondary structure for DNA that has since gained wide acceptance. Figure 10–24 illustrates the proposed structure, in which two polynucleotides are arranged in a double helix stabilized by hydrogen bonding between the base groups opposite to each other in the two chains. RNA is generally a single strand of helical polynucleotide.

Figure 10–23 Francis H. C. Crick (1916–) (*right*) and James D. Watson (1928–) (*left*), working in the Cavendish Laboratory at Cambridge, built scale models of the double helical structure of DNA based on the x-ray data of Maurice H. F. Wilkins. Knowing distances and angles between atoms, they compared the task to the working of a three-dimensional jigsaw puzzle. Watson, Crick, and Wilkins received the Nobel Prize in 1962 for their work relating to the structure of DNA.

Figure 10–24 (*a*) Double helix structure proposed by Watson and Crick for DNA. S-sugar, P-phosphate, A-adenine, T-thymine, G-guanine, C-cytosine. (*b*) Hydrogen bonds in the thymine-adenine and cytosine-guanine pairs stabilize the double helix. Adenine will also base pair with uracil in mRNA since there is no thymine in mRNA.

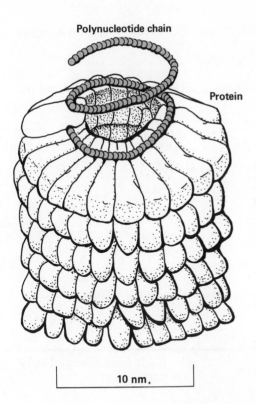
Polynucleotide chain

Protein

Figure 10–25 The imagined structure of the tobacco mosaic virus. The polynucleotide chain is coiled, and there is one protein unit for each three nucleotide units. Part of the polynucleotide chain is exposed for clarity. This structure is based on x-ray studies.

10 nm.

Virus Structure

A virus is a parasitic chemical complex that can reproduce only when it has invaded a host cell. It has the ability to disrupt the life processes of the host cell and order its cell contents to reproduce the virus structure. The isolated virus unit has neither the enzymes nor the smaller molecules necessary to reproduce itself alone.

A virus is a polynucleotide surrounded by a layer of protein. One virus that has been studied in detail is the tobacco mosaic virus, illustrated in Figure 10–25.

SELF-TEST 10-C

1. The best term to describe the general function of enzymes is (catalyst/intermediate/oxidant).

2. In the lock-and-key analogy of enzyme activity, the enzyme functions as the

   _____ while the substrate molecule serves as the _____.

3. Pellagra can be prevented by intake of the vitamin named _____.

4. The activation energy of many biological reactions is decreased if a(n)

   _____ is present.

5. Apoenzyme + coenzyme ⟶ _____.

6. Riboflavin is a vitamin that is needed because it is part of an essential

   _____.

7. The coenzyme nicotinamide adenine dinucleotide (NAD^+) cannot be made by the human body unless it has a supply of _____.

8. That portion of the enzyme at which the reaction is catalyzed is called the

   _____ _____.

9. (a) The sugar in RNA is _____,

   (b) while the one in DNA is _____.

10. A nucleotide contains _____, _____, and

    _____.

11. The secondary structure of DNA consists of a(n) _____ _____.

12. A virus is a chemical compound that can reproduce itself. (True/False)

MATCHING SET

_____ 1. D-glucose

_____ 2. linoleic acid

_____ 3. methionine

_____ 4. enzymes

_____ 5. carbohydrate storage in animals

_____ 6. starch

_____ 7. polypeptides

_____ 8. prostaglandins

_____ 9. DNA

_____ 10. fibrous protein

_____ 11. cellulose

_____ 12. vitamins

_____ 13. enzyme that splits sucrose into fructose and glucose.

(a) made from essential fatty acids

(b) polymer consisting of α-D-glucose units

(c) proteins

(d) sugar present in blood

(e) an essential amino acid

(f) sucrase

(g) an unsaturated fatty acid

(h) a polynucleotide

(i) biochemical catalysts

(j) coenzymes

(k) glycogen

(l) collagen

(m) polymer consisting of β-D-glucose units

QUESTIONS

1. Show the structure of the product that would be obtained if two alanine molecules (Table 10–2) react to form a dipeptide.

2. What is an essential amino acid?

3. The ketone structure of D-fructose has three asymmetric carbon atoms per molecule. How many isomers result from the asymmetric centers?

4. Name a polysaccharide that yields only D-glucose upon complete hydrolysis. Name a disaccharide that yields the same hydrolysis product.

5. What is the difference between the starch, amylopectin, and the "animal starch," glycogen?

6. What is the chief function of glycogen in animal tissue?

7. Explain the basic difference between starch, amylose, and cellulose.

8. What functional groups are always present in each molecule of an amino acid?

9. Give the name and formula for the simplest amino acid. What natural product has a high percentage of this amino acid?

10. If six different amino acids formed all the possible different tripeptides, how many would there be?

11. What is the meaning of the terms: primary, secondary, and tertiary structures of proteins?

12. What three molecular units are the parts of nucleotides?

13. Based on the structures in Figure 10–18, explain the meaning of the prefix *deoxy-* in deoxyribonucleic acid.

14. How many trinucleotides with the structure indicated in Figure 10–22 could be made with the nitrogenous bases listed in Figure 10–19?

15. What are the basic differences between DNA and RNA structures?

16. What stabilizing forces hold the double helix together in the secondary DNA structure proposed by Watson and Crick?

17. Consult a medical dictionary and determine the difference between atherosclerosis and arteriosclerosis.

18. What two molecular structures are present in viruses?

19. Distinguish between the three types of RNA as to molecular weight, number of mononucleotide units per molecule, and function.

20. Using Figure 10–14, explain the concept of activation energy for a chemical reaction and show the effect of an enzyme on the activation energy.

21. Point out the three important similarities between the lock-and-key analogy and enzymatic activity.

22. What is a coenzyme and why is one sometimes necessary?

BIO-CHEMICAL PROCESSES

CHAPTER 11

In Chapter 10 we examined the structures of some molecules that are important in biochemical systems. We shall now look at some of the reactions by which such molecules are constructed, serve their function, and are subsequently destroyed. There are three aspects of biochemical systems which should be kept in mind in this study. First, the contents of a living cell are in a dynamic state; the molecules are constantly being synthesized and degraded. However, the healthy living cell is characterized by a "steady state" condition in which the rates of buildup and breakdown are nearly the same at any time. Second, biochemical processes are very general in that the same basic chemistry is employed by a wide variety of cells. For example, the same types of reactions utilized to obtain energy from a compound in human beings also occur in simple unicellular organisms. Third, biochemical systems are composed of literally thousands of different kinds of molecules. Out of such an apparent chaos comes an ordered array of reactions that supports all life forms. In view of this complexity, it is evident that an elementary treatment of what is presently known about biochemical reactions will have to deal only with selected highlights.

Similar biochemical reactions occur in a wide variety of living cells.

BIOCHEMICAL ENERGY AND ATP

Most useful biochemical processes require energy to drive the reactions toward the desired products. This energy comes from the ultimate energy source in our environment, the sun. How is the energy of sunlight stored in various compounds, and in turn supplied to feed the life processes? Energy is stored primarily by *adenosine triphosphate,* often written ATP (Figure 11–1). The energy obtained by the oxidation of foods is mostly used to synthesize ATP. ATP is the immediate source of energy in muscular contraction, and the energy released by it allows many energy-requiring biochemical reactions to occur.

ATP is an energy-rich molecule that furnishes the energy required by many biochemical processes.

Each ATP molecule contains two so-called high-energy phosphate bonds. These are marked by wiggle bonds ($\sim$) in Figure 11–1. In the presence of a suitable catalyst, ATP will undergo a three-step hydrolysis. The

Figure 11–1 Molecular structure of adenosine triphosphate (ATP). Note the similarity between ATP and the fundamental unit of a nucleic acid, the nucleotide (see Chapter 10).

hydrolysis of ATP to adenosine diphosphate (ADP) and phosphoric acid releases about 12 kilocalories per mole. The second hydrolysis of ADP to adenosine monophosphate (AMP) also produces about 12 kilocalories of energy per mole. Finally, the hydrolysis of AMP to adenosine, which involves a low energy bond, releases only about 2.5 kilocalories per mole.

$$\boxed{\text{ATP}} + \text{HOH} \xrightarrow{\text{Catalyst}}$$

$+ \ H_3PO_4 + 12 \ \text{kcal}$
(APPROX.)

ADENOSINE DIPHOSPHATE
(ADP)

$$\boxed{\text{ADP}} + \text{HOH} \xrightarrow{\text{Catalyst}}$$

$+ \ H_3PO_4 + 12 \ \text{kcal}$
(APPROX.)

ADENOSINE MONOPHOSPHATE
(AMP)

$$\boxed{\text{AMP}} + \text{HOH} \xrightarrow{\text{Catalyst}} \text{ADENOSINE} + \text{H}_3\text{PO}_4 + 2.5\text{ kcal}$$
$$\text{(APPROX.)}$$

ADENOSINE

The energies for these three reactions are given only to show the relatively large amount of energy associated with the hydrolysis of the first two phosphate units in the ATP structure; however, not all of the energy released by chemical reactions is available to initiate other chemical changes. A more meaningful value is the energy that *is* available to initiate other chemical change; this is termed the change in **free energy**. For the three hydrolytic reactions of ATP, the approximate amounts of free energy released are:

ATP hydrolysis: 7.4 kcal per mole released
ADP hydrolysis: 6.8 kcal per mole released
AMP hydrolysis: 2.2 kcal per mole released

A visual conception of the releases of free energy by hydrolysis of ATP and ADP is shown in Figure 11–2.

These free energy values indicate the amount of chemical potential energy in the phosphate bonds in the ATP molecule. Such energy-producing reactions provide the energy necessary for energy-requiring

The change in free energy is generally represented by the symbol ΔG, and is the energy provided by a reaction to do work. The symbol, G, honors J. Willard Gibbs, who did much to establish the concept of free energy.

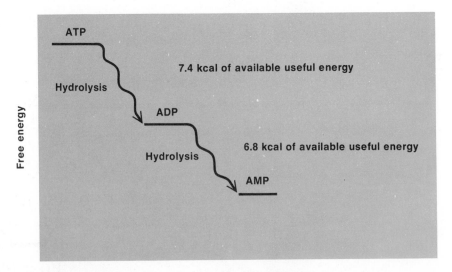

Figure 11–2 Free energy furnished by successive hydrolyses of phosphate groups from ATP. This energy is available to bring about other useful chemical reactions.

biochemical reactions to occur, and thus supply the energy necessary for the basic life processes.

PHOTOSYNTHESIS

Photosynthesis is a very complex process that produces the relatively simple overall reaction in which carbon dioxide and water are converted into energy-rich carbohydrates by solar energy.

$$6CO_2 + 6H_2O + 688\ kcal \longrightarrow C_6H_{12}O_6 + 6O_2$$

CARBON   WATER   ENERGY   GLUCOSE   OXYGEN
DIOXIDE   (SUNLIGHT)

In photosynthesis carbon dioxide is ***reduced*** to form a sugar,

$$6CO_2 + 24H^+ + 24e^- \longrightarrow C_6H_{12}O_6 + 6H_2O$$

and water is ***oxidized*** to oxygen

$$12H_2O \longrightarrow 6O_2 + 24H^+ + 24e^-$$

Note that these two half reactions, the first reduction and the second oxidation, give the overall reaction when added together.

Not all the details of photosynthesis are fully understood. However, some aspects are presented here because it is the beginning of the energy flow through biochemical systems.

Photosynthesis is generally considered in terms of the ***light reaction,*** which can occur only in the presence of light energy, and the ***dark reaction,*** which can occur in the dark, but feeds on the high energy structures produced in the light reaction. Actually, both the light and dark reactions are a series of reactions, all occurring simultaneously in the green plant cell. The light reaction is unique to green plants while the dark reaction is characteristic of both plant and some animal cells.

The Light Reaction

Photosynthesis is initiated by a quantum of light energy. The green plant contains certain pigments that readily absorb light in the visible region of the spectrum. The most important of these are the chlorophylls, ***chlorophyll a*** and ***chlorophyll b.*** Note that both chlorophylls are compounds of magnesium and both have complex ring systems. Such ring systems usually absorb light in the visible region of the spectrum; consequently, they are colored. For example, chlorophyll is green because it absorbs light in the violet region (about 400 nanometers) and the red region (about 650 nanometers) and allows the green light between those wavelengths to be reflected or transmitted.

When chlorophyll absorbs photons of light, electrons are raised to higher energy levels. As these electrons move back down to the ground state, very efficient subcellular components of the plant cell known as chloroplasts grab this energy and, through a series of steps that are not all completely known, store the energy as chemical potential energy. As

---

Marginal notes:

Reduction: gain of electrons or hydrogen.
Oxidation: loss of electrons or hydrogen.

Photosynthesis involves a number of different steps and is a very complex process.

The energy of a photon is captured by chlorophyll by raising an electron to a higher energy state.

CH$_2$
‖
CH   H
      C
H$_3$C ─────── N   N ── CH$_2$CH$_3$ CH$_3$
HC ─── Mg ── CH
H   N   N
H$_3$C ─────────── CH$_3$
H   C
CH$_2$HC─C═O
CH$_2$ COOCH$_3$
COOC$_{20}$H$_{39}$

CHLOROPHYLL A

CH$_2$
‖
CH   H
      C
H$_3$C ─────── N   N ── CH$_2$CH$_3$ HC═O
HC ─── Mg ── CH
H   N   N
H$_3$C ─────────── CH$_3$
H   C
CH$_2$HC─C═O
CH$_2$ COOCH$_3$
COOC$_{20}$H$_{39}$

CHLOROPHYLL B

shown in Figure 11–3, one of the chemicals used to store this energy is nicotinamide adenine dinucleotide phosphate (NADP⁺).

Energy is absorbed by NADP⁺ in the process of being reduced by a hydrogen donor to NADPH + H⁺.

The structure of NADP⁺ is on page 255.

$$NADP^+ + 2H + energy \longrightarrow NADPH + H^+$$

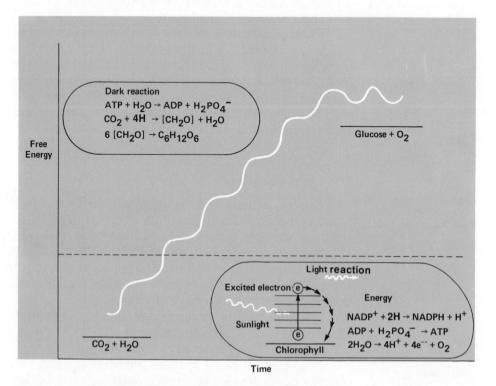

Figure 11–3 Free energy within the chemical system is increased as carbon dioxide and water are converted into glucose and oxygen by photosynthesis. This results in stored, useful energy in glucose. A very simplified mechanism for photosynthesis is discussed in the text.

NADPH eventually transfers its precious energy to the universal store-house of biochemical energy, adenosine triphosphate, ATP.

Green plants pro-duce oxygen by oxi-dizing water.

Thus far little mention has been made of the electrochemical charge balance and the oxygen produced by photosynthesis, both of which are important parts of the light reaction. Water, in the presence of chloro-plasts and light, is decomposed into oxygen, hydrogen ions, and electrons:

$$2H_2O \longrightarrow 4H^+ + 4e^- + O_2$$

The hydrogen ions and electrons are available for maintaining the balance of charge, and oxygen is liberated from the plant cell.

At this point we no longer need the energy of the sun. We have the energy necessary to run biochemical systems stored in the ATP structure. If the plant cell is given the minerals along with carbon dioxide and water, it, or subsequent living cells, can use the energy in ATP in the dark to pro-vide energy for the complex biochemical reactions that take place.

The Dark Reaction

The dark reaction is responsible for the ultimate conversion of gaseous carbon dioxide to glucose. It was discovered by Melvin Calvin (1911–   ; Nobel Prize in 1961). Calvin studied the uptake of radioactive carbon in carbon dioxide by plant cell chloroplasts. He illuminated the plants for definite short periods of time and then analyzed the plant cells to determine which compounds contained the most radioactive carbon. As the time periods were reduced, more of the radioactive carbon was found in those compounds into which it had been initially incorporated and less in compounds that had been formed in subsequent reactions. For example, after only five seconds' illumination, radioactive carbon is found in the compound 3-phosphoglyceric acid:

The mechanisms of many reactions can be studied by the use of radioactive atoms known as tracers.

3-PHOSPHOGLYCERIC ACID

This compound is apparently formed in the initial reaction in which CO_2 from the air reacts with some molecule present in the plant. Calvin discovered that the key was the reaction of atmospheric CO_2 with ribulose 1,5-diphosphate to give two molecules of 3-phosphoglycerate:

$$
\left.
\begin{array}{c}
^*CO_2 \\
\text{CARBON DIOXIDE} \\
+ \\
CH_2OPO_3{}^{2-} \\
| \\
C{=}O \\
| \\
HCOH \\
| \\
HCOH \\
| \\
CH_2OPO_3{}^{2-}
\end{array}
\right\}
\longrightarrow
\begin{array}{c}
^*COO^- \\
| \\
HCOH \\
| \\
CH_2OPO_3{}^{2-} \\
+ \\
COO^- \\
| \\
HCOH \\
| \\
CH_2OPO_3{}^{2-}
\end{array}
$$

RIBULOSE 1,5-DIPHOSPHATE   3-PHOSPHOGLYCERATE

*C indicates radioactive carbon (Calvin's experiments).

Figure 11—4 Abbreviated version of the dark reaction of photosynthesis. The energy needed to carry out the dark reactions is furnished from the high energy compounds produced in the light reaction. (P) indicates a phosphate group.

The 3-phosphoglycerate is then transformed into other carbohydrates in reactions that produce glucose and regenerate ribulose 1,5-diphosphate (for further uptake of more atmospheric CO_2). The energy needed to carry out these reactions is furnished by the NADPH and ATP generated from the light reaction.

Figure 11–4 presents the cyclic character of the dark reaction of photosynthesis. Note that carbon dioxide enters the cycle at the upper left and that sugars are removed at the lower right. Since just one carbon atom enters the cycle per revolution, only one fructose 6-phosphate molecule out of seven is converted to glucose.

When the light and dark reactions are considered, one can readily see that the net equation for photosynthesis, given at the beginning of this section, is a tremendous simplification of the actual process. According to the latest research, the light reaction may be summarized as follows:

Oxygen and ATP are the products of the light reaction.

$$12H_2O + 12NADP^+ + 24ADP + 24H_3PO_4 \longrightarrow$$
$$24ATP + 12H^+ + 12NADPH + 6O_2 + 24H_2O$$

Water has been placed on both sides of the equation because water is part of the input of the reaction (each $NADP^+$ requires one H_2O for reaction) and is part of the output of the reaction (the formation of each ATP splits out one H_2O). The coefficients of the equation are in the correct ratios to give $6O_2$, the production of which provides the electrons to form one glucose molecule ($C_6H_{12}O_6$) in the dark reaction. The dark reaction (carbon-fixation) may be represented as follows:

Glucose is a product of the dark reaction.

$$6CO_2 + 18ATP + 12NADPH + 12H^+ + 12H_2O \longrightarrow$$
$$C_6H_{12}O_6 + 18ADP + 12NADP^+ + 18H_3PO_4$$
GLUCOSE

---

1. The source of energy for photosynthesis is _____.

2. Most of the energy obtained by food oxidation is used immediately to synthesize the molecule _____.

3. The hydrolysis of ATP results in the molecules _____ and _____. The other "product" is _____.

4. Energy available to do work is called _____ energy.

5. The reactants in the photosynthesis process are _____ and _____; _____ must also be supplied.

6. Energy absorbed by chloroplasts in the green cells of a plant is transferred by means of the molecule _____ to ATP.

7. The first chemical product of the dark reaction, as determined by Calvin, is _____ .

8. The final product of the dark reaction is the substance _____ .

DIGESTION

From a chemical point of view, digestion is the breakdown of ingested foods through hydrolysis. The products of these hydrolytic reactions are relatively small molecules that can be absorbed through the intestinal walls into the body fluids where they are used for metabolic processes. The hydrolytic reactions of digestion are catalyzed by enzymes, there being a specific enzyme for each hydrolysis. The hydrolysis of carbohydrates ultimately yields simple sugars, proteins yield amino acids, and fats yield fatty acids. The activities of some of the more important enzymes are described in Table 11–1. Note that the pH of the stomach fluid is much lower than the pH of the intestines.

Digestion is the hydrolysis of carbohydrates, fats, and proteins to provide small molecules that can be absorbed.

TABLE 11—1 Principal Digestive Enzymes

Enzyme	Source	Substrate	Products	Optimal pH
Ptyalin	Salivary glands	Starch	Smaller carbohydrate polymers (minor physiologic role)	6–7
Pepsin	Chief cells of stomach	Protein	Polypeptides	1.6–2.4
Gastric lipase	Stomach	Fat	Glycerides, fatty acids (minor physiologic role)	–
Enterokinase	Duodenal mucosa	Trypsinogen	Trypsin	–
Trypsin		Denatured proteins and polypeptides	Small polypeptides (also activates chymotrypsinogen to chymotrypsin)	8.0
Chymotrypsin	Exocrine pancreas	Proteins and polypeptides	Small polypeptides	8.0
Nucleases		Nucleic acids	Nucleotides	–
Carboxypeptidases		Polypeptides	Smaller polypeptides*	–
Pancreatic lipase		Fat	Glycerides, fatty acids, glycerol	8.0
Pancreatic amylase		Starch	Maltose units	6.7–7.0
Aminopeptidases		Polypeptides	Smaller polypeptides†	8.0
Dipeptidase		Dipeptide	Amino acids	–
Maltase		Maltose	Hexoses	5.0–7.0
Lactase		Lactose	(glucose, galactose	5.8–6.2
Sucrase	Intestinal glands	Sucrose	and fructose)	5.0–7.0
Nucleotidase		Nucleotides	Nucleosides, phosphoric acid	–
Nucleosidase		Nucleosides	Purine or pyrimidine base, pentose	–
Intestinal lipase		Fat	Glycerides, fatty acids and glycerol	8.0

*Removal of C-terminal amino acid.
†Removal of N-terminal amino acid.

Carbohydrate Digestion and Absorption

The principal forms of carbohydrates in our food are (1) high molecular weight polymers such as starch and glycogen, (2) disaccharides such as sucrose and lactose, and (3) simple sugars such as glucose and fructose. The first enzyme capable of acting on ingested food is furnished by the saliva and is named salivary amylase, or ptyalin. Its action on starch or glycogen produces limited amounts of the disaccharide maltose. Ptyalin is inactivated by the high acidity in the stomach. The stomach furnishes no enzymes that can catalyze the splitting of carbohydrate polymers. Indeed, the high acidity of the gastric juice would destroy or inactivate all such complex structures.

When food passes from the stomach, the acidity is neutralized by a secretion of the pancreas. The enzymes secreted by the pancreas can split some of the polysaccharides to maltose and maltose to glucose, and can catalyze other hydrolytic reactions. The final result is a mixture of simple sugars such as glucose, fructose, and galactose. These simple sugars are then absorbed into the blood stream, where the concentration of blood sugar is regulated by the hormone insulin. If the sugar level is too high, the simple sugars are converted into the polysaccharide glycogen in the liver; if the blood sugar level is too low, the stored glycogen is hydrolyzed to raise the level. Malfunctions of our biological systems can lead to too much blood sugar (hyperglycemia) or too little blood sugar (hypoglycemia); either condition, if sustained, indicates one type or another of diabetes.

Acidic solutions: pH below 7 Basic solutions: pH above 7

Human blood normally contains between 0.08 and 0.1% glucose.

Insulin is a protein.

Fat and Oil Digestion and Absorption

The term *lipid* denotes a group of compounds that includes fats, oils, and other substances whose solubility characteristics are similar to those of fats and oils. These compounds are not all structurally related, and we shall consider only the triglycerides (triesters of fatty acids and glycerol) in this chapter. A typical triglyceride is palmitooleostearin. Its structure and its hydrolytic products are shown in the following equation:

$$H_2C-O-\overset{\displaystyle O}{\overset{\|}{C}}-(CH_2)_{14}CH_3$$

$$HC-O-\overset{\displaystyle O}{\overset{\|}{C}}-(CH_2)_7CH=CH(CH_2)_7CH_3 + 3HOH$$

$$H_2C-O-\overset{\displaystyle O}{\overset{\|}{C}}-(CH_2)_{16}CH_3$$

PALMITOOLEOSTEARIN (A TRIGLYCERIDE)

WATER

$$CH_3(CH_2)_{14}COOH$$ PALMITIC ACID

$$CH_3(CH_2)_7CH=CH(CH_2)_7COOH$$ OLEIC ACID

$$CH_3(CH_2)_{16}COOH$$ STEARIC ACID

$$HO-CH_2$$
$$HO-CH$$
$$HO-CH_2$$
GLYCEROL

Digestion of fats and oils occurs primarily in the intestinal tract where bile salts secreted by the liver aid in the process. The enzyme that aids in the hydrolysis of the fatty acid esters is water-soluble, while the fats and oils are insoluble in water. Bile salts emulsify the oil, that is, they break up the oil into very tiny drops and prevent the drops from recombining readily. The tiny drops provide more surface area for the enzyme to attack so digestion can occur. The bile salts form an interface between the nonpolar oil and the polar water and make it possible for the oil to "dissolve" in water. For a molecule to be an emulsifier between polar and nonpolar molecules, the emulsifier must have characteristics of both. One of the principal bile salts is derived from glycocholic acid.

Bile salts act much like detergent molecules.

$$CH_3 \quad\quad O \quad H \quad\quad O$$
$$OH \quad CH(CH_2)_2\overset{\|}{C}-\overset{|}{N}-CH_2\overset{\|}{C}-O^- \quad Na^+$$

SODIUM SALT OF GLYCOCHOLIC ACID

Notice that the bulky hydrocarbon groups of this bile salt are compatible with oil or fat and that the —OH and ionic groups anchor to water molecules. The bile salts emulsify oil in a manner similar to the action of a soap or detergent during the cleaning process (Chapter 8).

Protein Digestion and Absorption

The hydrolysis of proteins begins in the stomach and continues in the small intestine. Several different types of enzymes are known to be involved. These enzymatic systems must be controlled very carefully, for they have the potential of digesting the walls of the stomach and intestines. A number of these enzymes are secreted in an inactive form. For example, pepsin, which is secreted in the stomach, is first present in a form called pepsinogen. The molecular weight of pepsinogen is 42,600 amu. In the presence of the acid of the stomach, pepsinogen is broken down to **pepsin.** The molecular weight of pepsin is 34,500 amu. It is reasonable to believe that pepsinogen, pepsin, other enzymes, and the stomach acid normally have no effect on the stomach wall. However, pepsin would have considerable action on the stomach protein if it were formed under the mucous lining, a lining which is constantly sloughing off like the outer skin.

The stomach is protected from protein-splitting enzymes by a mucous lining.

Pepsin facilitates the breakdown of only about 10 percent of the bonds in a typical protein, leaving polypeptides with molecular weights from 600 up to 3000 amu. In the small intestine, hydrolysis is completed to amino acids, which are absorbed through the intestinal wall.

Some protein enzymes are sold commercially. Meat tenderizers are proteases, materials that speed up partial digestion of meat. Enzymes are used as stain removers in detergents, although they may irritate the skin of some individuals. Related enzymes are also used to free the lens of the eye prior to cataract surgery.

Enzymes and Heredity

Genetic effects are often observed in the pattern of enzymes produced by individuals or races. An example of this is found in "lactose intolerance," common in certain peoples of Asia (e.g., Chinese and Japanese) and Africa (many black tribes), whose diets have traditionally contained little milk after the age of weaning. While infants, such people manufacture the enzyme **lactase** that is necessary to digest lactose, a sugar occurring in all mammals' milk. As they grow older their bodies stop producing this enzyme because their diets normally contain no milk, and the ingestion of milk products containing lactose can lead to considerable discomfort in the form of stomach aches and diarrhea. People whose ancestral adult diets contained substantial amounts of milk or milk products (African tribes such as the Masai, Mongois, Caucasians, etc.) continue to produce lactase as adults and can eat such foods and digest the lactose they contain. It is quite possible that this is only one of several similar cases in which a traditional tribal diet has altered the pattern of production of digestive enzymes.

GLUCOSE METABOLISM

The biochemical process by which living cells obtain energy from glucose and similar compounds involves many chemical reactions. The initial sequence of reactions can follow two courses; one of these does not use oxygen (***anaerobic***) while the other one does (***aerobic***).

Aerobic: use oxygen. Anaerobic: use no oxygen.

When a muscle is used, glucose is anaerobically converted by a series of steps known as the ***Embden-Meyerhof pathway*** to lactic acid (Figure 11–5). We are concerned here only with the starting material, the energy flow, and the final products. The overall reaction can be represented by the equation:

$$C_6H_{12}O_6 + 2ADP + 2H_3PO_4 \longrightarrow 2CH_3-\overset{\displaystyle H}{\underset{\displaystyle OH}{C}}-\overset{\displaystyle O}{C}-OH + 2ATP + 2H_2O$$

GLUCOSE   LACTIC ACID

Muscular activity converts glucose to lactic acid, which produces fatigue in muscles as the lactic acid accumulates.

If the muscle is used strenuously for a sufficiently long period of time, the lactic acid buildup will produce tiredness and a painful sensation in the muscles. The blood stream eventually carries away the lactic acid, but time and oxygen are needed to convert the lactic acid to carbon dioxide and water, which are excreted. This is accomplished by the slower, aerobic (with air) process known as the ***Krebs cycle.*** As you look at Figure 11–6, imagine how a molecule of citric acid reacts with the enzyme aconitase. The citric acid has —OH and —H exchange sites which exchange and produce isocitric acid. The isocitric acid dissociates from aconitase and goes to the next enzyme, isocitric dehydrogenase, where two hydrogen atoms are removed by $NADP^+$, and so on around the cyclic pathway. Hydrogen atoms are removed, carboxyl groups are destroyed as carbon dioxide is formed, and hydrolysis occurs. Since two carbon atoms are fed into the cycle at oxaloacetic acid by acetyl coenzyme A, two carbon atoms in the form of two molecules of carbon dioxide (CO_2) must be eliminated before the cycle returns to oxaloacetic acid. This is exactly what happens. See whether you can find the two molecules of CO_2 formed in the Krebs cycle. Figure 11–7 will help.

If the Krebs cycle is aerobic, where does the oxygen enter the cycle? Oxygen is required to regenerate the coenzyme $NADP^+$ from NADPH. It is also required to remove the hydrogen atoms from NADH and $FADH_2$. There are about seven known steps involved in the removal of hydrogen

THE EMBDEN-MEYERHOF PATHWAY

Figure 11–5 The Embden-Meyerhof pathway—anaerobic oxidation of glucose and glycogen. The (P) indicates inorganic phosphate, PO_4^{3-}. (Adapted from Routh, J. I., *Introduction to Biochemistry,* W. B. Saunders Co., Philadelphia, 1971, p. 100.)

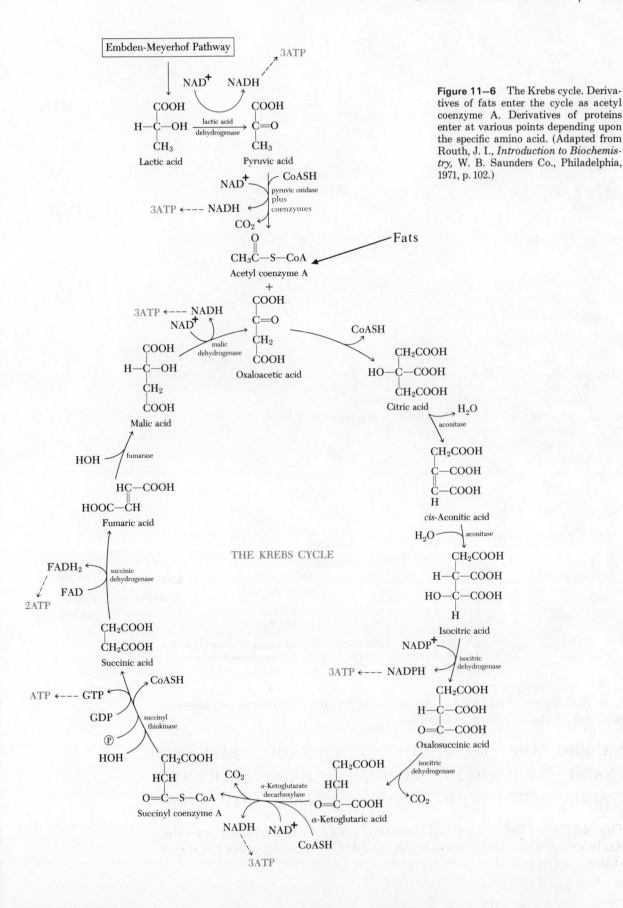

Figure 11–6 The Krebs cycle. Derivatives of fats enter the cycle as acetyl coenzyme A. Derivatives of proteins enter at various points depending upon the specific amino acid. (Adapted from Routh, J. I., *Introduction to Biochemistry,* W. B. Saunders Co., Philadelphia, 1971, p. 102.)

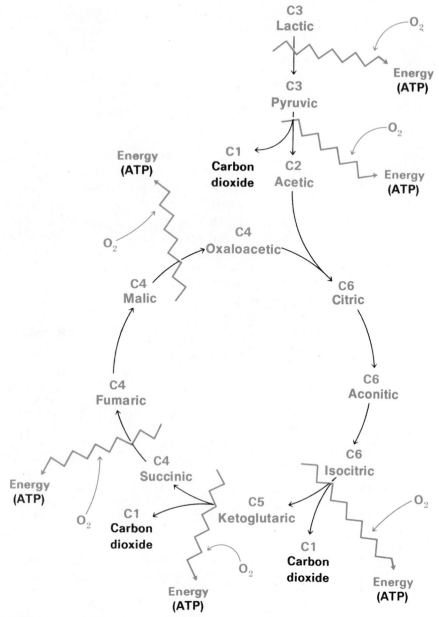

Figure 11–7 A summary of the Krebs cycle showing the number of carbon atoms per molecule for the principal substances in the cycle.

from NADH and NADPH and six steps for $FADH_2$. The steps can be summarized by the following equations:

$$NADPH + 3ADP + 3H_3PO_4 + H^+ + \tfrac{1}{2}O_2 \longrightarrow NADP^+ + 3ATP + 4H_2O$$

$$NADH + 3ADP + 3H_3PO_4 + H^+ + \tfrac{1}{2}O_2 \longrightarrow NAD^+ + 3ATP + 4H_2O$$

$$FADH_2 + 2ADP + 2H_3PO_4 \qquad + \tfrac{1}{2}O_2 \longrightarrow FAD + 2ATP + 3H_2O$$

The $NADP^+$, NAD^+, and FAD are now ready to be fed back into the Krebs cycle, where the process is continued by extracting other hydrogen atoms.

When these equations are combined with the principal sequence of reactions for the Krebs cycle, we have an equation for the conversion of lactic acid into carbon dioxide and water aerobically.

$$\underset{\text{LACTIC ACID}}{C_3H_6O_3} + 18ADP + 18H_3PO_4 + 3O_2 \longrightarrow 3CO_2 + 21H_2O + 18ATP$$

The key products of glucose metabolism are ATP (energy), CO_2, and H_2O.

The key product of this oxidation reaction is ATP. Energy derived from glucose (and lactic acid) is stored and transported in the phosphorus-oxygen bonds of ATP. This is the usable product of the Krebs cycle. Waste products are carbon dioxide and water. Since two lactic acid molecules are formed from one glucose molecule, 36 ATP's are formed per glucose molecule oxidized via the Krebs cycle. Two more ATP's are formed in the Embden-Meyerhof pathway, making a total of 38 ATP molecules formed in the complete oxidation of a glucose molecule. If we burn a mole of glucose to carbon dioxide and water outside the body, 688,000 calories of energy are released. This is the total energy available. Earlier in this chapter, the point was made that breaking a mole of $P-O-P$ bonds and forming $H_2PO_4^-$ gives up about 7,400 calories of free energy (ΔG). Thirty-eight moles of ATP molecules would provide 281,200 calories of free energy.

$$38 \text{ moles ATP} \times 7,400 \frac{\text{calories}}{\text{mole ATP}} = 281,200 \text{ calories}$$

This is an efficiency of 41 percent for obtaining stored energy from the total usable energy available.

$$\frac{281,200 \text{ calories} \times 100\%}{688,000 \text{ calories}} = 41\%$$

This figure is remarkable when you consider that the efficiency of the automobile engine is only about 20 percent, and real heat engines of any size seldom go above 35 percent.

When ATP converts back to ADP by losing a phosphate group, the energy is used to move muscles, such as those that provide the pumping action of the heart, or moving the diaphragm so we can breathe, or hundreds of other movements required for everyday life.

Coupled reactions allow the energy from one reaction to be used to run another reaction.

The energy is stored in ATP and released from ATP by means of **coupled reactions.** In Figures 11–6 and 11–7, curved arrows denote reactions in which sufficient free energy is provided by one reaction to drive another reaction. These are examples of coupled reactions. In fact, the very first step in the oxidation of glucose is a specific example of a coupled reaction. The substitution of a phosphate group for a hydrogen on glucose is an unfavorable reaction, as far as energy is concerned.

(UNFAVORABLE—REQUIRES ENERGY)

Glucose + H_3PO_4 + about 3,000 cal
of free energy
per mole of glucose $\rightleftharpoons$ Glucose-6-PO_4 + H_2O

If this reaction is coupled with the favorable hydrolysis of ATP,

(FAVORABLE—RELEASES ENERGY)
$$ATP + H_2O \rightleftharpoons ADP + H_3PO_4 + \text{ about 7,400 cal of free energy}$$
$$\text{per mole of ATP}$$

the net reaction is favorable:

Glucose + $\cancel{H_3PO_4}$ + 3,000 cal
of free energy $\rightleftharpoons$ Glucose-6-PO_4 + $\cancel{H_2O}$

ATP + $\cancel{H_2O}$ $\rightleftharpoons$ ADP + $\cancel{H_3PO_4}$ + 7,400 cal
of free energy

Glucose + ATP $\rightleftharpoons$ Glucose-6-PO_4 + ADP + 4,400 cal of
free energy, net

Even with this favorable free energy change, the enzyme glucokinase is required to hasten the process.

Fats and proteins (as amino acids) can enter the Krebs cycle also. Amino acids can enter at several places after first being converted by a series of reactions into one of the compounds in the cycle. Each amino acid has its particular point of entry. For example, aspartic acid is converted into α-ketoglutaric acid and enters at that point in the Krebs cycle.

Fatty acids go through a series of at least five reactions in which the hydrocarbon chain of the fatty acid is decreased by a two-carbon fragment. The fragment completes a molecule of acetyl coenzyme A, which is important in the Krebs cycle (Figure 11–6). Palmitic acid, a 16-carbon fatty acid (see Chapter 8), would do seven turns around the cycle to form eight acetyl CoA molecules.

In conclusion, you now have seen some of the detailed chemistry involved in simply raising your arm, and you are now aware of what happens to some of the sugars and starches that disappear down the hatch. Of course, there is much more known than is presented here and there appears to be no end to what is left to be discovered.

Reactions that are favored energetically may still require a catalyst to proceed at a reasonable rate.

SELF-TEST 11-B
1. Digestion is the breakdown of foodstuffs by _____.

2. Substances whose solubility characteristics are similar to those of fats and oils are termed _____.

3. Bile salts act as (catalysts/emulsifying agents/enzymes).

4. The two products of the Embden-Meyerhof pathway are _____ and _____.

5. The end products of the Krebs energy cycle are _____ and _____.

6. Energy is stored and released from ATP by chemical reactions called _____ reactions.

SYNTHESIS OF LIVING SYSTEMS

In the quest for a molecular understanding of living systems, theories must finally be put to the ultimate test of synthesis. If some of the complex biochemical substances can be synthesized and can successfully participate in the life processes, we can be reassured that we are on the right track. It should be emphasized that the biochemist is presently working at the molecular level, and as yet only a relatively few of the giant molecules have been characterized. The syntheses in a living cell are far too complex for our present methods to duplicate. However, in spite of the enormousness of this undertaking, remarkable strides have been made recently, and interest in current research in this area is intense.

Our current knowledge is far from sufficient to allow the synthesis of a living cell.

Replication of DNA

Almost all of the cells of one organism contain the same chromosome structure in their nuclei. This structure remains constant regardless of whether the cell is starving or has an ample supply of food materials. Each organism begins life as a single cell with this same chromosome structure;

The DNA molecule is capable of causing the synthesis of its duplicate.

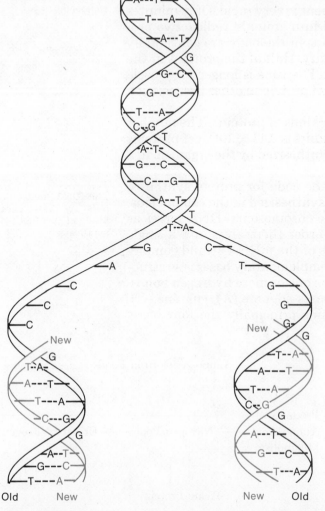

Old   New   New   Old

Figure 11–8 Replication of DNA structure. When the double helix of DNA (black) unwinds, each half serves as a template on which to assemble subunits (color) from the cell environment.

in sexual reproduction one half of this structure comes from each parent. These well known biological facts, along with recent discoveries concerning polynucleotide structures, lead to the conclusion that the DNA structure is faithfully copied during normal cell division (mitosis—both strands) and only one half is copied in cell division producing reproductive cells (meiosis—only one strand).

A prominent theory of DNA replication, based on verifiable experimental facts, suggests that the double helix of the DNA structure unwinds and each half of the structure serves as a template or pattern to reproduce the other half from the molecules in the cell environment (Figure 11–8). Replication of the DNA occurs in the nucleus of the cell. (The components of a typical cell are illustrated in Figure 10–20 on page 257.)

To make a replicate is to make a complement (something that fits) of the original.

Replicates:
object—cast of object
screw—hole for
    screw
bolt—bolt hole
adenine—thymine

Natural Protein Synthesis

The proteins of the body are being replaced and resynthesized continuously from the amino acids available to the body. The amino acids and proteins in the body can be considered as constituents of a "nitrogen pool"; additions to and losses from the pool are shown in Figure 11–9.

The use of isotopically labeled amino acids has made possible studies of the average lifetimes of amino acids as constituents in proteins—that is, the time it takes the body to replace a protein in a tissue. For a process that must be extremely complex, replacement is very rapid. Only minutes after radioactive amino acids are injected into animals, radioactive protein can be found. Although all the proteins in the body are continually being replaced, the rates of replacement vary. Half of the proteins in the liver and plasma are replaced in *six days*. The time is longer for muscle proteins, about 180 days, and replacement of protein in other tissues, such as bone collagen, takes even longer.

The proteins in the human body are being continuously replaced.

Recall that each organism has its own kinds of proteins. The number of possible arrangements of 20 amino acid units is 2.43×10^{18}, yet proteins characteristic of a given organism can be synthesized by the organism in a matter of a few minutes.

The DNA in the cell nucleus holds the code for protein synthesis. Messenger RNA, like all forms of RNA, is synthesized in the cell nucleus. The sequence of bases in one strand of the chromosomal DNA serves as the template for monoribonucleotides to order themselves into a single strand of mRNA (Figure 11–10). The bases of the mRNA strand complement those of the DNA strand. A pair of complementary bases is so structured that each one fits the other and forms one or more hydrogen bonds. Messenger RNA contains only the four bases adenine (A), guanine (G), cytosine (C), and uracil (U). DNA contains principally the four bases

The DNA molecule tells the cell what kind of protein to synthesize.

Figure 11–9   The nitrogen pool.

Amino Acids from Foods

↓

Products of
Amino Acid ⟵ Nitrogen Pool ⟶ Urinary Nitrogen
Metabolism

↑ ↓

Tissue Protein

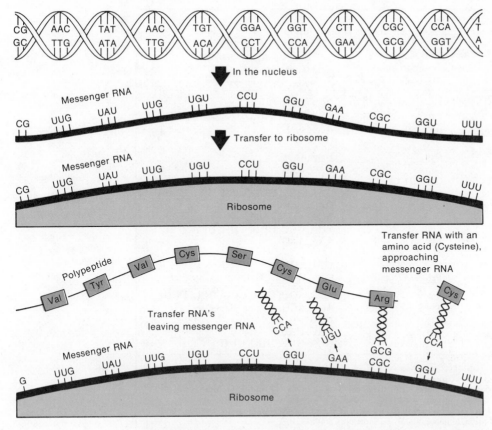

Figure 11–10 A schematic illustration of the role of DNA and RNA in protein synthesis. A, C, G, T, and U are nitrogen bases characteristic of the individual nucleotides. See Figure 10–19 for structures of the bases, and Table 11–2 for abbreviations of the amino acids used.

adenine (A), guanine (G), cytosine (C), and thymine (T). The base pairs are as follows:

DNA	mRNA
A	U
G	C
C	G
T	A

This means that every place a DNA has an adenine base (A), the mRNA will transcribe a uracil base (U), and so on, provided the necessary enzymes and energy are present.

After transcription, mRNA passes from the nucleus of the cell to a ribosome, where it serves as the template for the sequential ordering of amino acids during protein synthesis. As its name implies, messenger RNA contains the sequence message, in the form of a three-base code, for ordering amino acids into proteins. Each of the thousands of different proteins synthesized by cells is coded by a specific mRNA or segment of a mRNA molecule.

Transfer RNA's carry the specific amino acids to the messenger RNA. Each of the 20 amino acids found in proteins has at least one corresponding tRNA, and some have multiple tRNA's (Table 11–2). For example,

Transfer RNA carries specific amino acids to the messenger RNA.

TABLE 11–2   Messenger RNA Codes for Amino Acids*

Amino Acid	Shortened Notation Used for Amino Acids in Figs. 11–10 and 11–12	Base Code on mRNA
Alanine	Ala	GCA, GCC, GCG, GCU
Arginine	Arg	AGA, AGG, CGA, CGG, CGC, CGU
Asparagine	Asp-NH_2	AAC, AAU
Aspartic acid	Asp	GAC, GAU
Cysteine	Cys	UGC, UGU
Glutamic acid	Glu	GAA, GAG
Glutamine	Glu-NH_2	CAG, CAA
Glycine	Gly	GGA, GGC, GGG, GGU
Histidine	His	CAC, CAU
Isoleucine	Ileu	AUA, AUC, AUU
Leucine	Leu	CUA, CUC, CUG, CUU, UUA, UUG
Lysine	Lys	AAA, AAG
Methionine	Met	AUG
Phenylalanine	Phe	UUU, UUC
Proline	Pro	CCA, CCC, CCG, CCU
Serine	Ser	AGC, AGU, UCA, UCG, UCC, UCU
Threonine	Thr	AGA, ACG, ACC, ACU
Tryptophan	Try	UGG
Tyrosine	Tyr	UAC, UAU
Valine	Val	GUA, GUG, GUC, GUU

*In groups of three (called codons), bases of mRNA code the order of amino acids in a polypeptide chain. A, C, G, and U represent adenine, cytosine, guanine, and uracil, respectively. Some amino acids have more than one codon, and hence more than one transfer RNA can bring the amino acid to messenger RNA. The research on this coding was initiated by Nirenberg.

Adapted from J. I. Routh, D. P. Eyman, and D. J. Burton, *Essentials of General, Organic, and Biochemistry,* Third Edition, Philadelphia, W. B. Saunders Company, 1977.

there are five distinctly different tRNA molecules specifically for the transfer of the amino acid leucine in cells of the bacterium *Escherichia coli.* At one end of a tRNA molecule is a trinucleotide base sequence (called the "anticode") that fits a trinucleotide base sequence on mRNA. At the other end of a tRNA molecule is a specific base sequence of three terminal nucleotides—CCA—with a hydroxyl group on the sugar exposed on the terminal adenine nucleotide group. This hydroxyl group reacts with a specific amino acid by an esterification reaction with the aid of enzymes.

$$\underset{\text{tRNA}}{\text{(MONONUCLEOTIDES)}_{75-90}\text{CCA}-\text{OH}} + \text{HOCCH}(NH_2)R \overset{O}{\overset{\|}{\longrightarrow}}$$

$$\underset{\text{tRNA-AMINO ACID}}{\text{(MONONUCLEOTIDES)}_{75-90}\text{CCA}-\text{OCCH}(NH_2)R} + H_2O$$

The P—O bonds of adenosine triphosphate (ATP) provide energy for the reaction between an amino acid and its transfer RNA. A molecule of ATP first activates an amino acid.

Figure 11–11 Bonding of activated amino acid to transfer RNA. AMP is adenosine monophosphate.

$$\text{ATP} + \text{Amino acid} \longrightarrow \text{ATP–amino acid-activated species}$$

The activated complex then reacts with a specific transfer RNA and forms the products shown in Figure 11–11.

The ribosome is the part of the cell in which protein synthesis takes place.

The transfer RNA and its amino acid migrate to the ribosome, where the amino acid is used in the synthesis of a protein. The transfer RNA is then free to migrate back and repeat the process.

Messenger RNA is used only once, or at most a few times, before it is depolymerized. While this may seem to be a terrible waste, it allows the cell to produce different proteins on very short notice. As conditions change, a different type of messenger RNA comes from the nucleus, a different protein is made, and the cell adequately responds to a changing environment.

Synthetic Protein

A mixture of amino acids can react to form polypeptide structures outside of a living cell. Only relatively simple catalytic agents are required, and the complex enzymatic environment of cytoplasm is not necessary. However, a mixture of only a few amino acids will result in a multitude of different protein structures. The synthesis of a particular sequence of amino acids presents many problems.

The proteins in living things are not a random sequence of amino acids.

Methods too complex to be detailed here have been devised to construct protein structures with a desired sequence of amino acids. In order to obtain the desired bond between an amino acid and the peptide already constructed, it is necessary to block all the functional groups except those which are to undergo the peptide reaction. With the chemical blocking groups in place, the particular amino acid is added and the peptide bond is made. This is followed by removal of the blocking groups. To build a polypeptide with 20 amino acid units, this complicated process would have to be repeated 19 times. As the peptide chain grows, it becomes increasingly difficult to carry out the chemical operations without disturbing the bonds previously formed.

In spite of the difficulties, the customized synthesis of a prescribed protein has progressed steadily. In 1953, Vincent du Vigneaud synthesized a hormone from 9 amino acids. For his work he received a Nobel Prize in 1955. Other, more lengthy molecules have been synthesized similarly by starting with the individual amino acids and adding them in proper se-

Figure 11–12 The structure of ribonuclease A, an enzyme. This protein structure is composed of 124 amino acid units. The structure is partially explained by sulfur-sulfur bonds between some of the acid units—for example, between 40 and 95. Refer to Table 11–2 for the names of the amino acids abbreviated in this figure.

quence to make the desired protein. Notable hormone syntheses were those of β-corticotropin with 39 amino acid units, and insulin, with 51 units. In 1969, two teams of researchers synthesized the first enzyme, ribonuclease (Figure 11–12), to be assembled outside the living cell from individual amino acids. Ribonuclease contains 124 amino acid units. One team was led by Rockefeller University's Robert B. Merrifield and Bernd Gutte, and the other team at Merck Sharp & Dohme research laboratories was led by Robert G. Denkewalter and Ralph F. Hirschmann.

Many large protein molecules have now been made in the laboratory.

Merrifield's method has had wider application. In this automated technique, an insoluble solid support, polystyrene, acts as an anchor for the peptide chain during the synthesis (Figure 11–13). The first amino acid is firmly bonded to a small polystyrene bead, and each of the other amino acids is then added one at a time in a stepwise manner. The synthesis of ribonuclease required 369 chemical reactions and 11,931 steps in a continuous operation on a machine developed for this purpose.

Figure 11–13 Solid phase method of synthesizing a protein is carried out stepwise from the carboxyl end toward the amino end of the peptide. An aromatic ring of the polystyrene (1) is activated by attaching a chloromethyl group (2). This first amino acid (black), protected by a butyloxycarbonyl (Boc) group (black box), is coupled to the site (3) by a benzyl ester bond and is then deprotected (4). Subsequent amino acid units are supplied in one of two activated forms; a second unit is shown in one of these forms, the nitrophenyl ester of the amino acid (5). The ester (colored box) is eliminated as the second unit couples to the first. Then the second unit is deprotected, leaving a dipeptide (6). These processes are repeated to lengthen the peptide chain.

Figure 11—14 Dr. Li (*center*) checks experiment with Dr. Richard Noble (*left*) and Dr. Donald Yanashiro (*right*). This group was the first to synthesize the HGH polypeptide. (From *Chemical and Engineering News,* January 11, 1971.)

In 1970, a group led by Choah Hao Li at the University of California used Merrifield's method to synthesize human growth hormone (HGH) (Figure 11–14). HGH has 188 amino acid units and a molecular weight of about 21,500 amu. This hormone is produced naturally by the front lobe of the pituitary gland, a pea-sized body located at the base of the brain. In humans, it has a dual function—it controls milk formation and it regulates many aspects of growth. In childhood, excess secretions of HGH can cause giantism; too little HGH causes dwarfism.

These synthetic proteins are identical in every respect to the natural products. They perform the same functions and give the same results on analysis. The limits of the present methods are unknown; in time even longer molecules are expected to be duplicated.

Synthetic Nucleic Acids

Progress in the synthesis of polynucleotides has been difficult, principally because of the difficulties involved in determining the proper blocking groups. Progress, although slow, is being made.

In 1959, Arthur Kornberg synthesized a DNA-type polynucleotide, for which he received a Nobel Prize. He used natural enzymes as templates to arrange the nucleotides in the order of a desired polynucleotide. His product was not biologically active. In 1965, Sol Spiegelman synthesized the polynucleotide portion of an RNA virus. This polynucleotide was biologically active and reproduced itself readily when introduced into living cells. In 1967, Mehran Goulian and Kornberg synthesized a fully infectious virus of the more complicated DNA type.

In 1970, Gobind Khorana synthesized a complete, double-stranded, 77-nucleotide gene. He, too, used natural enzymes to join previously synthesized, short, single-stranded polynucleotides into the double-stranded gene.

If scientists can construct DNA, can they then control the genetic code? Genes are the submicroscopic, theoretical bodies proposed by early geneticists to explain the transmission of characteristics from parents to progeny. It was thought that genes composed the chromosomes, which are large enough to be observed through the microscope as the central figures in cell division. It is now generally believed that DNA structures carry the message of the genes; hence, DNA contains the *genetic code.* If scientists can construct DNA, they could very well alter its structure and thereby control the genetic code.

A *mutation* occurs whenever an individual characteristic appears that has not been inherited but is duly passed along as an inherited factor to the next generation. A mutation can readily be accounted for in terms of an alteration in the DNA genetic code; that is, some force alters the nucleotide structure in a reproductive cell. Some sources of energy, such as gamma radiation, are known to produce mutations. This is entirely reasonable because certain kinds of energy can disrupt some bonds, which can re-form in another sequence.

A mutation results when there has been an alteration of the genetic code contained within the DNA molecule.

If scientists can control the genetic code, can they control hereditary diseases such as sickle cell anemia, gout, some forms of diabetes, or mental retardation? If the understanding of detailed DNA structure and the enzymatic activity in building these structures continues to grow, it is reasonable to believe that some detailed relationships between structure and gross properties will emerge. If this happens, it may be possible to build compounds which, when introduced into living cells, can combat or block inherited characteristics.

---

1. (a) The basic code for the synthesis of protein is contained in the _____ molecule.

   SELF-TEST 11-C

   (b) The synthesis of a protein is carried out when _____ molecules bring up the required amino acids to messenger RNA.

2. When DNA replicates itself, each nitrogenous base in the chain is matched to

   another one via _____ bonds.

3. The energy for DNA replication and natural protein synthesis is supplied by

   substances such as _____.

4. What nitrogenous base complements (matches through hydrogen bonding)

   adenine (A)? _____ cytosine (C)? _____

   guanine (G)? _____ thymine (T)? _____

   uracil (U)? _____

5. At this point in time, a gene has been synthesized from individual nucleotides in the laboratory without the aid of natural enzymes. (True/False)

6. Replication means the same as duplication. (True/False)

MATCHING SET

_____ 1. energy "cash" in the living cell

_____ 2. mutation

_____ 3. enzyme that splits polysaccharides in the mouth

_____ 4. natural protein

_____ 5. first oxidation product of glucose

_____ 6. occurs under aerobic (with air) conditions

_____ 7. reaction utilizing ATP

_____ 8. product of ATP hydrolysis

_____ 9. molecule synthesized in a laboratory

_____ 10. molecules that absorb light energy

_____ 11. due to inadequate supply of lactase

(a) ptyalin

(b) ADP + energy

(c) Krebs cycle

(d) chlorophylls

(e) coupled reaction

(f) structure determined by DNA and RNA

(g) lactic acid

(h) altered DNA

(i) ATP

(j) insulin

(k) lactose intolerance

QUESTIONS

1. Write an equation for the digestion of:
   (a) starch to a disaccharide
   (b) a disaccharide to a simple sugar
   (c) a protein to amino acids
   (d) a triglyceride to fatty acids

2. What is the metal in chlorophyll?

3. If you were to "feed" radioactive carbon dioxide to a green plant, what would be the first radioactive carbon compound formed? Who made this discovery?

4. Give the structure of ATP and point out the region of the molecule that contains bonds which are hydrolyzed in coupled reactions.

5. What is meant by coupled reactions? Give an example. Why do the energetics of biochemical systems make coupled reactions necessary?

6. What type of compound first absorbs light energy in photosynthesis? Give an example.

7. In photosynthesis, why is it partially correct to say that light is an oxidizing agent?

8. What are the two major divisions in photosynthesis? Express in words what is accomplished in each.

9. What is the source of oxygen in photosynthesis?

10. What part of photosynthesis could even take place in an animal cell?

11. Since chlorophyll loses electrons because of light, it must subsequently gain electrons from somewhere. Where do they come from?

12. What is the basic nature of the digestion processes for large molecules?

13. The chemical changes in the Krebs cycle can be classified as dehydrogenation (removal of hydrogens, a type of oxidation), dehydration (removal of water), hydrolysis (reaction with water in which water loses its molecular identity), decarboxylation (removal of COOH group and formation of CO_2), and phosphorylation (adding a phosphate group, such as $H_2PO_4^-$). Beginning with citric acid and progressing around the cycle to oxaloacetic acid, determine the total number of each kind of chemical change.

14. What compound produces soreness in the muscles after a period of vigorous exercise?

15. Which compounds in the CO_2 fixation scheme in photosynthesis could enter directly the reactions of the Embden-Meyerhof pathway or the Krebs cycle?

16. If protein digestion is facilitated by enzymes, and these enzymes are produced in body organs made of proteins, explain why the enzymes do not cause rapid digestion of the organs themselves.

17. What is pyruvic acid? Why is it so important in getting energy from sugars?

18. What happens if an amino acid is needed for protein synthesis and the amino acid can neither be made by the body nor obtained from the diet? Does the modern theory of protein synthesis include an explanation of the role of essential amino acids? (These are amino acids which cannot be manufactured by the human body; they must be obtained in the diet.)

19. Does a strand of DNA actually duplicate itself base for base in the formation of a strand of messenger RNA? Explain.

20. (a) Describe the general method of synthesizing DNA *in vitro* (in a test tube) at present.
    (b) Why would the synthesis of a polynucleotide from the individual phosphoric acid, sugar, and nitrogenous bases be a breakthrough in controlling the genetic code?

21. Discuss some aspect of the feasibility of solving sociological and psychological problems via religion, DNA alteration, chemical suppressants, political pressures, and/or persuasive dialogue.

22. Check the recent issues of *Science* or other scientific news publications to find out about more recent work done on synthesis of proteins and polynucleotides.

SCIENCE AND TECHNOLOGY AS A NEW PHILOSOPHY

A PHILOSOPHICAL AND HISTORICAL BACKGROUND

The various factors that led to the systematic development of science and its associated technologies in the Western World are both numerous and complex. Rather than attempt to develop an overall "complete" view, we shall discuss only a few of the empirical results. One of the key points seems to be the development of *printing techniques,* which enormously increased the availability of the accumulated knowledge of mankind (see Figure 12–1). Another seems to be the continuous *refusal to accept authority of a person or a group of people as the ultimate judge of truth* in certain areas; this indeed still seems to be one of the major differences between mathematics and the sciences on the one hand, and other areas of human knowledge. No one ever states, as proof of the validity of a piece of scientific information, that so-and-so said this and therefore it must be true. This skeptical attitude is to be found throughout all nations at all times; however, it was only in Western Europe that such a frame of mind came to be cultivated systematically and ultimately formed the basis of some intellectual organizations whose primary purpose was the search for and propagation of a kind of **objective knowledge.**

The most important aspects of this type of knowledge are the ways in which it changes. With the passage of time, it becomes *more extensive, more accurate, more broadly disseminated, and,* most important of all, *more concisely summarized* in terms of generalizations or their equivalent: mathematical equations giving the quantitative relationships connecting a set of properties.

Because the physical sciences deal with our universe, our world, and

Direct observation and re-observation, not who did it, are the ultimate bases of truth in the sciences.

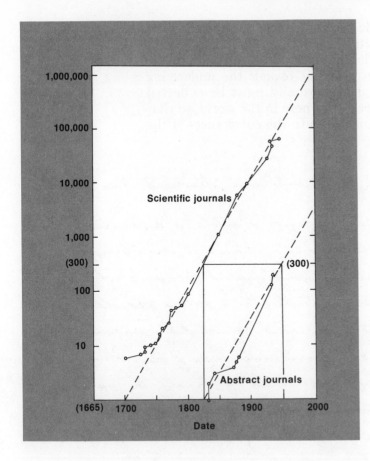

Figure 12–1 Rate of increase in number of scientific journals since 1665. (Reproduced with permission of Yale University Press.)

Technology: the manipulation of our environment.

Mankind is in the midst of a knowledge explosion.

our environment, this growth of knowledge leads to an ever increasing understanding of the physical world and the ways it can be manipulated to obtain desired ends. This manipulation of our environment can attain a high degree of sophistication. Presently, modern **technology** has two types of roots. One derives from the practical knowledge of craftsmen and frequently has been transmitted orally or by apprenticeships. The other is in the sciences and has been transmitted by publications of various sorts (Figure 12–1) and formal education. In practice the dual roots tend to fuse and reinforce each other. The older method of oral transmission had its ups and downs, as is especially obvious in certain areas where knowledge of techniques has died out (e.g., the early Middle Ages in Europe as contrasted with the Roman Empire). The newer method of transmission by publication tends to make such knowledge available to all who can read and who have the required training for understanding (see Figure 12–2).

As time has passed, the effectiveness of this new method of collecting and extending knowledge has become more apparent to the ordinary citizen. It has also become accepted by scholars in more disciplines and has led to the extension of this method to new areas of knowledge which previously had not been the subject of systematic study.

At this point one might well ask for a definition of **science** or the scientific method. While it is difficult to give such definitions that are acceptable to all, a close approximation may be achieved. First of all, "science" consists of that which has been collected by the use of the *scientific method.* The term "scientific method" refers more to a mental attitude than to a procedure. The attitude is one that requires the strictest intellectual honesty in the collection of observable facts, and in the arrangement of these facts into a pattern that reveals the underlying basis of the *observed* behavior. The data normally must be collected under conditions that can be reproduced anywhere in the world, so that new data can be obtained to confirm or to refute the correctness of the

Figure 12–2 A page from the notes of Dr. Bruce Merrifield (Professor of Biochemistry at Rockefeller University), the designer of a protein-making machine. Progress in science is made through disciplined human thought, taking advantage of previous knowledge and useful theories. See Figure 11–13. (From *Chemical and Engineering News*, August 2, 1971.)

suggested pattern (geologists and astronomers are excused from these rigid requirements). The results obtained are thus independent of differences in language, culture, religion, or economic status in the various observers, and so represent a unique type of objective truth.

The methods of collecting observed, objective data can be grouped generally into about four classes: trial and error, planned research, accident, and "let's do some experiments and see what happens."*

Discovery by trial and error begins by having a problem to solve and doing various experiments in the hope that something desirable will emerge. The next set of experiments then depends on the results obtained. The discovery of the Edison battery by Thomas Edison's group is an example of discovery by trial and error. His group performed more than 2000 experiments, each guided by the last, before settling on the composition of Edison's battery.

Discovery by planned research comes from plotting out specific experiments to test a well defined hypothesis. The carcinogenic nature of some compounds is determined by progressing through a set pattern of experimental tests.

Discovery by accident may be a misnomer. The investigator is usually actively involved in investigating nature through experimentation, but "accidentally" finds some phenomenon not originally imagined or conceived. Thus the "accident" has an element of serendipity, and is not seen unless the investigator is a trained observer. Pasteur's statement sums it up: "chance favors the prepared mind." The discovery of penicillin by Fleming was a type of "accidental" scientific discovery.

---

*A. B. Garrett, "The Discovery Process and the Creative Mind," *Journal of Chemical Education,* Vol. 41, pages 479–482 (1964).

TABLE 12–1 How Long It Has Taken Some Fruitful Ideas to Reach Technological Realization

Innovation	Conception	Realization	Incubation interval
Antibiotics	1910	1940	30 yr
Automatic transmission	1930	1946	16
Ballpoint pen	1938	1945	7
Cellophane	1900	1912	12
Dry soup mixes	1943	1962	19
Filter cigarettes	1953	1955	2
Heart pacemaker	1928	1960	32
Hybrid corn	1908	1933	25
Instant camera	1945	1947	2
Instant coffee	1934	1956	22
Liquid shampoo	1950	1958	8
Long-playing records	1945	1948	3
Nuclear energy	1919	1965	46
Nylon	1927	1939	12
Photography	1782	1838	56
Radar	1907	1939	32
Roll-on deodorant	1948	1955	7
Self-winding wrist-watch	1923	1939	16
Video tape recorder	1950	1956	6
Xerox copying	1935	1950	15
Zipper	1883	1913	30

By permission from *Industrial Research and Development,* October, 1978.

The method of "let's try some experiments and see what happens" lacks the formality of the other methods. Experimentation evolves with vague hypotheses or no specific hypotheses because no well defined pathway is obvious. Joseph Priestley, who discovered oxygen in 1774, utilized this approach on occasion.

Regardless of the method of inquiry and discovery, there is often a period of incubation time between the discovery of a usable idea supported by evidence and the fruition of the idea through application of technology. Some examples of such incubation times are listed in Table 12–1.

Over the last 200 years, accumulated scientific and technological knowledge has been put to use on a very extensive scale in Europe and in those areas of the world that have had the means and the will to follow the examples set by England, which underwent this "industrial revolution" first (see Figure 12–3). The result has been the development of a society largely dependent upon and supported by a technology which is itself undergoing constant change. The first consequence of this technology has been to increase the *rate* at which things can be produced. This in turn has continually changed the occupational patterns of millions of human beings, and has brought forcefully to mind the persistence of *change* in our pattern of life.

> Technology increases the rate at which things can be produced.

These changes have influenced profoundly the way in which people think about their material wants and the ways in which they can be satisfied. For example, there seems to be little argument with the statement, "If the number of human beings on the earth could be stabilized, a much higher standard of living could prevail over most of the earth." A statement such as this would have been greeted with widespread derision 500 years ago. While depending on technology, people today are beginning to doubt its ability to solve both personal and social problems on a long-range basis. It is obvious that confusion exists on this point since the cries about the curses of technology come from people who are highly dependent on it and who are constantly asking for even more help from technology.

TECHNOLOGY: ITS TRIUMPHS AND PROBLEMS

Almost as soon as the industrial revolution began in England, the public realized that technological progress brought with it a series of prob-

Figure 12–3 Name some discoveries that led to this change in transportation. How many of them are chemical in nature?

lems. The first to be noticed was the necessity for progress to be accompanied by changing patterns of employment.

Technological unemployment results when men are displaced from their jobs by machines.

It is obvious that if a machine makes as much thread as 100 men can make, the men are released to do other work. The 100 men, however, do not look on this as an advantage, especially if they are settled in their place of employment with their families. The new opportunities that result from such a machine are rarely of benefit directly to the men displaced. The wealth of their country is increased since there are now 100 men able to do other work. However, the initial reaction of the men in 18th century England was to riot and break up the machinery.

The increased use of fuels of all sorts, especially the introduction of coal and coke into metallurgical plants and then the use of coal to fuel steam engines, led to widespread problems with air pollution which were recognized and discussed over 200 years ago (Figure 12–4).

The most important point of these results is the realization that technological progress is always obtained at some cost, and the cost may not be obvious at the outset.

FUMIFUGIUM:

OR,

The Inconvenience of the AER,

AND

SMOAKE of LONDON

DISSIPATED.

TOGETHER

With some REMEDIES humbly proposed

By J. E. Esq;

To His Sacred MAJESTIE,

AND

To the PARLIAMENT now Assembled.

Published by His Majesties Command.

Lucret. l. 5.

Carbonumque gravis vis, atque odor infinuatur
Quam facile in cerebrum?——

LONDON:
Printed by W. GODBID, for GABRIEL BEDEL, and THOMAS
COLLINS; and are to be sold at their Shop at the Middle
Temple Gate, neer Temple Bar. M.DC.LXI.
Re-printed for B. WHITE, at Horace's Head, in Fleet-street.
M DCCLXXII.

Figure 12–4 Title page from J. Evelyn, F.R.S., *The Smoake of London*. The Latin quotation is from the Roman poet Lucretius (97–53 B.C.). It may be translated, "How easily the heavy potency of carbons and odors sneaks into the brain!" (Courtesy A. E. Gunther and the University Press, Oxford.)

A very important technological development was recognized as necessary in 1890 by Sir William Crookes, who addressed the British Association for the Advancement of Science on the problem of the fixed nitrogen supply (that is, nitrogen in a chemical form that plants are capable of using in growth). At that time, scientists recognized that nitrogen compounds were necessary in fertilizers and that the future food supply of mankind would be determined by the amount of nitrogen compounds that could be made available for this purpose. The source of these nitrogen supplies was then limited to rapidly-depleting supplies of guano (bird droppings) in Peru and to sodium nitrate in Chile. It was realized that when these were exhausted, widespread famine would result unless an alternative supply could be developed. This problem was recognized first by English scientists as a potentially acute one, because by the 1890's England had become very dependent upon imported food supplies. The Industrial Revolution allowed the population to grow rapidly, so that it soon outstripped the domestic food supply.

The amount of food that can be grown is related to the nitrogen content of the soil.

Widespread interest in this problem led to research on a number of chemical reactions by which nitrogen can be obtained from the relatively inexhaustible supply present in the air. Air is 21 percent oxygen (O_2) and 79 percent nitrogen (N_2). The nitrogen in the air is present as the rather unreactive molecule N_2, and in this form it can be used as a source of other nitrogen compounds only by a few kinds of bacteria. Some is also transformed into NO by lightning, and when this is washed into the soil by rain it can be utilized by plants. Needless to say, the amount of nitrogen transformed into chemical compounds useful to plants by these processes is quite limited and cannot be increased easily.

The N_2 molecule is rather unreactive because of its strong triple bond ($N\equiv N$).

Several chemical reactions were developed to form useful compounds from atmospheric nitrogen, but the best known and most widely used one has an ironic history. While England was interested in nitrogen for fertilizers, Germany was interested in nitrogen for explosives. The German General Staff realized that the British Navy could blockade German ports and cut them off from the sources of nitrogen compounds in South America. As a consequence, when a German chemist named Fritz Haber showed the potential of an industrial process in which nitrogen reacts with hydrogen in the presence of a suitable catalyst to form ammonia, the German General Staff was quite interested and furnished support through the German chemical industry for the study of the reaction and

The Haber process transforms the nitrogen of the air into ammonia.

$$N_2(g) + 3H_2(g) \rightleftharpoons 2NH_3(g)$$

Figure 12–5 An early German ammonia plant. (From "The Realm of Chemistry," Econ-Verlag Gmb H, Düsseldorf, 1965.)

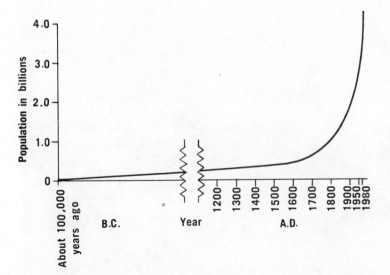

Figure 12–6 World population size. (Adapted from Turk, A., Turk, J., and Wittes, J. T.: *Ecology, Pollution and Environment.* Philadelphia, W. B. Saunders Co., 1972.)

the development of industrial plants based on it. The first such plant was in operation by 1911, and by 1914 such plants were being built very rapidly by Germany.

When the First World War broke out in August, 1914, many people thought that a shortage of explosives based on nitrogen compounds would force the war to end within a year. Unfortunately, by this time the nitrogen-fixing industry in Germany was capable of supplying the needed compounds in large amounts. This process thus prolonged the war considerably and resulted in an enormous increase in mortality. Subsequently, the ammonia process has been used on a huge scale to prepare fertilizers and now is largely responsible for the fact that the earth can support a population of more than four billion (see Figure 12–6). Ammonia production by this process exceeds 40,000 tons per day in the United States alone.

Ammonia is the number four commercial chemical.

The same type of problem seems to arise from the development of many technological processes. The utilization of nuclear energy brings with it the ability to make nuclear explosives from one of the byproducts. The development of rapid and convenient means of transportation, such as the automobile and the airplane, also brings forth new weapons of war and air pollution problems. We must learn to control our technology in such a manner as to maximize its benefits and minimize its disadvantages. These problems arise with *all* technological developments, even the most primitive. The discovery of the techniques necessary to the manufacture of iron led first to the development of weapons (swords) by their discoverers, the Hittites, who then proceeded to conquer their neighbors and lead the first successful invasion of Egypt (ca. 1550 B.C.).

Understanding nature does not necessarily lead to a better world.

TECHNOLOGY AND THE HUMAN ENVIRONMENT

Technology usually drastically alters the environment.

The growth in large scale technology has a very large number of effects, both direct and indirect, on the human environment. The examination of a few of these shows just how complex these consequences can be.

An obvious case is the development of atomic energy. When the

Figure 12–7 Land burial trench at the Oak Ridge National Laboratory reservation. Each day's accumulation of waste containers is buried under three or more feet of earth. (From *Radioactive Wastes*. Courtesy of the U.S. Atomic Energy Commission, Washington, D.C.)

incredibly large amounts of energy released by nuclear reactions were first recognized, the development of such energy sources was placed on a top priority basis. After some nuclear reactors had been built and actually placed in operation, it was evident that their operation was accompanied by some serious potential risks to their human users. The first was the danger of some potential disaster such as the explosion of a boiler, with the consequent dispersal of radioactive material. The second danger was in the generation of radioactive products as the nuclear reaction proceeded. The uranium used in such reactors was transformed into a wide variety of fission products, which made the operation of the pile more difficult as time went on. The repurification of the uranium could be accomplished chemically, but what was to be done with the radioactive wastes generated in the process? Obviously they could not be dumped into a sewer, since many of them have relatively long half-lives. Present practice calls for solid radioactive wastes to be placed in vaults and buried at carefully chosen sites (Figure 12–7). Liquid wastes are placed in underground storage tanks. The level of radioactivity of such wastes as a function of time is shown in Figure 12–8. The problem that faces us here is obtaining the benefits of the technology with minimum disruption of our own environment.

Nuclear wastes constitute an unsolved problem. See Chapter 17.

The same type of problem arises whenever we introduce a specific chemical compound into our environment to accomplish one single thing.

Figure 12—8 Disposal of radioactive wastes, for a hypothetical waste product with a one-month half-life. (From Turk, A., Turk, J., and Wittes, J.: *Ecology, Pollution and Environment.* Philadelphia, W. B. Saunders Co., 1972.)

Technological procedures have both obvious and not so obvious consequences.

General structure of PCB's, Z is either an H atom or a Cl atom.

The compound is often capable of a variety of actions and can lead to consequences undreamed of by those who introduce it. Many such examples can be found in the fields of drugs, industrial chemicals, and insecticides.

Thalidomide was designed to be an effective tranquilizer, and for this purpose it was an unqualified success. Unfortunately, when taken by pregnant women, it often prevents the normal development of the children they bear.

Polychlorinated biphenyls (PCB's) have been manufactured in the U.S. since 1929 and reached a maximum domestic production of 85 million pounds in 1970. There are 209 possible PCB's, and about half have been synthesized and characterized. As a class of compounds, PCB's are very stable to heat, acids, and bases. Their stability and insulating ability led to wide uses as insulating fluids and coolants in electrical transformers, commercial capacitors, hydraulic systems, gas turbines, and vacuum pumps, and as fire retardants and plasticizers. More than 95% of all power capacitors contain PCB's, and approximately 5% of the transformers are PCB filled. While very desirable in some uses, the great stability of PCB's is very undesirable in other situations. When the human body is exposed to PCB vapor or fluid, the PCB received by the body is poorly metabolized and resides for the most part in fatty tissues. As much as 29 parts of PCB per billion have been found in tissues of the general population. The known toxic effects of PCB's in humans include an acne-like skin eruption (chloracne), pigmentation of the skin and nails, excessive eye discharge, swelling of eyelids, and distinctive hair follicles. PCB's are known to cause cancer in test animals and are suspected of being carcinogenic in humans. Since 1971, the Monsanto Company, the major domestic manufacturer of PCB's, has voluntarily restricted domestic sales to uses in closed systems.

It seems that we must always combat insects, since they and other biological pests still claim about 35% of our food supply (Table 12–2). If we could eliminate that loss, we could feed most of the world's hungry people. The most effective aid we now receive in this struggle against insects is from insecticides. Unfortunately, many insecticides are rather unspecific poisons.

TABLE 12–2   Estimated Annual
Crop Loss Due to Pests and
Disease (Percent of Total U.S.
Crop Destroyed)*

Crop	Insects	Weeds	Disease
Corn	12	10	12
Rice	4	17	7
Wheat	6	12	14
Potatoes	14	3	19
Cotton	19	8	12

*From Scientific Aspects of Pest Control, National
Academy of Science Publication No. 1402.

Arsenic compounds were used on a very large scale as insecticides, but a realization of their great toxicity for human beings has led to their replacement by other compounds equally effective but usually less dangerous. However, some of these compounds which are most effective in protecting crops are nevertheless toxic to humans and are even capable of causing death when improperly handled or ingested.

Even those insecticides that are not obviously harmful to us often have side effects which make them unattractive. DDT, for example, is very effective for the control of a wide variety of insects and, as far as can be ascertained, is nontoxic to humans. Its widespread use from 1943 to 1965 virtually eliminated plague from the world population by controlling fleas and lice. The plague brought terrible devastation during the Middle Ages and persisted well into our century. DDT has also been responsible for huge increases in crops, has greatly improved the health of armies and refugees during the awful years of World War II, has improved the health of cattle, and has even eliminated the house fly from homes where it was used. Hence there were strong reasons for using DDT; it was not a willful or arbitrary thing.

DDT is far more harmful to many species of wild animals than it is to human beings.

Because of the widespread use of DDT, and the fact that it is accumulated in fish and animal fats, it is found to concentrate enormously in birds of prey (eagles, pelicans, and ospreys), whose diet is principally fish, and to disrupt their reproductive cycles (Figure 12–9). A side effect such as this led to extensive agitation to replace DDT with other compounds that are equally effective as insecticides yet free from this particular side effect. It is possible, however, that the compounds used to replace DDT will have other side effects, which will not become obvious until after they have been used for some time. It would seem that the use of any insecticide will have a considerable effect upon the bird population of an area, if only through the changes that it introduces into the food supply of the birds. The perplexing problem in this area and others is to know and implement the balance between benefits and subsequently determined drawbacks. Most chemical substances, like DDT, cannot be labeled "good" or "bad." Ozone, for another example, is necessary in our upper atmosphere to filter out most of the incoming ultraviolet light, but it is dangerous for us to breathe ozone.

Chemistry is not, however, the only science that can furnish technological means for the control of insects. There is a wide variety of bio-

Figure 12—9 The concentration of DDT in birds of prey, such as hawks, results in poor egg shell formation along with other abnormalities in the reproductive cycle.

Most technical problems require the efforts of several disciplines for a solution.

logical processes that can be used for the same purposes. One is to furnish assistance to the biological species that destroy insect pests, such as parasitic insects. This can lead to problems with the thriving new species, too. Another method is the introduction of large numbers of sterile insects. When these mate with normal insects, no offspring are produced; the numbers of a type of undesirable insect can be reduced considerably by this process, at least temporarily. These examples are given to emphasize the fact that there are usually several very different kinds of processes that can be used to solve any given practical problem. In the future these will be assessed, at least in part, on their ability to leave the environment undamaged as well as on their monetary cost.

TECHNOLOGICAL DEVELOPMENT AND ITS ENVIRONMENTAL CONSEQUENCES

A very good example of a highly desirable technological development which has consequences that are obviously not so desirable is seen in the development of fertilizers. Most of us agree that an abundant food supply is both necessary and good. Most would admit, too, that every crop harvested from a field removes essential nutrients from that field: nitrogen, phosphorus, potassium, and so forth. By replacing these lost elements with fertilizer, we can restore or enhance the amount of food we obtain from the field. The increased yields obtained with increased application

Figure 12–10 This pond was fertilized with chemicals to produce an excessive growth of algae. Part *A* shows the water covered with the plant growth: Part *B* shows the decay that follows when the concentrated form of life cannot be sustained. (Courtesy of Dr. D. L. Brockway, Federal Water Quality Administration.)

of fertilizers have been well established and commonly held to be desirable. It is also well established that increased fertilization of a field increases the concentration of these essential nutrients in the rainwater that runs off such land.

The change in the mineral concentration in the runoff water is capable of causing drastic changes in the rivers and lakes into which it drains (Figure 12–10). By increasing the amount of minerals available, we have greatly stimulated the growth of surface algae in rivers and lakes, and as these grow they choke out other forms of life. The growth of such slimes makes the water much less capable of supporting its normal population of fish, and ultimately they die off. The algae also make the water more difficult to purify for drinking purposes. After a time the increased food supply is paid for by a general disruption of the biology of the waterways that drain an area. This disruption has occurred in many areas in the United States where the movement of water through lakes and rivers is slow.

Fertilizers increase the growth of both desirable and undesirable organisms.

Other examples of this same kind of process can be seen in the development of energy sources, the use of antibiotics, weed killers, and, in fact, any kind of process that releases chemical compounds into the environment in sufficient quantities to cause an appreciable percentage increase in its concentration in the environment. This holds for the hydrocarbons released from automobile engines and for the carbon dioxide and sulfur dioxide released by electric power plants. The changes that increased concentrations of various compounds cause cannot always be estimated on the basis of experiments covering only a short period of time, since some of the effects may be very long range ones that build up slowly.

What are the questions that should be asked before a new chemical is widely used?

IS TECHNOLOGY ESCAPING CONTROL?

It is quite easy to imagine a situation in which a technological process can introduce into our environment drastic and irreversible changes, which set a chain of events into action before we can stop them. This is

INDUSTRIAL
CO₂ EMISSIONS TO U.S. ATMOSPHERE
1890–2000

PROJECTED FIGURES

**GREATEST GROWTH
RATE (1965–1985)**

Figure 12–11 United States' industrial emissions of CO_2 into the atmosphere in millions of tons per year.

especially easy to visualize in the case of changes that might be provided by the continuous increase in the carbon dioxide content of our atmosphere (Figure 12–11). The increased emission of CO_2 is caused by the increased use of "fossil" fuels, such as coal and petroleum, to furnish power for energy generation and transportation. What will be the long range effects, if any, from this steadily increasing amount of carbon dioxide in our atmosphere? At the present time, no one really knows.

There are other situations where the consequences of technology seem more obviously to be moving beyond the control of man. The ability to build nuclear weapons and power plants is now spreading quite rapidly, and soon may be well within the reach of all but the smallest nations. How are these to be controlled? What mechanisms can be developed to prevent mankind from destroying itself in an atomic holocaust triggered by some insignificant local dispute? This is obviously a very urgent problem.

The application of technology to the problems of warfare has already produced some frightening developments, including chemical and biological warfare agents whose discovery has closely followed the extension of human knowledge in these sciences. The same kind of knowledge that allows more effective drugs and insecticides to be synthesized facilitates the synthesis of more effective agents for gas warfare. The understanding of cellular behavior that allows us to produce new varieties of high yielding grains also can be used to develop new techniques of biological warfare.

Obviously, a distinction must be made between scientific knowledge, which can be used for good or evil purposes, and the types of technology that are developed specifically for destructive purposes. The kind of emotional thinking that puts all technological developments under some kind of moral ban can only lead us rapidly to a new Dark Age. Most of the materials of the world can be used for a variety of purposes, which are ethical to varying degrees. Selection must be made between good developments, the indifferent ones, and the bad ones. This has *always* been a problem for mankind and probably always will be. In such a situation, ignorance can be catastrophic; only by a study of such processes and an evaluation of their probable consequences can rational selections be made. In this case it makes more sense to ask if we are being given control over things beyond our understanding than to ask, "Is technology escaping control?" Technology is always *initiated* by human beings and is under some sort of actual or potential control by us. Whatever the minds of human beings can devise, the minds of human beings can control. Only the will to do so is needed. Our worst enemies are ignorance and selfishness.

Technology is always initiated by human beings and is therefore under our control in principle.

Figure 12—12 Human thought versus chemicals. Will mankind ever completely control chemicals—or will they control us? (Courtesy of Varian, Palo Alto, California.)

There is still another manner in which fears arise over the developing course of technology and our ability to control it. The vast majority of mankind has no knowledge of the scientific principles upon which technology is based and accordingly must accept the opinions of others on technological matters. This dependence upon experts, in turn, arouses fears of these experts and the damage they can do either by error or by evil intent. This fear, in some people, is an unreasoning, blind, driving force which leads them to condemn all technology and, indeed, all thinkers and all knowledge. (How do you feel about someone smarter than you?) The only antidote to such fear is an understanding of technology based on study. Mankind's perennial enemy has been ignorance and the prejudice it generates. Human progress has always been the result of activities of that small percentage of people who accept neither the ignorance nor the popular prejudices of their fellow human beings.

In all fairness, it must be noted that the benefits of technology far outweigh its disadvantages, and also that its transformation of the human condition is still in its infancy. There is, even now, an enormous number of practical problems facing mankind that can be solved only by new scientific discoveries and the technological advances that they will make possible. It is the responsibility of all of us to learn about the sciences and to develop an understanding of the possible ways in which this knowledge may be developed to the advantage of mankind. In these learning processes, we shall discover an avenue to responsible participation in decision making, an understanding of technological advances, and a thorough enjoyment of our investigations.

No person should hesitate to probe into technology to find how it affects his life. This is necessary if he or she is to make good decisions on many basic questions.

---

1. The ultimate test of a scientific theory is its agreement with _____. **SELF-TEST 12-A**

2. Different workers, in different countries, who carry out a particular laboratory experiment in exactly the same way should get _____ result.

3. Technology allows us to produce things at a greater _____.

4. We now expect that over the period of our own lifetime our way of life will

    _____.

5. Nuclear power plants produce competitively priced energy but dangerously

    _____ by-products.

6. The principal justification for the use of chemical fertilizers and pesticides is

    that they allow us to produce more _____.

Rate the following on a scale: GOOD, BAD, INDIFFERENT. Be prepared to state your reasons. **SUBJECTIVE TEST**

1. DDT _____

2. Nuclear energy _____

3. Coal as a fuel _____

4. Petroleum as a fuel _____

5. Fertilizers _____

6. Birth control pills _____

7. Plastic containers _____

8. Synthetic foods _____

9. Solar energy _____

10. Genetic manipulation _____

MATCHING SET

_____ 1. transmission of scientific knowledge

_____ 2. allowed World War I to be prolonged

_____ 3. source of the element N_2

_____ 4. alternative to insecticides

_____ 5. unsatisfactory way to establish scientific truth

_____ 6. insecticide harmful to wildlife

_____ 7. increases rate of change in life

_____ 8. used in explosives

_____ 9. needed to grow crops

_____ 10. scientific methods

_____ 11. basis of scientific truth

(a) appeal to authority

(b) DDT

(c) technology

(d) nitrogen compounds

(e) used to study nature

(f) printed journals

(g) Haber process (ammonia production)

(h) air

(i) observation

(j) biological controls

TOPICS FOR THEMES

Suggested topics for short themes to express your views. The depth of research is to be determined by you and (or) your instructor. If a topic is to be researched, the *Readers' Guide* is a good place to start, and *Scientific American* (well-indexed!) is a good place to continue.

1. Contrast knowledge gained by scientific methods with knowledge in philosophy, religion, art and (or) some other discipline.

2. What role should scientists play in deciding moral uses of their discoveries?

3. Would it be good to limit the number of people on earth in order to have more material and energy per person?

4. Explore some ways in which technology has changed the occupational patterns of people.

5. Find examples in which a technological development at first was considered more good than bad, but with passage of time was looked upon as more bad than good.

6. Is technological "progress" always obtained at some cost? Explain.

7. Discuss ways in which side effects of chemicals can be determined prior to putting a product on the market.

8. Should every conceivable side effect of a product be tested for and eliminated before the product is brought on the market? Justify your answer.

9. Techniques have been developed to analyze for some elements when only a nanogram (10^{-9} gram) is present in the sample. What effects may this have on the consciousness of side effects of chemicals?

10. Modern society demands more energy with less pollution and less expenditure of money. From your point of view, is this a "pipe dream"? Discuss reasons for your answer.

11. To what degree should food chains be considered when eradication of insects via insecticides seems to be the only practical way to protect a food supply and health?

12. What should be done with radioactive wastes? What problems are involved?

13. Discuss the need to recycle chemicals in view of limited resources.

14. What can one do to make a better, safer, more livable world for everybody with regard to waste materials?

15. To what extent should political decisions influence scientific investigation?

16. Should scientists be morally responsible in their investigations of nature?

17. Resolved: Education is the solution to problems concerning energy, pollution, and the proper use and reuse of materials. Defend or deny.

18. Discuss one or more ways in which food production creates a pollution problem and wastes energy.

19. Cite a technological advance that was not given specifically in this chapter; state its benefits, and briefly discuss problems arising from it.

20. How should wastes be dealt with? In your discussion consider minimum use of energy, minimum pollution, and maximum recycling.

21. Discuss significant differences and similarities between science and technology.

22. Discuss an important practical problem facing mankind in terms of how science and technology have helped to solve the problem and, in turn, have created other problems.

23. Describe evidence of technological advances that caused a decline of civilization. (Consider the article in the *Journal of Occupational Medicine,* Vol. 7, pages 53–60 (1965).)

24. Technology is more beneficial to mankind than detrimental. Defend or deny.

25. Discuss areas or problems in which technology is most needed today.

26. Discuss the limitations of science.

27. Discuss the limitations of technology.

28. What factors should be considered before a chemical is banned for a particular use? Defend each factor.

CHAPTER 13

TOXIC SUBSTANCES IN OUR ENVIRON- MENT

Toxic substances are materials that upset the incredibly complex system of chemical reactions occurring in the human body. Sometimes toxic substances cause mere discomfort; sometimes they cause illness, sometimes disability, or even death. Toxic symptoms can be caused by very small amounts of extremely toxic materials (an example is sodium cyanide) or larger amounts of a less toxic substance. The term *toxic substances* usually is limited to materials that are dangerous in small amounts. However, as most of us know, ill effects can be caused by excessive intake of substances normally considered harmless (eating too much candy, for example). Fortunately, in most cases the human body is capable of recognizing "foreign" chemicals and ridding itself of them. In this chapter, we shall focus on the chemical mechanisms by which toxic substances act.

A large enough dose of any compound can result in poisoning.

DOSE

Lethal doses of toxic substances are customarily expressed in milligrams (mg) of substance per kilogram (kg) of body weight of the subject. For example, the cyanide ion (CN^-) is generally fatal to human beings in a dose of 1 mg of CN^- per kg of body weight. For a 200-pound (90.7 kg) person, about one tenth of a gram of cyanide is a lethal dose. Examples of somewhat less toxic substances and the range of lethal doses for human beings are the following:

"Dosis sola facit venenum"—the dose makes the poison.

Morphine	1–50 mg per kg
Aspirin	50–500 mg per kg
Methyl alcohol	500–5,000 mg
Ethyl alcohol	5,000–15,000 mg

312 / TOXIC SUBSTANCES IN OUR ENVIRONMENT

TABLE 13–1 Approximate Comparison of LD_{50} Values to Lethal Doses for Human Adults

Oral LD_{50} for Any Animal (mg/kg)	Probable Lethal Oral Dose for Human Adult
Less than 5	a few drops
5 to 50	"a pinch" to 1 teaspoonful
50 to 500	1 teaspoonful to 2 tablespoonsful
500 to 5,000	1 ounce to 1 pint (1 pound)
5,000 to 15,000	1 pint to 1 quart (2 pounds)

A quantitative measure of toxicity is obtained by introducing various dosages of substances to be tested into laboratory animals (such as rats). That dosage that is found to be lethal in 50 percent of a large number of the animals under controlled conditions is called the LD_{50} (lethal dosage—50%) and is reported in milligrams of poison per kilogram of body weight. Thus, if a statistical analysis of data on a large population of rats showed that a dosage of 1 mg per kg was lethal to 50 percent of the population tested, the LD_{50} for this poison would be 1 mg per kg. Obviously, metabolic variations and other differences between species will produce different LD_{50} values for a given poison in different kinds of animals. For this reason such data cannot be extrapolated to human beings with any assurance, but it is safe to assume that a substance with a low LD_{50} value for several animal species will also be quite toxic to humans (Table 13–1).

Toxic substances can be classified according to the way in which they disrupt the chemistry of the body. Some of the modes of action of toxic substances can be described as *corrosive, metabolic, neurotoxic, mutagenic,* and *carcinogenic,* and these will serve as the bases of our discussion.

Metabolism (from the Greek, *metaballein,* to change or alter) is the sum of all the physical and chemical changes by which living organisms are produced and maintained.

CORROSIVE POISONS

Toxic substances that actually destroy tissues are corrosive poisons. Examples include strong acids and alkalies, and many oxidants such as those found in laundry products, which can destroy tissues. Sulfuric acid (found in auto batteries) and hydrochloric acid (also called muriatic acid; used for cleaning purposes) are very dangerous corrosive poisons. So is sodium hydroxide, used in clearing clogged drains. Death has resulted from the swallowing of 1 ounce of concentrated (98 percent) sulfuric acid, and much smaller amounts can cause extensive damage and severe pain.

Concentrated mineral acids such as sulfuric acid act by first dehydrating cellular structures. The cell dies because its protein structures are destroyed by the acid-catalyzed hydrolysis of the peptide bonds.

In the early stages of this process there will be a large proportion of larger fragments present. Subsequently, as more bonds are broken, smaller and smaller fragments result, leading to the ultimate disintegration of the tissue.

Some poisons act by undergoing chemical reaction in the body to produce corrosive poisons. Phosgene, the deadly gas used during World War I, is an example. When inhaled, it is hydrolyzed in the lungs to hydrochloric acid, which causes pulmonary edema (a collection of fluid in the lungs) due to the dehydrating effect of the strong acid on tissues. The victim dies of suffocation because oxygen cannot be absorbed effectively by the flooded and damaged tissues.

Chemical "warfare gases," such as phosgene, were outlawed by an international conference in 1925.

$$\begin{array}{c} O \\ \parallel \\ C \\ Cl \quad Cl \end{array} + H_2O \longrightarrow 2HCl + CO_2$$

PHOSGENE   HYDROCHLORIC ACID   CARBON DIOXIDE

Sodium hydroxide, NaOH (caustic soda—a component of drain cleaners), is a very strongly alkaline, or basic, substance that can be just

TABLE 13–2 Some Corrosive Poisons

Substance	Formula	Toxic Action	Possible Contact
Hydrochloric acid	HCl	Acid hydrolysis	Tile and concrete floor cleaner; concentrated acid used to adjust acidity of swimming pools
Sulfuric acid	H_2SO_4	Acid hydrolysis, dehydrates tissue—oxidizes tissue	Auto batteries
Phosgene	ClCOCl	Acid hydrolysis	Combustion of chlorine-containing plastics (PVC or Saran)
Sodium hydroxide	NaOH	Base hydrolysis	Caustic soda, drain cleaners
Trisodium phosphate	Na_3PO_4	Base hydrolysis	Detergents, household cleaners
Sodium perborate	$NaBO_3 \cdot 4H_2O$	Base hydrolysis—oxidizing agent	Laundry detergents, denture cleaners
Ozone	O_3	Oxidizing agent	Air, electric motors
Nitrogen dioxide	NO_2	Oxidizing agent	Polluted air, automobile exhaust
Iodine	I_2	Oxidizing agent	Antiseptic
Hypochlorite ion	OCl^-	Oxidizing agent	Bleach
Peroxide ion	O_2^{2-}	Oxidizing agent	Bleach, antiseptic
Oxalic acid	$H_2C_2O_4$	Reducing agent, precipitates Ca^{2+}	Bleach, ink eradicator, leather tanning, rhubarb, spinach, tea
Sulfite ion	SO_3^{2-}	Reducing agent	Bleach

as corrosive to tissue as strong acids. The hydroxide ion also catalyzes the splitting of peptide linkages:

$$R-\overset{\overset{\text{O}}{\|}}{C}-\overset{\overset{\text{H}}{|}}{N}-R' + H_2O \xrightarrow[\text{base}]{OH^-} R-\overset{\overset{\text{O}}{\|}}{C}-OH + H-\overset{\overset{\text{H}}{|}}{N}-R'$$

Both acids and bases, as well as other types of corrosive poisons, continue their action until they are consumed in chemical reactions.

Some corrosive poisons destroy tissue by oxidizing it. This is characteristic of substances such as ozone, nitrogen dioxide, and possibly iodine, which destroy enzymes by oxidizing their functional groups. Specific groups, such as the —SH and —S—S— groups in the enzyme, are believed to be converted by oxidation to nonfunctioning groups; alternatively, the oxidizing agents may break chemical bonds in the enzyme, leading to its inactivation.

A summary of some common corrosive poisons is presented in Table 13–2.

METABOLIC POISONS

Metabolic poisons are more subtle than the tissue-destroying corrosive poisons. In fact, many of them do their work without actually indicating their presence until it is too late. Metabolic poisons can cause illness or death by interfering with a vital biochemical mechanism to such an extent that it ceases to function or is prevented from functioning efficiently.

Carbon Monoxide

The interference of carbon monoxide with extracellular oxygen transport is one of the best understood processes of metabolic poisoning. As early as 1895, it was noted that carbon monoxide deprives body cells of oxygen (asphyxiation), but it was much later before it was known that carbon monoxide, like oxygen, combines with hemoglobin:

$$O_2 + \text{hemoglobin} \longrightarrow \text{oxyhemoglobin}$$

$$CO + \text{hemoglobin} \longrightarrow \text{carboxyhemoglobin}$$

Laboratory tests show that carbon monoxide reacts with hemoglobin to give a compound (carboxyhemoglobin) that is 140 times more stable than the compound of hemoglobin and oxygen (oxyhemoglobin). (See Figure 13–1.) Since hemoglobin is so effectively tied up by carbon monoxide, it cannot perform its vital function of transporting oxygen.

An organic material that undergoes incomplete combustion will always liberate carbon monoxide. Sources include auto exhausts, smoldering leaves, lighted cigars or cigarettes, and charcoal burners. In the United States alone, combustion sources of all types dump about two hundred million tons of carbon monoxide per year into the atmosphere.

While the best estimates of the maximum global background level of carbon monoxide are of the order of 0.1 ppm, the background concen-

ppm—parts per million—a measure expressing concentration. 50 ppm CO means 50 ml CO for every million ml of air.

Figure 13–1 Structure of the heme portion of hemoglobin. (*a*) Normal acceptance and release of oxygen. (*b*) Oxygen blocked by carbon monoxide.

To convert ppm to percent, divide by 10,000.

tration in cities is higher. In heavy traffic, sustained levels of 100 or more ppm are common; for offstreet sites an average of about 7 ppm is typical for large cities. A concentration of 30 ppm for eight hours is sufficient to cause headache and nausea. Breathing an atmosphere that is 0.1 percent (1,000 ppm) carbon monoxide for four hours converts approximately 60 percent of the hemoglobin of an average adult to carboxyhemoglobin (Table 13–3), and death is likely (Figure 13–2).

Since both the carbon monoxide and oxygen reactions with hemoglobin involve easily reversed reactions, the concentrations, as well as relative strengths of bonds, affect the direction of the reaction. In air that contains 0.1 percent CO, oxygen molecules outnumber CO molecules 200 to 1. The larger concentration of oxygen helps to counteract the greater combining power of CO with hemoglobin. Consequently, if a carbon monoxide victim is exposed to fresh air or, still better, pure oxygen (provided he is still breathing), the carboxyhemoglobin (HbCO) is gradually decomposed, owing to the greater concentration of oxygen:

Air is 21% O_2 by volume; in 1,000,000 "air molecules" there would be 210,000 O_2 molecules.

$$HbCO + O_2 \rightleftarrows HbO_2 + CO$$

REACTION SHIFTED TO RIGHT BECAUSE OF GREATER CONCENTRATION OF OXYGEN

TABLE 13–3   Concentration of CO in Atmosphere versus Percentage of Hemoglobin (Hb) Saturated*

CO concentration in air	0.01% (100 ppm)	0.02% (200 ppm)	0.10% (1,000 ppm)	1.0% (10,000 ppm)
Percentage of hemoglobin molecules saturated with CO†	17	20	60	90

*A few hours of breathing time is assumed.
†Normal human blood contains up to 5% of the hemoglobin as carboxyhemoglobin (HbCO).

100ppm   1,000ppm   1,300ppm   >2,000ppm

Figure 13—2 A healthy adult can tolerate 100 ppm carbon monoxide in air without suffering ill effect. A one-hour exposure to 1,000 ppm causes a mild headache and a reddish coloration of the skin develops. A one-hour exposure to 1,300 ppm turns the skin cherry red and a throbbing headache develops. One-hour exposure to concentrations greater than (>) 2,000 ppm will likely cause death. (From E. Meyer: *Chemistry of Hazardous Materials*. Prentice-Hall, Inc., Englewood Cliffs, NJ, 1977.)

Although carbon monoxide is not a cumulative poison, permanent damage can occur if certain vital cells (e.g., brain cells) are deprived of oxygen for more than a few minutes.

Individuals differ in their tolerance of carbon monoxide, but generally those with anemia or an otherwise low reserve of hemoglobin (e.g., children) are more susceptible. No one is helped by carbon monoxide, and smokers suffer chronically from its effects. It is a subtle poison, since it is odorless and tasteless.

Cyanide

The cyanide ion, CN^-, is the toxic agent in cyanide salts such as sodium cyanide used in electroplating. Since the cyanide is a relatively strong base, it reacts easily with many acids (weak and strong) to form volatile hydrogen cyanide, HCN:

$$CH_3COOH + Na^+CN^- \rightleftharpoons HCN + Na^+CH_3COO^-$$

ACETIC ACID   SODIUM CYANIDE   HYDROGEN CYANIDE   SODIUM ACETATE

Since HCN boils at a relatively low temperature (26°C), it is a gas at temperatures slightly above room temperature. It is often used as a fumigant in storage bins and holds of ships because it is toxic to most forms of life and, in gaseous form, can penetrate into tiny openings, even into insect eggs.

Natural sources of cyanide ions include the seeds of the cherry, plum, peach, apple, and apricot fruits. Hydrogen cyanide is produced by hydrolysis of certain compounds, such as amygdalin, contained in the seeds:

$$O\!-\!\boxed{C_6H_{10}O_4 \cdot C_6H_{11}O_5}$$

SUGAR UNITS

$$C\equiv N + 2H_2O \longrightarrow HCN + 2C_6H_{12}O_6 +$$

AMYGDALIN   HYDROGEN CYANIDE   GLUCOSE   BENZALDEHYDE

Figure 13–3 The mechanism of cyanide (CN^-) poisoning. Cyanide binds tightly to the enzyme cytochrome C, an iron compound, thus blocking the vital ADP-ATP reaction in cells.

The cyanide is not toxic as long as it is tied up in the amygdalin, but presumably if enough apple or peach seeds were hydrolyzed in warm acid, sufficient HCN would result to cause considerable danger. There are a few recorded instances of humans poisoned by eating large numbers of apple seeds. Amygdalin is not confined to the seeds; amounts as high as 66 mg per 100 g have been reported in peach leaves.

The cyanide ion is one of the most rapidly working poisons. Lethal doses taken orally act in minutes. Cyanide poisons by asphyxiation, as does carbon monoxide, but the mechanism of cyanide poisoning is different (Figure 13–3). Instead of preventing the cells from getting oxygen, cyanide interferes with oxidative enzymes, such as cytochrome oxidase. Oxidases are enzymes containing a metal, usually iron or copper. They catalyze the oxidation of substances such as glucose:

> A metabolite is any substance in a metabolic process.

$$\text{Metabolite (H)}_2 + \tfrac{1}{2}O_2 \xrightarrow{\text{Oxidase}} \text{Oxidized metabolite} + H_2O + \text{Energy}$$

The iron atom in cytochrome oxidase is oxidized from Fe^{2+} to Fe^{3+} to provide electrons for the reduction of O_2. The iron regains electrons from other steps in the process. The cyanide ion forms stable cyanide complexes with the metal ion of the oxidase and renders the enzyme incapable of reducing oxygen or oxidizing the metabolite.

> $Fe^{2+} \longrightarrow Fe^{3+} + e^-$
> Oxidation

$$\text{Cytochrome oxidase (Fe)} + CN^- \longrightarrow \underbrace{\text{cytochrome oxidase } (Fe)\cdots CN^-}_{\text{complex}}$$

In essence the electrons of iron are "frozen"—they cannot participate in the oxidation-reduction processes. Plenty of oxygen gets to the cells, but the mechanism by which the oxygen is used in the support of life is stopped. Hence the cell dies, and if this occurs fast enough in the vital centers, the victim dies.

The body has a mechanism for ridding itself slowly of cyanide ions. The cyanide-oxidative enzyme reaction is reversible, and other enzymes such as rhodanase, found in almost all cells, can convert cyanide ions to relatively harmless thiocyanate ions. For example,

> The body can rid itself of a large number of toxic substances if the dose is small enough and sufficient time is allowed.

$$\underset{\text{THIOSULFATE}}{CN^- + S_2O_3^{2-}} \xrightarrow{\text{Rhodanase}} \underset{\text{THIOCYANATE}}{SCN^- + SO_3^{2-}}$$

This mechanism is not as effective in protecting a cyanide-poison victim as it might appear, since there is only a limited amount of thiosulfate available in the body at a given time.

Fluoroacetic Acid

Nature has used the synthesis of fluoroacetic acid as part of a defense mechanism for certain plants. Native to South Africa, the *gilfbaar* plant contains lethal quantities of fluoroacetic acid. Cattle that eat these leaves usually sicken and die.

$$F-\overset{\overset{\displaystyle H}{|}}{\underset{\underset{\displaystyle H}{|}}{C}}-C\overset{\displaystyle O}{\underset{\displaystyle O-H}{}}$$

FLUOROACETIC ACID

Sodium fluoroacetate, the sodium salt of this acid (Compound 1080), is a potent rodenticide (rat poison). Because it is odorless and tasteless it is especially dangerous, and its sale in this country is strictly regulated by law.

Fluoroacetate is toxic because it enters the Krebs cycle, producing

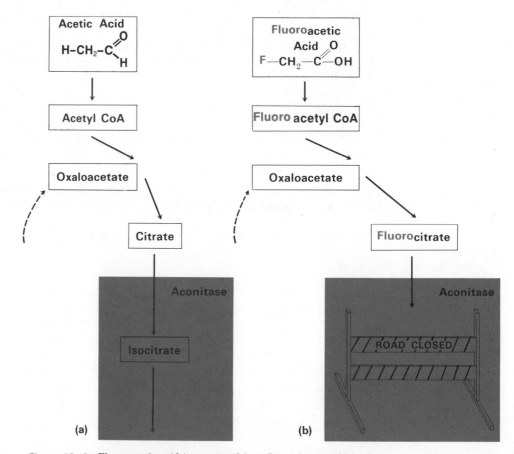

Figure 13–4 Fluoroacetic acid is converted into fluorocitrate, which then forms a stable bond with the enzyme aconitase. This blocks the normal Krebs cycle, a portion of which is shown.

fluorocitric acid, which in turn blocks the Krebs cycle (Figure 13–4). The C—F linkage apparently ties up the enzyme aconitase, thus preventing it from converting citrate to isocitrate.

Some poisons can act by mimicking other compounds.

In this instance, the poison is similar enough to the normal substrate to compete effectively for the active sites on the enzyme. If a poison has sufficient affinity for the active site on the enzyme, it blocks the normal function of the enzyme. The blocking of the Krebs cycle by fluorocitrate is a typical example of this affinity. If fluoroacetate is not present in excessive amounts, its action can be reversed simply by increasing the concentration of available citrate.

Heavy Metals

Heavy metals are perhaps the most common of all the metabolic poisons. These include such common elements as lead and mercury, as well as many less common ones such as cadmium, chromium, and thallium. In this group we should also include the infamous poison, *arsenic,* which is really not a metal but is metal-like in many of its properties, including its toxic action.

Arsenic, a classic homicidal poison, occurs naturally in small amounts in many foods. Shrimp, for example, contain about 19 ppm arsenic, while corn may contain 0.4 ppm arsenic. Some agriculturial insecticides contain arsenic (Table 13–4), and so some arsenic is observed in very small amounts on some fruits and vegetables. The Federal Food and Drug Administration (FDA) has a limit of 0.15 mg of arsenic per pound of food, and this amount apparently causes no harm. Several drugs, such as arsphenamine, which has found some use in treating syphilis, contain covalently bonded arsenic. In its ionic forms, arsenic is much more toxic.

Arsenic and heavy metals owe their toxicity primarily to their ability to react with and inhibit sulfhydryl (—SH) enzyme systems, such as those involved in the production of cellular energy. For example, glutathione (a tripeptide of glutamic acid, cysteine, and glycine) occurs in most tissues; its behavior with metals illustrates the interaction of a metal with sulfhydryl groups. The metal replaces the hydrogen on two sulfhydryl groups on adjacent molecules (Figure 13–5), and the strong bond that results effectively eliminates the two glutathione molecules from further reaction. Glutathione is involved in maintaining healthy red blood cells.

TABLE 13–4 Some Arsenic-Containing Insecticides

Name	Formula
Lead arsenate	$Pb_3(AsO_4)_2$
Monosodium methanearsenate	$CH_3-\overset{\displaystyle O}{\underset{\displaystyle OH}{As}}-O^-Na^+$
Paris green (copper acetoarsenite)	$3CuO \cdot 3As_2O_3 \cdot Cu(C_2H_3O_2)_2$

$$2 \text{ Glutathione} + \text{Metal ion (M}^{2+}) \longrightarrow \text{M (Glutathione)}_2 + 2\text{H}^+$$

Figure 13–5 Glutathione reaction with a metal (M).

Glutathione-Metal Complex

The typical forms of toxic arsenic compounds are inorganic ions such as arsenate (AsO_4^{3-}) and arsenite (AsO_3^{3-}). The reaction of an arsenite ion with sulfhydryl groups results in a complex in which the arsenic unites with two sulfhydryl groups, which may be on two different molecules of protein or on the same molecule:

ARSENITE   SULFHYDRYL   ARSENIC
            GROUPS   COMPLEX

The problem of developing a compound to counteract *Lewisite,* an arsenic-containing poison gas used in World War I, led to an understanding of how arsenic acts as a poison and subsequently to the development of an antidote. Once it was understood that Lewisite poisoned people by the reaction of arsenic with protein sulfhydryl groups, British scientists set out to find a suitable compound that contained highly reactive sulfhydryl groups which could compete with sulfhydryl groups in the natural substrate for the arsenic, and thus render the poison ineffective. Out of this research came a compound now known as British Anti-Lewisite (BAL).

$$\begin{array}{ccc} CH_2 & CH & CH_2 \\ | & | & | \\ OH & SH & SH \end{array}$$
BAL
British Anti-Lewisite

The BAL, which bonds to the metal at several sites, is called a ***chelating agent*** (Greek, *chela,* claw), a term applied to a reacting agent that envelops a species such as a metal ion. BAL is one of a large number of compounds that can act as chelating agents for metals (Figure 13–6).

A chelating agent encases an atom or ion like a crab or an octopus surrounds a bit of food.

With the arsenic or heavy metal ion tied up, the sulfhydryl groups in vital enzymes are freed and can resume their normal functions. BAL is a standard therapeutic item in a hospital's poison emergency center and is used routinely to treat heavy metal poisoning.

Mercury deserves some special attention because it has a rather peculiar fascination for some people, especially children, who love to

$$\begin{array}{c} \text{CH}_2\text{—OH} \\ | \\ \text{CH—SH} \\ | \\ \text{CH}_2\text{—SH} \end{array} + \text{M}^{2+} \longrightarrow \begin{array}{c} \text{CH}_2\text{—OH} \\ | \\ \text{CH—S} \\ \qquad\qquad \text{M} \\ \text{CH}_2\text{—S} \end{array} + 2\text{H}^+$$

BAL   Heavy metal ion   Chelated metal ion

Figure 13–6 BAL chelation of arsenic or a heavy metal ion such as lead.

A vivid description of the psychic changes produced by mercury poisoning can be found in the Mad Hatter, a character in Lewis Carroll's *Alice in Wonderland.* The fur felt industry once used mercury(II) nitrate, $Hg(NO_3)_2$, to stiffen the felt. Chronic mercury poisoning accounted for the Mad Hatter's odd behavior; it also gave the workers in hat factories symptoms known as "hatter's shakes."

Amalgam: any mixture or alloy of metals in which mercury is a constituent.

touch it (Figure 13–7). It is poisonous and, to make matters worse, mercury and its salts accumulate in the body. This means the body has no quick means of ridding itself of this element and there tends to be a build-up of the toxic effects leading to ***chronic*** poisoning.

While mercury is rather unreactive compared to other metals, it is quite volatile and easily absorbed through the skin. In the body the metal atoms are oxidized to Hg_2^{2+} [mercury(I) ion] and Hg^{2+} [mercury(II) ion]. Compounds of both Hg_2^{2+} and Hg^{2+} are known to be toxic.

Today mercury poisoning is a potential hazard to those working with or near this metal or its salts, such as dentists (who use it in making amalgams for fillings), various medical and scientific laboratory personnel (who routinely use mercury compounds or mercury pressure gauges), and some agricultural workers (who employ mercury salts as fungicides).

Mercury can also be a hazard when it is present in food. It is generally believed that mercury enters the food chain through small organisms that feed at the bottom of bodies of water that contain mercury from industrial waste or mercury minerals in the sediment. These in turn are food for bottom-feeding fish. Game fish in turn eat these fish and accumulate the largest concentration of mercury, the accumulation of poison building up as the food chain progresses.

Lead is another widely encountered heavy-metal poison. The body's method of handling lead provides an interesting example of a "metal

Figure 13–7 The first step to coating a coin with mercury. Children love to coat coins with metallic mercury—a very dangerous practice since mercury is easily inhaled and also passes through the skin.

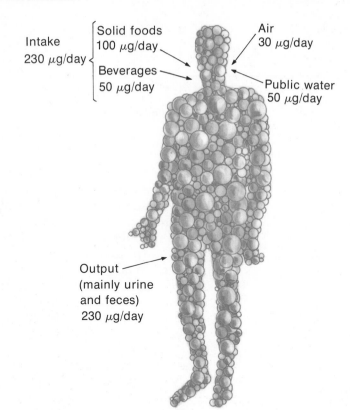

Intake 230 μg/day
- Solid foods 100 μg/day
- Beverages 50 μg/day
- Air 30 μg/day
- Public water 50 μg/day

Output (mainly urine and feces) 230 μg/day

Figure 13–8 Lead equilibrium in man. Figures chosen for intake are probable upper limits.

equilibrium" (Figure 13–8). Lead often occurs in foods (100 to 300 μg per kg), beverages (20 to 30 μg per liter), public water supplies (100 μg per liter, from lead-sealed pipes), and even air (2.5 μg per cubic meter, from lead compounds in auto exhausts). With this many sources and contacts per day, it is obvious that the body must be able to rid itself of this poison; otherwise everyone would have died long ago of lead poisoning! The average person can excrete about two milligrams of lead a day through the kidneys and intestinal tract; the daily intake is normally less than this. However, if intake exceeds this amount, accumulation and storage results. In the body lead not only resides in soft tissues but is also deposited in bone. In the bones lead acts on the bone marrow, while in tissues it behaves like other heavy-metal poisons, such as mercury and arsenic. Lead, like mercury and arsenic, can also affect the central nervous system.

1 μg (microgram) = 10^{-6} gram

1 mg (milligram) = 10^{-3} gram = 1,000 μg

Lead salts, unless they are very insoluble, are always toxic, and their toxicity is directly related to the salt's solubility. One common covalent lead compound, tetraethyllead, $Pb(C_2H_5)_4$, until recently a component of some gasolines, is different from most other metal compounds in that it is readily absorbed through the skin. Even metallic lead can be absorbed through the skin; cases of lead poisoning have resulted from repeated handling of lead foil, bullets, and other lead objects.

One of the truly tragic aspects of lead poisoning is the fact that even though lead-pigmented paints have not been used in this country for interior painting during the past 30 years, children are still poisoned by lead from old paint. Health experts estimate that up to 225,000 children become ill from lead poisoning each year, with many experiencing mental retardation or other neurological problems. The reason for this is twofold. Lead-based paints still cover the walls of many older dwellings. Coupled

Figure 13-9 The structure of the chelate formed when the anion of EDTA envelops a lead (II) ion.

with this is the fact that many children in poverty-striken areas are ill-fed and anemic. These children develop a peculiar appetite trait called *pica,* and among the items that satisfy their cravings are pieces of flaking paint, which may contain lead. In 1969, about 200 children in the United States alone died of lead poisoning and untold thousands suffered permanent damage.

Toxicologists have discovered a very effective chelating agent to remove lead from the human body—ethylenediaminetetraacetic acid, also called EDTA (Figure 13-9).

$$\begin{matrix} \text{HOOCCH}_2 & & \text{CH}_2\text{COOH} \\ & \diagdown \qquad \diagup & \\ & \text{N}-\text{CH}_2-\text{CH}_2-\text{N} & \\ & \diagup \qquad \diagdown & \\ \text{HOOCCH}_2 & & \text{CH}_2\text{COOH} \end{matrix}$$

EDTA
(ETHYLENEDIAMINETETRAACETIC ACID)

The calcium disodium salt of EDTA is used in the treatment of lead poisoning because EDTA by itself would remove too much of the blood serum's calcium. In solution EDTA has a greater tendency to complex with lead (Pb^{2+}) than with calcium (Ca^{2+}). As a result, the calcium is released and the lead is tied up in the complex:

$$[\text{CaEDTA}]^{2-} + Pb^{2+} \longrightarrow [\text{PbEDTA}]^{2-} + Ca^{2+}$$

The lead chelate is then excreted in the urine.

---

1. Corrosive poisons such as sulfuric acid destroy tissue by _____ **SELF-TEST 13-A**

   followed by _____ of proteins.

2. Corrosive poisons, such as ozone, nitrogen dioxide, and iodine, destroy tissue

   by _____ it.

3. Carbon monoxide poisons by forming a very strong bond with iron in

   _____ and thus prevents the transport of _____
   from the lungs to the cells throughout the body.

4. CO is a cumulative poison. (True/False.)

5. The cyanide ion has the formula _____ . It poisons by complexing with iron in the enzyme _____ _____ , and thus prevents the use of _____ in the oxidative processes in the cells.

6. Give an example of a metabolic poison that is toxic because its structure is so similar to a useful substance that it can mimic the useful substance.

   _____

7. BAL is an antidote for _____ . BAL is effective because its sulfhydryl (—SH) groups _____ arsenic and heavy metals and render them ineffective toward enzymes.

8. Mercury is a cumulative poison. (True/False.)

NEUROTOXINS

Some metabolic poisons are known to limit their action to the nervous system. These include poisons such as strychnine and curare (a South American Indian dart poison), as well as the dreaded nerve gases developed for chemical warfare. The exact modes of action of most neurotoxins are not known for certain, but investigations have discovered the action of a few.

A nerve impulse or stimulus is transmitted along a nerve fiber by electrical impulses. The nerve fiber connects with either another nerve fiber or with some other cell (such as a gland, or cardiac, smooth, or skeletal muscle) capable of being stimulated by the nerve impulse (Figure 13–10). Neurotoxins often act at the point where two nerve fibers come together, called a *synapse.* When the impulse reaches the end of certain nerves, a small quantity of *acetylcholine* is liberated. This activates a receptor on an adjacent nerve or organ. The acetylcholine is thought to activate a nerve ending by changing the permeability of the nerve cell membrane. The method of increasing membrane permeability is not clear, but it may be related to an ability to dissociate fat-protein complexes or to penetrate the surface films of fats. Such effects can be brought about by as little as 10^{-6} mole of acetylcholine, which could alter the permeability of a cell so ions can cross the cell membrane more freely.

To enable the receptor to receive further electrical impulses, the enzyme *cholinesterase* breaks down acetylcholine into acetic acid and choline (Figure 13–11):

Investigations of the actions of neurotoxins have provided insight into how the nervous system works.

$$\underset{\text{Acetylcholine}}{\overset{O \qquad\quad CH_3}{CH_3\overset{\|}{C}OCH_2CH_2N^+CH_3, OH^-}}$$

Permeability: the ability of a membrane to let chemicals pass through it.

10^{-6} of a mole of acetylcholine is 6×10^{17} molecules.

$$CH_3\overset{\|}{\underset{O}{C}}OCH_2CH_2\overset{CH_3}{\underset{CH_3}{-N^+-}}CH_3, OH^- + H_2O \xrightarrow{\text{Cholinesterase}} CH_3\overset{O}{\overset{\|}{C}}OH + HOCH_2CH_2\overset{CH_3}{\underset{CH_3}{-N^+-}}CH_3, OH^-$$

ACETYLCHOLINE   WATER   ACETIC ACID   CHOLINE

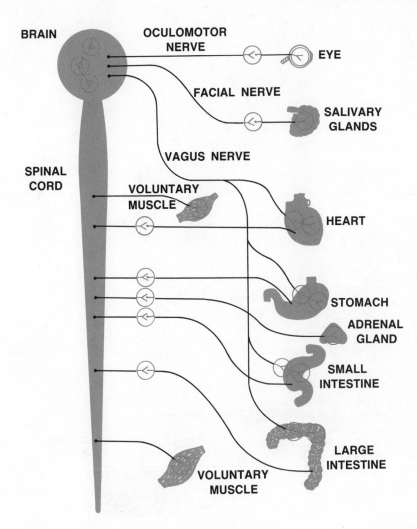

Figure 13–10 "Cholinergic" nerves, which transmit impulses by means of acetylcholine, include nerves controlling both voluntary and involuntary activities. Exceptions are parts of the "sympathetic" nervous system that utilize norepinephrine instead of acetylcholine. Sites of acetylcholine secretion are circled in color; poisons that disrupt the acetylcholine cycle can interrupt the body's communications at any of these points. The role of acetylcholine in the brain is uncertain, as is indicated by the broken circles.

In the presence of potassium and magnesium ions, other enzymes, such as acetylase, resynthesize new acetylcholine from the acetic acid and the choline within the incoming nerve ending:

$$\text{Acetic acid} + \text{Choline} \xrightarrow{\text{Acetylase}} \text{Acetylcholine} + \text{H}_2\text{O}$$

The new acetylcholine is available for transmitting another impulse across the gap.

Neurotoxins can affect the transmission of nerve impulses at nerve endings in a variety of ways. The ***anticholinesterase poisons*** prevent the breakdown of acetylcholine by deactivating cholinesterase. These poisons are usually structurally analogous to acetylcholine, so they bond to the enzyme cholinesterase and deactivate it (Figure 13–12). The cholinesterase molecules bound by the poison are held so effectively that the restoration of proper nerve function must await the manufacture of new cholinesterase. In the meantime, the excess acetylcholine overstimulates nerves, glands, and muscles, producing irregular heart rhythms, convulsions, and death. Many of the organic phosphates that are widely used as insecticides are metabolized in the body to produce anticholinesterase

Figure 13—11 The acetylcholine cycle, a fundamental mechanism in nerve impulse transmission, is affected by many poisons. An impulse reaching a nerve ending in the normal cycle (*a*) liberates acetylcholine, which then stimulates a receptor. To enable the receptor to receive further impulses, the enzyme *cholinesterase* breaks down acetylcholine into acetic acid and choline; other enzymes resynthesize these into more acetylcholine. (*b*) Botulinus and dinoflagellate toxins inhibit the synthesis, or the release, or acetylcholine (*1*). The "anticholinesterase" poisons inactivate cholinesterase, and therefore prevent the breakdown of acetylcholine (*2*). Curare and atropine desensitize the receptor to the chemical stimulus (*3*).

Figure 13–12 (*a*) The mechanism of cholinesterase breakdown of acetylcholine. (*b*) The tie-up of cholinesterase by an anticholinesterase poison like the nerve gas DFP blocks the normal hydrolysis of acetylcholine since the acetylcholine cannot bind to the enzyme.

Curare, used by South American Indians in poison darts, was brought to Europe by Sir Walter Raleigh in 1595. It was purified in 1865, and its structure was determined in 1935.

poisons. For this reason, they should be treated with extreme care. Some poisonous mushrooms also contain an anticholinesterase poison. Figure 13–13 contains the structures of some anticholinesterase poisons.

Neurotoxins such as *atropine* and *curare* are able to occupy the receptor sites on nerve endings of organs that are normally occupied by the impulse-carrying acetylcholine. When atropine or curare occupies the receptor site, no stimulus is transmitted to the organ. Acetylcholine in excess causes a slowing of the heart beat, a decrease in blood pressure, and excessive saliva, whereas atropine and curare produce excessive thirst

Name	Structure	LD_{50} (rat;oral), mg/kg	Use
Sarin	CH₃—C(H)(CH₃)—O—P(=O)(CH₃)—F	0.55	World War II nerve gas
Tabun	CH₃—N(CH₃)—P(=O)(CN)—O—CH₂—CH₃	3.7	World War II nerve gas
Parathion	CH₃—CH₂—O—P(=S)(O—CH₂—CH₃)—O—C₆H₄—NO₂	2.0	Insecticide
Paraoxon	CH₃—CH₂—O—P(=O)(O—CH₂—CH₃)—O—C₆H₄—NO₂	1.8	Insecticide
Carbyl (Sevin)	O—C(=O)—N(H)(CH₃) (naphthyl)	400	Insecticide

Figure 13–13 Some anticholinesterase poisons. In animals, parathion is converted into paraoxon in the liver. Carbyl does not bind to cholinesterase as strongly as the organophosphates.

and dryness of the mouth and throat, a rapid heart beat, and an increase in blood pressure. The normal responses to acetylcholine activation are absent, and the opposite responses occur when there is sufficient atropine present to block the receptor sites.

Neurotoxins of this kind can be extremely useful in medicine. For example, atropine is used to dilate the pupil of the eye to facilitate examination of its interior. Applied to the skin, atropine sulfate and other atropine salts relieve pain by deactivating sensory nerve endings on the skin. Atropine is also used as an antidote for anticholinesterase poisons. Curare has long been used as a muscle relaxant.

A well-known, natural organic compound that blocks receptor sites in a manner similar to that of curare and atropine is *nicotine.* This powerful poison causes stimulation and then depression of the central nervous system. The probable lethal dose for a 70-kilogram man is less than 0.3 gram. It is interesting to note that pure nicotine was first extracted from tobacco and its toxic action observed *after* tobacco use was established as an acceptable habit.

Natural or synthetic morphine is the most effective pain reliever

TABLE 13–5 Alkaloid Neurotoxins That Compete with Acetylcholine for the Receptor Site[*]

Name	Normal Contact	Lethal Dose (for a 70 kg human)	Formula
Atropine	Dilation of pupil of the eye	0.1 g	(structure)
Curare	Muscle relaxant	20 mg	(structure)
Nicotine	Tobacco, insecticide	75 mg	(structure)
Caffeine	Coffee, tea, cola drinks	13.4 g (One cup of coffee contains about 40 mg caffeine)	(structure)
Morphine	Opium— pain killer	100 mg	(structure)
Codeine	Opium— pain killer	0.3 g	(structure)
Cocaine	Leaves of *Erythroxylon coca* in South America	1 g	(structure)

[*]*Alkaloid* is broadly defined as a physiologically active compound found in plants and containing amino nitrogen atoms (consequently it has basic properties). The nitrogen atom, or atoms, are frequently found as part of rings.

known. It is widely used to relieve short-term acute pain resulting from surgery, fractures, burns, etc., as well as to reduce suffering in the later stages of terminal illnesses such as cancer. The manufacture and distribution of narcotic drugs are stringently controlled by the federal government through laws designed to keep these products available for legitimate medical use. Under federal law, some preparations containing small amounts of narcotic drugs may be sold without a prescription (for example, cough mixtures containing codeine), but not many.

In spite of stringent controls, drugs like **morphine, heroin, meperidine,** and **methadone** are abused and illicitly used. Heroin is prepared from morphine, which is derived from sap in the opium poppy. It takes about 10 pounds of opium to prepare one pound of morphine. Morphine reacts with acetic anhydride in a one-to-one reaction to form heroin. Street-grade heroin is only 9 to 10% pure.

Meperidine and **methadone** are products of chemical laboratories rather than of poppy fields. Meperidine was claimed to be nonaddictive when first produced. Experience, however, proved otherwise (as it did with morphine and heroin). A major difference between methadone and morphine and heroin is that when methadone is taken orally, under medical supervision, it prevents withdrawal symptoms for approximately 24 hours.

Some other natural products that affect the central nervous system and can be neurotoxic in comparatively small amounts are listed in Table 13–5.

Morphine is the most effective pain killer known.

Meperidine

Methadone

MUTAGENS

Mutagens are chemicals capable of altering the genes and chromosomes sufficiently to cause abnormalities in offspring. Chemically, mutagens alter the structures of DNA and RNA, which compose the genes (and, in turn, the chromosomes) that transmit the traits of parent to offspring. Mature sex, or germinal, cells of humans normally have 23 chromosomes; body, or somatic, cells have 23 *pairs* of chromosomes.

A mutagen is a chemical that can change the hereditary pattern of a cell.

Although many chemicals are under suspicion because of their mutagenic effects on laboratory animals, it should be emphasized that no one has yet shown conclusively that any chemical induces mutations in the germinal cells of man. Part of the difficulty of determining the effects of mutagenic chemicals in man is the extreme rarity of mutation. A specific genetic disorder may occur as infrequently as only once in 10,000 to 100,000 births. Therefore, to obtain meaningful, statistical data, a carefully controlled study of the entire population of the United States would be required. In addition, the very long time between generations presents great difficulties, and there is also the problem of tracing a medical disorder to a single, specific chemical out of the tens of thousands of chemicals with which we come in contact.

If there is no direct evidence for specific chemical mutagenic effects in human beings, why, then, the interest in the subject? The possibility of a deranged, deformed human race is frightening; the chance for an improved human body is hopeful; and the evidence for chemical mutation in plants and lower animals is established. A wide variety of chemicals is known to alter chromosomes and to produce mutations in rats, worms,

TABLE 13–6   Mutagenic Substances as Indicated by
Experimental Studies on Plants and Animals

Substance	Experimental Results
Aflatoxin (from mold, *Aspergillus flavus*)	Mutations in bacteria, viruses, fungi, parasitic wasps, human cell cultures, mice
Benzo(α)pyrene (from cigarette and coal smoke)	Mutations in mice
Caffeine	Chromosome changes in bacteria, fungi, onion root tips, fruit flies, human tissue cultures
Captan (a fungicide)	Mutagenic in bacteria and molds; chromosome breaks in rats and human tissue cultures
Dimethyl sulfate (used extensively in chemical industry to methylate amines, phenols, and other compounds)	Methylates DNA base guanine; potent mutagen in bacteria, viruses, fungi, higher plants, fruit flies
LSD (lysergic acid diethylamide)	Chromosome breaks in somatic cells of rats, mice, hamsters, white blood cells of humans and monkeys
Maleic hydrazide (plant growth inhibitor; trade names Slo-Gro, MH-30)	Chromosome breaks in many plants and in cultured mouse cells
Mustard gas (dichlorodiethyl sulfide)	Mutations in fruit flies
Nitrous acid (HNO_2)	Mutations in bacteria, viruses, fungi
Ozone (O_3)	Chromosome breaks in root cells of broadleaf plants
Solvents in glue (glue sniffing) (toluene, acetone, hexane, cyclohexane, ethyl acetate)	4% more human white blood cells showed breaks and abnormalities (6% versus 2% normal)
TEM (triethylenemelamine) (anticancer drug, insect chemosterilants)	Mutagenic in fruit flies, mice

bacteria, fruit flies, and other plants and animals. Some of these are listed in Table 13–6.

The horrible thalidomide disaster in 1961 focused worldwide attention on chemically induced birth defects. Thalidomide, a tranquilizer and sleeping pill, caused gross deformities (flipper-like arms, shortened arms, no arms or legs, and other defects) in children whose mothers used this drug during the first two months of pregnancy. The use of this drug resulted in more than 4000 surviving malformed babies in West Germany, more than 1000 in Great Britain, and about 20 in the United States. With shattering impact, this incident demonstrated that a compound can appear to be remarkably safe on the basis of animal studies (so safe, in fact, that thalidomide was sold in West Germany without prescription) and yet cause catastrophic effects in humans. While the tragedy focused attention on chemical mutagens, thalidomide presumably does not cause genetic damage in the germinal cells, and is really not mutagenic. Rather,

Thalidomide
(A teratogen)

Figure 13–14 Reaction of nitrous acid (HONO) with nitrogenous bases of DNA. Nitrogen gas (N_2) and water are also products of each reaction.

thalidomide, when taken by a woman during early pregnancy, causes direct injury to the developing embryo. A substance that acts in this manner is called a ***teratogen*** (Greek *teras,* monster). Even though thalidomide is generally believed not to be mutagenic, the disaster spotlighted the serious potential hazards of supposedly nontoxic chemicals in our environment.

Experimental work on the chemical basis of the mutagenic effects of nitrous acid (HNO_2) has been very revealing. Repeated studies have shown that nitrous acid is a potent mutagen in bacteria, viruses, molds, and other organisms. In 1953, at Columbia University, Dr. Stephen Zamenhof demonstrated experimentally that nitrous acid attacks DNA. Specifically, nitrous acid reacts with the adenine, guanine, and cytosine bases of DNA by removing the amino group of each of these compounds. The eliminated group is replaced by an oxygen atom (Figure 13–14). The changed bases may garble a part of DNA's genetic message, and in the next replication of DNA the new base may not form a base pair with the proper nucleotide base.

For example, adenine (A) typically forms a base pair with thymine (T) (Figure 13–15). However, when adenine is changed to hypoxanthine, the new compound forms a base pair with cytosine (C). In the second replication, the cytosine forms its usual base pair with guanine (G). Thus, where an adenine-thymine (A-T) base pair existed originally, a guanine-cytosine (G-C) pair now exists. The result is an alteration in the DNA's genetic coding, so that a different protein is formed later.

Do all of these findings mean that nitrous acid is mutagenic in

Figure 13–15 Alteration of DNA genetic code by base pairing nitrous acid-converted nitrogenous bases. The bases are adenine (A), cytosine (C), guanine (G), hypoxanthine (Hx), and thymine (T).

humans? Not necessarily. We do know that **sodium nitrite** has been widely used as a preservative, color enhancer, or color fixative in meat and fish products for at least the past 30 years. It is currently used in such foods as frankfurters, bacon, smoked ham, deviled ham, bologna, Vienna sausage, smoked salmon, and smoked shad. The sodium nitrite is converted to nitrous acid by hydrochloric acid in the human stomach:

$$NaNO_2 + HCl \longrightarrow HNO_2 + NaCl$$

The Food and Drug Administration (FDA) now considers the mutagenic effects of nitrous acid in lower organisms sufficiently ominous to suggest strongly that the use of sodium nitrite in foods be severely curtailed, and a complete ban of this use of sodium nitrite is being considered. A number of European countries already restrict the use of sodium nitrite in foods. The concern is that this compound, after being converted in the body to nitrous acid, may cause mutation in somatic cells (and possibly in germinal cells) and thus could possibly produce cancer in the human stomach. Other scientists doubt that nitrous acid is present in germinal cells and, therefore, seriously question whether this compound could be a cause of genetically produced birth defects in man. The uncertainty of extrapolating results obtained in animal studies to human beings hovers over the mutagenic substances.

Thus far research has concentrated on the action of chemicals in causing mutations in bacterial viruses, molds, fruit flies, mice, rats, human white blood cells, and so on. Perhaps in the next 10 to 20 years it will be demonstrated that these chemicals can produce transmissible alteration of chromosomes in human germinal cells. Meanwhile, many scientists are pressing for a more vigorous research effort to expand our knowledge of chemically induced mutations and of their potentially harmful effects in man. One intriguing theory that will surely invoke experimental examination is the belief that some compounds cause cancer because they are first and foremost mutagenic. The supporting evidence at present is still extremely inconclusive.

CARCINOGENS

Carcinogens are chemicals that cause cancer. ***Cancer*** is an abnormal growth condition in an organism that manifests itself in at least three ways. The rate of cell growth (that is, the rate of cellular multiplication) in cancerous tissue differs from the rate in normal tissue. Cancerous cells may divide more rapidly or more slowly than normal cells. Cancerous cells spread to other tissues; they know no bounds. Normal liver cells divide and remain a part of the liver. Cancerous liver cells may leave the liver and be found, for example, in the lung. Most cancer cells show partial or complete loss of specialized functions. Although located in the liver, cancer cells no longer perform the functions of the liver.

Attempts to determine the cause of cancer have evolved from early studies in which the disease was linked to a person's occupation. It was first noticed in 1775 that persons employed as chimney sweeps in England had a higher rate of skin cancer than the general population. It was not

Cancer of the epithelial tissue—*carcinoma*.

Cancer of the connective tissue—*sarcoma*.

TABLE 13—7   A Sample of Carcinogenic Compounds*

Compound	Use or Source	Formula of a Representative Compound
Vinyl chloride	Monomer for making polyvinyl chloride	
Benzene	Solvent in chemical industry, starting material for many other chemicals	
Aromatic amines		
2-Naphthylamine	Optical bleaching agent	
Benzidine	Polymer formation, manufacture of dyes, stain for tissues, rubber compounding	
2-Acetylaminofluorene	Herbicide	
Azo dyes		
4-Dimethylamino-azobenzene	Yellow dye (once used as a food color)	
Nitrosoamines		
N-Ethyl-N-nitroso-n-butylamine	Insecticide, gasoline and lubricant additive	
Polynuclear compounds		
Benzo(α)pyrene	Cigarette and coal smoke	
Aflatoxin B_1	Mold on peanuts and other plants (*Aspergillus flavus*)	

*Most of the evidence is based on studies with mice.

ANILINE   2-NAPHTHYLAMINE

BENZO(α)PYRENE

until 1933 that **benzo(α)pyrene,** $C_{20}H_{12}$, (a 5-ringed aromatic hydrocarbon) was isolated from coal dust and shown to be metabolized in the body to produce one or more carcinogens. In 1895, the German physician Rehn noted three cases of bladder cancer, not in a random population, but in employees of a factory which manufactured dye intermediates in the Rhine Valley. Rehn attributed these cancers to his patients' occupation. These and other cases confirmed that at times as many as 30 years passed between the time of the initial employment and the occurrence of bladder cancer. The principal product of these factories was aniline. While aniline was first thought to be the carcinogenic agent, it was later shown to be noncarcinogenic. It was not until 1937 that continuous, long-term treatment with **2-naphthylamine,** one of the suspected dye intermediates, in dosages of up to 0.5 g per day produced bladder cancer in dogs. Since then other dye intermediates have been shown to be carcinogenic.

A vast amount of research effort has verified the carcinogenic behavior of a large number of diverse chemicals. Some of these are listed in Table 13–7. This research has led to the formulation of a few generalizations concerning the relationship between chemicals and cancer. For example, carcinogenic effects on lower animals are commonly extrapolated to man. The mouse has come to be the classic animal for studies of carcinogenicity. Strains of inbred mice and rats have been developed that are uniform and show a standard response.

Some carcinogens are relatively nontoxic in a single, large dose, but may be quite toxic, often increasingly so, when administered continuously. Thus, much patience, time, and money must be expended in carcinogenic studies. The development of a sarcoma in humans, from the activation of the first cell to the clinical manifestation of the cancer, takes from 20 to 30 years. With life expectancy of an average person in the United States now set at about 70 years, it is not surprising that the number of deaths due to cancer is increasing.

Cancer does not occur with the same frequency in all parts of the world. Breast cancer occurs with lower frequency in Japan than in the United States or Europe. Cancer of the stomach, especially in males, is more common in Japan than in the United States. Cancer of the liver is not widespread in the Western Hemisphere but accounts for a high proportion of the cancers in the Bantu in Africa and in certain populations in the Far East. The widely publicized incidence of lung cancer is higher in the industrialized world and is increasing at an appreciable rate.

Some compounds cause cancer at the point of contact. Other compounds cause cancer in an area remote from the point of contact. The liver, the site at which most toxic chemicals are removed from the blood, is particularly susceptible to such compounds. Since the original compound does not cause cancer on contact, some other compound made from it must be the cause of cancer. For example, it appears that the substitution of an NOH group for an N—H group in an aromatic amine derivative produces at least one of the active intermediates for carcinogenic amines. If R denotes a two- or three-ring aromatic system, then the process can be represented as follows:

Cancer spreads from one tissue to another via metastases.

An abnormal growth is classified as cancerous or malignant when examination shows it is invading neighboring tissue. A growth is benign if it is localized at its original site.

$$\underset{\substack{\text{INACTIVE} \\ \text{ON CONTACT}}}{\overset{\overset{\displaystyle H}{|}}{RNCOCH_3}} \longrightarrow \underset{\substack{\text{ACTIVE ON} \\ \text{CONTACT}}}{\overset{\overset{\displaystyle OH}{|}}{RNCOCH_3}} \longrightarrow \underset{\substack{\text{OTHER UNKNOWN} \\ \text{INTERMEDIATES}}}{RX? \longrightarrow RY?} \xrightarrow{\text{Tissue}} \text{Tumor cell}$$

As indicated by the variety of chemicals in Table 13–7, many molecular structures produce cancer while closely related ones do not. The 2-naphthylamine mentioned earlier is carcinogenic, but repeated testing gives negative results for 1-naphthylamine.

1-NAPHTHYLAMINE
(NONCARCINOGENIC)

2-NAPHTHYLAMINE
(CARCINOGENIC)

For some types of cancer there are distinct stages that ultimately result in cancer. These may be identified as the *initiation period,* the *development* or *promotion period,* and the *progression period.* A single, minute dose of a carcinogenic polynuclear aromatic hydrocarbon, such as benzo(α)pyrene, applied to the skin of mice produces the permanent change of a normal cell to a tumor cell. This is the initiation step. No noticeable reaction occurs unless further treatment is made. If the area is painted repeatedly with noncarcinogenic croton oil, even up to one year later, carcinomas appear (Figure 13–16). This is the development period. Additional fundamental alterations in the nature of the cells occur during the progression period. If there is no initiator, there are no tumors. If there is initiator but no promoter, there are no tumors. If the initiator is followed by repeated doses of promoter, tumors appear. This seems to indicate that cancer cannot be contracted from chemicals unless repeated doses are administered or applied.

Just how do these toxic substances work? Cancer might be caused if the carcinogen combines with growth control proteins, rendering them inactive. During the normal growth process the cells divide and the organism grows to a point and stops. Cancer is abnormal in that cells

> Almost all chemical carcinogens have an induction period.

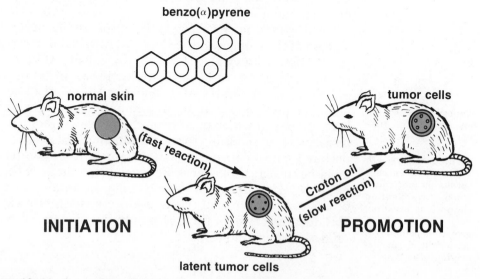

Figure 13–16 A second chemical can promote tumor growth in mice after an initiation period. Treatment of mice with croton oil produces no tumor nor does treatment with small quantities of benzo (α) pyrene alone. Croton oil is an irritant oil similar to castor oil. Both are derived from plants.

continue to divide and portions of the organism continue to grow. One or more proteins are thought to be present in each cell with the specific duty of preventing replication of DNA and cell division. Virtually all of the carcinogens bind firmly to proteins, but so do some similar compounds that are noncarcinogenic. The specific growth proteins involved are not yet known for any of the carcinogens, despite considerable efforts to find them.

Another theory suggests that carcinogens react with and alter nucleic acids so the proteins ultimately formed on the messenger RNA are sufficiently different to alter the cell's function and growth rate. The carcinogen may be included in the DNA or RNA strands by covalent bonding or it may be entangled in the helix and held by weak van der Waals attractions. The carcinogenic compounds nitrosodimethylamine and mustard gas have been shown to react with nucleic acids.

While researchers collect data in their laboratories and speculate on the theoretical structural causes of cancer, we can studiously avoid compounds known to cause cancer in man. It has been proposed that as much as 80% of all human cancer has its origin in carcinogenic chemicals.

HALLUCINOGENS

Hallucinogens can produce temporary changes in perception, thought, and mood. These substances include *mescaline, lysergic acid*

TABLE 13—8 Some Hallucinogenic Chemicals

Chemical	Aliases	Notes	Active Compound	Formula
LSD	Acid Crackers The chief The hawk	Very powerful hallucinogen	D-Lysergic acid diethylamide	
Mescaline	Peyote	Extracted from mescal buttons from peyote cactus in South and Central America	3,4,5-Trimethoxy-phenethylamine	
Marihuana	Grass Pot Reefers Locoweed Hash Mary Jane	Leaves of the 5-leaf-per-frond marihuana plant (*Cannabis sativa*)	3,4-Trans-tetra-hydrocannabinol	
Amphetamines	Pep pills Speed Bennies Ups	Eighteen times more active than mescaline	Benzedrine (for example)	
Phencyclidine	PCP Hog Elephant Peace pills Angel dust	Powerful analgesic-anesthetic, used in veterinary medicine	Phencyclidine	

diethylamide (LSD), tetrahydrocannabinol (the active component of marihuana), and a broadening field of more than 50 other substances. They are included in this chapter as a separate section because they can be toxic in comparatively small doses.

Several characteristics of some of the more famous hallucinogens are given in Table 13–8. Each of these substances is capable of disturbing the mind and producing bizarre and even colored interpretations of visual and other external stimuli. Mescaline is one of the oldest known hallucinogens, having been isolated from the peyote plant in 1896 by Heffter. Indeed, as early as 1560 the Mexican Indians who ate or drank the peyote were described by Spanish explorers as experiencing "terrible or ludicrous visions; the inebriation lasting for two or three days and then disappearing."

Although known for nearly 5,000 years, marihuana is one of the least understood of all natural drugs. Very early in China's history it was used as a medicine, and in the United States it had early use as an analgesic and a poultice for corns. Marihuana has little acceptable medical use in this country at present; experimental use is presently being studied, such as the treatment of glaucoma.

The body does not become dependent on the continuing use of marihuana as it does with heroin or other narcotics. However, reliable scientific data are not available with regard to chronic toxicity resulting from long-term use of the drug. It is known to cause dryness of the mouth, leading to dental problems.

Today there are many substances that produce hallucinogenic experiences; the most powerful one known is LSD. Our brief discussion will be restricted to this compound.

Lysergic Acid Diethylamide (LSD)

Literally hundreds of scientists are doing research on the effects of hallucinogens, including LSD. They are investigating the influence of these drugs on nerve and brain function, the possibility of chromosome alteration, tissue damage, psychological changes, and a host of other physiological and psychological effects. Scientists are not near a consensus on the toxic effects that LSD can cause. Collecting valid data is very difficult; many users of LSD overestimate the quantity of the drug they have used, the purity is often in question, and some take other drugs in addition. Even if the purity of the LSD is known, the results are difficult to interpret. For example, a study in which mice were given 0.05 to 1.0 microgram of LSD on the seventh day of pregnancy showed a 5 percent incidence of badly deformed mouse embryos. Another study, probably equally valid and meticulous, showed no unusual fetal damage to rat embryos when 1.5 to 300 micrograms of LSD was administered during the fourth or fifth day of pregnancy. Babies with depressed skulls were aborted from two women who had taken LSD during the early weeks of pregnancy. Two other women, also LSD users, delivered full term, apparently normal babies.

There are, however, some dangers that are well documented. These drugs destroy one's sense of judgment. Such things as height, heat, or even a moving truck may seem to hold no danger for the person under the influence of a hallucinogen. A dose of 50 to 200 micrograms will take the user on a "trip" for approximately eight to sixteen hours. Excessive or prolonged use of LSD can cause a person to "freak out" and possibly to sus-

All of the hallucinogenic drugs are toxic.

LSD has been linked with birth defects.

CH₂CH₂NH₂

Serotonin

Norepinephrine

tain permanent brain damage. After one "trip," a user can experience another "trip" some time later, unexpectedly, without taking any more of the drug. The debate over LSD's dangers is continuing and is not likely to be resolved soon.

Out of the darkness of LSD-induced suicides and brain damage has come some new understanding of how the brain works. It is an interesting coincidence that soon after LSD was found localized in areas of the brain responsible for eliciting a human being's deep-seated emotional reactions, the compounds were discovered that are probably responsible for transmitting the impulse across synapses in the brain. Serotonin and norepinephrine are thought to act in a manner similar to acetylcholine, which was discussed earlier. These compounds carry the message from the end of one neuron to the end of another across the synapse. Somewhat later Dilworth Wooley discovered that LSD blocks serotonin action (Figure 13–17). This has led to an interesting theory to explain the LSD trip.

The theory of the hallucinogenic mechanism is, roughly, as follows. The LSD releases in some chemical way those emotional experiences that are generally hidden away, or chemically stored in the lower midbrain and the brain stem. Serotonin or norepinephrine, or both, inhibit the escape of these experiences into consciousness, as they turn off certain excitations. LSD interacts with serotonin or norepinephrine at the synapse and nullifies their blocking effect. Thus, these stored, previously rejected thoughts are allowed to enter the conscious part of the brain where the imaginary trip occurs.

This theory is built on the experimental fact that LSD blocks the action of serotonin, but there are other theories as well as some unexplained phenomena involved. Psychiatry is fraught with theories of why LSD causes an uplifting experience in some people and a frightening one in others. There are theories about dosage, purity of the compound, psychiatric state of the tripper, and conditions of administration. If there is anything close to a consensus, it is that the more controlled and relaxed the surroundings, the "better" the trip will be; even this, however, is debatable. Many trips, even under medical supervision, have not been satisfactory.

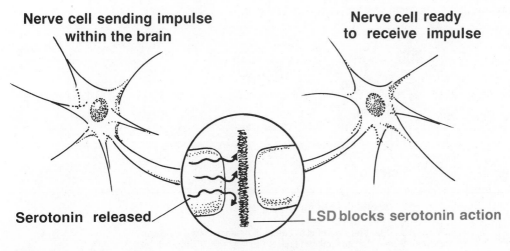

Nerve cell sending impulse within the brain

Nerve cell ready to receive impulse

Serotonin released

LSD blocks serotonin action

Figure 13–17 LSD blocks the flow of serotonin from one brain nerve cell to another.

ALCOHOLS

Alcohols have some well known toxic effects, but a complete chemical explanation has eluded scientists so far.

Methyl alcohol (methanol or wood alcohol) is highly poisonous, and unlike the other simple alcohols, it is, in effect, a cumulative poison in human beings. It has a specific toxic effect on the optic nerve, causing blindness with large doses. After its rapid uptake by the body, oxidation occurs in which the alcohol is first converted into formaldehyde and then to formic acid, which is eliminated in the urine.

$$CH_3OH \xrightarrow{\text{Oxidation}} H-\overset{\overset{\displaystyle O}{\|}}{C}-H \xrightarrow{\text{Oxidation}} H\overset{\overset{\displaystyle O}{\|}}{C}OH$$

METHYL ALCOHOL   FORMALDEHYDE   FORMIC ACID

This is a slow process and, for this reason, daily exposure to methyl alcohol can cause an extremely dangerous buildup of the alcohol in the body. The toxic effect on the optic nerve is thought to be caused by the oxidative products.

(a)

(b)

Figure 13–18   The blocking of ethyl alcohol oxidation by disulfiram (Antabuse). When the normal oxidative process (a) is stopped by blockage of the oxidative enzyme (b), acetylaldehyde builds up in the body (c).

CH₃CH₂   CH₂CH₃
　　　＼　／
　　　　N
　　　　|
　　　C＝S
　　　　|
　　　　S
　　　　|
　　　　S
　　　　|
　　　C＝S
　　　　|
　　　　N
　　　／　＼
CH₃CH₂   CH₂CH₃

Disulfiram

Ethyl alcohol (ethanol, grain alcohol) is found in alcoholic beverages, yet it is toxic like the other simple alcohols and is quantitatively absorbed by the gastrointestinal tract. About 58 percent of a dose is absorbed in 30 minutes, 88 percent in 1 hour, and 93 percent in 90 minutes. Over 90 percent of the ethyl alcohol is then slowly oxidized to carbon dioxide and water, mainly in the liver.

The intoxicated person's staggering gait, stupor, and nausea are caused by the presence of acetaldehyde, but the chemical reactions involved are not fully understood. The compound disulfiram (Antabuse) is sometimes given as a treatment for chronic alcoholism because it blocks the oxidative steps beyond acetaldehyde (Figure 13–18). The accumulation of acetaldehyde causes nausea, vomiting, blurred vision, and confusion. This is supposed to encourage the partaker to avoid this severe sickness by avoiding alcohol. Interestingly, this drug was discovered by two researchers who took a dose for another purpose and got violently ill that evening at a cocktail party.

---

1. Substrates that poison the nervous system are called _____.　　**SELF-TEST 13-B**

2. Most neurotoxins affect chemical reactions that occur in the opening between two nerve cells. These openings are called _____.

3. The electrical impulse is carried across a synapse by the chemical

   _____.

4. Mutagens alter the structures of _____ or _____.

5. If a substance is mutagenic in test animals, particularly dogs, it must necessarily be mutagenic in human beings. (True/False)

6. The first occupation definitely linked to cancer was _____

   _____.

7. An active component of marihuana is _____.

8. The nausea and stupor of drunkenness from ethyl alcohol are caused by

   _____ and not by the alcohol itself.

_____	1.	metabolic poison	(a)	lysergic acid diethylamide
_____	2.	metabolism	(b)	cyanide ion
_____	3.	corrosive poison	(c)	cancer in connective tissue
_____	4.	neurotoxin	(d)	benzo(α)pyrene metabolite
_____	5.	mutagen	(e)	cancerous growths of lung tissue located in liver

MATCHING SET

_____	6. carcinogen	(f)	use of chemicals in the body
_____	7. carcinoma	(g)	cancer of skin
_____	8. metastases	(h)	sodium hydroxide (caustic soda)
_____	9. sarcoma	(i)	atropine
_____	10. chelating agent	(j)	alters DNA
_____	11. hallucinogen	(k)	EDTA

QUESTIONS

1. Give an example of a toxic substance that is toxic as a result of

   (a) binding to an oxygen-carrying molecule.
   (b) disguising itself as another compound.
   (c) attack on an enzyme.
   (d) hydrolysis.

2. True/False. Explain each answer concisely.

   (a) Lead is a corrosive poison.
   (b) Carbon monoxide and cyanide poison in the same way.
   (c) There are no known chemical compounds that cannot be toxic under some circumstances.

3. The application of a single, minute dose of a fused ring hydrocarbon such as benzo(α)pyrene fails to produce a tumor in mice. Does this mean this compound is definitely noncarcinogenic? Give a reason for your answer.

4. Describe the chemical mechanism by which the following substances show their toxic effects.

   (a) fluoroacetic acid
   (b) phosgene
   (c) curare

5. Should any laws and regulations be placed on the use of any of the following? Justify your answers.

   (a) LSD
   (b) marihuana
   (c) ethanol

6. Discuss some of the pros and cons of testing toxic substances on animals.

7. Give chemical reactions in words for

   (a) action of NaOH on tissue.
   (b) action of carbon monoxide in blood.
   (c) reaction of EDTA and lead ion.

8. What questions do you think need to be answered before the action of ethyl alcohol is understood?

9. Assume a normal diet has the quantity of lead in a given quantity of food stated in the text. What would a person's total food intake of lead be per day?

10. What is the meaning of the symbolism LD_{50}?

11. Write chemical equations for:

   (a) the hydrolysis of acetylcholine.
   (b) acid hydrolysis of a protein having a glycine-glycine primary structure.

CHAPTER 14

WATER: ITS USE AND MISUSE

Water is certainly one of the most important of all chemical compounds. Its intimate participation in life processes marks it as indispensable to all known organisms. The rapidly expanding scope of human life, with its associated industrial development, is increasing the demand for water supplies so quickly that future needs may easily be underestimated. The time has passed when we can take an abundant water supply for granted (Table 14–1).

Over the centuries the world water supply has remained essentially constant. However, the availability of relatively pure water is being diminished by pollution. When the Pilgrims landed at Plymouth Rock, drinkable water could be obtained directly from most of the streams in the Colonies without endangering health. Now there are only a very few streams with drinkable water, and some bodies of water, especially those near large cities, are not even fit for swimming.

Pollution causes something that is normally clean to be unfit for direct use.

TABLE 14–1   Federal Laws Relating to Water Quality

	Name	Number	Year
1.	Federal Water Pollution Control Acts*	92-500, 95-217	1972–
2.	Atomic Energy Act (radioactive wastes)	83-703	1954
3.	Marine Protection Act	92-532	1972
4.	Federal Insecticide, Fungicide, and Rodenticide Act	92-516	1972
5.	Ports and Waterways Safety Act	92-340	1972
6.	Safe Drinking Water Act	93-523	1974
7.	Hazardous Materials Transportation Act	93-633	1974
8.	Resource Conservation and Recovery Act	94-580	1976
9.	Toxic Substances Control Act	94-469	1976

*The principal laws, which have been amended numerous times. The other laws listed are supportive in water quality control.

Water pollution arises from two main sources. First, water, being an unusually good solvent, readily dissolves materials from the soil. Second, and by far the most serious, are the waste materials of our culture; mercury and raw sewage effluents are but two examples.

It is important to realize at the outset that the technical know-how is available to solve water pollution problems, and much of this know-how is patterned after natural purification processes. Pure water, then, becomes a question of costs.

Water—the universal solvent.

WATER REUSE

It is estimated that an average of 4350 billion gallons of rain (and snow) fall on the contiguous United States each day. Of this amount, 3100 billion gallons return to the atmosphere by evaporation and transpiration. The discharge to the sea and underground reserves amounts to 800 billion gallons daily, leaving 450 billion gallons of surface water each day for domestic or commercial use. The 48 contiguous states withdrew from natural sources 40 billion gallons per day in 1900 and 325 billion gallons per day in 1960, and it is estimated that the demand will be at least 900 billion gallons per day by the year 2000. It is evident that the reuse of surface water is the only way that human needs for water can be met in the future. Much progress can be, and some has been, made in the reuse of water. For example, a Dow Chemical plant in Pittsburg, California used 20,000 gallons per minute (gpm) prior to its water reuse study. After its reuse plans were activated, the same plant required only 1000 gpm fresh water.

The increasing reuse of surface water by cities contributes to the pollution problem. The fact that rivers serve as a source of water for many cities, as well as receivers for their sewage and industrial waste effluents, means that a series of cities on the banks of a river will use some of the same water over and over again. If we choose not to purify our water after use or, in some cases, do not give nature enough time to do the job, the river becomes progressively more polluted downstream. The polluted water may contain heat, toxic substances, dissolved chemicals, and an undesirable population of microorganisms, along with silt. However, it should be noted that what is pollution to one user may not be to the next. The industrial user who needs water for cooling may primarily be concerned with heat content. The director of the municipal water plant is likely to be concerned with microorganisms, chemicals, and then heat, in that order.

Nature recycles water—why can't we?

Since the reuse of water is necessary, and since the purification of used water before reuse is often necessary, the question arises as to where the purification should occur. Purification immediately prior to reuse actually may lead to greater chemical problems because of the complex mixtures that result. Purification immediately after use is likely to involve only one type of problem characteristic of that use.

WATER PURIFICATION IN NATURE

Water is a natural resource which, within limitations, is continuously renewed. The familiar water cycle (Figure 14–1) offers a number of oppor-

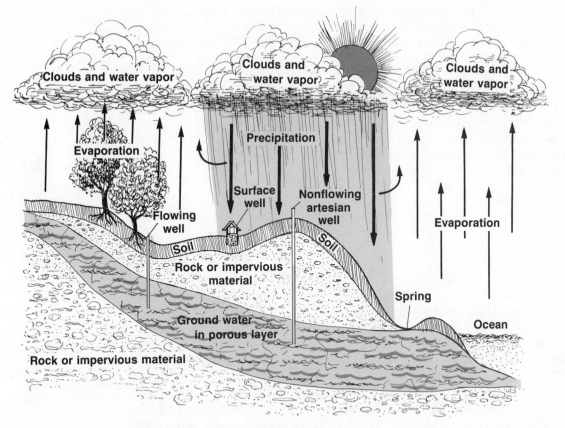

Figure 14–1 The water cycle in nature.

Rainwater in clean air is very pure.

tunities for nature to purify its water. The worldwide *distillation* process results in rainwater containing only traces of nonvolatile impurities, along with gases dissolved from the air. *Crystallization* of ice from ocean salt water results in relatively pure water in the form of icebergs. *Aeration* of groundwater as it trickles over rock surfaces, as in a rapidly running brook, allows volatile impurities to be released to the air. *Sedimentation* of solid particles occurs in slow-moving streams and lakes. *Filtration* of water through sand rids the water of suspended matter such as silt and algae. Next, and of very great importance, are the *oxidation processes*. Essentially all naturally occurring organic materials—plant and animal tissue, as well as their waste materials—are changed through a complicated series of oxidation steps in surface waters to simple substances common to the environment. Finally, another process that is used by nature is *dilution*. Most, if not all, pollutants found in nature are rendered harmless if reduced below certain levels of concentration by dilution with pure water.

Volatile: goes easily into the gaseous state.

Pure water:
Chemist: "Pure H$_2$O —no other substance."
Housewife: "Nothing harmful to human beings."
Game and Fish Commission: "Nothing harmful to animals."
Sunday boater: "Pleasing to the eye and nose, no debris."
Ecologist: "Natural mixture containing necessary nutrients."

Before the advent of the exploding human population and the industrial revolution, natural purification processes were quite adequate to provide ample water of very high purity in all but desert regions. Nature's purification processes can be thought of as massive but somewhat delicate. In many instances the activities of humans push the natural purification processes beyond their limit, and polluted water accumulates.

A simple example of nature's inability to handle man-made pollution comes from dragging gravel from stream beds. This excavation leaves large amounts of suspended matter in the water. For miles downstream from a source of this pollutant, aquatic life is destroyed. Eventually, the solid matter settles, and normal life can be found again in the stream.

A more complex example, one for which there is not nearly so much reason to hope for the eventual solution by natural purification, is the degradability of organic materials. A ***biodegradable*** substance is composed of molecules which are broken down to simpler ones in the natural environment. For example, cellulose suspended in water will eventually be converted to carbon dioxide and water.

$$(C_6H_{10}O_5)_x \xrightarrow[\text{Biochemical catalysts}]{\text{Oxidizing agents}} CO_2 + H_2O$$

CELLULOSE   CARBON   WATER
                          DIOXIDE

The overall oxidation process is a complicated one, involving microorganisms and their enzymatic secretions. Many reaction steps are required; some are not yet understood. Some organic compounds, notably some of those synthetically produced, are not easily biodegradable; these substances simply stay in the natural waters or are absorbed by life forms and remain intact for long periods of time.

Some man-made molecules are not biodegradable and therefore are very persistent in natural waters.

Detergents offer an interesting contrast in biodegradability. Recall that detergents are composed of long organic molecules, one end of which is "soluble" in water and the other in oil. Branched-chain detergent molecules, such as

$$CH_3-CH-CH_2-CH-CH_2-CH-CH_2-CH-\langle\bigcirc\rangle-SO_3^{\ominus}\,Na^{\oplus}$$
$$\qquad |\qquad\qquad |\qquad\qquad |\qquad\qquad |$$
$$\qquad CH_3\qquad\;\; CH_3\qquad\;\; CH_3\qquad\;\; CH_3$$

are not readily biodegradable. The enzymes of most of the bacteria present do not catalyze this decomposition. When such detergents were widely used, foaming problems occurred in sewers, streams, and, in extreme cases, in the purified fresh water supply (Figure 14–2). With the subsequent development of straight-chain detergents, such as

$$CH_3CH_2CH_2CH_2CH_2CH_2CH_2CH_2CH_2CH_2CH_2CH_2OSO_3^{\ominus}\,Na^{\oplus}$$

biodegradability was readily achieved. Even so, a major river can still be overloaded and caused to foam from detergent for a mile or more downstream from a raw sewage exit.

Bacteria are capable of breaking detergent molecules down quite rapidly if the detergents have only straight chains of carbon atoms present.

Some chemically stable materials, such as DDT and radioisotopes like strontium-90, are made in such large quantities that they cannot be dealt with effectively by dilution. If such materials are judged to be harmful to the natural environment, either we must choose not to make or use them, or we must assist nature in dealing with these species. No one really knows what catastrophic effects would be caused by the sinking of a ship loaded with DDT or the release of strontium-90 that would be produced in a nuclear war.

Figure 14–2 Foam in natural waters. A high concentration of detergent materials dissolved in water causes foam because of the lowered surface tension of the water. If the water is badly polluted, the foam may exist miles from the pollution source. (From Singer, S. F., "Federal interest in estuarine zones builds," *Environmental Science and Technology,* Vol. 3, p. 2, 1969.)

Acid rain in Pasadena, California (1976–77) contained:
(1) Acid (pH = 4.06)
(2) NH_4^+ ⎫
(3) K^+ | Concentrations in 10^{-5} to 10^{-6} molar range.
(4) Ca^{2+} |
(5) Mg^{2+} |
(6) Cl^- |
(7) NO_3^- |
(8) SO_4^{2-} ⎭

Even nature's pure rainwater is in jeopardy in isolated instances. In areas in which heavy concentrations of automobile fumes collect, poisonous lead compounds have been found in rainwater in concentrations many times higher than the 0.01 ppm generally allowed in drinking water. The concentration of the lead can be correlated with the concentration of exhaust fumes in the air. Fortunately, lead does not long remain in water, since it generally forms insoluble compounds.

THE SCOPE OF WATER POLLUTANTS

When natural purification processes cannot cope with materials added to water, pollution results.

There was a time when polluted water could be thought of in terms of dissolved minerals, natural silt, and contaminants associated with the natural wastes of animals and humans. As our use of water has increased, the pollution has become more diversified. The U.S. Public Health Service now classifies water pollutants into the eight broad categories listed in Table 14–2.

Biochemical Oxygen Demand

The way in which organic materials are oxidized in the natural purification of water deserves special attention. The process opposes *eutrophication.* Even in the natural state, living organisms found in

TABLE 14–2   Classes of Water Pollutants,
with Some Examples

1. Oxygen-demanding wastes	Plant and animal material
2. Infectious agents	Bacteria and viruses
3. Plant nutrients	Fertilizers, such as nitrates and phosphates
4. Organic chemicals	Pesticides, such as DDT, detergent molecules
5. Other minerals and chemicals	Acids from coal mine drainage, inorganic chemicals such as iron from steel plants
6. Sediment from land erosion	Clay silt on stream bed may reduce or even destroy life forms living at the solid-liquid interface
7. Radioactive substances	Waste products from mining and processing of radioactive material, radioactive isotopes after use
8. Heat from industry	Cooling water used in steam generation of electricity

natural waters are constantly discharging organic debris into the water. To change this organic material into simple inorganic molecules (such as CO_2 and H_2O) requires oxygen. The amount of oxygen required to oxidize a given amount of organic material is called the ***biochemical oxygen demand (BOD).*** The oxygen is required by microorganisms, such as many forms of bacteria, to metabolize the organic matter that constitutes their food. Ultimately, given near normal conditions and enough time, the microorganisms will convert huge quantities of organic matter into the following end-products:

$$\text{Organic carbon} \longrightarrow CO_2$$
$$\text{Organic hydrogen} \longrightarrow H_2O$$
$$\text{Organic oxygen} \longrightarrow H_2O$$
$$\text{Organic nitrogen} \longrightarrow NO_3^-$$

Eutrophication is the enrichment of a body of water with nutrients such that growth of organisms in the water makes it unfit for human purposes.

One way to determine the amount of organic pollution is to determine how much oxygen a given sample of polluted water will require. For example, a known volume of the polluted water is diluted with a known volume of standardized sodium chloride solution of known oxygen content. This mixture is then held at 20°C for five days in a closed bottle. At the end of this time the amount of oxygen that has been consumed is taken to be the biochemical oxygen demand.

Highly polluted water often has a very high concentration of organic material, with resultant very large biochemical oxygen demand (Figure 14–3). In extreme cases, more oxygen is required than is available from the environment, and putrefaction results. Fish and other freshwater aquatic life can no longer survive. The aerobic bacteria (those that require oxygen for the decomposition process) die. As a result of the death of these organisms, even more lifeless organic matter results and the BOD soars. Nature, however, has a back-up system for such conditions. A whole new set of microorganisms (anaerobic bacteria) takes over; these organisms take oxygen from oxygen-containing compounds to convert organic matter to CO_2 and water. Organic nitrogen is converted to elemental nitrogen by these bacteria. Given enough time, enough oxygen may become available, and aerobic oxidation will then return.

A stream containing 10 parts per million (ppm) by weight (just 0.001 percent) of an organic material, the formula of which can be represented

A quantitative relationship exists between oxygen needs and organic pollutants to be destroyed. This is BOD.

A standardized solution is one of known concentration.

Fish cannot live in water that has less than 0.004 gram O_2 per liter (4 ppm).

Temperature °C	Solubility g_{O_2}/liter H_2O
0	0.0141
10	0.0109
20	0.0092
25	0.0083
30	0.0077
35	0.0070
40	0.0065

These data are for water in contact with air at 760 mm mercury pressure.

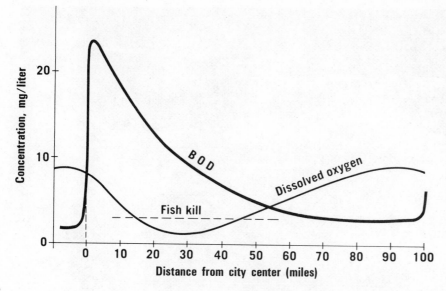

Figure 14–3 Graph showing oxygen content and oxidizable nutrients (BOD) as a result of sewage introduced by a city. The results are approximated on the basis of a river flow of 750 gallons per second. Note that it takes 70 miles for the stream to recover from a BOD of .023 gram oxygen per liter. (From Turk, A., et al.: *Environmental Science.* Philadelphia, W. B. Saunders Co., 1974.)

by $C_6H_{10}O_5$, will contain 0.01 g of this material per liter. The calculation used to obtain this is:

$$?g = 1 \text{ liter of water}$$

$$?g = 1 \text{ liter} \times \frac{1000 \text{ ml}}{1 \text{ liter}} \times \frac{1 \text{ g}}{\text{ml}} = 1000 \text{ g}$$

0.001 percent of this is the pollutant:
0.001 percent of 1000 g = (0.00001)(1000 g) = 0.010 g

Characteristic BOD Levels

	g_{O_2}/liter
Untreated municipal sewage	0.1 to 0.4
Run-off from barnyards and feed lots	0.1 to 10
Food processing wastes	0.1 to 10

High concentration of organic pollutants
↓
Low oxygen concentration
↓
Dead organisms
↓
Higher concentration of organic pollutants
↓
Lower oxygen concentration
↓
Anaerobic conditions

To transform this pollutant to CO_2 and H_2O, the bacteria present use oxygen as described by the equation:

$$C_6H_{10}O_5 + 6O_2 \longrightarrow 6CO_2 + 5H_2O$$

RELATIVE WEIGHT 162   RELATIVE WEIGHT 192

The 0.010 g of pollutant requires 0.012 g of dissolved oxygen.

$$\frac{?g \text{ oxygen}}{\text{liter}} = \frac{0.010 \text{ g pollutant}}{\text{liter}} \times \frac{192 \text{ g oxygen}}{162 \text{ g pollutant}} = \frac{0.012 \text{ g oxygen}}{\text{liter}}$$

At 68°F (20°C), the solubility of oxygen in water under normal atmospheric conditions is 0.0092 g of oxygen per liter.

Since the BOD (0.012 g per liter) is greater than the equilibrium concentration of dissolved oxygen (0.0092 g per liter), as the bacteria utilize the dissolved oxygen in this stream, the oxygen concentration of the water will soon drop too low to sustain any form of fish life. Life forms can survive in water where the BOD exceeds the dissolved oxygen if the water is flowing very vigorously in a shallow stream (this facilitates the absorption of more oxygen from the air via aeration).

BOD values can be greatly reduced by treating industrial wastes and sewage with oxygen and/or ozone. Numerous commercial clean-up opera-

Figure 14—4   Fish kills can be caused by the lack of a substance necessary for life, such as oxygen, or by the presence of toxic materials that interfere with the life processes. A heavy concentration of organic matter in a stream may depress the oxygen concentration below that required to support fish life. (From Tidwell, P.: Anti-pollution: A Fish Management Priority. The Tennessee Conservationist, *37*:4, 1971.)

tions now being developed and used employ this type of "burning" of the organic wastes. Another benefit of treating waste water with oxygen is that some of the non-biodegradable material becomes biodegradable as a result of partial oxidation.

Thermal Pollution

Thermal pollution results when water is used for cooling purposes and in the process has its own temperature raised. Water has a very high *heat capacity* (the heat required to raise a unit weight of water one degree) and a high *heat of vaporization* (the heat required to change a unit weight of liquid water to gaseous water); this combination makes water an ideal cooling fluid for thermal power stations, nuclear energy generators, and industrial plants. The extent to which thermal pollution can occur is illustrated by the Thames River, the center of an industrial complex in England. The yearly average temperature in this river, corrected for changes in climatic conditions, rose from 53°F in 1930 to 60°F in 1950. Heat was released into this river in 1930 at the average rate of 132,000 kilocalories per second. By 1950 the rate became 888,000 kilocalories per second. The river, even with various means of cooling (evaporation, conduction, and discharge to the sea), simply could not dissipate

Water will absorb large amounts of heat before it is vaporized.

Vast amounts of heat are released to nature by industrial processes.

Figure 14—5 Thermal pollution. Plant effluent causes the Calumet River to "steam." Such heat discharges can change the temperature of a natural body of water enough to vastly alter the ecological balances. (From "Technology," *Chemical and Engineering News,* Feb. 24, 1969.)

The solution of oxygen in water is facilitated by:
1. exposed surface area of water;
2. low temperature;
3. low concentration of oxygen in the water.

Aquatic life is very sensitive to temperature. Lethal temperatures for various species of fish in Wisconsin and Minnesota:

trout	77°F
white sucker	84–85°F
walleye	86°F
yellow perch	84–88°F
fathead minnow	93°F

the heat as fast as it was fed into the river. Table 14—3 lists the types of contributors to this heat inflow.

The most obvious result of thermal pollution is to make the water less efficient for further cooling applications. Far more important, however, are the biological and biochemical implications.

The oxygen content of water in contact with air is dependent on the temperature of the water, since more oxygen can dissolve in a quantity of cold water than in the same quantity of warm water. Also, the *rate* at which water dissolves oxygen is directly proportional to the difference between the actual concentration of oxygen present and the equilibrium value. This is extremely fortunate, since it means the rate of solution of oxygen increases sharply as the oxygen is consumed. Since a larger surface area will allow quicker absorption of oxygen, the rate of absorption of oxygen from the air is much greater in a shallow, cold mountain stream than in a deep lake behind a dam in a warm river.

The increased temperature of the Thames River decreased the oxygen content by four percent over what it otherwise would have been. However, the biochemical results of this factor alone could not be determined, since the river was anaerobic in 1950 as a result of other pollutants.

TABLE 14—3 Contributors to Increased Heat Content of Thames River in 1950

	Percent
Fossil fuel power stations	75
Sewage effluents	9
Industrial effluents	6
Fresh water discharge	6
Biochemical activity	4

TABLE 14–4   A Projection of the Thermal
Pollution Problem in the United States*

Year	Per Capita Electrical Energy Consumption in U.S. (48 States, Actual or Projected)	Cooling Water Needs (Billion Gallons per Day)
1950	2,000 Kwh	∼30
1968	6,500 Kwh	∼100
1980	11,500 Kwh	200†
2000	24,000 Kwh	450†

*Notes: a. Surface runoff is 1250 billion gallons per day.
b. Generation of electrical energy has been doubling every 10 years.
c. 500 new power stations will be needed by the year 2000, at the present growth rate.
d. Kwh is kilowatt-hour, an energy unit equivalent to 860 kilocalories.

†Projected.

Even though more is to be learned about thermal pollution, two conclusions seem obvious: (1) thermal pollution aggravates the problem of oxygen supply; and (2) a significant rise in the temperature of a stream can drastically change or even destroy entire biological populations.

Solutions to thermal pollution involve cooling the water in evaporation towers or storage lakes before returning it to the natural body of water. Cycling the water for reuse after cooling has obvious advantages. Table 14–4 shows a quantitative forecast of the thermal pollution problem in the United States.

Fertilizers

Fertilizers contain such chemicals as ammonium nitrate, NH_4NO_3, and "superphosphate," $CaH_4(PO_4)_2 \cdot CaSO_4$. These substances are soluble enough to be carried away by the surface run-off during and after a hard rain. Contamination of water by fertilizers leads to very undesirable effects. In a large measure, this contamination results from the phosphate (PO_4^{3-}) and nitrate (NO_3^-) present in fertilizers. It is generally believed that phosphate and nitrate encourage the growth of large amounts of algae. It is certain that nitrate in sufficient concentration is toxic to most higher organisms.

Limnology is the study of physical, biochemical, and biological aspects of freshwater systems. One of the conclusions of limnology is that massive but relatively delicate balances exist in such systems. The growth of a limited kind and quantity of algae is good for a lake; such growth is a source of needed oxygen via photosynthesis:

Limnology—
the study of fresh-
water systems.

$$6CO_2 + 6H_2O \xrightarrow[\text{Catalysts}]{\text{Sunlight}} C_6H_{12}O_6 + 6O_2$$

**CARBON
DIOXIDE**   **WATER**   **SUGAR**   **OXYGEN**

However, in waters that contain excessive amounts of phosphate and nitrate leached from fertilized farm land, algae growths are sometimes

Leached—
dissolved by rain-
water and carried
into streams or lakes.

so massive that they tend to choke out desirable life forms. Such "blooms" of algae may lead to eutrophication. In this state, the dead algae actually give rise to an oxygen-deficient environment, especially at lower depths. This is because the decaying algae quickly consume the available oxygen produced by the algae; the BOD rises sharply.

As yet it is not clear that the nitrate and phosphate are the only or even the primary causes of the algae blooms. Some have suggested the primary cause could be an excessive organic pollution. Bodies of water in which the phosphate concentrations were kept abnormally low—less than 10 parts per billion—produced algae blooms. It was reasoned that the higher-than-normal levels of carbon dioxide resulting from the bacterial oxidation of organic materials caused the excessive algae growth. It appears at this stage of studying algae blooms that there can be causes other than excess nitrate and phosphate. The confusion exists because the limnological problem is so complex and our present knowledge is so limited.

Is there a danger of too much phosphate (PO_4^{3-}) and nitrate (NO_3^-)?

Detergent Builders

Detergents contain ***detergent builders,*** which are additives that make the surfactant (detergent) molecules more efficient and effective. In some detergents, as much as 40 percent of the total weight is in the form of detergent builders. A common detergent builder is sodium tripolyphosphate.

$$3Na^+, \left[\begin{array}{c} O \quad\quad O \\ \diagdown \quad \diagup \\ P \\ \diagup \quad \diagdown \\ O \quad\quad O \\ | \quad\quad\quad | \\ O-P \quad P-O \\ \diagup \quad \diagdown \diagup \quad \diagdown \\ O \quad O \quad\quad O \end{array} \right]^{3-}$$

Many polyphosphate structures exist, and a commercial preparation is likely to contain several. At least one positive action of detergent builders is to tie up metal ions such as Fe^{3+}.

$$Fe^{3+} + 2PO_4^{3-} \longrightarrow Fe(PO_4)_2^{3-}$$

IRON-PHOSPHATE COMPLEX

Detergents and detergent builders both may present water pollution problems.

Because of the use of phosphorus in detergent builders, more than 500 million pounds of this element end up in our waste waters each year in the form of phosphate. A typical detergent contains about nine percent phosphorus by weight.

By 1979 six states had passed laws banning the retail sale of phosphate detergents. The Environmental Protection Agency reported in 1978 that the phosphate concentration was greatly reduced in Lake Michigan along the shore of Indiana after that state banned phosphate detergents.

No substitute with qualities equal to those of phosphates has yet been found. Sodium citrate is somewhat effective and is biodegradable. Synthetic zeolites are currently being introduced and show promise. Natural zeolites have been used for many years to remove metal ions from hard water (Figure 14–6).

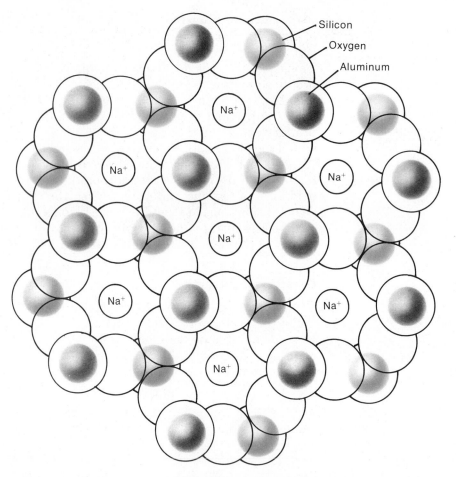

Figure 14—6 The structure of natural zeolite. The aluminum, oxygen, and silicon atoms form a macromolecular structure that is negatively charged. The positive sodium ions are loosely held in the holes within the lattice. Water, containing other metal ions, will exchange some of these ions for sodium ions when passed through the zeolite.

SELF-TEST 14-A

1. What liquid is sometimes called the universal solvent? _____

2. Which process does nature not use in purifying natural water: distillation, crystallization, sedimentation, chlorination, or aeration? _____

3. BOD stands for _____

4. BOD is the amount (grams) of _____ per _____ required to oxidize a given amount of organic material.

5. _____ is the study of physical, biochemical, and biological aspects of freshwater systems.

6. Detergents contain detergent builders, such as _____, in addition to the detergent chemical itself.

Pesticides

The use of synthetic insecticides increased enormously on a world-wide basis after World War II. As a result, insecticides such as DDT have found their way into lakes and rivers. There is a great variety of pesticides, and their use frequently leads to severe damage to other forms of animal life, such as fish and birds. The toxic reactions and peculiar biological side effects of many of the pesticides were not thoroughly studied or understood prior to their widespread use.

A good case in point is **DDT.** This insecticide, which had not been shown to be toxic to man in doses up to those received by factory workers involved in its manufacture (400 times the average exposure), does have peculiar biological consequences. The structure of DDT is such that it is *not* metabolized (broken down) very rapidly by animals; it is deposited and stored in the fatty tissues. The biological half-life of DDT is about eight years; that is, it takes about eight years for an animal to metabolize one-half of an amount it assimilates. The enzymes are just not present for rapid breakup of this molecule. If ingestion continues at a steady rate, it is evident that DDT will build up within the animal over a period of time. For many animals this is not a problem, but for some predators, such as eagles and ospreys, which feed on other animals and fish, the consequences are disastrous. The DDT in the fish eaten by such birds is concentrated in the bird's body, which attempts to metabolize the large amounts by an alteration in its normal metabolic pattern. This alteration involves the use of compounds that normally regulate the calcium metabolism of the bird and are vital to its ability to lay eggs with thick shells. When these compounds are diverted to their new use, they are chemically modified and are no longer available for the egg-making process. As a consequence, the eggs the bird does lay are easily damaged, and the survival rate decreases drastically. This process has led to the nearly complete extinction of eagles and ospreys in some parts of the United States where formerly they were numerous.

Since DDT is not readily biodegradable, there is a resultant buildup of this substance in natural waters. However, it is not an irreversible process; the Environmental Protection Agency reported a 90% reduction of DDT in Lake Michigan fish by 1978 as a result of the ban on the use of this insecticide. DDT and other insecticides such as **dieldrin** and **heptachlor** are referred to as **persistent pesticides.** Substitutions of other substances with biodegradable structures are now made where possible. The compound **chlordan** is an example of just such a substitution. It is interesting to note that the structural differences between heptachlor (persistent) and chlordan (short-lived) are relatively slight (look at the chlorine atom on the lower five-membered ring).

DDT

Half-life is the time required for half of the substance to disappear (see Chapter 17).

Do we really have to decide between pests and eagles?

City water treatment generally does not remove DDT.

Dieldrin

and aldrin

were banned by the courts from their major uses (such as on corn) in 1974.

HEPTACHLOR   CHLORDAN

There are many other insecticides that are actually much more toxic to humans than is DDT. These include inorganic materials based on arsenic compounds, as well as a wide variety of phosphorus derivatives based on structures of the type

Heptachlor and chlordan were banned for most garden and home use in December, 1975.

$$R \diagdown \atop R' \diagup \!\!\!> P \overset{\overset{\displaystyle Z}{\|}}{} - X$$

where Z is oxygen or sulfur, R and R' are alkyl, alkoxy, alkylthio or amide groups, and X is a group which can be easily split from the phosphorus. Insecticides of this type include parathion, which is effective against a very large number of insects but is also *very* poisonous to human beings. These compounds are anticholinesterase poisons. One of their most important properties, however, is that they are readily hydrolyzed to less toxic substances that are not residual poisons.

$$C_2H_5O \diagdown \atop C_2H_5O \diagup \!\!\!> P \overset{\overset{\displaystyle S}{\|}}{} - O - \bigcirc - NO_2$$

Parathion

Perhaps the most publicized case of a poorly handled insecticide was the manufacture of Kepone ($C_{10}Cl_{10}O$, a very complex molecule containing several fused rings) in a converted filling station. The resulting contamination of the James River was extensive, as was the personal suffering of the workers. The dredging costs to clean the river are estimated by the EPA to be one billion dollars, taking one dredge 120 years to complete the job. Total clean-up costs are estimated at $7.2 billion. Obviously, there are no immediate plans to clean up this chemical spill.

While it might occur to us to forbid completely the use of such insecticides, we would soon find that agricultural production would drop precipitously were this to be done. An enlarged human population requires that crops be protected against the ever-present threat of destruction by insects. Furthermore, the use of insecticides is essential if we are to control malaria, plague, and other diseases. It must be recognized that a single insect species, the tsetse fly, alone has retarded the development of over 4,000,000 square miles of Africa by means of the diseases it transmits to people and cattle.

The goal of the insecticide quest: a selectively toxic chemical that is quickly biodegradable.

The choice of solutions to our problems with insects is not an easy one. The use of insecticides introduces them into our environment and our water supplies. A refusal to use insecticides means that we must tolerate malaria, plague, sleeping sickness, and consumption of a large part of our food supply by insects. It is obvious that continuing research is needed on new methods and materials for the control of insect populations.

Industrial Wastes

Industrial wastes can be an especially vexing sort of pollution problem because they often are not removed or are removed very slowly by naturally occurring purification processes and are generally not removed at all by a typical municipal water treatment plant. Table 14–5 lists some of the large variety of industrial pollutants that have been put into our natural waters or buried in the ground on a large-scale basis.

Chemicals buried in dumps have in some cases infiltrated groundwater or surface water supplies. The Love Canal chemical landfill subsequently caused near panic conditions in Niagara Falls, N.Y. because of polluted ground- and surface waters. The water in a private well near

TABLE 14—5 A Partial List
of Industrial Wastes That
Have Been Dumped into
Streams and Lakes

Acids
Bases
Salts
Various metal solutions
Oils
Bacteria
Emulsifying agents
Grease
Plant and animal wastes
Dyes
Waste solvents
Poisons such as cyanides and mercury
Numerous chemicals from washing operations

Toone, Tennessee was shown to contain carbon tetrachloride, chloroform, benzene, heptachlor and chlordan along with other chemicals. The well, which "smells like shoe polish," is $1\frac{1}{4}$ miles from a chemical dump that was closed six years prior to the testing of the well. The chemical dump is estimated to contain 16.5 million pounds of chemical wastes from a pesticide plant.

Chemical destruction by burning has been of limited success in getting rid of unwanted chemicals because of air pollution. Ships have been fitted out as floating incinerators, but pollution of the surrounding water and air has followed.

Waste liquid from chromium plating is an example of industrial pollution. It has been a common practice to dump waste water containing chromate (CrO_4^{2-}) and cyanide (CN^-) ions into our rivers. The method for chemical clean-up of such wastes has long been known, and involves three steps:

(1) Alkaline oxidation with sodium hypochlorite (laundry bleach) to oxidize the CN^- ion.

$$2CN^- + 5OCl^- + H_2O \longrightarrow 2HCO_3^- + N_2 + 5CL^-$$
CYANIDE   HYPOCHLORITE   BICARBONATE   NITROGEN   CHLORIDE

(2) Reduction of CrO_4^{2-} to Cr^{3+} by sulfur dioxide (SO_2); the Cr^{3+} salts are soluble.

$$2CrO_4^{2-} + 3SO_2 + 2H_2O \longrightarrow 2Cr^{3+} + 3SO_4^{2-} + 4OH^-$$
CHROMATE   SULFUR   CHROMIUM(III)   SULFATE   HYDROXIDE
           DIOXIDE

(3) Alkaline precipitation of other metal ions.

$$M^{2+} + 2OH^- \longrightarrow M(OH)_2$$

Essentially all metal ions except the alkali metals and the alkaline earths are insoluble in a strongly basic solution.

TABLE 14—6   The Fate of Spent Pickle Liquor

Content of pickle liquor:	From 0.5 up to 10% acid, 12% iron
Pollution problem:	Lowers pH of stream and produces solid slime
Scope of problem:	8 to 15 gallons produced per ton of steel; 50 million tons of steel produced in U.S. per year.
Method of disposal used:	1. Discharge into natural waterway. 2. Discharge into deep well. 3. Neutralization of acid before release to waterway. Sludge is produced and iron is released. 4. Hauling away liquor. 5. Recovery. Cool liquid from 180°F. Iron crystallizes as $FeSO_4 \cdot 7H_2O$. Acid is concentrated and recycled. 6. Regeneration: one step beyond step 5. Commercial HCl and Fe_2O_3 recovered.
Immediate costs:	$6 > 5 > 4 > 3 > 2 > 1$
Long-range costs:	$1 > 2 > 3 > 4 > 5 > 6$

Another example of industrial pollution is hydrochloric acid, which is used to clean steel. Table 14—6 outlines the various solutions, with relative costs, to this type of pollution.

Heavy metals such as lead, mercury, cadmium, chromium, manganese, copper, and zinc are generally toxic to life forms in greater than trace amounts. Here we will discuss pollution of water by one of these metals, mercury (Table 14—7).

About 10,000 tons of mercury are mined annually worldwide, and about 5,000 tons are somehow lost. Because of the hazards associated with it, world mercury consumption is going down, having decreased by 15% from 1974 to 1975. Some of the loss is discharged as waste effluent from manufacturing plants, and some of the rest is incorporated as traces into products in which it does not belong. For example, one chlorine plant is known to have been discharging as much as 200 pounds of mercury a day. In the production of chlorine, mercury is used as an electrode in the electrolysis of a solution of sodium chloride (brine). When the salt solution becomes too dilute, it is discarded, and this discharge carries some mercury along with it.

About 5000 tons of mercury are lost into the environment each year.

TABLE 14—7   Mercury Consumption in the United States*

1975 Consumption (tons)	
Electrical apparatus (batteries)	644
Electrolytic chlorine	578
Paint	263
Instruments	174
Dental preparations	88
Agriculture	22
Pharmaceuticals	16
All other uses	146
Total	1931

*Mineral Yearbook, 1975.

Fish and other aquatic organisms concentrate the mercury which is present in the organisms they eat.

Mercury in solution is often in the form of dimethylmercury, $(CH_3)_2Hg$. This form is especially hazardous. Research indicates that metallic mercury and inorganic mercury compounds can be converted to dimethylmercury by anaerobic bacteria in the mud of lake bottoms, as well as by fish and mammals.

One way to remove mercury from industrial wastes before they are discharged into public waters is by the process developed by the FMC Corporation. The effluent is run into a large concrete basin where sodium sulfide is added. Insoluble mercury sulfide is precipitated and collected by filtration:

This reaction is profit-making and provides economic incentives for pollution control.

$$Hg^{2+} \quad + \quad S^{2-} \quad \longrightarrow \quad HgS\downarrow$$

MERCURY(II)   SULFIDE   MERCURY(II) SULFIDE
SOLUBLE   INSOLUBLE

The filtrate, which contains less than 3 parts per billion (ppb) mercury, is discharged into public waters. Less than 0.1 pound mercury is discharged per day.

The next step is to solubilize the mercury sulfide. The mud is stirred in a solution of sodium hydroxide and chlorine. The sulfide ion is oxidized by the hypochlorite ion (formed by the sodium hydroxide and chlorine), and mercury ions go into solution.

$$HgS \quad + \quad OCl^- \quad + \quad H_2O \longrightarrow \quad Hg^{2+} \quad + \quad Cl^- \quad + \quad S \quad + \quad 2OH^-$$

INSOLUBLE   HYPOCHLORITE   WATER   MERCURY(II)   CHLORIDE   SULFUR   HYDROXIDE
MERCURY(II)
SULFIDE

The mercury ions are then reduced to elemental metallic mercury.

Industrial wastes—only a cost problem; the solution is clear.

The technology necessary to remove industrial wastes from the water before it is returned to natural waters is available now. The limiting factor is the cost. We will simply have to decide whether we are willing to spend a greater fraction of our resources in order to have a less polluted environment. The control of this pollution is made easier by the use of

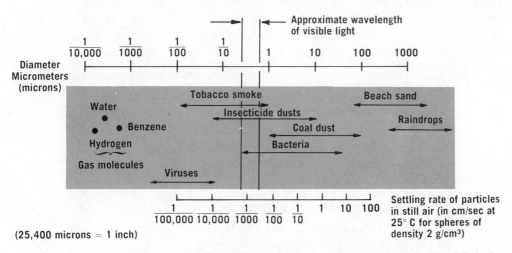

(25,400 microns = 1 inch)

Figure 14–7 The settling rates in air of particles as a function of their size. Note that very long times are needed for the smaller particles. For practical purposes, colloidal particles will not settle. Settling rates in water are slower in comparison with the rates in air. (From Turk, A., et al.: *Environmental Science*. Philadelphia, W. B. Saunders Co., 1974.)

Figure 14—8 Silt particles are stabilized by kinetic energy of the water and repulsive electrostatic forces.

national and even international standards for all parties to observe. For example, steel mills have been and, to a lesser extent, still are notorious polluters of water. Some companies would find it very difficult to compete if they had to clean up all of their wastes while other companies did not. Concerted action, whether voluntary or enforced, is necessary. The problem has not been completely solved in fact as it has been in theory. It only requires the will and commitment of the people.

Chemical methods capable of eliminating industrial water pollution are known for most pollutants.

Silt

Silt, which is finely divided solid material, is picked up and carried along by flowing water until the velocity of the water is reduced as it enters a lake, dam area, or the sea. The silt is stabilized by two factors: (1) the kinetic energy of the moving water keeps at least some of the silt moving against gravity, and (2) charges build up on the silt particles, causing them to repel each other (Figures 14—7 and 14—8). In the quiet water of a lake, silt tends to settle out. The saltwater of the sea is very effective in removing silt; hence, the great river deltas at the river mouths. The ions in saltwater neutralize the electrostatic charge on the particles and then collisions between them result in coagulation. The larger particles settle readily.

A sticky precipitate will settle silt.

Silt is undesirable in municipal water because it is esthetically unappealing and also because the silt particles can be carriers of chemicals or microorganisms, or both, that are harmful to humans. Silt is readily removed in holding tanks by sedimentation and by filtration beds. Sedimentation is greatly enhanced if aluminum sulfate and calcium hydroxide (slaked lime) are added to the settling (holding) tank; this results in the formation of aluminum hydroxide, which is a very sticky precipitate. The salt solution tends to neutralize static charges, and the sticky $Al(OH)_3$ collects the silt particles by physical attraction. Most of it settles and filtration removes the rest.

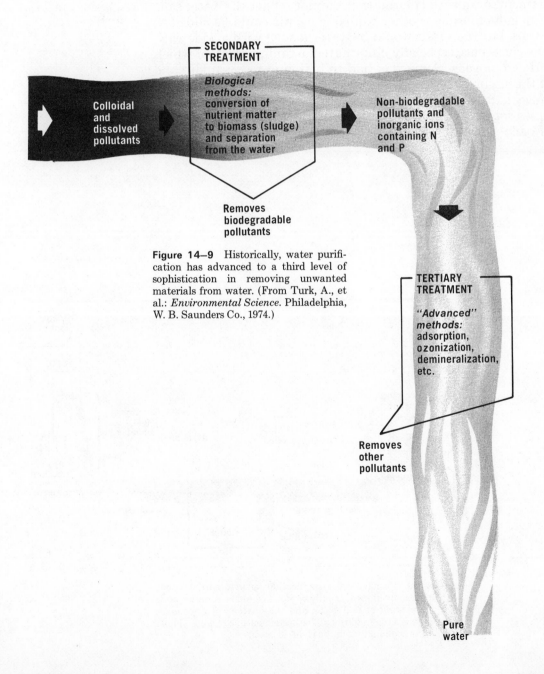

Figure 14—9 Historically, water purification has advanced to a third level of sophistication in removing unwanted materials from water. (From Turk, A., et al.: *Environmental Science.* Philadelphia, W. B. Saunders Co., 1974.)

WATER PURIFICATION:
CLASSICAL AND MODERN PROCESSES

The outhouses of some rural dwellers had their counterparts in city cesspools. The terrible job of cleaning led to the development of cesspools that could be flush-cleaned with water, followed by a connecting series of such pools that could be flushed from time to time. City sewer systems with no holding of the wastes were the next step.

Since there were not enough pure wells and springs to serve the growing population, water purification techniques were developed. The classical method, which is now termed *primary water treatment,* involved settling and filtration (Figure 14–10).

If the intake water is polluted enough with biological wastes, the primary treatment, even with chlorination, cannot render the water safe. To be sure, enough chlorine or other oxidizing agents could be added to kill all life forms, but the result would be water loaded with a wide variety of noxious chemicals, especially chlorinated organics, many of which are suspected carcinogens. Some way had to be found to coagulate and separate out the organic material that passed through the primary filters.

Secondary water treatment revives the old cesspool idea under a more controlled set of conditions and acts only on the material that will not settle or cannot be filtered (Figure 14–11). It operates in an oxygen-

Cesspools were an early and crude form of the modern activated sludge process.

Sewage is still 99.9 percent water!

Figure 14–10 Primary water purification removes particles that will settle or can be filtered, usually by sandbed filters. Aeration adds oxygen and gets rid of foul gases, and chlorination kills microbes. These are more recent additions. The system outlined here is almost as common as communities in the civilized world.

Figure 14–11 Sewage plant schematic, showing facilities for primary and secondary treatment. (From *The Living Waters*. U.S. Public Health Service Publication No. 382.)

rich environment (aerobic), whereas the cesspool operates in an oxygen-poor environment (anaerobic). The results are the same: the organic molecules that will not settle are consumed by organisms; the resulting sludge will settle. Bacteria and even protozoa are introduced into the oxygen-rich environment for this purpose. Two techniques, the trickle filter (Figure 14–12) and the activated sludge method (Figure 14–13), have been widely used in secondary water treatment.

Figure 14–12 Picture of a trickle filter with a section removed to show construction details. Rotating pipes discharge the water over a bed of stones. As a result, the organic molecules are "eaten" by microorganisms. (From Warren: *Biology and Water Pollution Control*. Philadelphia, W. B. Saunders Co., 1971. Photo courtesy of Link-Belt/FMC.)

Figure 14–13 The activated sludge process provides a closed-system environment in which organic molecules can be consumed by organisms that will readily settle. (From Turk, A., et al.: *Environmental Science.* Philadelphia, W. B. Saunders Co., 1974.)

Primary and secondary water treatment systems will not remove dissolved inorganic materials such as poisonous metal ions or even residual amounts of organic materials. These materials are removed by a variety of ***tertiary water treatments.***

Filtration through activated carbon, carbon that has been heated to high temperatures to drive off volatile material, adsorbs most organic and odor-causing compounds. Recently it has been shown that many metal ions are removed by this tertiary treatment as well. The method practically eliminates carcinogens from drinking water.

Since oceans are an "infinite" water supply, we shall examine some of the tertiary water purification treatments of seawater. Of course, these same techniques serve as the tertiary stage in reclaiming sewage or any other polluted water supply.

> Primary, secondary, and tertiary water treatment methods can be used in both the purification of water to be consumed and the preparation of sewage to be sent back into a stream.

Fresh Water from the Sea

As we look for new supplies of water, attention is inevitably drawn to the sea, as well as to the vast supplies of brackish water frequently found by drilling in arid lands. Seawater contains 3.5 percent salts; brackish waters have smaller salt concentrations. If we can convert such water to potable (drinkable) water inexpensively, we have a potentially valuable process. Thus far, the cost of purification of saltwater is high in comparison with the cost of water supplied by a typical European or American municipal water system. However, one can expect the cost of fresh water purification to increase if pollution continues to increase, and the cost of purifying saltwater to decrease as technology is improved.

> The sea is a largely untapped source of water.

The purification of saltwater is at present approached in two general ways. The first way involves separation of the water from the salt by evaporation or by freezing. A second approach is to use the influence of electrical charge, chemical attraction, or selective membranes to cause the ions in the saltwater to move out of the stream of water flow. If this process continues long enough, pure water results. Changing the state of water is expensive because of the large amount of energy that must move either into or out of the water. It takes 540 cal to vaporize 1 gram of water, and 80 cal must be removed to freeze each gram of water. The removal of ions from water is expensive because it is relatively slow for large flows of water and involves costly electric power.

Distillation

Distillation processes for the purpose of obtaining fresh water from seawater have been developed to a considerable state of refinement. In view of the large amounts of heat required, various methods have been devised to preheat the water with "waste heat." Since the steam has to lose 540 cal per gram in order to liquefy, this energy can be used to heat up the raw water to near the boiling point. Solar and nuclear energy have also been used to heat the water. Distillation plants are now the most common method in use to refine seawater and can produce water at a cost of less than $1 per 1000 gallons.

Because water needs much heat for vaporization, distillation requires a great deal of energy and is expensive.

Freezing

When cold seawater is sprayed into a vacuum chamber, the evaporation of some of the water cools the remainder, and ice crystals form in the brine. When the crystals of ice form, they tend to exclude the salt ions. Any solid separating from a liquid will tend to take only the molecules or ions that fit into the particular solid pattern. Recall that this generalization is the basis for purification by recrystallization. Even though the separation of salt and water is not complete in one step, the ice has less salt than the same weight of liquid solution. The ice crystals can be collected on a filter, washed with a small amount of fresh water, and then melted to obtain purified water. The process is repeated until the desired degree of purity is achieved. Plants have been built and successfully operated that produce as much as 250,000 gallons of pure water per day by this purification technique.

The ice in icebergs can be melted and drunk by humans.

Hydrate Formation

Propane (C_3H_8), a simple hydrocarbon, offers an interesting approach to the purification of water. In comparison to water, propane molecules have relatively little attraction for each other. Recall that water molecules interact through hydrogen bonding, while weaker van der Waals forces are the only attractive forces available between propane molecules. Consequently, the heat changes in the freezing and melting of propane are considerably less than for water. Experiment shows that when propane is solidified in the presence of water, a hydrate is formed which is a crystalline solid containing water molecules trapped in the "holes" between the molecules of the solid propane. The salt ions that are attached to a hydrated sphere of water are too large to fit into the "holes." If the solid hydrate is initially separated from the saltwater and then heated, it first melts; the propane then vaporizes, leaving the purified water behind. This purification method for saltwater has not been used widely because of high costs that result from the relatively complex equipment required and the lack of experience in this new technology.

Ion Exchange

In the process called ion exchange, brackish water or seawater is first passed through a cation exchange resin to replace the cations with H^+, and then through an anion exchange resin to replace the anions with OH^-; these two ions then neutralize each other.

Modern ion-exchange resins are high molecular weight polymers that contain firmly bonded functional groups which can exchange one ion for another as an ionic solution is passed over the giant molecules. ***Cation exchangers,*** or positive-ion exchangers, swap their hydrogen ions for the positive ions present in solution. They are usually organic derivatives of sulfuric acid, and their action can be depicted as:

The chloride ion is shown on both sides of this reaction to indicate that it is not removed.

$$\text{Polymer-SO}_3\text{H} + \quad \text{Na}^+ + \text{Cl}^- \longrightarrow$$
$$\text{BRACKISH WATER}$$

$$\text{Polymer-SO}_3^-\text{Na}^+ + \qquad \text{H}^+ + \text{Cl}^-$$
$$\text{FROM CATION EXCHANGER}$$

When the polymer is saturated with cations, it can be regenerated by treatment with strong acid, which reverses the above reaction. Because they generate weakly acidic solutions, cation exchange resins themselves do not do a complete job. When they are followed by ***anion exchange resins,*** they provide a good route to very pure water (Figure 14–14). An anion exchange resin can replace the anions in solution by hydroxide ions. These resins are again high molecular weight polymers, but now the polymer contains a nitrogen atom bonded to the polymer and three other groups. The reaction is:

$$\text{Polymer} - \overset{|}{\underset{|}{\text{N}}}{}^+\text{OH}^- + \quad \text{H}^+ + \text{Cl}^- \longrightarrow$$
$$\text{FROM CATION}$$
$$\text{EXCHANGER}$$

$$\text{Polymer} - \overset{|}{\underset{|}{\text{N}}}{}^+\text{Cl}^- + \qquad \text{H}_2\text{O}$$
$$\text{PURE WATER FROM}$$
$$\text{CATION AND ANION}$$
$$\text{EXCHANGERS}$$

When the anion exchange resin has exchanged all its hydroxide, it can be regenerated by treatment with a strong solution of sodium hydroxide:

Ion exchange resins can be regenerated.

$$\text{Polymer} - \overset{|}{\underset{|}{\text{N}}}{}^+\text{Cl}^- + \text{Na}^+ + \text{OH}^- \longrightarrow$$

$$\text{Polymer} - \overset{|}{\underset{|}{\text{N}}}{}^+\text{OH}^- + \text{Na}^+ + \text{Cl}^-$$

Unfortunately, the amount of seawater that can be purified by a given amount of ion exchange resin is quite small, and the resulting water is relatively costly.

Electrodialysis

We have learned that in the electrolysis of saltwater, positive ions migrate toward the negative electrode and negative ions move toward the positive electrode. If an electrolysis cell is divided into three compartments by semipermeable membranes, one permeable to positive ions and the other permeable to negative ions (Figure 14–15), the resulting ionic separation is called ***electrodialysis.*** Dialysis is the passage of selected

In electrodialysis, ions are attracted to oppositely charged electrodes on the other side of a membrane.

Water + H$^+$ + X$^-$

CATION EXCHANGER
R–H + M$^+$ → R–M + H$^+$

ANION EXCHANGER
N–OH + X$^-$ → N–X + OH$^-$

H$^+$ + OH$^-$ → H$_2$O

WATER (with M$^+$, X$^-$)

PURE WATER

Figure 14–14 The tank on the left contains a cation exchange resin which replaces metal ions in solution with hydrogen ions. The tank on the right contains an anion exchange resin which replaces nonmetal ions in solution with hydroxide ions. One might well ask why this kind of process cannot be used to turn seawater into pure water. The answer is that the process can be used, but its cost is much *greater* than that of distillation. It is best suited to the removal of small amounts of salts from water which is already quite pure.

Electrode

Membrane permeable to cations only

Membrane permeable to anions only

Electrode

All three compartments filled with brackish water

Figure 14–15 The essential features of the electrodialysis process. Each compartment of the cell contains brackish water. Application of an electrical potential across the cell causes the cations to move from the center into the left compartment and the anions to move into the right compartment. The salt content of the water in the center compartment is thus reduced, and raised in the end compartments.

— Current +

Salts removed from water in center compartment

species in solution through membranes while other species are excluded. Electrodialysis is the special case in which the passage of ions is influenced by an electrical field.

In Figure 14–15, note that positive ions (cations) can move to the left out of the center compartment but cannot move through the negative ion (anion) membrane from the right compartment to the center one. In a similar way, anions move out of the center compartment to the right. Thus, the ionic concentration of the water in the central compartment is reduced. If the process is continued long enough, the water in the central compartment loses most of its salt content.

Reverse Osmosis

Osmosis is the process whereby water moves through a semipermeable membrane from a region of relatively pure water into a region containing a concentrated solution. For example, water will move through a living cell membrane into the cell's protoplasm, causing the cell to become turgid. The resulting pressure inside the cell is often very high. If enough pressure is brought to bear on the solution inside the membrane (in other words, if a pressure greater than the osmotic pressure is applied in reverse), the water can actually be made to flow from the concentrated solution inside to the region outside (Figure 14–16). This is called *reverse osmosis.*

Reverse osmosis is a very promising process. Costs are down to 25¢ per 1000 gal for a plant producing 30,000,000 gal of pure water per day. This compares to 7.5¢ for the typical municipal treatment and 11¢ for the activated sludge treatments. Membranes used thus far are mostly cellu-

Figure 14–16 Reverse osmosis. (*a*) Mechanical pressure forces water against osmotic pressure to region of pure water. (*b*) Enlargement of individual membrane. (*c*) Mass of many membranes. (*d*) Industrial unit; feed water (salt) that passes through membranes collects at left end (permeate). The more concentrated salt solution flows out to the right as the reject.

lose acetate and work very well for saltwater. Much effort is being made in developing new films for a wide variety of applications of reverse osmosis.

It has been demonstrated that membrane processes are very effective in removal of polychlorinated biphenyls (PCB's), phenol, chromium, lead, and silver. It is probable that membrane processes can be developed to remove any dissolved species from water.

Softening of Hard Water

The presence of Ca^{2+}, Mg^{2+}, Fe^{3+}, or Mn^{2+} will impart "hardness" to waters. Hardness in water is objectionable because (1) it causes precipitates (scale) to form in boilers and hot water systems, (2) it causes soaps to form insoluble curds (this reaction does not occur with most synthetic detergents), and (3) it can impart a disagreeable taste to the water.

Hardness consisting of calcium or magnesium, present as their bicarbonates, is produced when water containing dissolved carbon dioxide trickles through limestone or dolomite:

> Hard water contains metal ions that react with soaps and give precipitates.

$$CaCO_3 \underset{\text{LIMESTONE}}{} + CO_2 + H_2O \longrightarrow Ca^{2+} + 2HCO_3^-$$

$$\underset{\text{DOLOMITE}}{CaCO_3 \cdot MgCO_3} + 2CO_2 + 2H_2O \longrightarrow Ca^{2+} + Mg^{2+} + 4HCO_3^-$$

Such "hard water" can be softened by removing these compounds. The principal methods for achieving this are (1) the lime-soda process and (2) ion-exchange processes.

> Water softeners that act like ion exchange resins are used to make soft water. They remove the hard water ions, Ca^{2+}, Mg^{2+}, and Fe^{3+}, and put Na^+ ions in the water in exchange.

The lime-soda process is based on the fact that calcium carbonate ($CaCO_3$) is much less soluble than calcium bicarbonate [$Ca(HCO_3)_2$] and that magnesium hydroxide is much less soluble than magnesium bicarbonate. The raw materials added to the water in this process are hydrated lime [$Ca(OH)_2$] and soda (Na_2CO_3). In the system, several reactions take place, which can be summarized:

$$HCO_3^- + OH^- \longrightarrow CO_3^{2-} + H_2O$$
$$Ca^{2+} + CO_3^{2-} \longrightarrow CaCO_3\downarrow$$
$$Mg^{2+} + 2OH^- \longrightarrow Mg(OH)_2\downarrow$$

(The downward arrow indicates that the chemical falls out as a precipitate.)

The overall result of the lime-soda process is to precipitate all the calcium and magnesium ions and to leave sodium ions as replacements.

Iron present as Fe^{2+} and manganese present as Mn^{2+} can be removed from water by oxidizing them with air (aeration) to higher oxidation states. If the pH of the water is 7 or above (either naturally or by adding lime), the insoluble compounds $Fe(OH)_3$ and $MnO_2(H_2O)_x$ are produced.

Chlorination

Chlorine is introduced into water as the gaseous free element (Cl_2), and it acts as a powerful oxidizing agent for the purpose of killing bacteria that remain in water after preliminary purification. The great majority of the bacteria in a water supply are removed before the addition of chlorine, by aluminum hydroxide precipitation. When the filtered water is allowed to stand, the bacterial population often decreases still more since its food supply is inadequate.

When water is heavily contaminated by sewage, the possibility exists that preliminary treatment will not be sufficient to remove *all* of the bacteria capable of causing disease. As cities grew during the late nineteenth century, their water supplies came more and more from waters that had already been used by other cities or from wells that were very close to cesspools or sewers. This meant that the possibility increased for an epidemic caused by a water-borne disease. The principal diseases that spread in this fashion include cholera, typhoid, paratyphoid, and dysentery.

Chlorine—an oxidizing agent—destroys microbes in water. As an oxidizing agent, chlorine is reduced: $Cl_2 + 2e^- \longrightarrow 2Cl^-$

The fact that cholera is caused by drinking polluted water was proved before the discovery of bacteria. An outbreak of cholera in London in 1854 was examined in great detail by Dr. John Snow. This was a relatively "small" epidemic, in which about 700 people died. Dr. Snow made a very careful study of the last few weeks in the lives of about 600 of those killed in the epidemic. Where information was available, he was able to show that in every case the victims of the plague had drunk water from a particular well. An examination of the well itself showed a crack in its wall through which sewage from persons sick with cholera had drained before the outbreak of the epidemic. This study showed that polluted water is a source of disease.

Subsequent identification of bacteria and proof that they can cause disease led to the chemical treatment of drinking water to kill bacterial populations. The chemical first used for this purpose was chloride of lime [a crude form of calcium hypochlorite, $Ca(OCl)_2$]. This chemical was first used to treat the water supply of Louisville, Kentucky, in 1896. Within a decade and a half, the use of hypochlorites of various sorts spread and became quite common. In 1910, Major C. R. Darnall of the U.S. Army Medical Corps demonstrated that the addition of small amounts of chlorine to partially purified water brought about almost complete elimination of bacteria. Because of the convenience of using elemental chlorine and its ability to remove bacteria even from waters heavily polluted with sewage, chlorine is now widely used in the purification of municipal water supplies. Both bromine and iodine can also be used for this purpose, but these are usually less convenient and more expensive.

Chlorine was used in military water supplies in World War I and thus became more acceptable for domestic use.

Chlorine acts as an oxidizing agent, not only on bacteria and other living microscopic inhabitants of water, but also on dead organic matter. The presence of a small excess amount is a guarantee that the bacterial population has been destroyed. Although there is no direct evidence on this point, it is possible that chlorine is also capable of destroying many, if not all, potentially disease-causing viruses that may be present in water drawn from a polluted source.

Chlorination of industrial wastes and city water supplies presents a potential threat because of residual concentrations of chlorinated organic compounds that are formed and are not removed by traditional purification procedures. City water supplies have been shown to contain such chloro-organics as dibromodichloromethane, trichloroethylene, bromoform, bromodichloromethane, chlorobenzene, and chloroform in concentrations of a few parts per million or less. A number of these chemicals in the same concentration range have been shown to be mutagenic to sal-

Figure 14—17 This apparatus adds chlorine in sufficient amounts to meet health standards (1 ppm residual) for a 60-million gallon per day water treatment plant. (Courtesy of the Robert L. Lawrence, Jr., Filtration Plant.)

monella bacteria. The effect of such water on humans is still a matter of conjecture. According to the Environmental Protection Agency, mutagenic or carcinogenic chemicals have been found in fourteen major river basins in the United States. If chlorination turns out to do more harm than good, which is not established at this point, other oxidizing agents such as ozone, chlorine dioxide, and bromine chloride have shown promise. However, potential problems with these chemicals are yet to be explored.

Fluoridation

Fluoride, too, is a poison in large concentrations, but a valuable additive in trace quantities.

Fluorine is added to water in the form of sodium fluoride to introduce the fluoride ion, F^-, for the purpose of reducing tooth decay.

Widespread fluoridation of water followed the discovery that certain cities in the western United States, where the water supplies contained abnormally high concentrations of natural fluoride, had populations with a less than average incidence of tooth decay. A thorough investigation showed that the ingestion of such fluorides by young children was especially effective in protecting their teeth against decay. As a consequence, the addition of small amounts of fluoride to drinking water to bring the fluoride concentration up to about 1 part per million was instituted on an experimental basis in some cities. The results bore out the original observation, that small amounts of fluoride in the drinking water could be effec-

Figure 14—18 Structure of hydroxyapatite and fluorapatite. The dark circles denote Ca^{2+} ions, the groups of five circles tied together by lines represent the PO_4^{3-} groups, and the largest circles represent OH^- groups in hydroxyapatite and F^- ions in fluorapatite.

PO$_4$$^{3-}$

Ca^{2+}

F$^-$ or OH$^-$

tive in reducing cavities to about 40 percent of their normal incidence. Excessive amounts of fluoride, however, can cause mottling of teeth.

The mineral content, or the hard part, of bones and teeth consists of two compounds of calcium. Calcium carbonate ($CaCO_3$) is present in bones and teeth in a crystalline form known to mineralogists as aragonite. The second compound is calcium hydroxyphosphate, $[Ca_5(OH)(PO_4)_3]$, called hydroxyapatite, and it is this crystalline structure that is modified and made more inert by the introduction of fluoride ions. Both aragonite and hydroxyapatite are susceptible to acid attack from food decay, but hydroxyapatite is the more so because of its basic characteristics associated with the OH^- ion. If fluoride ions are present in sufficient concentrations during the formative period of the teeth, fluorapatite $[Ca_5F(PO_4)_3]$ is laid down instead of hydroxyapatite. The two apatites have the same general structure, so the tooth structure is not physically changed (Figure 14–18). Since fluorapatite does not have the base properties of hydroxyapatite and is not readily attacked by mouth acids, the teeth are less likely to develop cavities.

SELF-TEST 14-B

1. Pick the pesticide with the shortest half-life in nature: DDT, dieldrin, heptachlor, or chlordan. _____

2. Heavy metal ions in more than trace concentrations are usually _____ _____ to life forms.

3. Silt is often stabilized in suspension by _____ forces.

4. Select the ions that may cause water to be hard: sodium, calcium, magnesium, potassium. _____

5. The element _____ is added to water to kill microbes in water, and a compound of _____ is added to protect the teeth against cavities.

6. Electrodialysis involves ion migration in a(n) _____ field.

7. Ion-exchange systems for removing salts from water involve _____ (how many) types of resins.

8. Ice formed from impure water will likely be (more pure than, as pure as, or less pure than) the original water. _____

9. Primary water treatment involves _____ and _____ of particles.

10. Tertiary water treatment removes _____ ions and trace amounts of _____ .

_____ 1.	sedimentation	(a)	tertiary purification process
_____ 2.	biodegradable	(b)	a measure of organic material in water
_____ 3.	detergent	(c)	widely used as a detergent builder
_____ 4.	BOD	(d)	results mostly from water used to cool a process
_____ 5.	thermal pollution	(e)	caused by metal ions such as Ca^{2+} and Mg^{2+} in solution
_____ 6.	phosphate	(f)	primary purification process
_____ 7.	zeolite	(g)	persistent pesticide
_____ 8.	mercury	(h)	secondary purification process
_____ 9.	activated sludge process	(i)	naturally reducible to simpler compounds
_____ 10.	reverse osmosis	(j)	soap substitute
_____ 11.	water hardness	(k)	poison metal
_____ 12.	DDT	(l)	experimental detergent builder

QUESTIONS

1. What are some processes that *decrease* the amount of dissolved oxygen in a stream? What are some processes that *increase* the amount of dissolved oxygen in a stream? Which ones are most readily subject to man's control?

2. Explain why each of the following introduces a pollution problem when its wastes are emptied into a stream:

   (a) a chlorine-producing plant.
   (b) a steel mill.
   (c) an electricity-generating plant burning oil or coal.
   (d) an agricultural area that is intensively cultivated.

3. An old rule of thumb is, "Water purifies itself by running two miles from the source of incoming waste." What processes are active in purifying the water? Is this adage foolproof? Explain.

4. What is a detergent builder? What property should it have to increase the cleansing power of the detergent? What property must it have to avoid environmental problems?

5. What are some consequences of requiring by law that all water put into our national waters be clinically and chemically pure H_2O? Do you think this is the proper solution to water pollution? If not, what is the proper compromise?

6. What are some ecological consequences of thermal pollution?

7. From your experience, add one additional example for each of the classes of pollutants listed in Table 14–2.

8. What pertinent facts would you try to gather if it were your responsibility to vote on a bill to regulate water pollution?

9. At what point should pollutants be removed from used water? Who should be responsible for this removal? Would you distinguish between industrial wastes and household wastes?

10. The most abundant elements in organic compounds are carbon, hydrogen, oxygen, and nitrogen. What are the oxidation products for these elements in the decomposition that occurs in nature?

11. Relate molecular structure to biodegradability for detergent molecules.

12. Classify water pollutants into as few major groups as you can. Describe some effects of each group and a removal process.

13. From your study of biochemistry, Chapters 10 and 11, explain why there is no question concerning the biodegradability of the proposed detergent builder, sodium citrate.

14. In your judgment, what are the most serious water pollution problems? Be ready to defend your points in class discussion.

15. What is natural osmosis? Explain the significance of the word "reverse" in reverse osmosis.

CHAPTER 15

AIR POLLUTION

Prior to 1960 there was little national or international concern about air pollution. Most smoke, carbon monoxide, sulfur dioxide, nitrogen oxides, and organic vapors were emitted into the air with little apparent thought of these pollutants being harmful as long as they were scattered

Figure 15–1 Air pollution covers the base of the St. Louis arch. (World Wide Photos, Inc.)

Figure 15–2 Region of Donora, Pennsylvania. In late October, 1948, Donora had a five-day siege of extreme air pollution. Before rain cleansed the air, over 800 domestic animals died and 43 percent of the population, or 5910 people, became ill. Eighteen died; the normal rate was two deaths every five days. (From Turk, A., et al.: *Environmental Science*. Philadelphia, W. B. Saunders Co., 1974.)

Figure 15–3 Coppertown Basin (Ducktown), Tennessee, as photographed in 1943. Copper ore (principally copper sulfide, Cu_2S) had been mined and smelted in this area since 1847. In the early years, large quantities of sulfur dioxide, a by-product, were discharged directly into the atmosphere and killed all vegetation for miles around the smelter. Today the sulfur is reclaimed in the exhaust stacks to make sulfuric acid, but the denuded soil remains a monument to the misuse of the atmosphere.

TABLE 15–1 Some Air Pollution Disasters*

Location and Date	Pollutants Type	Range	Meteorology	Health Effects Diseases and Symptoms	Excess Deaths
Meuse Valley, Belgium Dec., 1930	H_2SO_4 HF NO_2 CO CO_2	Concentrations unknown	Fog Wind speed—0.62 mph Ceiling—246 ft Temp. inversion	Cardiovascular hypotension Alkalosis Sore throat Cough Nausea Vomiting	60 to 80
St. Louis, Mo. Nov. 28, 1939 ("Black Tuesday")	Smoke	(Nighttime darkness through the day)	Temp. inversion Windless period		
Donora, Pa. Oct. 27–31, 1948	SO_2 (last day) H_2SO_4 mist Other sulfur compounds O_3 NOx Organic compounds Smoke	Concentrations unknown	Temp. inversion Fog Stagnant anticyclone	Cough Respiratory irritation Sore throat Chest constriction Dyspnea Eye irritation Vomiting Nausea Heart diseases Bronchitis (43% of population became ill)	18
London Nov. 26–Dec. 1, 1948	Black suspended matter SO_2	200 to 2800 mg/m^3 0.09 to 0.75 ppm	Fog Wind speed—0 to 4.6 mph Visibility—27 to 440 yd		700 to 800
London Dec. 5–9, 1952	SO_2 Black suspended matter	0.09 to 1.34 ppm 40 to 4500 mg/m^3	Wind vel.—0 to 5.8 mph Visibility—22 to 240 yd Fog	Bronchitis Emphysema Cardiovascular (ischemic heart disease) Acute wheezy chests Dyspnea Fever Yellow-black sputum	4000
London Jan. 3–6, 1956	SO_2 Black suspended matter	0.19 to 0.55 ppm 700 to 2400 mg/m^3	Wind speed—0 to 8.1 mph Visibility—5 to 12,000 yd		1000
Rotterdam Dec. 2–7, 1962	SO_2	5 times normal			Increase in mortality noted
London Dec. 5–10, 1962	Smog SO_2	1.98 ppm, highest hourly concentration		Wheezy chests Bronchitis Dyspnea Fever Yellow-black sputum	700
Osaka, Japan Dec. 7–10, 1962	Pollution levels high				60
New York Jan. 29–Feb. 12, 1963	SO_2 Smoke shade	0.2 to 0.5 ppm 1.5 to 7 Coh units†	Few winds <5 mph Few inversions Stagnating anticyclones were absent	Influenza Pneumonia Vascular lesions Cardiac arrests	200 to 400
New York Feb. 27–Mar. 10, 1964	SO_2 Smoke shade	0.0 to 7 ppm 2 to 5 Coh units†	A few inversions Few winds <5 mph		168

*From U.S. Environmental Protection Agency.
†Coh is the abbreviation for coefficient of haze, a unit of measurement of visibility interference.

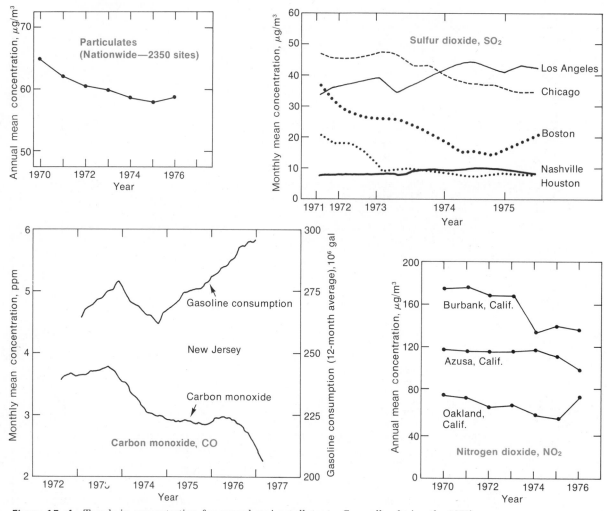

Figure 15—4 Trends in concentration for several major pollutants. Generally, during the 1970's concentrations have been decreasing despite increasing population pressures. [μg/m³ means micrograms (10^{-6} grams) per cubic meter; ppm is parts per million parts of air.]

into the atmosphere. Few controls were used to curtail the increasing pollution of the air brought on by increased urbanization, industrialization, and transportation via the automobile. Yet the worst air pollution disasters of all time occurred prior to 1960 (Table 15–1).

During the 1960's more public and governmental concern about air pollution led to controls being placed on emissions from automobiles, industries, and power companies. These efforts caused a general downward trend in most air pollutants (Figure 15–4).

The energy crisis of the 1970's has caused a relaxation of air quality standards (Table 15–2). Obviously a choice had to be made—do we want cheap energy and the attendant air pollution, or are we willing to pay for higher priced energy in order to obtain abatement of air pollution? A serious study of the material in this chapter, combined with a study of the chapter on energy (Chapter 17), provides a sound basis for the decisions society must make concerning energy and pollution, and its willingness to pay for them.

TABLE 15–2   Federal Air Quality Standards
(August 7, 1977)

	Concentration
Sulfur Dioxide	
Arithmetic mean (annual)	0.03 ppm
24-hour concentration not to be exceeded more than once per year	0.14 ppm
Suspended Particulates	
Geometric mean (annual)	80 $\mu g/m^3$*
24-hour concentration not to be exceeded more than once per year	365 $\mu g/m^3$
Carbon Monoxide	
8-hour concentration not to be exceeded more than once per year	9 ppm
1-hour concentration not to be exceeded more than once per year	35 ppm
Photochemical Oxidants	
1-hour concentration not to be exceeded more than once per year	0.08 ppm
Hydrocarbons	
3-hour concentration not to be exceeded more than once per year	0.24 ppm
Oxides of Nitrogen (NO_x)	
Arithmetic mean (annual)	0.05 ppm

*1 $\mu g = 10^{-6}$ g

Air pollutants tend to attack the site where they first enter the body, i.e., the lungs.

The long-range effects of air pollution on materials and the health of plants, animals, and human beings are beginning to emerge. The lung cancer rate in large metropolitan areas is twice as great as the rate in rural areas, even after full allowance is made for differences in cigarette smoking habits. The incidence of the serious pulmonary disease emphysema shot up eightfold during the decade of the sixties. However, only after we comprehend the short- and long-range effects of air pollution can we evaluate wisely its relative importance. A logical beginning is to understand some basic facts about our atmosphere.

THE ATMOSPHERE

The atmosphere is a mixture of an estimated 5500 trillion tons of gases, mostly nitrogen (78 percent) and oxygen (21 percent) with small quantities of argon, water vapor and carbon dioxide and still smaller quantities of other materials. About 99 percent of the total mass is below an altitude of 19 miles, and there is sufficient oxygen to sustain life only up to 4 miles above sea level (Figure 15–5). The region that contains most of the oxygen and supplies most of our weather is the *troposphere,* which extends to an average altitude of 7 miles.

ppm = number of molecules of pollutant per 1,000,000 molecules of air.

To change from percent to ppm multiply by 10,000.

A representative composition of dry air near sea level is given in Table 15–3. Near industrial complexes and communities some of the substances, normally present in very small amounts (less than 1 part per million), often increase to abnormal amounts of 50 or more ppm.

Pollutants litter the troposphere, move both vertically and horizontally, often reacting chemically with themselves or with materials on the surface of the earth, and in due time generally return to land or water. The troposphere—the rug in the sky under which we try to hide our wastes—receives more than 140,000 tons of combustion products per day from the United States alone. How pollutants behave chemically while in the atmosphere, and what their ultimate fate is, are interesting chemical

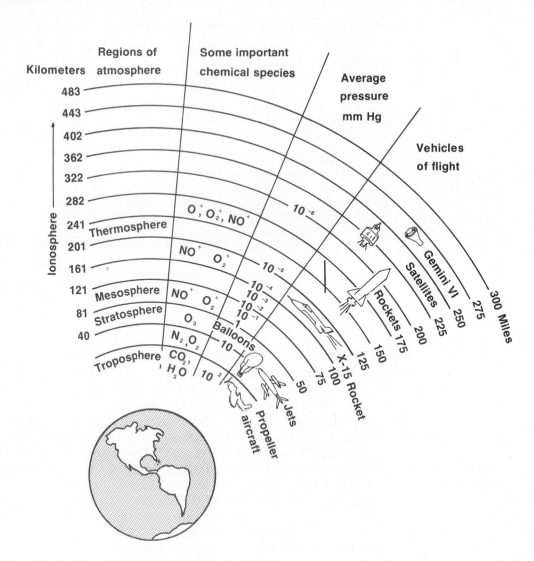

Figure 15—5 Some facts about our limited atmosphere. The troposphere was named by British meteorologist Sir Napier Shaw from the Greek word *tropos,* meaning turning. The stratosphere was discovered by the French meteorologist Leon Philippe Teisserenc de Bort, who believed that this region consisted of an orderly arrangement of layers with no turbulence or mixing. The word *stratosphere* comes from the Latin word *stratum,* meaning layer.

TABLE 15—3 Composition of Clean, Dry Air Near Sea Level (ppm)

Component	Content	Component	Content
Nitrogen	780,900	Nitrous oxide	0.25
Oxygen	209,400	Carbon monoxide	0.10
Argon	9,300	Xenon	0.08
Carbon dioxide	318	Ozone	0.02
Neon	18	Ammonia	0.01
Helium	5.2	Acetone	0.001
Methane	1.5	Nitrogen dioxide	0.001
Krypton	1.0	Sulfur dioxide	0.0002
Hydrogen	0.5	Lead	0.00000013

TABLE 15—4 Composition of Inhaled and Exhaled Air

	Inhaled Air (%)	Exhaled Air (%)
Oxygen	20.96	15.8
Carbon dioxide	.04	4.0
Nitrogen and other gases	79.00	80.2

questions with which we shall deal later. Except for the freons, pollutants do not seem to go beyond the heights of the troposphere.

The most important part of the atmosphere to us is the part we actually breathe. With every breath, the average adult inhales (and exhales) about one half liter of air. This quantity contains about 1.3×10^{22} (13 sextillion) molecules. Most of these molecules leave the lungs just as they entered. Only those that linger can influence our health.

Although pollution occurs in relatively small amounts compared to oxygen, nitrogen, and carbon dioxide, about 200 quadrillion *pollutant* molecules are inhaled per breath on a *clear* day in Los Angeles, where the average breath would contain:

A quadrillion is 10^{15}.

Carbon monoxide	175,000,000,000,000,000 molecules
Hydrocarbons	10,000,000,000,000,000
Peroxides	5,000,000,000,000,000
Nitrogen oxides	4,000,000,000,000,000
Lower aldehydes	3,500,000,000,000,000
Ozone	3,000,000,000,000,000
Sulfur dioxide	2,500,000,000,000,000

On a smoggy day, the numbers increase by a factor of five or more. The breath you are now inhaling could contain pollutant molecules in comparable amounts, give or take a few quadrillion.

TABLE 15—5 Air Pollutant Emissions in the United States in 1976 (millions of tons per year)*

	Totals	% of Totals	Carbon Monoxide	Sulfur Oxides	Hydro-carbons	Nitrogen Oxides	Particulates
Transportation	92.6	52	69.7	0.8	10.8	10.1	1.2
Fuel Combustion In Stationary Sources	40.9	23	1.2	21.9	1.4	11.8	4.6
Industry	28.3	16	7.8	4.1	9.4	0.7	6.3
Solid Waste Disposal	4.1	2	2.8	0	0.8	0.1	0.4
Miscellaneous	12.5	7	5.7	0.1	5.5	0.3	0.9
Totals:	178.4	100	87.2	26.9	27.9	23.0	13.4

*From U.S. Environmental Protection Agency, December, 1977.

SOURCES OF POLLUTANTS IN THE AIR

Automobiles, industry, and electric power plants are the main sources of air pollutants from human-controlled processes. Volcanic action, forest fires, dust storms and even growing plants are natural sources of air pollutants, but these contribute very little compared to the man-made sources. A summary of the principal sources of emissions in the United States in 1976 is shown in Table 15–5.

POLLUTANT PARTICLE SIZE

Pollutant particles may be grouped in three sizes: fundamental (such as single molecules, ions, or atoms), aerosol (ranging from a thousand to about a trillion atoms, ions, or small molecules per particle), and particulate (suspended particles composed of more than a trillion atoms, ions, or molecules). Aerosols range in size from 1 to about 10,000 nanometers in diameter; molecules are smaller and particulates are larger. Aerosols and particulates serve as collectors of the chemically active sulfur oxides, nitrogen oxides, ozone, hydrocarbons, and other pollutant molecules.

1 nanometer = 10^{-9} meter.

Aerosols

Aerosols can be either liquid or solid particles, and are small enough to remain suspended in the atmosphere for long periods of time. Smoke, dust, clouds, fog, mist, and sprays are typical aerosols. Aerosols have much greater surface area than would a single lump of the material of the same mass. Consider a cube 1 centimeter on an edge. Such a cube would have a surface area of 6 square centimeters. If this cube is subdivided into a sextillion (1,000,000,000,000,000,000,000) smaller cubes, each cube would be 1 nanometer on an edge, about the size of an aerosol particle. The total surface area is 60,000,000 square centimeters—about 1.5 acres! Because of this large surface area, aerosols have enormous capacities to adsorb and concentrate gases on their surfaces. Many times aerosols absorb toxic gases, provide the water medium for a reaction to occur, and then cause devastating results when breathed. Specific examples and the important role that aerosols play in smog formation will be cited later.

Aerosol particles are intermediate in size between small molecules and easily visible small particles.

Adsorption is the attachment of particles to a surface.

Figure 15–6 A dust sample collector for use in the backyard. The open jar containing water is exposed to the air for a known interval. Fifteen days is sufficient exposure time. The water is then evaporated and the residue weighed. Balances sensitive to 0.001 g should be used; the area (cm²) of the opening must be known.

Insect screen

Distilled water

Absorption is pulling particles inside.

1 micron = 10^{-6} meter.

Particulates

Particulates are generally large enough to be seen. They range in size from 1 to 10 microns in diameter. Over 26 million tons of soot, dust, and smoke particulates were deposited into the atmosphere of the United States in 1970. Average suspended particulate concentrations in the United States range from 0.00001 gram per cubic meter of air (g per m³) in remote rural areas to about six times that value in urban locations. In heavily polluted areas concentrations up to 0.002 g per m³, or 200 times the usual value, have been measured.

Particulates may cause physical damage to certain materials. Particles whipped by the wind grind exposed materials by abrasive action. Particles settling in electrical equipment can cause short circuits and foul contacts and switches. Dust settling in paint detracts from its beauty and hastens its deterioration by allowing water to reach the surface underneath. Particles may interfere physically with one or more of the clearance mechanisms in the respiratory tract of man and animals (inhibiting the ciliary transport of mucus, for example). People who have asthma or emphysema know that heavy concentrations of particles in the air increase discomfort. In extremely polluted regions, these diseases often lead to death.

Particulate effects depend heavily on the chemical nature of the particle.

Particulates may also injure human beings or animals because they are intrinsically toxic. The effects of lead and arsenic were described in Chapter 13. Lead compounds are emitted in automobile exhaust. Arsenic compounds are used as insecticides to dust plants. Particles containing fluorides, commonly emitted from aluminum-producing and fertilizer factories, have caused weakening of bones and loss of mobility in animals

Figure 15—7 The General Electric CF6-6D engines that power this McDonnell Douglas DC-10 were designed to eliminate exhaust smoke and reduce by one half the noise level at takeoff. American Airlines advertises that it spent in a ten year period an amount equal to 43 percent of its profits on noise and air pollution control systems. (Courtesy of American Airlines.)

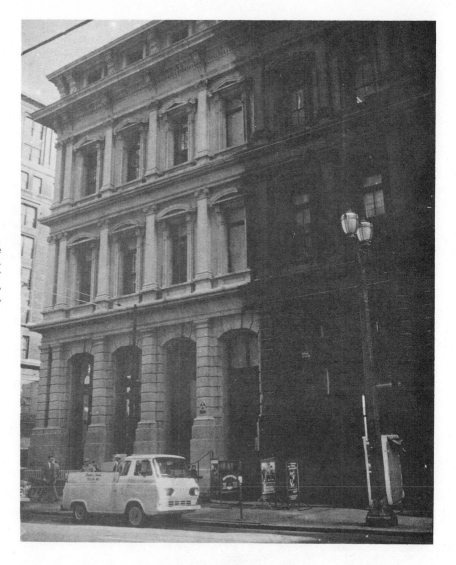

Figure 15–8 Old post office building being cleaned in St. Louis, Mo., 1963. (Photo by H. Neff Jenkins. From Stern, A. C.: *Air Pollution.* New York, Academic Press, 1968.)

that have eaten plants covered with the dust. Leaves of bean plants dusted with cement kiln particles (0.00047 g per cm² per day) wilt significantly.

Like aerosol particles, larger particulates may cause damage because they adsorb toxic substances. Sulfur dioxide, nitrogen oxides, hydrocarbons, and carbon monoxide do their greatest damage when concentrated on the surface of particulates.

Figure 15–9 Diagram of a small section of the lung. The walls of the airways are lined with cells, some of which secrete mucus. The walls are also lined with hairlike projections called cilia, which wave in rhythm and serve to propel mucus and impurities of the air out of the lungs. (From Turk, A., et al.: *Environmental Science.* Philadelphia, W. B. Saunders Co., 1974.)

Removing particulates by filtration

Removing particulates by centrifugal separation

Removing particulates and aerosols by scrubbing. Schematic drawing of a spray collector, or scrubber.

Particles are removed naturally from the atmosphere by gravitational settling and by rain and snow. They can be prevented from entering the atmosphere by treating industrial emissions by one or more of a variety of physical methods such as filtration, centrifugal separation, spraying, ultrasonic vibration, and electrostatic precipitation.

Glass fiber or silicone treated textile bags filter hot exhaust gases (290°C maximum); cotton, nylon, acrylics, wool, and felts are used to filter out cold aerosols.

Centrifugal separators whirl the gases and sling the particles against the outer walls, where they collect. The same result is obtained when fast-moving exhaust is made to change direction suddenly by impinging on a baffle. The dust particles bump against the wall, where they collect.

Spray treatment (scrubbing) is utilized by the petroleum industry, among others. The smoke from a waste gas burner is collected in the water from several small spray nozzles set a few inches above the tip of the burner.

Ultrasonic vibration operates on the principle that increased collisions of the particles in aerosols cause adherence and condensation of the suspended particles. The ultrahigh frequency vibrations cause the increased collisions.

Electrostatic precipitators can remove aerosols and dust particles smaller than 1 micron from plant exhaust gases. A diagram of a Cottrell electrostatic precipitator is shown in Figure 15–10. The central wire is connected to a source of direct current at high voltage (about 50,000 volts). As dust or aerosols pass through the strong electrical field, the particles attract ions that have been formed in the field, become strongly charged,

Figure 15–10 The Cottrell electrostatic precipitator.

and are attracted to the electrodes. The precipitated solid falls to the bottom where it is collected. Electrostatic precipitators are better than 98% effective in removing most particulates and aerosols from exhaust gases.

SMOG—INFAMOUS AIR POLLUTION

The poisonous mixture of smoke, fog, air, and other chemicals was first called *smog* in 1911 by Dr. Harold de Voeux in his report on a London air pollution disaster that caused the deaths of 1150 people. Through the years, smog has been a technological plague in many communities and industrial regions.

Industrial or London-type smog:
fog + SO_2
Photochemical smog:
fog + NO_x + hydrocarbons

Two general kinds of smog have been identified. One is the chemically reducing type that is derived largely from the combustion of coal and oil, and contains sulfur dioxide mixed with soot, fly ash, smoke, and partially oxidized organic compounds. This is the *London type,* which is diminishing in intensity and frequency as less coal is burned and more controls are installed. A second type of smog is the chemically oxidizing type, typical of Los Angeles; it is called *photochemical* smog because light—in this instance sunlight—is important in initiating the photochemical process. This smog is practically free of sulfur dioxide but contains substantial amounts of nitrogen oxides, ozone, ozonated olefins, and organic peroxide compounds, together with hydrocarbons of varying complexity.

"Olefin" is another name for an unsaturated hydrocarbon.

What general conditions are necessary to produce smog? Although the chemical ingredients of smogs often vary, depending upon the unique sources of the pollutants, certain geographical and meteorological conditions exist in nearly every instance of smog.

There must be a period of windlessness so that pollutants can collect

Figure 15—11 Smog over New York City. A heavy haze hangs over Manhattan Island, viewed from the roof of the RCA Building. The Empire State Building is barely visible in the background. (Wide World Photos, Inc.)

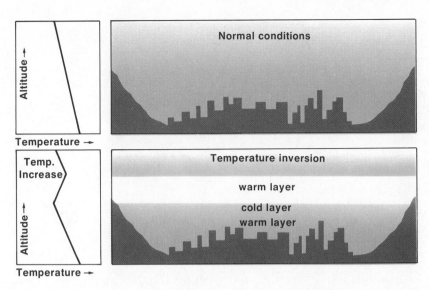

Thermal inversion: mass of warmer air over a mass of cooler air.

without being dispersed vertically or horizontally. This lack of movement in the ground air can occur when a layer of warm air rests on top of a layer of cooler air. This sets the conditions for a ***thermal inversion,*** which is an abnormal temperature arrangement for air masses. Normally the warmer air is on the bottom nearer the warm earth, and this warmer, less dense air rises and transports most of the pollutants to the upper troposphere where they are dispersed. In a thermal inversion the warmer air is on top, and the cooler, more dense air retains its position nearer the earth. The air becomes stagnated. If the land is bowl-shaped (surrounded by mountains, cliffs, or the like), this stagnant air mass can remain in place for quite some time.

When these natural conditions exist, man supplies the pollutants by combustion and evaporation in automobiles, electrical power plants, space heating, and industrial plants. The chief pollutants are sulfur dioxide (from burning coal and some oils), nitrogen oxides, carbon monoxide, and hydrocarbons (chiefly from the automobile). Add to these ingredients the radiation from the sun, and a massive smog is in the offing.

Photochemical Smog

Primary pollutants: pollutants emitted into the air.

Secondary pollutants: pollutants formed in the air by chemical reaction.

A chemical radical is a species with an unpaired electron. They are usually very reactive and short-lived.

A city's atmosphere is an enormous mixing bowl of frenzied chemical reactions. Ferreting out the exact chemical reactions that produce smog has been a tedious job, but in 1951, insight into the formation process was gained when smog was first duplicated in the laboratory. Detailed studies have subsequently revealed that the chemical reactions involved in the smog-making process are photochemical and that aerosols serve to keep the reactants together long enough to form secondary pollutants. Ultraviolet radiation from the sun is the energy source for the formation of this ***photochemical smog.***

The exact reaction scheme by which primary pollutants are converted into the secondary pollutants found in smog is still not completely understood, but the reactions shown in Figure 15–13 account for the major secondary pollutants. The process is thought to begin with the absorption of a quantum of light by nitrogen dioxide, which causes its breakdown into nitrogen oxide and atomic oxygen, a chemical radical. The very reac-

$$
\begin{aligned}
NO_2 + light &\longrightarrow NO + O\cdot \\
O\cdot + O_2 + M &\longrightarrow O_3 + M \\
O_3 + NO &\longrightarrow NO_2 + O_2 \\
O\cdot + Hc &\longrightarrow HcO\cdot \\
HcO\cdot + O_2 &\longrightarrow HcO_3\cdot
\end{aligned}
$$

$$
\begin{aligned}
HcO_3\cdot + Hc &\longrightarrow RCHO \text{ or } R\overset{O}{\overset{\|}{C}}R \\
HcO_3\cdot + NO &\longrightarrow HcO_2\cdot + NO_2 \\
HcO_3\cdot + O_2 &\longrightarrow O_3 + HcO_2\cdot
\end{aligned}
$$

$$
HcO_3\cdot + NO_2 \longrightarrow R\overset{O}{\overset{\|}{C}}\text{—O—O—N}\overset{O}{\underset{O}{}} + \text{other products}
$$
(PAN)

Figure 15–13 Simplified reaction scheme for photochemical smog. Ultraviolet light initiates the process to produce oxygen atoms. Hc is hydrocarbon (olefin or aromatic); M is a third body to absorb the energy released from forming the ozone. Among many possibilities, M could be a N_2 molecule, an O_2 molecule, or a solid particle. A species with a dot, as $HcO\cdot$, is a chemical radical. R is a saturated hydrocarbon group; PAN is peroxy acyl nitrate, a very reactive secondary pollutant that causes eyes to water and lungs to hurt.

tive atomic oxygen reacts with molecular oxygen to form ozone (O_3), which is then consumed by reacting with nitrogen oxide to form the original reactants—nitrogen dioxide and molecular oxygen. Atomic oxygen, however, also reacts with reactive hydrocarbons—olefins and aromatics—to form other chemical radicals. These radicals, in turn, react to form other radicals and secondary pollutants such as aldehydes (e.g., formaldehyde). About 0.2 ppm of nitrogen oxides and 1 ppm of reactive

Figure 15–14 Photochemical smog (a brown haze) enveloping the city of Los Angeles. (Los Angeles Times photo.)

TABLE 15–6 Approximate Concentrations of
Pollutants in Los Angeles on a Clear Day
(Visibility 7 Miles)
and on a Smoggy Day
(Visibility 1 Mile)

	Concentration (ppm)		
Pollutant	Clear Day	Smoggy Day	Increase
Carbon monoxide	3.5	23.0	×6.5
Hydrocarbons	0.2	1.1	×5.5
Peroxides	0.1	0.5	×5.0
Oxides of nitrogen	0.08	0.4	×5.0
Lower aldehydes	0.07	0.4	×6.0
Ozone	0.06	0.3	×5.0
Sulfur dioxide	0.05	0.3	×6.0

hydrocarbons are sufficient to initiate these reactions. The hydrocarbons involved come mostly from unburned petroleum products like gasoline.

Measurements made in several U.S. cities reveal that the amount of airborne aerosols increased tenfold during the 1960's. Despite the stricter controls effected during the last decade, Los Angeles did not show a downward trend in the occurrence and amount of smog, which is present about 300 days a year. A comparison of the levels of pollutants on a clear day and on a smoggy day is presented in Table 15–6.

Industrial Smog

The type of smog formed in London and around some industrial and power plants is thought to be caused by sulfur dioxide. Laboratory experiments have shown that sulfur dioxide increases aerosol formation, particularly in the presence of mixtures of hydrocarbons, nitrogen oxides, and air energized by sunlight. For example, mixtures of 3 ppm olefin, 1 ppm NO_2, and 0.5 ppm SO_2 at 50 percent relative humidity form aerosols which have sulfuric acid as a major product. Even with 10 to 20 percent relative humidity, sulfuric acid is a major product. Sulfuric acid, which is formed in this kind of smog, is very harmful to people suffering from bronchial diseases such as asthma or emphysema. At a concentration of 5 ppm for one hour, SO_2 can cause constriction of bronchial tubes. A level of 10 ppm for one hour can cause severe distress. In the 1962 London smog, readings as high as 1.98 ppm of SO_2 were recorded. The sulfur dioxide and sulfuric acid are thought to be the primary causes of deaths in the London smogs.

We shall next examine some of the principal components of air pollution and their chemistry. Emphasis will be placed on how the compounds are produced and how they can be eliminated from our atmosphere.

Relative humidity is a measure of the amount of water vapor air contains compared to the maximum amount it can contain.

1. _____ _____ _____ is abbreviated ppm. To change   SELF-TEST 15-A

from percent to ppm, multiply by _____ . Thus 0.092 percent

is _____ ppm.

2. Three substances normally considered primary pollutants are _____,

   _____, and _____.

3. In what region of the atmosphere does most air pollution exist?

   _____

4. Particles that can remain suspended in air for long periods of time and are intermediate in size between individual molecules and particulates are called

   _____.

5. Much of the effect of aerosols is due to their large _____

   _____.

6. A device that imparts charges to dust and aerosol particles so they can be

   attracted out of a gaseous effluent stream is known as a(n)_____

   _____.

7. The initial step in the formation of photochemical smog occurs when ultra-

   violet light decomposes a molecule of _____ _____.

8. During a thermal inversion, a (warm/cool) mass of air is above a mass of (warm/cool) air.

9. The ingredient that distinguishes industrial (or London-type) smog from

   the other kind is the substance _____.

SULFUR DIOXIDE

Sulfur dioxide is produced when sulfur or sulfur-containing substances are burned in air:

$$S + O_2 \longrightarrow SO_2(g)$$

Most of the sulfur dioxide in the atmosphere comes from sulfur-containing fuels burned in power plants, from smelting plants treating sulfide ores, and from sulfuric acid plants. The average sulfur content of all coal shipped from mines in the United States is about 2.0%, although most coal has either less than 1% or more than 3% sulfur. Coal contains sulfur in several forms: as elemental sulfur (S), as iron pyrite (FeS_2), and as organic compounds, such as mercaptans (compounds containing $-SH$ groups). According to the U.S. Environmental Protection Agency, the burning of fuels by the nation's power plants accounted for 80 percent of the SO_2 emitted into the atmosphere in 1976. One study found a concentration of 2200 ppm SO_2 in the stack of a coal-burning power plant. Concentrations as high as 2.9 ppm have been recorded up to half a mile from power plants. If coal and petroleum containing up to 5 percent sulfur are

$CH_3—SH$
Methyl mercaptan

Figure 15–15 Comparison of SO_2 trends at Bayonne, New Jersey, with regulations governing percent sulfur content in fuel. (From U.S. Environmental Protection Agency.)

A large modern power station (e.g., a station of 2,000 megawatts capacity) annually produces about the same amount of SO_2 as an industrial city of a million inhabitants.

More than 26 million tons of SO_2 were put into the air in 1976.

burned, a 1000-megawatt electric power plant could emit 600 tons of SO_2 per day. Note how the SO_2 concentration in the air has decreased in Bayonne, New Jersey, as the sulfur content of the fuel has been reduced (Figure 15–15). Much of the SO_2 pollution in this country is isolated in seven industrialized states (New York, Pennsylvania, Michigan, Illinois, Indiana, Ohio, and Kentucky), which account for almost 50 percent of the nation's total SO_2 output.

In spite of the large output of SO_2 in the United States per year, its concentration rarely exceeds a few parts per million. This is because SO_2 has a relatively short atmospheric life. In the presence of oxygen, sunlight, and water vapor, SO_2 is oxidized to SO_3, which is converted to sulfuric acid (H_2SO_4):

$$2SO_2 + O_2 \longrightarrow 2SO_3$$

$$SO_3 + H_2O \longrightarrow H_2SO_4$$

Sulfurous acid, H_2SO_3, is a weak acid; sulfuric acid, H_2SO_4, is a strong acid. Both can make the water in streams and lakes too acidic for fish.

The SO_2 and sulfuric acid are readily dissolved in rivers, lakes, and streams and can increase the acidity considerably.

$$H_2O + SO_2 \rightleftharpoons H_2SO_3 \rightleftharpoons H^+ + HSO_3^-$$
$$\text{WEAK ACID}$$

and

$$H_2SO_4 \longrightarrow H^+ + HSO_4^-$$
$$\text{STRONG ACID}$$

TABLE 15–7   Physiologic and Corrosive Effects of SO_2.*

SO_2 Exposure, ppm	Duration	Effect	Comment
0.03 to 0.12	Annual average	Corrosion	Moist temperate climate with particulate pollution
0.3	8 hr	Vegetation damage (alfalfa mostly, but many other species are similarly sensitive)	Laboratory experiment; other environmental factors optimal. Field studies are consistent but dose is difficult to estimate
0.47	<1 hr	Odor threshold (50% of subjects detect)	May be higher for many persons or when other methods are used
0.2	Daily average	Respiratory symptoms	Community exposure exceeding 0.2 ppm more than 3% of the time
>0.05	Long-term average	Respiratory symptoms	With particulates >100 $\mu g/m^3$
0.2	Daily average	Respiratory symptoms	With particulates
0.9	Hourly average	Respiratory symptoms	With particulates
>0.05	Monthly average	Respiratory symptoms, including impairment of lung function in children	With particulates

*From U.S. Environmental Protection Agency.

Often streams contain ions that help to offset the acidity increase. Sulfite is one such ion.

$$SO_3^{2-} + H^+ \longrightarrow HSO_3^-$$

SULFITE   BISULFITE

If the pH of streams varies too much, aquatic life suffers. Salmon, for example, cannot survive if the pH is as low as 5.5. The lower limit of tolerance for most organisms is a pH of 4.0. In the late 1950's and early 1960's, certain sections of the Netherlands had precipitation with a pH less than 4.

The corrosion of steel is promoted by sulfur dioxide. In a study involving 100-gram panels of steel exposed at several sites in Chicago in 1964, the amount of weight loss by corrosion was very definitely related to the SO_2 concentration in the air. Corrosive action is also promoted by particulate matter, high humidity, and high temperatures.

An interesting example of the action of sulfur dioxide in the air is the alarming disintegration of marble statues and buildings on the Acropolis in Athens, Greece. As all coatings have failed to protect the marble adequately, the only known solution is to bring the prized objects into air-conditioned museums.

The effects of SO_2 on vegetation are manifested as bleached spots, suppression of growth, leaf drop, and reduction in yield. Plant damage has been noticed 52 miles downwind from a smelting operation that emitted large quantities of SO_2. The SO_2, and SO_3 formed from it, enter the stomata of the leaves, H_2SO_4 is produced, and leaf tissue is hydrolyzed and dehydrated, causing yellowing and leaf drop (Figure 15–16).

There are many ways to control SO_2 pollution of the air. For ex-

Sulfur dioxide in the air is harmful to people, animals, plants, and buildings.

Figure 15–16 Rose leaves in Independence, Missouri, show marginal and interveinal necrotic injury from sulfur dioxide. (From National Air Pollution Control Administration, HEW Public Health Service, 1970. Publication No. AP-71.)

ample, tall stacks (Figure 15–17) emit the SO_2 into the higher atmosphere and away from the ground. The results of a study in Great Britain showed that tall stacks effectively reduce ground level SO_2 concentrations. During a 10-year period, the SO_2 emissions from power plants *increased* by 35 percent, but, as a result of the construction of tall stacks, the ground level concentrations *decreased* by as much as 30 percent.

Figure 15–17 World's largest chimney (as of 1972) standing 1,250 feet high (as tall as the Empire State Building) was built at a cost of $5.5 million for the Copper Cliff smelter in the Sudbury District of Ontario, Canada. (From Turk, A., et al.: *Environmental Science.* Philadelphia, W. B. Saunders Co., 1974.)

Chemical means are more efficient than high stacks in removing SO_2. One method involves heating limestone to produce lime (calcium oxide) and allowing this to react with the SO_2 to produce calcium sulfite, which can be removed from the stack by an electrostatic precipitator.

$$CaCO_3 \xrightarrow{\text{Heat}} CaO + CO_2$$
LIMESTONE   LIME

$$CaO + SO_2 \longrightarrow CaSO_3 \text{ (solid)}$$
CALCIUM SULFITE

Another efficient method involves passing the combustion gases through molten Na_2CO_3 (sodium carbonate), in which solid sodium sulfite is formed.

$$SO_2 + Na_2CO_3 \xrightarrow{800°C} Na_2SO_3 + CO_2$$

The use of sulfur-free fuels is one obvious method of eliminating SO_2 emissions (Figure 15–15). However, most low-sulfur coals are mined far from the major metropolitan areas where they are most needed for power generation.

Most of the sulfur in coal is contained in organic matter or pyrite (FeS_2, fool's gold). Pulverized coal, with the consistency of talcum powder, can be partially cleansed of pyrite by magnetic separation, electrostatic separation, froth flotation, and dry centrifugation. But none of these methods completely eliminates the sulfur dioxide emissions. Removal of the sulfur from coal can cost up to one dollar per ton.

Much of the petroleum used in the eastern United States, where SO_2 pollution is a problem, is Caribbean residual fuel oil, which has an average sulfur content of 2.6 percent. While technology has developed refining techniques for reducing the sulfur content to 0.5 percent, this procedure would increase the cost of the oil by about 35 percent. The process involves the formation of hydrogen sulfide by bubbling hydrogen through the oil in the presence of metallic catalysts (platinum-palladium, for example). Residual fuel oil produces less than 10 percent of the total utility power. In 1968, New York's Consolidated Edison switched to low-sulfur oil from Liberia and Nigeria at a cost of $3 million to change over from coal and an increase of $7.5 million in a $63 million fuel bill.

NITROGEN OXIDES

There are eight known oxides of nitrogen, two of which are recognized as important components of the atmosphere: dinitrogen oxide (N_2O) and nitrogen dioxide (NO_2).

Fixed nitrogen (nitrogen oxides are one type) is necessary to perpetuate nature's nitrogen cycle (Figure 15–18). In this respect, nitrogen oxides are useful. However, too large a quantity of nitrogen oxides in the air can lead to photochemical smog and bronchial problems. In these respects and many others, nitrogen oxides are harmful.

Most of the nitrogen oxides emitted are in the form of NO, a colorless

Fixed nitrogen is nitrogen chemically bonded to another element.

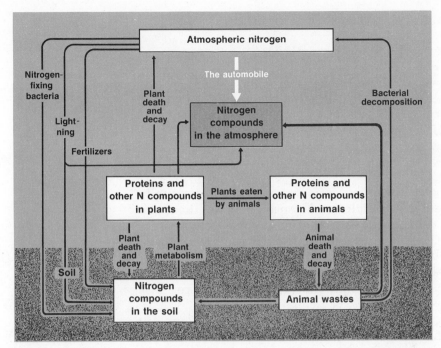

Figure 15–18 The nitrogen cycle.

reactive gas. In a combustion process involving air as the oxidant, some of the atmospheric nitrogen reacts with oxygen to produce NO:

$$N_2 + O_2 + Heat \longrightarrow 2NO$$

Nitrogen oxide is formed in this manner during electrical storms. Since the formation of nitrogen oxide requires heat, it follows that a higher combustion temperature would produce relatively more NO. This will be an important point to consider later in our discussion of the automobile and its nitrogen oxide emissions, since one way to achieve greater burning efficiency of fuels in automobile engines is to operate at higher temperatures.

In the atmosphere NO reacts rapidly with atmospheric oxygen to produce NO_2:

$$2NO + O_2 \longrightarrow 2NO_2$$
$$\text{NITROGEN DIOXIDE}$$

Nitrogen dioxide, a brown, choking gas, is a necessary component of photochemical smog (Figure 15–13). Normally the atmospheric concentration of NO_2 is a few parts per billion (ppb) or less.

If NO_2 does not react photochemically, it can react with water vapor in the air to form nitric and nitrous acids:

$$2NO_2 + H_2O \longrightarrow HNO_3 + HNO_2$$
$$\text{NITRIC ACID} \quad \text{NITROUS ACID}$$

In addition, nitrogen dioxide and oxygen yield nitric acid:

$$4NO_2 + 2H_2O + O_2 \longrightarrow 4HNO_3$$

These acids in turn can react with ammonia or metallic particles in the atmosphere to produce nitrate or nitrite salts. For example,

$$\underset{\text{AMMONIA}}{NH_3} + HNO_3 \longrightarrow \underset{\substack{\text{AMMONIUM NITRATE} \\ \text{(A SALT)}}}{NH_4NO_3}$$

Nitrates are important components of fertilizers.

The acids or the salts, or both, ultimately form aerosols, which eventually settle from the air or dissolve in raindrops. Nitrogen dioxide, then, is a primary cause of haze in urban or industrial atmospheres because of its participation in the process of aerosol formation. Normally nitrogen dioxide has a lifetime of about three days in the atmosphere.

At present, the emission of nitrogen oxides by human activities and our machines is only a minor part of that emitted by natural processes (Figure 15–18).

In laboratory studies, nitrogen dioxide in concentrations of 25 to 250 ppm inhibits plant growth and causes defoliation. The growth of tomato and bean seedlings is inhibited by 0.3 to 0.5 ppm NO_2 applied continuously for 10 to 20 days.

In a concentration of 3 ppm for 1 hour, nitrogen dioxide causes bronchioconstriction in man, and short exposures at high levels (150 to 220 ppm) produce changes in the lungs that produce fatal results. A seemingly harmless exposure one day can cause death a few days later.

CARBON MONOXIDE

Carbon monoxide (CO) is the most abundant and widely distributed air pollutant found in the atmosphere. It is produced in combustion processes when carbon or some carbon-containing compound is burned in an insufficient amount of oxygen:

The toxic effects of carbon monoxide are discussed in Chapter 13.

$$2C + \underset{\substack{\text{LIMITED} \\ \text{SUPPLY}}}{O_2} \longrightarrow 2CO$$

For every 1000 gallons of gasoline burned, 2300 pounds of CO are emitted if no emission controls are placed on the engine.

The background concentration of CO in cities is high (Figure 15–4); in heavy traffic sustained levels of 50 or more ppm are common, and instantaneous concentrations of 150 ppm and higher have been found. The variation of CO levels over a 24-hour period in an urban atmosphere is shown in Figure 15–19. However, in spite of increased emission of CO (enough to raise the concentration by 0.03 ppm yearly), the global level is not increasing at all. This is because the natural cycle dominates the pollutant emissions.

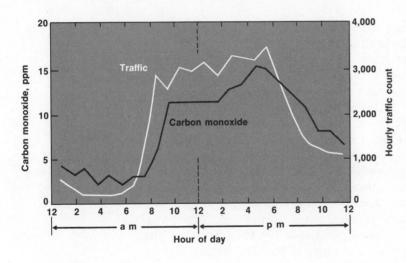

Figure 15-19 Hourly average carbon monoxide concentration and traffic count in midtown Manhattan. (From Johnson, K. L., Dworetzky, L. H., and Heller, A. N., "Carbon Monoxide and Air Pollution from Automobile Emissions in New York City." *Science,* Vol. 160, No. 3823, p. 67 [1968].)

HYDROCARBONS

Toluene

Hydrocarbons enter the atmosphere from natural sources as well as from man's activities.

Benzo(α)pyrene, a carcinogenic poly-nuclear aromatic hy-drocarbon found in smoke.

As we have seen in Chapter 7, hydrocarbons come in all shapes and sizes, beginning with methane, CH_4, and continuing to molecules containing hundreds of carbon atoms. Some have all single bonds, some have double bonds, and a few have triple bonds. Some are aromatic hydrocarbons and some are non-aromatic. Literally hundreds of these hydrocarbons and their oxygen, sulfur, nitrogen, and halogen derivatives find their way into the atmosphere.

Trees and other plants silently release turpentine, pine oil, and thousands of other hydrocarbons into the air. Bacterial decomposition of organic matter emits very large amounts of marsh gas, principally methane. We contribute our share of 15 percent (of the total global emissions; a greater quantity in urban areas) through incomplete incineration, through leakage of industrial solvents, through unburned fuel from the automobile, through evaporation of gasoline from tanks, through incomplete combustion of coal and wood, and through petroleum processing, transfer, and use.

In Chapter 13 we saw that polynuclear aromatic hydrocarbons like benzo(α)pyrene are capable of causing cancer in mice and in humans. In the late 1950's, the U.S. Public Health Service, Division of Air Pollution, surveyed 103 urban and 28 nonurban areas of the United States and found that the air in all of the 103 urban areas contained benzo(α)pyrene (BaP). Concentrations ranged from 0.11 to 61 micrograms per 1000 cubic meters of air, with the average concentration being 6.6 micrograms. In 1967, the estimated annual emission of BaP in the United States was 422 tons from burning coal, oil, and gas, 20 tons from refuse burning, 19 tons from industries (petroleum catalytic cracking, asphalt road mix, and the like), and 21 tons from motor vehicles. British researchers report that lung cancer in nonsmokers closely parallels the 10 times greater amount of BaP in city air than in rural air; there is 9 times more lung cancer in cities than in rural areas. A resident of a large town may inhale 0.20 g BaP a year. If he is a heavy smoker (two packs a day without filters), add another 0.15 g for a total of 0.35 g. This is about 40,000 times the amount of BaP necessary to produce cancer in a mouse. Coal smoke contains about 300 ppm BaP. Every million tons of coal burned in England in 1958 produced smoke laden with 750 tons of BaP. Many authorities attribute England's high lung cancer rate to this enormous production of BaP.

Other polynuclear aromatics have also shown carcinogenic activity. Particulates from the atmosphere around Los Angeles, London, Newcastle, Liverpool, and eight other urban sites were extracted with organic solvents. The extracts produced cancer in mice.

In the section on smog we discussed the role of olefins and aromatic hydrocarbons in photochemical smog formation. In a study made in Los Angeles in 1970, an average of 0.106 ppm (maximum of 0.33 ppm) aromatics was found in that city's atmosphere. (About 38 percent of the total was toluene and 40 percent the more reactive dialkyl- and trialkyl-benzenes.) These compounds are about as reactive as propylene and higher molecular weight olefins in causing smog formation. The automobile is responsible for emitting most of these hydrocarbons to the atmosphere. In fact, the automobile emits more than 200 different hydrocarbons and hydrocarbon derivatives.

Alkyl: hydrocarbon group such as ethyl, $-C_2H_5$, and octyl, $-C_8H_{17}$.

SELF-TEST 15-B

1. When sulfur is burned in air, the major product is _____.

2. The major source of sulfur dioxide is the burning of _____.

3. How does the production of electrical power rank as a producer of sulfur dioxide? _____

4. Fixed nitrogen is nitrogen combined with _____ _____.

5. The reaction of nitrogen with oxygen is (endothermic/exothermic).

6. In all combustion processes in air, some nitrogen _____ are formed.

7. Man's emission of nitrogen oxides has greatly disrupted the nitrogen cycle. (True/False)

8. Nitrogen oxides remain in the atmosphere for an indefinite length of time. (True/False)

9. Carbon monoxide is always formed when hydrocarbons are burned in a (limited/plentiful) amount of air.

10. In the human body, carbon monoxide is a formidable competitor with oxygen for _____.

11. The amount of carbon monoxide is definitely increasing in the atmosphere. (True/False)

12. A polynuclear aromatic hydrocarbon commonly found in the air and in cigarette smoke that is known to cause cancer is _____.

OZONE

Ozone is a pungent smelling gas that can be detected by the human nose at concentrations as low as 0.02 ppm. Sparking electrical appliances, lightning, and even silent electrical discharges convert oxygen into ozone, which is a more reactive form of oxygen.

$$\text{ENERGY} + 3O_2 \longrightarrow 2O_3$$
$$\text{OZONE}$$

Pure oxygen can be breathed for weeks by humans and animals without apparent injurious effects. Several studies have shown that concentrations of 0.3 to 1.0 ppm ozone, well within the recorded range of photochemical oxidant levels, after 15 minutes to 2 hours cause marked respiratory irritation accompanied by choking, coughing, and severe fatigue. For these reasons, outdoor recreation classes in Los Angeles public schools are cancelled on days when the ozone level reaches 0.35 ppm. Ozone at these levels for 1 hour depresses the body temperature, perhaps by an impairment of the brain center that regulates body temperature or by opening the pores of the skin. These levels (0.2 to 0.5 ppm) cause a considerable decrease in night vision in addition to other effects on vision.

Ozone attacks mercury and silver, which are not affected by molecular oxygen at room temperature. A typical reaction might be

$$6Ag + O_3 \longrightarrow 3Ag_2O$$

but

$$Ag + O_2 \longrightarrow \text{No reaction}$$

The stratosphere contains ozone. This "ozone layer" can be destroyed by certain man-made pollutants.

Even with all of the electrical sparks from lightning, electric motors and such, very little ozone is emitted into the air. It decomposes into molecular oxygen or reacts with other molecules too quickly to leave its source.

Ozone is found in the lower troposphere only as a secondary pollutant; that is, it is formed from other substances, as in photochemical smog (Figure 15–13). When sunlight impinges on automobile exhaust fumes, a considerable amount of ozone is produced. The stratosphere contains about 10 ppm of ozone in a layer that has the important function of filtering out some of the sun's ultraviolet light and providing an effective shield against radiation damage to living things.

Halogenated Hydrocarbons and the Ozone Layer

Most pollutants are adsorbed on surfaces or react with other pollutants in the troposphere, owing to their high concentration or high reactivity, and eventually wash out in the rain. One group of unreactive pollutants, the halogenated hydrocarbons, does not react quickly enough to be consumed in the troposphere, so they eventually mix with air in the stratosphere (see Figure 15–5). The common halogenated hydrocarbon pollutants are listed in Table 15–8.

In the stratosphere, where the ultraviolet radiation from the sun is most intense, these halogenated hydrocarbons are broken down into

TABLE 15—8   The Major Halogenated Hydrocarbons

Name	Formula	Uses	World Production (1973, millions of kg)	Emission into Atmosphere
Fluorocarbon-11	CCl_3F	Aerosol propellant	302	89%
Fluorocarbon-12	CCl_2F_2	Propellant, refrigerant	441	76%
Carbon tetrachloride	CCl_4	Making fluorocarbons	941	4%
Methyl chloroform	CH_3CCl_3	Metal cleaning	405	93%
Perchloroethylene	C_2Cl_4	Dry cleaning	743	83%

The U.S. accounts for about half the emissions of each of these chemicals. Released, they become atmospheric pollutants.

chemical radicals. The most common radical produced is the chlorine atom ($Cl \cdot$), as the breakdown of fluorocarbon-11 illustrates.

$$F-\underset{\underset{Cl}{|}}{\overset{\overset{Cl}{|}}{C}}-Cl + light \longrightarrow F-\underset{\underset{Cl}{|}}{\overset{\overset{Cl}{|}}{C}}\cdot + Cl \cdot$$

The chlorine atom then reacts with ozone (O_3), which is present in high concentration in the stratosphere.

$$O_3 + Cl \cdot \longrightarrow ClO \cdot + O_2$$

Since there are many oxygen atoms in the stratosphere because of the ozone production reaction,

$$O_2 + light \longrightarrow O + O$$
$$O + O_2 \longrightarrow O_3$$

the $ClO \cdot$ species can react with an oxygen atom to release the $Cl \cdot$ radical

$$ClO \cdot + O \longrightarrow O_2 + Cl \cdot$$

to react with another ozone molecule.

The net effect of these reactions is the destruction of ozone in the stratosphere. The total number of ozone molecules destroyed by a single chlorine atom can run into the thousands. Eventually the chlorine atom reacts with a water molecule to form HCl, which mixes into the troposphere and washes out in rain.

In 1974, M. J. Molina and F. S. Roland of the University of California predicted that the increasing use of halogenated hydrocarbons could seriously deplete the stratospheric ozone, with a corresponding increase in ultraviolet rays reaching the earth's surface. Since living things are very sensitive to ultraviolet rays (for instance, they increase skin cancer in humans), the depletion theory was judged to suggest a serious threat to our safety and health. Later research has shown that this theory is probably correct in its predictions if halogenated hydrocarbon emissions go unchecked. The EPA has now limited the emissions of these compounds.

CARBON DIOXIDE

Carbon dioxide is not usually considered a pollutant because it is a normal component of air and is directly involved in the give-and-take between animal life and plant life (a product of respiration, a reactant in photosynthesis). There is concern, however, because the concentration in the atmosphere is increasing (Figure 15–20). Studies at Scripps Institute of Oceanography have shown that the concentration of CO_2 in the atmosphere increased by 1.3 percent (or 3.7 ppm) from 1958 to 1962. Electric power plants, internal combustion engines, and the manufacture of cement are the principal technological sources of CO_2 emission, but there are numerous other sources—home heating, trash burning, and bacterial oxidation of soil humus, for example. These emissions have been entering the atmosphere faster than oceans and plants can remove them.

CO_2 is a necessary ingredient for photosynthesis.

In some ways an increase in CO_2 can be very beneficial. Experiments at Michigan State University showed an increase of 115 percent in the dry weight of lettuce and tomatoes when the CO_2 was increased from the normal 325 ppm (0.0325 percent by volume) to 800 to 2000 ppm during daylight hours. Fruit size increased and quality was enhanced; reproduction improved; tomatoes had higher vitamin C and sugar content; and plants were more resistant to some fungi, virus diseases, and insects. Plants have even responded favorably to 20,000 to 30,000 ppm CO_2 levels.

Carbon dioxide in the atmosphere produces a greenhouse effect.

A substantial increase of CO_2 in the air is potentially detrimental in two ways: it can increase the temperature of the atmosphere as well as increase the acidity of the oceans.

Carbon dioxide in the atmosphere acts similarly to the glass (or plastic) on a greenhouse. Both let the shorter wavelengths of light through but absorb the longer wavelengths as they are emitted from the surface of the earth. The carbon dioxide (and the glass) in turn emit the

Figure 15–20 Atmospheric carbon dioxide. It is estimated that doubling the concentration would increase the average world temperature by 3.8°C.

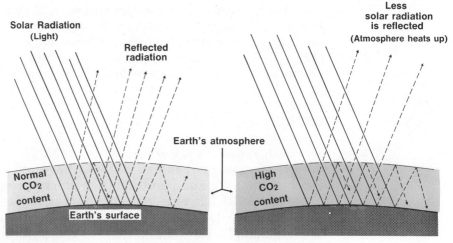

Figure 15—21 The greenhouse effect. Owing to a balance of incoming and outgoing energy in the earth's atmosphere, the mean temperature of the earth is 14.4°C (58°F). Carbon dioxide permits the passage of visible radiation from the sun to the earth but traps some of the heat radiation attempting to leave the earth.

absorbed energy as heat radiation, some of which returns to earth and some of which energizes other molecules via collisions. The net effect is an increase in temperature of the lower atmosphere, a phenomenon known as the *"greenhouse effect"* (Figure 15–21).

In contrast, solid and liquid aerosols scatter incoming sunlight of all wavelengths, thus decreasing the amount of solar energy that reaches the earth. The net effect is a cooling of the lower atmosphere. Calculations show that a 25 percent increase in aerosols (turbidity) would counteract a 100 percent increase in carbon dioxide concentration.

It is generally agreed that the global temperature increased about 0.4°C between 1880 and 1940 and then decreased by nearly 0.2°C by 1967. The CO_2 concentration is believed to have increased by 7.4 percent during a comparable interval of time—from 296 ppm in 1900 to 318 ppm in 1969. The cooling during the 1950's possibly was due in part to the particulates emitted by gigantic volcanic eruptions of Mt. Spurr in Alaska in 1953 and the Bezymyannaya (Kamchatka, U.S.S.R.) eruption in 1956. While an increasing CO_2 concentration could explain the increase in temperature and the increasing turbidity could explain the decrease in temperature, other factors may also have contributed significantly. For example, it is assumed in this explanation that solar radiation has been constant over this period of time. Actually, it may not have been; we simply do not know. (Carbon dioxide is the only substance whose global background concentration is known to be rising.)

Three mechanisms account for the removal of CO_2 from the atmosphere. These are photosynthesis, dissolving in surface water (principally the oceans), and weathering of silicate rocks. None of these seems adequate since the CO_2 level is on the rise. While a higher level of CO_2 increases plant activity and growth, it is apparent that the photosynthesis process has not picked up enough to maintain the status quo. It is known that pollutants such as SO_2 hinder photosynthesis and that much of our forest area is being cleared for housing and development. Perhaps these factors partially explain why photosynthesis cannot absorb enough CO_2. At the rates prevailing in 1962, we would need to add 2.7 billion acres of new, growing forest (a little more than the area of the United States) to absorb the CO_2 generated by the combustion of fossil fuel.

Perhaps as much as half of the CO_2 emitted by burning is absorbed by the surface waters. The oceans serve as a huge sink for carbon dioxide, which reacts with water to form carbonic acid. Carbon dioxide is acidic and sea water is slightly alkaline.

Some CO_2 is consumed by weathering silicate rocks to carbonates:

$$CaAl_2(SiO_3)_4 + 4CO_2 \longrightarrow CaCO_3 + Al_2(CO_3)_3 + 4SiO_2$$

This process is very slow and could not help to relieve a relatively sudden increase in CO_2 emissions.

When the CO_2 level increases, it does so at the expense of the oxygen supply. While this is another disadvantage of an increasing CO_2 concentration, the increase is far more detrimental than the equivalent decrease in oxygen. For example, approximately 60 percent of our oxygen need is being restored by photosynthesis within the United States. The rest comes from bacteria that reduce sulfates in anaerobic environments and from plankton upwellings in the sea. Some upwellings—containing literally millions of tons of diatoms, bacteria, and algae (called plankton)—are found off the coasts of California, Peru, Morocco, and Southwest Africa.

A major consumer of CO_2 (through photosynthesis) is phytoplankton (small ocean plants).

Carbon dioxide is not normally considered detrimental to man's health. Obviously our systems contain it at all times as a product of respiration. Prolonged exposure to about 5000 ppm CO_2 (15 times present levels), however, is considered unsafe primarily because the greater CO_2 concentration diminishes the O_2 concentration in our lungs.

THE AUTOMOBILE—A SPECIAL CASE IN AIR POLLUTION

Automobiles are the major emitters of air pollutants in the United States.

More than 100 million automobiles jam our roadways, each adding its share of pollutants to the atmosphere. Like it or not, the automobile, as we know it, is *the* major source of air pollution (see Table 15–5).

Automobile air pollutants enter the atmosphere in three major ways: from the exhaust, from the crankcase blowby (gases that escape around the piston rings), and by evaporation from the fuel tank and carburetor.

Gasoline is a mixture of hydrocarbons (isomers of octane, for example), tetraethyllead, 1,2-dibromoethane, 1,2-dichloroethane (in a molar ratio of 2 to 1 in ethyl gasoline) and various "additives" such as TCP (tricresyl phosphate). When this mixture is compressed with air in an automobile cylinder and ignited (Figure 15–23), many of the hydrocarbon molecules combine with oxygen to form CO, CO_2, and H_2O. Some hydrocarbons react with oxygen, other hydrocarbons, or both, to form a large variety of organic substances. Other hydrocarbons do not react at all.

Nitrogen combines with oxygen to form a variety of nitrogen oxides, principally NO and a little NO_2. The tetraethyllead, $Pb(C_2H_5)_4$, forms $PbClBr$, $PbCl_2$, and $PbBr_2$ by reacting with 1,2-dibromoethane and 1,2-dichloroethane. These combustion products, added to the evaporation losses of gasoline from the carburetor and gas tank, make up the noxious waste products of the automobile.

Fuel tank and carburetor evaporation
(15% of hydrocarbons)

Crankcase blowby
(20% of hydrocarbons)

Exhaust
100% of Pb
100% of CO
100% of NO
(65% of hydrocarbons)

Figure 15—22 Sources of automobile pollutant emissions. The percentages show the relative amounts of the various types of pollutants from the three major automobile sources for automobiles which have no pollution controls.

Tetraethyllead is used as a component of gasoline because it causes the mixture to burn evenly and all at once, thus raising the octane rating. When the burning of the gasoline-air mixture is too rapid or irregular, preignition occurs in the combustion chamber and causes a sharp explosion, which is heard as a "knock" in the engine. Straight-chain hydrocarbons can react with oxygen in an uncontrollably fast reaction to give the products carbon dioxide and water.

The octane rating of a fuel is numerically equal to the percent of 2,2,4-trimethylpentane (isooctane) in a mixture with heptane that has the same knocking characteristics as the fuel. The octane rating of n-heptane is 0 and the value is set at 100 for 2,2,4-trimethylpentane. $Pb(C_2H_5)_4$ added to gasoline raises its octane rating.

Ways to Control Emissions

What can be done to abate the emissions of the automobile? For one thing, governments, both state and federal, have recently passed legislation regulating automobile emissions and establishing standards. Sig-

Figure 15—23 Diagram of combustion chamber and some of the reactions that occur in it.

Spark plug

Compression chamber

Gasoline and air enter here

Waste gases

Crankcase

$O_2 + N_2 \rightarrow NO, NO_2$

$C_8H_{18} + O_2 \rightarrow CO + H_2O$
$\rightarrow CO_2$
$\rightarrow CH_3COCH_3, HCHO$
+ some unburned hydrocarbons
(and many other compounds)

TABLE 15–9   Levels of Auto Emissions in Grams per Mile

	Hydrocarbons, HC	Carbon Monoxide, CO	Nitrogen Oxides, NO_x
Prior to control	11	80	4.0
EPA standards for 1977 model cars	1.5	15.0	2.0
California standards, 1978	0.41	9.0	1.5
EPA standards for 1980 model cars*	0.41	9.0	2.0

*Proposed.
From U.S. Environmental Protection Agency.

nificant data regarding effects of governmental laws and control devices on levels of auto emissions are summarized in Table 15–9.

Two approaches are presently being developed to meet these standards: mechanical changes in automobiles and changes in fuel. Mechanical changes are made at relatively low cost, but fuel modifications offer advantages over mechanical devices. The most obvious advantage is that fuel changes would apply to all cars, old and new.

Beginning with the 1963 models, all automobiles have been equipped with a PCV (positive crankcase ventilation) valve. Fresh air is drawn through the crankcase by the vacuum system, through the PCV valve and connecting hose into the engine. This system reduces crankcase blowby exhaust. Newer models of automobiles have gas tank caps that allow air to come into the gas tank but do not allow gasoline vapors to pass out through the cap.

Major causes of hydrocarbon and CO emissions from automobiles are bad spark plugs and engines not properly timed.

Methods for reduction of hydrocarbon and carbon monoxide emissions in the exhaust are being developed along three lines: (1) the adjustment of air-fuel ratio, spark timing, and other variables; (2) the injection of air into the hot exhaust gases; and (3) the use of a catalytic converter. The first two methods have physical limits that will prevent complete conversion of hydrocarbons and carbon monoxide to carbon dioxide and water. Ideally, if the stoichiometric amount of oxygen is present, complete conversion to CO_2 is expected. For example, the combustion of a mole of octane requires 12.5 moles of oxygen:

$$C_8H_{18} + 12.5O_2 \longrightarrow 8CO_2 + 9H_2O + \text{heat}$$

Note that 13.5 molecules of reactants produce 17 molecules of products (or 27 produce 34). The heat produced and the increased number of molecules cause the expansion that pushes the piston and thereby does the work that drives the car.

But the time for reaction in an automobile engine is so short that the octane and oxygen molecules cannot react completely. In the chaotic split second of ignition and reaction, many undesirable side reactions occur. These reactions produce the 200 or so hydrocarbons found in an automobile's exhaust that were not originally in the gasoline.

Adjustment of the air-fuel ratio can reduce CO and hydrocarbon emissions, but this can mean loss of power. For example, it may be seen in Figure 15–24 that maximum power is obtained when the air-fuel ratio is 12.5:1. As shown in Figure 15–25, this ratio does not give the lowest emissions of CO and hydrocarbons. A ratio of 15.1:1 would produce about eight

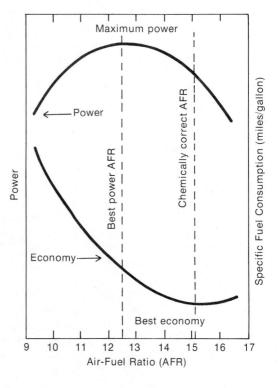

Figure 15—24 The effect of air-fuel ratio on power and economy for a particular engine. (From Wildeman, T. R.: "The Automobile and Air Pollution." *Journal of Chemical Education,* Vol. 51, No. 5, p. 290, 1974.)

times less CO (than an air-fuel ratio of 12.5:1) and slightly less hydrocarbons. Therefore, the 15.1:1 air-fuel ratio would give a more efficient use of the fuel—more complete burning. Why don't the air-fuel ratios for maximum power and most efficient use of fuel match? The answer is well-known to automotive engineers. All of the oxygen (not all of the fuel) must be consumed to produce maximum power. Part of the reason for this is that too much oxygen produces more NO_x, with a resultant loss of power.

Auto carburetors vaporize the fuel and mix it with air.

Figure 15—25 Exhaust gas composition versus air-fuel ratio for a spark-ignition engine with no pollution controls. The chemically correct air-fuel ratio is indicated by the colored arrow. (From Wildeman, T. R.: "The Automobile and Air Pollution," *Journal of Chemical Education,* Vol. 51, No. 5, p. 291, 1974.)

Endothermic:
$$N_2 + O_2 + 4.32 \text{ kcal}$$
$$\longrightarrow 2NO$$

Exothermic:
$$C + O_2 \longrightarrow CO_2$$
$$+ 94.1 \text{ kcal}$$

The formation of nitrogen oxides deducts heat energy that could be used for expansion. This reduces power.

The nitrogen oxide problem is simple to understand but hard to correct. Reactions of hydrocarbons and CO with oxygen are exothermic. Reactions of nitrogen with oxygen to form nitrogen oxides are endothermic. Therefore, if additional air is provided to burn the unburned hydrocarbons and CO, the heat produced provides the energy for the formation of nitrogen oxides. Thus, in the process of reducing the amounts of CO and hydrocarbons, nitrogen oxides are increased. Reducing the air from a ratio of 15.1:1 to a ratio of 12.5:1 can cause up to a 60% reduction in nitrogen oxide emissions.

Less nitrogen is oxidized to oxides of nitrogen if the temperature within the combustion chamber is reduced to the lowest practical limit. One way to do this is to recycle part of the exhaust gas back into the

Figure 15—26 Automotive hardware for pollution control.

Figure 15—27 Cutaway view of catalytic exhaust muffler showing catalyst pellets. (From Turk, A., et al.: *Environmental Science*. Philadelphia, W. B. Saunders Co., 1974.)

engine. This reduces the peak combustion temperature and the amount of nitric oxide that forms. Prototypes of such recycling systems have reduced nitrogen oxides emissions by 80 percent without causing the exhaust to exceed 275 ppm hydrocarbons and 1.5 percent CO. The recirculation of about 15 percent of the exhaust, with the reduction of nitrogen oxides by 80 percent, is accompanied by a cut in power output by 16 percent and a decrease in fuel economy by 15 percent. Changes in timing can increase power and economy, but the amount of nitrogen oxides produced also increases. Fuel injected specifically at the spark plug gap will produce low hydrocarbon, CO, and NO_x emissions, but then particulate emissions become a problem.

The amount of CO and unburned hydrocarbons in the exhaust can be reduced by injecting air into the hot gases just after they leave the combustion chamber—that is, at the exhaust valve. This is known as the "afterburner" technique; it does not affect the piston and drive train of the automobile. Some of the unburned hydrocarbons and carbon monoxide react with oxygen in the injected air and complete the combustion *outside* the power chamber. Although this reduces hydrocarbon and CO emissions, heat is evolved that must be dissipated without gaining any additional power to run the automobile. It also increases the nitrogen oxide problem.

Considerable research is being done to develop catalysts for converting CO, hydrocarbons, and NO_x to less harmful products. For example, in a laboratory mixture of 5 percent NO, 5 percent CO, and 90 percent helium run over a barium-promoted copper chromite catalyst at 200°C, 92 percent of the NO and CO was converted to N_2 and CO_2, neither of which is considered a pollutant in the ordinary sense. At 350°C, 98 percent conversion is effected.

$$2CO + 2NO \xrightarrow{\text{Ba-CuCrO}_2} 2CO_2 + N_2$$

Using the exhaust from a 1947 Ford V-8 engine, complete reaction to CO_2 and N_2 can be achieved with a copper chromite catalyst if the temperature is between 370° and 460°C and if the reactants are in the stoichiometric amounts—a most important condition. The perfect balance is difficult to achieve because an engine produces the most nitric oxide under conditions that make very little carbon monoxide (high temperatures, much oxygen). Furthermore, if too much oxygen is present in the exhaust, CO reacts with O_2 in preference to NO.

All post-1975 cars equipped with catalytic mufflers use a platinum-based catalyst that is inactivated by lead. For this reason they must use lead-free gasoline.

New Kinds of Cars?

Several thousand automobiles in this country are now burning natural gas instead of gasoline. Natural gas, which is mostly methane, CH_4, burns cleanly to CO_2 and water. A regulating device, positioned over

Batteries

Figure 15—28 Experimental electric car. Powered by conventional lead-acid batteries, it has a range of 50 miles. Lighter, more energetic batteries would increase its range significantly. Present-day batteries account for 30 percent of the total weight of the car.

Motor control

Coaxial D.C. motor, planetary gear and differential

the carburetor, costs about $350. The problem here is that our surplus of natural gas in this country is rapidly diminishing, a potential long-range problem.

At the turn of the century, a Baker Electric Coupe cost $2,600 and a Borland Electric Delux could be bought for $5,500.

One of the very few ways to eliminate combustion processes is to use electricity in some way to power our transportation devices. Early attempts to develop electric cars in the United States were made in Boston in 1888, 28 years after the development of the first storage battery in 1860. By 1912, about 6000 electric passenger and 4000 commercial vehicles were being manufactured annually in this country. The electric car lost out in competition with the internal combustion engine during the 1920's. The problems of short range (about 20 miles), low speeds (20 miles per hour maximum), 8 to 12 hours recharge time for the batteries, and relatively high price combined to eliminate the electric car from the race for leader in transportation.

At present, more than 100,000 rider-type, battery powered, materials-handling vehicles are operating in U.S. plants and warehouses where it is vital to avoid air pollution from internal combustion engines. Despite this positive start, attempts to employ electric vehicles for street and highway use generally have been commercial failures. The energy storage capacities of conventional batteries (lead-acid, nickel-iron, and silver-zinc) are too limited or expensive to provide an acceptable energy source for electric passenger cars (Figure 15–28). New battery designs may prove competitive for short-range urban travel.

Other developments currently either coming on the market or being seriously considered include the stratified charge engine and the turbine-generator engine. These relatively new inventions are reported to be improvements in transportation, fuel economy, and pollution control.

SELF-TEST 15-C

1. How many carbon atoms are in the most plentiful hydrocarbon molecule in gasoline? _____

2. The purpose of tetraethyllead in gasoline is to prevent _____.

3. What is the source of nitrogen used to form oxides of nitrogen in auto exhaust? _____

4. "Ethyl" gasoline necessarily contains _____.

5. What element found in "ethyl" gasoline poisons the catalysts in a catalytic muffler? _____.

6. Gasoline having the same knocking characteristics as a mixture of 7% n-heptane and 93% 2,2,4-trimethylpentane has an octane rating of _____.

7. In the process of adding more air to burn the residual CO and hydrocarbons, are more nitrogen oxides formed? _____

8. If all of the fuel is burned in a regular automobile engine, maximum power is produced. (True/False)

9. Injecting air into the hot exhaust gases reduces the amount of _____ and _____ but increases the amount of _____.

10. A metal catalyst that is capable of helping to convert CO and unburned hydrocarbons into CO_2 and H_2O is _____ (specific metal).

WHAT DOES THE FUTURE HOLD?

There will undoubtedly be an abatement of air pollution in the future; the sheer pressures of population increase will demand it. But life also will undoubtedly have to be different. Perhaps the first major change will be the disappearance of the automobile from the city, followed by a gradual modification of the power plant of the automobile until it is relatively nonpolluting. One interesting effect of the lower legal speed limit on our nation's highways has been the reduction of NO_x emissions due to the lower operating temperatures of the auto engines.

Most pollution exists because we demand the benefits of a technology that, for the most part, has given little consideration to the long-range effects of its products. When industry, automobile manufacturers, or power plant operators add equipment to stop noxious waste products from getting into the air, the costs are added to the already considerable manufacturing expense without adding one cent to the market value of the product being made. The cost to the consumer, however, will go up

At this time, it is uncertain whether the knowledge of the harmful, long-range effects of air pollution will bring people to the point where they are willing to give up the immediate activities that give rise to it. No one seems willing to use less energy or to give up the automobile.

and will be reflected in the increased price of consumer goods, automobiles, and electric power rates. This is a high price, of course, but what is the value of clean air?

In the final analysis, we all pollute the atmosphere, and much of the pollution is due to the misapplication of chemical techniques. Although we are becoming aware of the problems and although it is within the capabilities of chemical technology to eradicate most forms of air pollution, the process will be very slow.

(Use each choice no more than once)

MATCHING SET

_____ 1. smog

(a) heat absorbing

_____ 2. primary pollutant

(b) prevents lead from fouling spark plugs

_____ 3. secondary pollutant

(c) mixture of fog, SO_2, and (or) hydrocarbons and NO_2

_____ 4. aerosol

(d) parts per million

_____ 5. micron

(e) disease of lungs

_____ 6. ppm

(f) hydrocarbon with double bond

_____ 7. chemical radical

(g) eradication

_____ 8. olefin

(h) CO

_____ 9. photochemical

(i) molecular-sized species with an unpaired valence electron

_____ 10. abatement

(j) destroys stratospheric ozone

_____ 11. emphysema

(k) process of decreasing

_____ 12. brown gas

(l) intermediate in size between individual molecules and particulates

_____ 13. endothermic

(m) benzo(α)pyrene

_____ 14. polynuclear hydrocarbon

(n) peroxy acyl nitrate

_____ 15. chlorofluorocarbon

(o) chemical reaction energized by light

_____ 16. 1,2-dibromoethane

(p) 10^{-6} meter

(q) NO_2

QUESTIONS
1. The formation of photochemical smog involves very reactive chemical radicals. What structural feature makes these radicals reactive?

2. Write a balanced chemical equation for the burning of iron pyrite (FeS_2) in coal to sulfur dioxide and Fe_2O_3, iron(III) oxide.

3. What conditions are necessary for thermal inversion?

4. What are the basic chemical differences between industrial smog and Los Angeles-type smog?

5. What are the major sources of the following pollutants?

    (a) carbon monoxide
    (b) sulfur dioxide
    (c) nitrogen oxides
    (d) ozone

6. What is a photochemical reaction? Give an example.

7. If air pollutants rise from the earth into the atmosphere, why do they not continue on into space?

8. Why is it so difficult to avoid the formation of either carbon monoxide or nitrogen oxides in the combustion process in an automobile engine?

9. What effects does weather have on local air pollution problems? on regional air pollution problems?

10. Knowing the chemistry of photochemical smog formation, list some ways to prevent its occurrence.

11. What part do aerosols play in the formation of smogs?

12. Describe the antipollution devices on your car.

13. Describe an air pollution problem in your community. How can this problem be solved?

14. Discuss the merits of abatement versus eradication of air pollution.

15. Of the following air pollutants—particulates, sulfur dioxide, carbon monoxide, ozone, and nitrogen dioxide—

    (a) which is normally a secondary pollutant?
    (b) which is emitted almost exclusively by human-controlled sources?
    (c) which can be removed from emissions by centrifugal separators?
    (d) which two react with water to form acids?

16. Define:

    (a) micron
    (b) ppm

17. What compound gives "ethyl" gasoline its name?

18. What two products are produced by the perfect combustion of hydrocarbons in gasoline?

19. Which are more effective in producing smog, paraffinic (all single bonds) or olefinic (some double bonds) hydrocarbons? Why?

20. Why is natural gas a good substitute for oil and coal as far as air pollution is concerned? What problem is related to its substitution?

21. What is the approximate concentration of air pollutants in the atmosphere during smoggy conditions?

22. Describe an effective way of eliminating carbon monoxide from the exhaust of an automobile.

23. What is the chief source of air pollution in the United States?

CHAPTER 16

CONSUMER CHEMISTRY

A major portion of the household budget is spent on consumer products. We buy chemical products to feed, cleanse, disinfect, wax, deodorize, paint, remove spots, relieve pain, cure disease, fertilize, preserve, destroy, and do hundreds of other things for us. You already know about the undesirable qualities of certain chemicals as pollutants (Chapters 14 and 15) and as toxic substances (Chapter 13). You are also familiar with many of the chemical products that make our lives more pleasant, healthy, and convenient, provided we use them in the proper amounts and for their intended purposes. You are aware that some, if not all, consumer products ultimately become pollutants, and it is a question of values as to how we use them in our society.

Consideration of consumer products must involve disposal as well as preparation and use.

To reap the full benefit of the products we purchase, we should know the types of raw materials in the products, the ways in which products perform their jobs, and the precautions to be taken when using the products. Many formulations under different brand names are essentially the same and cost about the same to produce. Competitive products vie for sales through the cleverness of the name, the attractiveness of the container, and the effectiveness of advertisements as well as the performance of the product. The selling costs are generally added to the price without, of course, increasing the true value of the product. When you have some knowledge of the chemistry of the product, the small print on the label becomes important. A comparison of lists of ingredients sometimes uncovers better formulations, and even harmful substances. In this way, chemical knowledge can then protect the consumer.

Synergism—cooperative action so that the total effect is greater than the sum of the parts.

Most consumer products are mixtures, but their properties are not necessarily blends of individual properties of the components. In some formulations, the individual properties of certain chemicals are reinforced and enhanced by other chemicals in the mixture. This is called *synergism,* defined as the cooperative action of discrete agents such that the total effect is greater than the sum of the effects of each used alone. Citric acid has very little antioxidant effect on foods; butylated hydroxyanisole (BHA) has considerable effect on preventing oxidation of foods. When these two substances are used together in foods, the antioxidative powers of the BHA are increased several-fold. The citric acid is said to have synergistic action on BHA.

413

A closely related term is ***potentiation.*** Potentiators do not have a particular effect themselves but exaggerate the effect in other chemicals. The 5′-nucleotides, for example, have no taste, but they enhance the flavor of meat or the effectiveness of salt.

We cannot discuss all consumer products in this chapter, but a few topics will be explained at the molecular level. You can, if you wish, extend this approach to understand the products you use and make wise choices in your selection of them.

PART I—CHEMICALS IN FOODS

Preservation of Foods

Foods generally lose their usefulness and appeal a short time after harvest. Bacterial decomposition and oxidation are the prime reasons steps must be taken to lengthen the time that a foodstuff remains edible. Any process that prevents the growth of microorganisms and/or retards oxidation is generally an effective preservative process for food. Perhaps the oldest technique is the drying of grains, fruits, fish and meat. Water is necessary for the growth and metabolism of microorganisms, and it is also important in oxidation. Dryness then thwarts both the oxidation of food and the microorganisms that feed on it.

Dry foods tend to be stable.

Food additives include monosodium glutamate, benzoic acid, vitamins, butylated hydroxytoluene. Food additives prevent oxidation, add flavor, texture and color, kill bacteria, increase vitamin content, etc.

Figure 16—1 Between the harvested and the consumer-ready food one often finds the addition of a large variety of food additives.

A hypertonic solution is more concentrated than solutions in its immediate environment.

Osmosis is the flow of water from a more dilute solution through a membrane into a more concentrated solution.

Chemicals may also be added as preservatives. Salted meat, and fruit preserved in a concentrated sugar solution, are protected from microorganisms. The abundance of sodium chloride or sucrose in the immediate environment of the microorganisms forms a *hypertonic* condition in which water flows by *osmosis* from the microorganism to its environment. The salt and sucrose have the same effect on the microorganism as does dryness. Both dehydrate the microorganism.

The canning process for preserving food was developed around 1810, and involves first heating the food to kill all bacteria and then sealing it in bottles or cans to prevent access of other microorganisms and oxygen. Some canned meat has been successfully preserved for over a century. Newer techniques for the preservation of food include vacuum freezing, pasteurization, cold storage, irradiation, and chemical preservation.

Antimicrobial Preservatives

Food spoilage caused by microorganisms is a result of the excretion of toxins. A preservative is effective if it prevents multiplication of the microbes during the shelf-life of the product. Sterilization by heat or radiation, or inactivation by freezing, is often undesirable since the quality of the food is impaired. Chemical agents seldom achieve sterile conditions but can preserve foods for considerable lengths of time.

A preservative must interfere with microbes but be harmless to the human system—a delicate balance.

Antimicrobial preservatives are widely used in a large variety of foods. For example, in the United States sodium benzoate is permitted in nonalcoholic beverages and in some fruit juices, fountain syrups, margarines, pickles, relishes, olives, salads, pie fillings, jams, jellies, and preserves. Sodium propionate is legal in bread, chocolate products, cheese, pie crust, and fillings. Depending on the food, the weight of the preservative permitted ranges up to a maximum of 0.1 percent for sodium benzoate and 0.3 percent for sodium propionate.

SODIUM BENZOATE   SODIUM PROPIONATE

Postulated mechanisms for the action of food preservatives may be grouped into three categories: (1) interference with the permeability of cell membranes of the microbes for foodstuffs, so the bacteria die of starvation; (2) interference with genetic mechanisms so the reproduction processes are hindered; and (3) interference with intracellular enzyme activity so that metabolic processes such as the Krebs cycle cease.

Atmospheric Oxidation

Microbial activity results in oxidative decay of food, but it is not the only means of oxidizing food. The direct action of oxygen in the air, *atmospheric oxidation,* is the chief factor in destroying fats and fatty portions of foods. Chemically, oxygen reacts with the fat to form a hydro-peroxide (R—OOH).

PORTION OF AN UNSATURATED   HYDROPEROXIDE
FAT MOLECULE

The mechanism involves the formation of very reactive free radicals (species with one or more unpaired electrons) in a chain reaction process. For example:

Free radicals usually have short lives; in rare cases they have a stable structure.

$$RH + O_2 \longrightarrow R\cdot + HO_2\cdot$$

FAT   FREE RADICALS

$$R\cdot + O_2 \longrightarrow ROO\cdot$$

$$ROO\cdot + RH \longrightarrow ROOH + R\cdot$$

HYDROPEROXIDE

Foods kept wrapped, cold, and dry are relatively free of oxidation. An antioxidant added to the food can also hinder oxidation. Antioxidants most commonly used in edible products contain various combinations of butylated hydroxyanisole (BHA), butylated hydroxytoluene (BHT), or propyl gallate:

BHA

BHT   PROPYL GALLATE

To prevent the oxidation of fats, the antioxidant can donate the hydrogen atom in the —OH group to the free radicals and stop the chain reactions. The bulky aromatic radicals formed are relatively stable and unreactive; they add the unpaired electrons to their supply of delocalized electrons:

$$R \cdot + (CH_3)_3C-\underset{\underset{CH_3}{\overset{\overset{OH}{|}}{\bigcirc}}}{}-C(CH_3)_3 \longrightarrow RH + (CH_3)_3C-\underset{\underset{CH_3}{\overset{\overset{\dot{O}}{|}}{\bigcirc}}}{}-C(CH_3)_3$$

ANTIOXIDANT   STABLE FREE RADICAL WITH LITTLE TENDENCY TO REACT

If antioxidants are not present, the hydroperoxy group will attack a double bond. This reaction leads to a complex mixture of volatile aldehydes, ketones, and acids, which cause the odor and taste of rancid fat.

Sequestrants

Metals get into food from the soil and from machinery during harvesting and processing. Copper, iron, and nickel, and their ions, catalyze the oxidation of fats. However, a molecule of citric acid bonds with the metal ion, thereby rendering it ineffective as a catalyst. With the competitor metal ions tied up, antioxidants such as BHA and BHT can accomplish their task much more effectively.

Citric acid belongs to a class of food additives known as ***sequestrants.*** For the most part sequestrants react with trace metals in foods, tying them up in complexes so the metals will not catalyze the decomposition or oxidation of food. Sequestrants like sodium and calcium salts of EDTA (ethylenediaminetetraacetic acid) are permitted in beverages, cooked crab meat, salad dressing, shortening, lard, soup, cheese, vegetable oils, pudding mixes, vinegar, confectioneries, margarine, and other foods. The amount ranges from 0.0025 to 0.15 percent. The structural formula of EDTA bonded to a metal ion is shown in Figure 16–2.

To sequester means "to withdraw from use." The sequestering ability of EDTA accounts for its use in treating heavy metal poisoning (Chapter 13).

Figure 16–2 The structural formula for the metal chelate of ethylenediaminetetraacetic acid (EDTA).

Flavor in Foods

Flavors result from a complex mixture of volatile chemicals. Since we have only four tastes (sweet, sour, salt, bitter), much of the sensation of taste in food is smell. For example, the flavor of coffee is determined largely by its aroma, and this in turn is due to a very complex mixture of over 100 compounds, mostly volatile oils.

Most flavor additives originally came from plants. The plants are crushed, and the compound is extracted with various solvents such as ethanol or carbon tetrachloride. Sometimes a single compound is extracted; more often, a mixture of several compounds occurs in the residue. By repeated efforts, relatively pure oils are obtained. Oils of wintergreen, peppermint, orange, lemon, and ginger, among others, are still obtained this way. These oils, alone or in combination, are then added to foods to obtain the desired flavor. Gradually, analyses of the oils and flavor components of plants have revealed the active compounds responsible for the flavor. Today, the synthetic preparation of the same flavors actively competes with natural sources.

The Food and Drug Administration has recently banned some of the naturally occurring flavoring agents that were formerly used, including safrole, the primary root beer flavor, found in the root of the sassafras tree.

Safrole

Flavor Enhancers

Flavor enhancers have little or no taste of their own but amplify the flavors of other substances. They exert synergistic and potentiation effects. Potentiators were first used in meats and fish. Now they are also used to intensify the flavor and cover unwanted flavors in vegetables,

Figure 16–3 A stereochemical interpretation of the sensation of smell. The substance fits a cavity in the back of the oral cavity. If the atoms are properly spaced, they sensitize nerve endings which transmit impulses to the brain. The brain identifies these sensations as a particular smell. A complete explanation of smell is certainly more involved than the simple idea presented here.

O   CH₃
OH
O

Maltol
(from pine needles)

bread, cakes, fruits, nuts, and beverages. Three common flavor enhancers are *monosodium glutamate (MSG), 5'-nucleotides* (similar to inosinic acid; Chapter 10), and *maltol.*

MSG is a natural constituent of many foods, such as tomatoes and mushrooms.

In some people, MSG causes the so-called "Chinese restaurant syndrome," an unpleasant reaction that includes headaches, sweating, and other symptoms usually occurring after an MSG-rich Chinese meal. Tomatoes and strawberries affect some individuals in the same way.

O
C—CH₂CH₂C—C
HO   O⁻Na⁺
NH₂

MONOSODIUM GLUTAMATE

OH
N   N
N   N
O
HO—P—O—CH₂
OH   O
H   H
H   H
OH   OH

INOSINIC ACID
(A 5'-nucleotide)

Brain damage is caused when MSG is injected in very high dosage under the skin in 10-day-old mice. When these laboratory results were reported, considerable discussion ensued concerning the merits of MSG. National investigative councils have suggested that it be removed from baby foods since infants do not seem to appreciate enhanced flavor. However, in the absence of hard evidence that MSG is harmful in the amounts used in regular food, no recommendations were made relative to its use.

1. Butylated hydroxyanisole plus citric acid in food is an example of (synergism/potentiation) _____.   **SELF-TEST 16-A**

2. Name two of the oldest means for preserving food. _____

3. Antimicrobial preservatives make foods sterile. (True/False)

4. Flavors result from (volatile/non-volatile) compounds. _____

5. Antioxidants are (more/less) easily oxidized than the food into which they are placed. _____

6. Citric acid is an example of a(n) _____.
Such compounds tie up metals in stable complexes.

7. A flavor in a food can usually be traced to a single compound. (True/False)

8. Monosodium glutamate is a(n) _____ _____.

9. Salt is effective in preserving foods because it kills microorganisms by _____ them.

10. BHT serves as an antioxidant by destroying _____.

Sweetness

Sweetness is characteristic of a wide range of compounds, many of which are completely unrelated to sugars. Lead acetate, $Pb(CH_3COO)_2$, is sweet but poisonous. A number of *artificial sweeteners* are allowed in foods. These are primarily used for special diets such as those of diabetics. Artificial sweeteners have no known metabolic use in the body and do not require insulin.

Insulin is a hormone that regulates glucose metabolism.

Saccharin

The most common artificial sweetener is saccharin.

SACCHARIN

Saccharin is about 300 times sweeter than ordinary sugar (sucrose). When ingested, saccharin passes through the body unchanged. It therefore has no food value other than to render an otherwise bland mixture more tasty. Saccharin has a somewhat bitter aftertaste, which renders it unpleasant to some users. Glycine, the simplest amino acid, which is also sweet-tasting, is often added to counteract this bitter taste.

Saccharin is a synthetic chemical.

Laboratory studies have shown that high doses of saccharin cause cancer in mice. After months of consideration, the Institute of Medicine of the National Academy of Science joined the Food and Drug Administration in its 1978 statement that saccharin should be banned in U.S. foods.

Cyclamates

The sweetness of cyclamate salts was discovered accidentally. Sodium cyclamate is about 30 times sweeter than sugar. Because a cyclamate does not have the aftertaste of saccharin, and because it could be used in cooked or baked products, it rapidly replaced saccharin in a wide variety of dietary products.

SODIUM CYCLAMATE

In 1969, experimental evidence indicated that cyclamates, under certain conditions, caused cancer of the bladder in laboratory mice. When this evidence was coupled with the fact that 20 percent of all cyclamate users metabolized cyclamate to a metabolite that caused chromosome breakage in rat cells, the Food and Drug Administration banned cyclamates from foods.

Aspartame

Aspartame, a new entry into the sweetener market, was approved by the FDA in 1974 and subsequently withdrawn by its maker, G. D. Searle Co., when toxicity questions were raised. Aspartame is about 180 times sweeter than table sugar (sucrose). The caloric value of aspartame is similar to that of proteins. It is metabolized in the body as a peptide. The caloric intake of consumers using the product is greatly reduced, since much smaller amounts are needed to produce the same sweetening effect. Aspartame does not have the bitter aftertaste associated with other artificial sweeteners.

ASPARTAME

At present, there is no generally acceptable artificial sweetener for foods.

Food and Esthetic Appeal

Food Colors

Colored organic substances often are *conjugated* molecules, having alternating double and single bonds in the carbon chain or ring.

Food colors are generally large organic molecules having several double bonds and aromatic rings. These conjugated structures can absorb only certain wavelengths of light and pass the rest; this behavior gives the substance its characteristic color. β-Carotene is an orange-red substance that occurs in a variety of plants, the carrot in particular, and is commonly used as a food color. It has a conjugated system of delocalized electrons. Most of the other food colors have similar conjugated systems.

β-Carotene is a *precursor* of vitamin A. (It is changed into vitamin A in the body.)

β-CAROTENE

pH Control

Weak organic acids are added to such foods as cheese, beverages, and dressings to give a mild acidic taste. They often mask undesirable aftertastes. Weak acids and acid salts, such as tartaric acid and potassium acid tartrate, react with bicarbonate to form CO_2 in the baking process.

Some acid additives control the pH of food during various stages of processing as well as of the finished product. In addition to single substances, there are several combinations of substances that will adjust and then maintain a desired pH. These mixtures are called **buffers.** An example of one type of buffer is potassium acid tartrate, $KHC_4H_4O_6$.

Buffer solutions resist change in acidity and basicity; pH remains constant.

Adjustment of the pH of a fruit juice is allowed by the FDA. If the pH of the fruit is too high, it is permissible to add acid (called an **acidulant**). Citric acid and lactic acid are the most common acidulants used since they are believed to impart good flavor; but phosphoric, tartaric, and malic acids are also used. These acids are often added at the end of the cooking time to prevent extensive hydrolysis of the sugar. In the making of jelly they are sometimes mixed with the hot product immediately after pouring. To raise the pH of a fruit that is unusually acidic, buffer salts such as sodium citrate or sodium potassium tartrate are used.

Small amounts of certain acids are allowed to be added to some foods.

The versatile acidulants also function as preservatives to prevent growth of microorganisms, as synergists and antioxidants to prevent rancidity and browning, as viscosity modifiers in dough, and as melting point modifiers in such food products as cheese spreads and hard candy.

Anticaking Agents

Anticaking agents are added to hygroscopic foods—in amounts of 1 percent or less—to prevent caking in humid weather. Table salt (sodium chloride) is particularly subject to caking unless an anticaking agent is present. The additive (magnesium silicate, for example) incorporates water into its structure as water of hydration and does not appear wet as sodium chloride does when it absorbs water physically on the surface of its crystals. As a result, the anticaking agent keeps the surface of sodium chloride crystals dry and prevents crystal surfaces from co-dissolving, which would join the crystals together.

Hygroscopic substances absorb moisture from the air.

Stabilizers and Thickeners

Stabilizers and thickeners improve the texture and blends of foods. The action of carrageenin (a polymer from edible sea weed) is shown in Figure 16–4. Most of this group of food additives are polysaccharides (Chapter 10) having numerous hydroxyl groups as a part of their structure. The hydroxyl groups form hydrogen bonds with water to prevent the segregation of water from the less polar fats in the food, and to provide a more even blend of the water and oils throughout the food. Stabilizers and thickeners are particularly effective in icings, frozen desserts, salad dressing, whipped cream, confectionery, and cheeses.

Stabilizers and thickeners are types of emulsifying agents.

Surface Active Agents

Surface active agents are similar to stabilizers, thickeners, and detergents in their chemical action. They cause two or more normally incom-

Figure 16—4 The action of carrageenin to stabilize an emulsion of water and oil in salad dressing. An active part of carrageenin is a polysaccharide, a portion of which is shown above. The carrageenin hydrogen-bonds to the water, which keeps it dispersed. The oil, not being very cohesive, disperses throughout the structure of the polysaccharide. Gelatin (a protein) undergoes similar action in absorbing and distributing water to prevent ice crystals in ice cream.

patible (nonpolar and polar) chemicals to disperse in each other. If the chemicals are liquids, the surface active agent is called an *emulsifier.* If the surface active agent has a sufficient supply of hydroxyl groups, such as cholic acid has, the groups form hydrogen bonds to water. Cholic acid and its associated group of water molecules are distributed throughout dried egg yolk in a manner quite similar to that of carrageenin and water in salad dressing.

Some surface active agents have both hydroxyl groups and a relatively long nonpolar hydrocarbon end. Examples are diglycerides of fatty acids, polysorbate 80, and sorbitan monostearate. The hydroxyl groups on one end of the molecule are anchored via hydrogen bonds in the water, and the nonpolar end is held by the nonpolar oils or other substance in the food. This provides tiny islands of water held to oil. These islands are distributed evenly throughout the food.

Hydrogen bonding plays a major role in stabilizers, thickeners, surface active agents, and humectants.

Polyhydric Alcohols

Polyhydric alcohols are allowed in foods as humectants, sweetness controllers, dietary agents, and softening agents. Their chemical action is based on their multiplicity of hydroxyl groups that hydrogen-bond to water. This holds water in the food, softens it, and keeps it from drying out. Tobacco is also kept moist by the addition of polyhydric alcohols such as glycerol. An added feature of polyhydric alcohols is their sweetness. Two particularly effective alcohols added to sweeten sugarless chewing gum are mannitol and sorbitol. Compare the structures of these alcohols with the structure of glucose presented in Chapter 10. It is not surprising that the very similar structures of these polyhydric alcohols and the isomers of glucose produce a similar taste sensation.

Kitchen Chemistry

Leavened Bread

Leavened bread is as old as recorded history.

Sometimes cooking causes a chemical reaction that releases carbon dioxide gas. The carbon dioxide causes breads and pastries to rise. Yeast has been used since ancient times to make bread rise, and remains of bread made with yeast have been found in Egyptian tombs and the ruins of

Pompeii. The metabolic processes of the yeast furnish gaseous carbon dioxide, which creates bubbles in the bread and makes it rise:

$$C_6H_{12}O_6 \xrightarrow[\text{from yeast}]{\text{Zymase}} 2CO_2 + 2C_2H_5OH$$

GLUCOSE   (GAS)   ETHANOL

When the bread is baked, the CO_2 expands even more to produce a light, airy loaf.

Carbon dioxide can be generated in cooking by other processes. For example, baking soda (which is simply sodium bicarbonate, $NaHCO_3$, a base) can react with acidic ingredients in a batter to produce CO_2.

$$NaHCO_3 + H^+ \longrightarrow Na^+ + H_2O + CO_2(gas)$$

Baking powders contain sodium bicarbonate and an added acid salt or a salt which hydrolyzes to produce an acid. Some of the compounds used for this purpose are potassium hydrogen tartrate, $KHC_4H_4O_6$, calcium dihydrogen phosphate monohydrate, $Ca(H_2PO_4)_2 \cdot H_2O$, and sodium acid pyrophosphate, $Na_2H_2P_2O_7$. The reactions of these white, powdery salts with sodium bicarbonate are similar, although the compounds all have somewhat different appearances. For example:

$$KHC_4H_4O_6 + NaHCO_3 \xrightarrow{\text{Water}} KNaC_4H_4O_6 + H_2O + CO_2(gas)$$

Cooking and Precooking—"Preliminary Digestion"

The cooking process involves the partial breakdown of proteins or carbohydrates by means of heat and hydrolysis (Figure 16–5). The polymers that must be degraded if cooking is to be effective are the carbohydrate cellular wall materials in vegetables and the collagen or connective tissues in meats. Both types of polymers are subject to hydrolysis in hot water or moist heat. In either case, only partial depolymerization is required.

In recent years several precooking additives have become popular; the **meat tenderizers** are a good example. These are simply enzymes that catalyze the breaking of peptide bonds in proteins via hydrolysis at room temperature. As a consequence, the same degree of "cooking" can be obtained in a much shorter heating time. Meat tenderizers are usually plant products such as papain, a proteolytic (protein-splitting) enzyme from the unripe fruit of the papaw tree. Papain has considerable effect on connective tissue, mainly collagen and elastin, and shows some action on muscle

Cooking starts the digestive process, although it is rarely needed for this purpose. Since it may destroy nutrients, it is done mostly for pleasure.

Figure 16–5 The hydrolysis of starch during the cooking of foods such as potatoes and rice.

fiber proteins. On the other hand, microbial protease enzymes (from bacteria, fungi, or both) have considerable action on muscle fibers. A typical formulation for the surface treatment of cuts of beef contains 2 percent commercial papain or 5 percent fungal protease, 15 percent dextrose, 2 percent monosodium glutamate (MSG), and salt.

Vitamins and Minerals—Nutrient Supplements

A type of food additive that has received considerable attention is the nutrient supplement. The addition of iodide (as KI) to common table salt to prevent goiter is one example. Many food products now contain added vitamins: vitamin D is added to milk, the B vitamins are added to wheat flour, vitamin C is added to certain beverages, and minerals are added to many cereal products. Since these compounds are needed by the body, they fall into a somewhat different category from the other nonessential additives. The purpose of the various vitamins are given in Table 16–1.

The GRAS List

The GRAS list is a noble effort—but at the present time it is not foolproof.

The Food and Drug Administration lists about 600 chemical substances *"generally recognized as safe" (GRAS)* for their intended use. A small portion of this list is given in Table 16–2. It must be emphasized that an additive on the GRAS list is safe *only if it is used in the amounts and in the foods specified.* The GRAS list was published in several installments in 1959 and 1960. It was compiled by asking experts in nutrition, toxicology, and related fields to give their opinions about the safety of using various materials in foods. Since its publication, few substances have been

---

TABLE 16–1 The Functions of Vitamins*

Vitamin	Biochemical Function	Deficiency Effects
A	Regeneration of rhodopsin (visual purple)	Excessive light sensitivity; night blindness; increased susceptibility to infection
B_1 (Thiamine)	Nerve activity; carbohydrate metabolism	Beriberi; serious nervous disorders; muscular atrophy; serious circulatory changes
B_2 (Riboflavin	Coenzyme; affects sight	Sores on the lips; bloodshot and burning eyes; excessive light sensitivity
B_3 (Nicotinamide)	Metabolism of ATP	Stunted growth; pellagra
B_5 (Pantothenic acid)	Growth factor; component of coenzyme A	Retarded growth
B_6 (Pyridoxine)	Coenzyme; metabolism of fatty acids	Retarded growth; anemia; leukocytosis; insomnia; lesions about eyes, nose, mouth
B_7 (Biotin)	Growth factor; affects scaly and greasy skin; CO_2 fixation; coenzyme	Scaly and greasy skin
B_9 (Folic acid)	Coenzyme; tyrosine metabolism	Anemia
B_{12} (Cobalamin)	Growth factor; involved in synthesis of DNA, hemoglobin	Degeneration of the spinal cord, pernicious anemia
C (Ascorbic acid)	Coenzyme; reducing agent; cholesterol metabolism	Hemorrhages; lesions in the mouth; muscular degeneration; sterility; scurvy
D	Ca and P metabolism	Abnormal development of bones and teeth; rickets
E (α-Tocopherol)	Antioxidant; cofactor between cytochromes b and c	Sterility
K	Synthesis of prothrombin, coenzyme Q	Hemorrhages; slow clotting of blood

*The vitamins participate in biochemical changes concerned with the utilization of foodstuffs.

added to the GRAS list and some, such as the cyclamates, saccharin, and Red Dye No. 2, have been removed.

It is evident, in view of the more than 2500 known food additives, that many more chemicals than those that appear on the GRAS list are approved (or at least, not banned) for use as food additives by the FDA. It is

---

TABLE 16—2   A Partial List of Food Additives Generally Recognized as Safe*

Food Colors	Preservatives	Sequestrants
Annatto (yellow)	Benzoic acid	Citrate esters: isopropyl, stearyl
Carbon black	Na benzoate	Citric acid
Carotene (yellow-orange)	Methylparaben	EDTA, Ca^{2+} and Na^+ salts
Cochineal (red)	Oxytetracycline	Pyrophosphate, Na^+
Food dyes and colors	Propylparaben	Sorbitol
Titanium dioxide (white)	Propionic acid	Tartaric acid
	Ca propionate	Na tartrate
	Na propionate	
Acids, Alkalies, and Buffers	Sorbic acid	**Stabilizers and Thickeners**
	Ca sorbate	
	K sorbate	
Acetates: Ca, K, Na	Na sorbate	Agar-agar
Acetic acid	Sulfites, Na^+, K^+	Algins: NH_4^+, Ca^{2+}, K^+, Na^+
Calcium lactate		Carrageenin
Citrates: Ca, K, Na	**Antioxidants**	Gum acacia
Citric acid		Gum tragacanth
Fumaric acid	Ascorbic acid	Sodium carboxymethyl cellulose
Lactic acid	Ca ascorbate	
Phosphates, CaH, Ca_3, Na_2, Na_3, NaAl	Na ascorbate	**Flavorings**
	Butylated hydroxy-anisole (BHA)	
Potassium acid tartrate	Butylated hydroxy-toluene (BHT)	Acetanisole (slight haylike)
Sorbic acid	Lecithin	Allyl caproate (pineapple)
Tartaric acid	Propyl gallate	Amyl acetate (banana)
	Sulfur dioxide and sulfites	Amyl butyrate (pearlike)
	Trihydroxybuty-rophenone (THBP)	Bornyl acetate (piney, camphor)
Surface Active Agents		Carvone (spearmint)
		Cinnamaldehyde (cinnamon)
Cholic acid		Citral (lemon)
Glycerides: mono- and diglycerides of fatty acids	**Flavor Enhancers**	Ethyl cinnamate (spicy)
		Ethyl formate (rum)
Polyoxyethylene (20) sorbitan mono-palmitate	Monosodium glutamate (MSG)	Ethyl propionate (fruity)
	5'-Nucleotides	Ethyl vanillin (vanilla)
Sorbitan mono-stearate	Maltol	Eucalyptus oil (bittersweet)
		Eugenol (spice, clove)
		Geraniol (rose)
Polyhydric Alcohols		Geranyl acetate (geranium)
		Ginger oil (ginger)
Glycerol		Linalool (light floral)
Sorbitol		Menthol (peppermint)
Mannitol		Methyl anthranilate (grape)
Propylene glycol		Methyl salicylate (wintergreen)
		Orange oil (orange)
		Peppermint oil (peppermint) (menthol)
		Pimenta leaf oil (allspice) (eugenol, cineole)
		Vanillin (vanilla)
		Wintergreen oil (wintergreen) (methyl salicylate)

*For precise and authoritative information on levels of use permitted in specific applications, the regulations of the U.S. Food and Drug Administration and the Meat Inspection Division of the U.S. Department of Agriculture should be consulted.

If past history is a guide, at least some of these compounds will be taken off the GRAS list in the future.

quite expensive to introduce a new food additive with the approval of the FDA. Allied Chemical Corporation began research in 1964 on a new synthetic food color, Allura Red AC. It was approved by the FDA and went on the market in 1972. The cost for introducing this product was $500,000, and about half of this amount was spent on safety testing.

SELF-TEST 16-B

1. Which has the sweetest taste when an equal amount of each is tasted: table sugar, a cyclamate, or saccharin? _____

2. What molecular characteristic do most food colors have? _____ _____

3. What acids are added to foods to lower the pH? _____ and _____.

4. The gas released by leavening agents is _____.

5. Monosodium glutamate (MSG) is used in meat as a _____.

6. Cooking _____ some chemical bonds.

7. The GRAS list is the FDA list of food additives that are _____ _____.

8. Hydrogen bonding generally plays a very important role in the action of surface _____ agents.

9. In order to produce CO_2 in bread, sodium bicarbonate reacts with a(n) _____.

MATCHING SET

_____ 1. β-carotene   (a) nutrient supplement in food

_____ 2. monosodium glutamate   (b) food color

_____ 3. copper, nickel, and iron   (c) flavor enhancer

_____ 4. sodium benzoate   (d) catalyze oxidation of fats

_____ 5. potentiator   (e) antimicrobial preservative

_____ 6. mineral   (f) exaggerates some chemical effects

PART II—MEDICINES

The average life expectancy in the United States has risen from age 49 in 1900 to age 73 in 1976. It is expected to go beyond the age of 79 by the year 2000. The major contributing factor is the widespread use of a large assortment of new medicinal compounds.

The contents of the medicine cabinet have changed drastically in the past few decades. A survey of physicians shortly before World War I revealed the ten most essential drugs (or drug groups) to be ether, opium and its derivatives, digitalis, diphtheria antitoxin, smallpox vaccine, mercury, alcohol, iodine, quinine, and iron. When another survey was made at the end of World War II, at the top of the list were sulfonamides, aspirin, antibiotics, blood plasma and its substitutes, anesthetics and opium derivatives, digitalis, antitoxins and vaccines, hormones, vitamins, and liver extract. Today there is an even wider array of medicinal chemicals, but drugs for reducing fever, relieving pain, and fighting infection still head the list in all areas of medical practice.

Medicines rise and fall in popularity.

Many new drugs are tested each year, but few ever reach the market place.

Antacids

The walls of a human stomach contain thousands of cells that secrete hydrochloric acid, the main purposes of which are to suppress growth of bacteria and to aid in the hydrolysis (digestion) of certain foodstuffs. Normally, the stomach's inner lining is not harmed by the presence of this hydrochloric acid, since the mucosa, the inner lining of the stomach, is replaced at the rate of about a half million cells per minute. When too much food is eaten the stomach often responds with

The contents of the stomach are highly acidic.

TABLE 16–3   Widely Prescribed Drugs

Generic Name*	Medical Use
Tetracycline HCl	Antibiotic
Ampicillin	Antimicrobial (kills bacteria but is not derived from a plant, as is an antibiotic)
Phenobarbital	Sedative, hypnotic, anticonvulsant
Thyroid	Increases rates of metabolism
Prednisone	Antiinflammatory, antiallergic agent similar to cortisone
Digoxin	Decreases the rate of the heartbeat but increases the force of the heartbeat, similar to digitalis
Meprobamate	Tranquilizer
Erythromycin	Antimicrobial
Penicillin G potassium	Antibiotic
Nitroglycerin	Dilates the blood vessels of the heart
Penicillin VK	Antibiotic
Quinidine sulfate	Slows the heartbeat, also used for malaria and hiccups
Paregoric	A tincture (alcoholic solution) of camphorated opium used as an analgesic
Reserpine	Tranquilizer
Nicotinic acid (Niacin)	Dilates blood vessels; also, essential B vitamin with antipellagra activity

The generic name for a drug is its widely accepted chemical name.

*The generic name for a drug is its generally accepted chemical name rather than a specific brand name. For example, the generic drug, tetracycline HCl, is sold under such brand names as Achromycin V, Ambracyn, Artomycin, Diacycline, Quatrex, and others. Medical doctors can prescribe either the generic name or a brand name. If the generic name is used, the prescription often is cheaper, particularly if the drug is not protected by patents and can be manufactured and marketed competitively by several companies. Only about 12% of the prescriptions written in 1974 used generic names.

TABLE 16—4   The Chemistry of Some Antacids

Compound	Reaction in Stomach	Comments
Magnesium oxide MgO	$MgO + 2H^+ \longrightarrow Mg^{2+} + H_2O$	MgO is white and tasteless
Milk of magnesia $Mg(OH)_2$ in water	$Mg(OH)_2 + 2H^+ \longrightarrow Mg^{2+} + 2H_2O$	The water suspension has an unpleasant chalky consistency
Calcium carbonate $CaCO_3$	$CaCO_3 + 2H^+ \longrightarrow Ca^{2+} + H_2O + CO_2$	Calcium carbonate is purified limestone
Sodium bicarbonate $NaHCO_3$	$NaHCO_3 + H^+ \longrightarrow Na^+ + H_2O + CO_2$	Baking soda, like $CaCO_3$, produces CO_2 gas in the stomach
Aluminum hydroxide $Al(OH)_3$	$Al(OH)_3 + 3H^+ \longrightarrow Al^{3+} + 3H_2O$	$Al(OH)_3$ is a clear gel
Dihydroxyaluminum sodium carbonate $NaAl(OH)_2CO_3$	$NaAl(OH)_2CO_3 + 4H^+ \longrightarrow Na^+ + Al^{3+} + 3H_2O + CO_2$	Sold as Rolaids, will not ordinarily cause pH to go above 5
Sodium citrate $Na_3C_6H_5O_7 \cdot 2H_2O$	$Na_3C_6H_5O_7 \cdot 2H_2O + 3H^+ \longrightarrow 3Na^+ + H_3C_6H_5O_7 + 2H_2O$	Mild

an outpouring of acid, which lowers the pH to a point where discomfort is felt.

If the reduction of acidity is too great the stomach responds by secreting an excess of acid. This is "acid rebound."

Antacids are compounds used to decrease the amount of hydrochloric acid in the stomach. The normal pH of the stomach ranges from 1.2 to 3.0. Some alkaline compounds used for antacid purposes and their modes of action are given in Table 16—4.

Analgesics

Analgesics relieve pain, but they are harmful in large doses.

Analgesics are pain killers. Most people need these compounds at one time or another. When we have a headache we take aspirin. When we have a tooth filled or extracted, the dentist uses novocaine. Intense suffering requires a strong pain killer, such as codeine or morphine. While these compounds are immensely useful, they are nevertheless dangerous if taken or used improperly. They can even become killers if taken in overdose.

Early man may well have used opium. Although not all opium derivatives have therapeutic value, most of them are very efficient pain killers. Their chief disadvantage lies in their addictive properties.

Opium is obtained from the opium poppy by scratching the seed pod with a sharp instrument. From this scratch flows a sticky mass that contains about 20 different compounds called *alkaloids* (organic nitrogenous bases containing basic nitrogen atoms). About 10 percent of this mass is the alkaloid *morphine,* which is primarily responsible for opium's effects.

Morphine and its derivatives are addictive drugs.

MORPHINE

Two derivatives of morphine are of interest. One of these is ***codeine,*** a methyl ether of morphine, which is less addictive than morphine and is about as powerful an analgesic. The other compound is ***heroin,*** the diacetate ester of morphine. Heroin is much more addictive than morphine and for that reason finds no medical uses in the United States.

Analgesics may or may not be habit-forming.

CODEINE (METHYL ETHER)

Two acetate groups

HEROIN

One of the most effective substitutes for morphine is ***meperidine,*** first reported in 1931, and now sold as Demerol. It is less addictive than morphine. Two other relatively strong pain relievers used today are ***pentazocine*** (Talwin) and ***propoxyphene*** (Darvon). Talwin is slightly addictive, while Darvon has not been shown to be. However, Darvon has been much abused; there were approximately 600 Darvon-related deaths in 1977. Critics argue it to be more dangerous and less effective in killing

MEPERIDINE (DEMEROL)

TABLE 16—5   Some Local Analgesics

Cocaine		Probably the first local analgesic used
Procaine (Novocaine)		Often used in dental work
Lidocaine (Xylocaine)		More potent than procaine, can be applied to the skin

pain than other available opiates. Note that in the structures of these compounds there is a strong resemblance to the morphine structure.

Considerable progress has been made in understanding the drug action of the opiates. Solomon Snyder, along with coworkers, discovered in 1973 that the brain and spinal cord contain specific bonding sites into which the opiate molecules fit as a key fits a lock. John Hughes and Hans Kosterlitz followed in 1975 with the discovery that vertebrates produce their own opiates, which they named enkephalins (Greek: in the head). Individuals with a high tolerance for pain produce more enkephalins and consequently tie up more receptor sites; hence there is less pain. A dose of heroin would temporarily bond to a high percentage (or all) of the sites and there would be little or no pain. Continued use of heroin would cause the body to reduce or cease its production of enkephalins. If the use of the narcotic is stopped, the receptor sites are empty and withdrawal symptoms appear.

Many of the *local analgesics,* or local anesthetics, are nitrogen compounds, like the alkaloids (Table 16–5). Local analgesics include the naturally occurring *cocaine,* derived from the leaves of the coca plant of South America, and the familiar novocaine (procaine). All of these drugs act by some blockage of the nerves that transmit pain. Acetylcholine appears to be the "opener of the gate" for sodium (Na^+) and potassium (K^+) ions to flow into a nerve cell (Figure 16–6).

There are times when milder, general analgesics are required, and few compounds work as well for as many people as *aspirin.* Not only is aspirin an analgesic, but it is also an antipyretic, a fever reducer.

The synthesis of aspirin was outlined in Chapter 7. Each year, about 40 million pounds are manufactured in the United States.

The greatest danger presented by aspirin is that of stomach bleeding, caused when an undissolved aspirin tablet lies on the stomach wall. As the aspirin molecules pass through the fatty layer of the mucosa, they appear to injure the cells, causing small hemorrhages. The blood loss for most individuals taking two 5-grain tablets is between 0.5 ml and 2 ml. Some people are more susceptible. Early aspirin tablets were not particularly fast-dissolving, which aggravated this problem greatly. Today, aspirin

PENTAZOCINE
(TALWIN)

PROPOXYPHENE
(DARVON)

Many aspirin tablets contain starch to hasten their disintegration in the stomach.

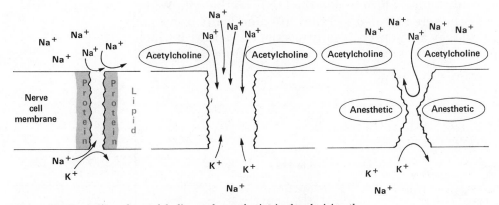

Figure 16–6 Action of acetylcholine and anesthetics in depolarizing the membrane of a nerve cell. Acetylcholine makes it possible for sodium and potassium ions to neutralize the negative charge associated with a nerve impulse so another impulse can be transmitted. Anesthetics block the action of acetylcholine and do not allow repetitive impulses to travel along the nerve.

tablets are formulated to disintegrate quickly, although crushing the tablet in a little water might not be a bad idea.

An alternative for pain-sufferers who have trouble with aspirin may be acetaminophen (Tylenol). Although known for about as long as aspirin (a century), acetaminophen is a relative newcomer to the market. Like aspirin, acetaminophen is both an analgesic and an antipyretic.

Acetylsalicylic acid (aspirin)

Antiseptics and Disinfectants

An antiseptic is a compound that prevents the growth of micro-organisms. It now has the legal meaning *"germicide,"* or a compound that *kills* microorganisms. A disinfectant is a compound that destroys pathogenic bacteria or microorganisms, but usually not bacterial spores. Disinfectants are generally poisonous, and, therefore, suitable only for external use as on the skin or a wound.

Pathogenic bacteria cause many illnesses.

Some common germicides are listed in Table 16–6. Some of these, such as the halogens, sodium hypochlorite, hydrogen peroxide, and potassium permanganate, are effective because of their oxidizing properties. This is a general property and allows them to oxidize any kind of cell, including human cells. For this reason they are used mostly as disinfectants in destroying the germs on nonliving objects. Phenol (carbolic acid) is readily absorbed by cells and is a general poison. The quaternary ammonium compounds are surface active agents, and their bactericidal effect seems to be related to their ability to weaken the cell wall so the cell contents cannot be contained.

A quaternary ammonium chloride.

One of the newer developments solves the problem of applying antiseptics to children. You may remember the sting of "iodine" when applied to a scratch or a wound. Old-fashioned iodine is a solution of iodine (I_2) in alcohol with a little potassium iodide (KI) to increase the solubility of the iodine. The alcohol causes most of the pain. Now there are polymers, such as polyvinylpyrrolidone, which complex iodine molecules (Figure 16–7); the products (iodophors) are soluble in water. The resultant solution is a very efficient and painless disinfectant. The iodophors are active ingredients in a popular mouthwash and in restaurant glassware disinfectants.

A tincture is an alcoholic solution. Tincture of iodine is a solution of water, iodine, and potassium iodide in ethanol.

Because such compounds are generally toxic to living matter, it is necessary to utilize only dilute solutions and then only on the skin. While they help to prevent the spread of disease, they are practically useless in its treatment because they act nonspecifically against all cells with which they come in contact. They are to be distinguished from antibiotics,

---

TABLE 16–6 Some of the More Common Antiseptics

Iodine	Phenols
Sodium hypochlorite	Mercurochrome
Potassium permanganate	Metaphen
Hydrogen peroxide	Merthiolate
Iodophors	Pine oil
Ethanol	Soap
Quaternary ammonium compounds	Hexylresorcinol
Chloramine-T	Mercuric chloride

Figure 16—7 An iodophor: complex of polyvinylpyrrolidone (PVP) and iodine.

which act more selectively against infecting bacteria than against "host" cells within the human body.

Antimicrobial Medicines

In our time, the quest for drugs to wipe out disease due to micro-organisms has been virtually fulfilled by the ***antibiotics*** (Table 16–7). In the original sense, an antibiotic is a substance such as penicillin, produced by a microorganism, that inhibits the growth of another organism. It has become common practice to include synthetic chemicals such as the sulfa drugs in a discussion of antibiotics.

Since the antibiotics are so efficient, they were the first of what came to be called "miracle" drugs. Their job generally is to aid the white blood cells by stopping bacteria from multiplying. When a person falls victim to or is killed by a disease, it means that the invading bacteria have multiplied faster than the white blood cells could devour them, and that the bacterial toxins increased more rapidly than the antibodies could neutralize them. The action of the white blood cells and antibodies plus an antibiotic is generally enough to repulse an attack of the disease germs.

An antibody is a specific protein produced to protect the organism from harmful invading molecules.

The Sulfa Drugs

Sulfa drugs represent a group of compounds discovered in a conscious search for antibiotics. In 1904, the German chemist Paul Ehrlich (1854–1915; Nobel prize in 1908) realized that infectious diseases could be conquered if toxic chemicals could be found that attacked parasitic organisms within the body to a greater extent than they did host cells. Ehrlich achieved some success toward his goal; he found that certain dyes that were used to stain bacteria for microscopic examination could also kill the

TABLE 16—7 Deaths per 100,000 Americans Due to Different Causes		
	1900	1970
Infectious diseases	500	50
Influenza and pneumonia	210	30
Diphtheria	40	fewer than 1
Typhoid and paratyphoid	30	fewer than 1
Whooping cough	10	fewer than 1
Gastrointestinal problems	150	fewer than 10

bacteria. This led to the use of dyes against African sleeping sickness and arsenic compounds against syphilis.

In 1935, after experimenting with several drugs, Gerhard Domagk, a pathologist in the I. G. Farbenindustrie Laboratories in Germany, found that Prontosil, a dye, was somewhat effective against bacterial infection in mice. *Prontosil* can be changed to ***sulfanilamide,*** which is very effective.

Large doses of sulfa drugs are required compared to the doses of "true antibiotics."

$$H_2N-\bigcirc-N=N-\bigcirc-SO_2NH_2 \xrightarrow{H_2O} H_2N-\bigcirc-SO_2-NH_2$$

NH₂ (on the first ring)

PRONTOSIL   SULFANILAMIDE

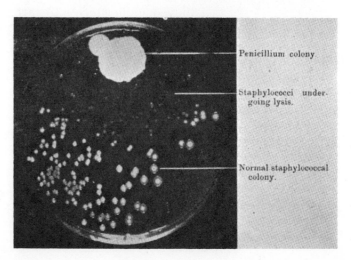

Figure 16–8 *A,* Sir Alexander Fleming. *B,* A culture-plate showing the dissolution of staphylococcal colonies in the neighborhood of a colony of penicillium. (From article by Fleming, *British Journal of Experimental Pathology,* June, 1929.)

Penicillium colony.

Staphylococci undergoing lysis.

Normal staphylococcal colony.

This discovery led to the synthesis and testing of a large number of related compounds in the search for drugs that are more effective or less toxic to the infected experimental animal. By 1964, more than 5000 sulfa drugs had been prepared and tested.

A sulfa drug mimics an essential compound.

Sulfa drugs inhibit bacteria by preventing the synthesis of folic acid, a vitamin essential to their growth. The drugs' ability to do this apparently lies in their structural similarity to a key ingredient in the folic acid synthesis, para-aminobenzoic acid.

SULFANILAMIDE,
A TYPICAL SULFA DRUG

p-AMINOBENZOIC ACID

The close structural similarity of sulfanilamide and p-aminobenzoic acid permits sulfanilamide to be incorporated into the enzymatic reaction sequence instead of p-aminobenzoic acid. By bonding tightly, sulfanilamide shuts off the production of the essential folic acid, and the bacteria die of vitamin deficiency. In humans and the higher animals, p-aminobenzoic acid is not necessary for folic acid synthesis, so sulfa drugs have no effect on this mechanism.

The Penicillins

Penicillin was first discovered in 1928 by Alexander Fleming, a bacteriologist at the University of London, who was working with cultures of *Staphylococcus aureus,* a germ that causes boils and some other types of infections. In order to examine the cultures with a microscope, he had to

Figure 16–9   Penicillin kills bacteria by interfering with the formation of cross links in the cell wall.

remove the cover of the culture plate for a while. One day as he started work he noticed that the culture was contaminated by a blue-green mold. For some distance around the mold growth, the bacterial colonies were being destroyed. Upon further investigation, Fleming found that the broth in which this mold had grown also had an inhibitory or lethal effect against many pathogenic (disease-causing) bacteria. The mold was later identified as *Penicillium notatum* (the spores sprout and branch out in pencil shapes, hence the name).

Penicillin, the name given to the antibacterial substance produced by the mold, apparently has no toxic effect on animal cells, and its activity is selective. The structure of penicillin (see margin) has now been determined. Many different penicillins exist, differing in the structure of the R group. Penicillin G is the most widely used in medicine. Several antibiotics, such as penicillin and bacitracin, are known to prevent cell-wall synthesis in bacteria, as shown in Figure 16–9.

Penicillin G

Streptomycin and the Tetracyclines

In 1937, following collaboration with René Dubos, Selman Waksman isolated a compound from a soil organism, *Streptomyces griseus,* which came to be known as **streptomycin** and was released to physicians in 1947. This compound was quite successful in controlling certain types of bacteria but later had to be withdrawn, because of its adverse side effects.

Streptomycin has many undesirable side effects.

In 1945, B. M. Duggan discovered that a gold-colored soil fungus, *Streptomyces aureofaciens,* produced a new type of antibiotic, **Aureomycin,** the first of the **tetracyclines.** Research then stepped up to a fever pitch. Pfizer laboratories tested 116,000 different soil samples before they discovered the next antibiotic, which they named **Terramycin.**

AUREOMYCIN

TERRAMYCIN

Tetracyclines get their names from their four-ring structures.

Compounds of the tetracycline family are so named because of their four-ring structure. One side effect of taking these drugs is diarrhea caused by the killing of the patient's intestinal flora (the bacteria normally residing in the intestines).

The Steroid Drugs

A large and important class of naturally occurring compounds is derived from the tetracyclic structure given below.

These compounds are known as ***steroids,*** and they occur in all plants and animals. The most abundant animal steroid is cholesterol, $C_{27}H_{46}O$. The human body synthesizes cholesterol and also readily absorbs dietary cholesterol through the intestinal wall. It is associated with gallstones and hardening of the arteries.

> Cortisone is a "powerful drug" having a major effect on biological systems.

Biochemical alteration or degradation of cholesterol leads to many steroids of great importance in human biochemistry. When ***cortisone,*** one of the adrenal cortex hormones, is applied topically or injected into a diseased joint, it acts as an antiinflammatory agent and is of great use in treating arthritis.

CHOLESTEROL

CORTISONE

Structurally related to cholesterol and cortisone are the sex hormones. One female sex hormone, ***progesterone,*** differs only slightly from an important male hormone, ***testosterone.***

PROGESTERONE

TESTOSTERONE

Other female hormones are estradiol and estrone, called ***estrogens.*** The estrogens differ from the other steroids discussed earlier in that they contain an aromatic A ring (in color).

ESTRONE

ESTRADIOL

Birth Control Pills

One of the most revolutionary medical developments of the 1960's was the worldwide introduction and use of "The Pill." More than ten million women in the United States use birth control pills. The basic feature of oral contraceptives for women is their chemical ability to simulate the hormonal processes resulting from pregnancy and, in so doing, prevent ovulation. Ovulation, the production of eggs by the ovary, ceases at the onset of pregnancy because of hormonal changes (Figure 16–10). This same result can be produced by the administration of a variety of steroids, some of which are effective when taken orally, although the mechanism of their action and their long-term effects are not known in detail.

The active ingredients of the Pill are the hormones progesterone and estrogen, or their derivatives.

Allergens and Antihistamines

A person may have an unpleasant physiological response to poison ivy, pollen, mold, food, cosmetics, penicillin, aspirin, and even cold, heat, and ultraviolet light. In the United States about 5000 people die yearly from bronchial asthma, at least 30 from the stings of bees, wasps, hornets, and other insects, and about 300 from ordinary doses of penicillin. The reason: ***allergy***. About one in 10 suffers from some form of allergy; more than 16 million Americans suffer from hay fever.

An allergy is an adverse response to a foreign substance or to a physical condition that produces no obvious ill effects in *most* other organisms, including humans. An ***allergen*** (the substance that initiates the allergic reaction) is, in most cases, a highly complex substance—usually a protein.

An allergy is a physiological response such as sneezing, runny nose, coughing, dermatitis, etc., to the introduction of a foreign substance. This foreign substance is called an allergen.

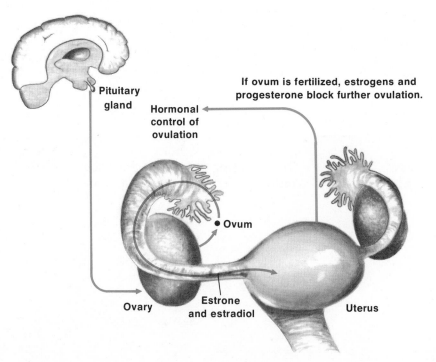

Figure 16–10 Some of the control processes in the female reproductive cycle. The pituitary gland, located in the brain area, sends out hormones which cause the ovary to release other hormones. Estrone and estradiol initiate the deterioration of the old wall of the uterus and the formation of a new wall. If the ovum is fertilized, it attaches to the uterine wall, and estrogens and progesterone are released to prevent further ovulation. The Pill serves the same function as the estrogens and progesterone in that it prevents the ovulation process.

Pituitary gland

If ovum is fertilized, estrogens and progesterone block further ovulation.

Hormonal control of ovulation

Ovum

Ovary

Estrone and estradiol

Uterus

Most allergens are high molecular weight substances.

Some are polysaccharides or compounds formed by combining a protein and a polysaccharide. Usually allergens have a molecular weight of 10,000 amu or more.

The principal allergen of ragweed pollen, a major allergy-producer, has been isolated and is named ragweed antigen E. It is a protein with a molecular weight of about 38,000 amu; it represents only about 0.5 percent of the solids in ragweed pollen, but contributes about 90 percent of the pollen's allergenic activity. A mere 1×10^{-12} gram of antigen E injected into an allergic person is enough to induce a response.

10^{-12} is .000000000001

The allergens come in contact with special cells in the nose and breathing passages to which a particular type of antibody is attached, the IgE antibody, which has a molecular weight of about 196,000. An allergic person has 6 to 14 times more IgE in his blood serum than a nonallergic person. The IgE is formed in the nose, bronchial tubes, and gastrointestinal tract, and binds firmly to specific cells, called *mast cells,* in these regions.

Antibodies are high molecular weight proteins, called immunoglobulins, which attack foreign proteins such as allergens.

Antigen E from ragweed reacts with the IgE antibody attached to the mast cells, forming antigen-antibody complexes. The formation of these antigen-antibody complexes leads to the release of so-called "allergy mediators" from special granules in the mast cells (Figure 16–11). The most potent of these mediators found so far is *histamine.* Although it is widely distributed in the body, it is especially concentrated in the 250 to 300 granules of the mast cells. Histamine accounts for many, if not most, of the symptoms of hay fever, bronchial asthma, and other allergies.

$$H_2NCH_2CH_2 \underset{\underset{H}{N}}{\overset{N}{\rule{0pt}{0pt}}}$$

HISTAMINE

Histamine causes runny noses, red eyes, and other hay fever symptoms.

The chemical mediators such as histamine must be released from the cell to cause the symptoms of allergy. The release mechanism is an energy-requiring process in which the granules may move to the outer edge of the living cell and, without leaving the cell, discharge their contents of hista-

Figure 16–11 A postulated mechanism for the cause of and relief from hay fever. The details are described in the text.

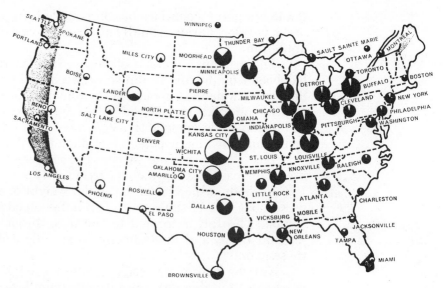

Figure 16–12 In this Abbott Laboratories map, the size of each circle represents the amount of all late-summer and fall pollens found in the air in each city. Dark portions show amount of ragweed pollen. Shaded areas are regions of low pollen count.

mine through a temporary gap in the cell membrane. This sends histamine on its way to produce the toxic effects of hay fever.

Treatment consists of three procedures: avoidance (Figure 16–12), desensitization, and drug therapy. Desensitization therapy is costly and inconvenient, since 20 or more injections are required to achieve what usually is a partial cure. One chemical idea of desensitization is to inject a blocking antibody that preferentially reacts with the allergen so that it cannot react with the IgE allergy-sensitizing antibody. This breaks the chain of events leading to the release of histamine or other allergy-producing mediators. Many small injections, spaced in time, are required to build up a sufficient level of the blocking antibody.

Epinephrine (Adrenalin), steroids, and antihistamines are effective drugs in treating allergies. The first two are particularly effective in treating bronchial asthma, while the ***antihistamines,*** introduced commercially in the United States in 1945, are the most widely used drugs for treating allergies. More than 50 antihistamines are offered commercially in the United States. Many of these contain, as does histamine, an ethylamine group, $(-CH_2CH_2N\!<)$:

PYRIBENZAMINE
(AN IMPORTANT ANTIHISTAMINE)

These drugs act competitively by occupying the receptor sites normally occupied by histamine on cells. This, in effect, blocks the action of histamine.

Diet Pills

The desired rate of weight loss for most overweight people ranges from about 1 to 2 pounds per week. For most women, diets ranging between 1200 and 1500 food calories per day will bring about a satisfactory weight loss. For men, the range is between 1500 and 2000 food calories per day. An intake of less than 1200 calories is generally not advised.

A variety of special, rather unusual diets have been reported to be effective in weight reduction. These diets have been known under such names as the grapefruit diet, the water diet, the 10-day diet, and others. The problem with the vast majority of these diets is that they are often made up of only a few foods, so that they are not nutritionally adequate. Five basic types of nutrients are essential: protein, fat, carbohydrate, vitamins, and certain minerals. They must be included in the diet for normal health. If the diet is lacking in any one category, the body suffers. Therefore, a balanced diet plus water is required, even if the amounts are reduced.

Many individuals have difficulty staying on a diet, and they must be highly motivated in order to maintain a caloric level low enough to bring about a loss of weight. Appetite-depressing drugs, known as *anorexigenic drugs,* have been used by some physicians to curb the appetite. More than two billion diet pills are distributed per year. The ingredients include amphetamines (which suppress the appetite), digitalis (which affects the heart), various diuretics (which increase the amount of water excreted), thyroid (which increases rates of metabolism), and prednisone (an anti-inflammatory and antiallergic agent). Some of the drugs are potentially addictive; others tax the heart or cause potassium loss from the body.

By themselves anorexigenic drugs will not control obesity. Reliance on such drugs, rather than proper diet, leads to failure in a weight-reduction plan, and their unsupervised use may be harmful.

In addition to the appetite-depressing drugs, there are many other products that are advertised to help overweight people lose weight. These products include methyl cellulose, vitamins, iron and calcium compounds, benzocaine (a local anesthetic), dextrose, potassium p-amino-benzoate, caffeine, flavorings, and a few other substances. By themselves, however, none of these products will cause a reduction in weight.

Drugs in Combination

Drugs, like some food additives, can have enhanced effects when placed in certain chemical environments. Sometimes the effects are harmful, sometimes helpful. Take the case of an aging business executive who took an antidepressant and then ate a meal that included aged cheese and wine. The antidepressant is an inhibitor of monoamine oxidase, an enzyme that helps to control blood pressure. Both the aged cheese he ate and the wine he drank contained pressor amines, which raise blood pressure. Without the controlling effect of the enzyme, these amines skyrocketed his blood pressure and caused a stroke. Neither the amines nor the antidepressant alone would likely cause the stroke, but the combination did.

Likewise, people who take digitalis for heart trouble and for reducing the sodium level in the blood should take aspirin only under medical supervision. Aspirin can cause a 50 percent reduction in salt excretion for three or four hours after it is taken.

A food calorie is actually a kilocalorie, or 1000 calories.

Many reducing diets are not nutritionally adequate.

Diet pills should be used under medical supervision.

Pressor amines tend to increase blood pressure.

Alcohol increases the action of many antihistamines, tranquilizers, and drugs such as reserpine (for lowering blood pressure) and scopolamine (contained in many over-the-counter nerve and sleeping preparations), making such combinations extremely dangerous. Staying away from dangerous alcohol-drug combinations is not as easy as it may seem. Many people fail to realize that a large number of over-the-counter preparations —such as liquid cough syrup and tonics—contain appreciable amounts of alcohol.

Not all drug combinations are bad. Doctors have been highly successful in prolonging the lives of leukemia and other cancer victims with combinations of drugs that individually could not do the job. Resistant kidney disease has also responded to drug combinations in cases in which single drugs were ineffective.

Perhaps the best advice is to take medicine only when you are seriously ill, making sure that a physician knows what you are taking.

SELF-TEST 16-C

1. What acid is secreted by the walls of the stomach? _____

2. Is an analgesic a painkiller or a germ fighter? _____

3. Select the natural compound from which the other is made: morphine, heroin. _____

4. A molecule of aspirin has two functional groups. What are they?_____

   and _____

5. An antipyretic reduces _____.

6. Penicillin was discovered by _____ _____.

7. Allergens are complex proteins with a usual molecular weight of 10,000 amu or more. (True/False)

8. Two groups of medicines that act by mimicking other compounds are

   _____, which mimic p-aminobenzoic acid, and

   _____, which mimic histamine.

MATCHING SET

____ 1. histamine	(a)	interferes with cell wall structure
____ 2. cortisone	(b)	antiseptic
____ 3. tetracycline	(c)	source of morphine
____ 4. penicillin	(d)	analgesic
____ 5. sulfa drug	(e)	antacid
____ 6. iodine	(f)	a steroid

_____ 7. procaine (novocaine)   (g) causes symptoms of hay fever

_____ 8. opium poppy   (h) antibiotic containing a four-ring
                                         structure

_____ 9. dihydroxyaluminum   (i) sulfanilamide
          sodium carbonate

PART III—BEAUTY AIDS

People find many reasons for applying various chemical preparations (***cosmetics***) to their skin and hair. We wish to be clean, beautiful, healthy, and pleasing to others. Considerable progress has been made in producing chemical products that color hair and skin, disinfect our body surfaces, and keep down unwanted body odor. The purpose here is not to enable you to make face creams or hair sprays, but to give you a better understanding of the basic chemistry involved. In fact, amateur chemical preparations should be avoided, since toxic reactions are often encountered when impure or otherwise harmful concoctions are used without proper testing.

Cosmetic products require careful testing before they are used on humans.

Skin and Hair

The skin, hair, and nails are protein structures. Skin (Figure 16–13), like the other organs of the body, is not uniform tissue. The outermost

Figure 16–13 Replica of the surface of human forearm skin, showing three hairs emerging from the skin (× 225). (Courtesy of E. Bernstein and C. B. Jones, *Science,* 166: 252–253, 10 October, 1969. Copyright 1969 by the American Association for the Advancement of Science.)

Figure 16—14 Electron micrograph of human hair. Note the layers of keratinized cells.

layer, called the **stratum corneum,** or **corneal layer,** is where most cosmetic preparations for the skin act. The corneal layer is composed principally of dead cells with low moisture content and a surface pH of about 4, slightly acidic. The principal protein of the corneal layer is **keratin,** which is composed of about 22 different amino acids. Its structure renders it insoluble in, but slightly permeable to, water. In order to control the moisture content of the corneal layer so that it does not dry out and slough off too quickly, moisturizers are added to the skin.

Hair (Figure 16–14) is composed principally of keratin. An important difference between hair keratin and other proteins is its high content of the amino acid **cystine.** About 16 to 18 percent of hair protein is cystine, while only 2.3 to 3.8 percent of the keratin in corneal cells is cystine. This amino acid plays an important role in the structure of hair.

The toughness of both skin and hair is due to the bridges between different protein chains, such as hydrogen bonds and $-S-S-$ linkages, called **disulfide** linkages.

Keratin is skin protein.

$$
\begin{array}{l}
\quad\quad\quad NH_2 \\
\quad\quad\quad | \\
HOOC-CH-CH_2-S \\
\quad\quad\quad\quad\quad\quad\quad | \\
HOOC-CH-CH_2-S \\
\quad\quad\quad | \\
\quad\quad\quad NH_2
\end{array}
$$

Cystine

$$
\begin{array}{llll}
O=C & & & N-H \\
| & & & | \\
H-CCH_2-S-S-CH_2C-H \\
| & & & | \\
H-N & & & C=O \\
| & & & |
\end{array}
$$

Disulfide linkage

Another type of bridge between two protein chains, which is important in keratin as well as in all proteins, is the **ionic bond.** Consider the interaction between a lysine $-NH_2$ group and a carboxylic $-COOH$

The structures of protein tissues are due in part to disulfide cross links and to ionic bonds between "molecules."

group of glutamic acid on a neighboring protein chain. At pH 4.1, the presence of a $-NH_3^+$ group and a $-COO^-$ group is most favorable for keratin. If the two charges approach closely, an ionic bond is formed.

$$HCCH_2CH_2CH_2CH_2NH_2 + HOOCCH_2CH_2CH \xrightarrow{\text{At pH 4.1}}$$
LYSINE   GLUTAMIC ACID

$$HCCH_2CH_2CH_2CH_2NH_3^+ \ ^-OOCCH_2CH_2CH$$
(IONIC BOND)

As the pH rises above 4, keratin will swell and become soft as these cross links are broken. This is an important aspect of hair chemistry.

Changing the Shape of Hair

Changing the shape of hair is a story of oxidation - reduction reactions.

When hair is wet, it can be stretched to one and a half times its dry length because water (pH = 7) weakens some of the ionic bonds and causes swelling of the keratin. Imagine the disulfide cross links remaining between two protein chains in hair as in Figure 16–15(a). Winding the hair on rollers causes tension to develop at the cross-links (b). In "cold" waving, these cross-links are broken by a reducing agent (c), relaxing the tension. Then, an oxidizing agent regenerates the cross-links (d) and the hair holds the shape of the roller. The chemical reactions in simplified form are shown in Figure 16–16.

$$CH_2-C\overset{\displaystyle O}{\underset{\displaystyle OH}{<}}$$
$$|$$
$$SH$$
THIOGLYCOLIC ACID

The most commonly used reducing agent is thioglycolic acid. The common oxidizing agents used include hydrogen peroxide, perborates ($NaBO_2 \cdot H_2O_2 \cdot 3H_2O$), and sodium or potassium bromate ($KBrO_3$). A typical neutralizer solution contains one or more of the oxidizing agents dissolved in water. The presence of water and strong base in the oxidizing

Figure 16–15 A schematic diagram of a permanent wave.

Figure 16—16 Structural changes that occur in hair during a permanent wave.

solution also helps to break and re-form hydrogen bonds between adjacent protein molecules. However, too-frequent use of strong base causes hair to become brittle and lifeless.

Various additives are present in both the oxidizing and the reducing solutions in order to control pH, odor, and color, and for general ease of application. A typical waving lotion contains 5.7 percent thioglycolic acid, 2.0 percent ammonia, and 92.3 percent water.

Hair can be straightened by the same solutions. It is simply "neutralized" (or oxidized) while straight (no rolling up).

Hair Coloring and Bleaches

Hair contains two pigments: brown-black melanin and an iron-containing red pigment. The relative amounts of each actually determine the color of the hair. In deep black hair melanin predominates, while in light-blond, the iron pigment predominates. The depth of the color depends upon the size of the pigment granules.

Melanin—black.

Iron pigment—red.

Formulations for dyeing hair vary from temporary coloring (removable by shampoo), which is usually achieved by means of a water-soluble dye that acts on the surface of the hair, to semipermanent dyes, which penetrate the hair fibers to a great extent (Figure 16–17). These often consist of cobalt or chromium complexes of dyes dissolved in an organic solvent. Permanent dyes are generally "oxidation" dyes. They penetrate the hair, and then are oxidized to give a colored product that is permanently attached to the hair by chemical bonds or that is much less soluble than the reactant molecule. Permanent hair dyes generally are derivatives of phenylenediamine.

NH_2

NH_2

p-PHENYLENEDIAMINE

Phenylenediamine dyes hair black. A blond dye can be formulated with *p*-aminodiphenylaminesulfonic acid,

$$NH_2 - \bigcirc - NH - \bigcirc - SO_3H$$

Hair

Dye adsorbed on surface of hair

Dye rinses

Dye enclosed inside of hair

Semipermanent dye

Figure 16–17 Methods of dyeing hair.

or *p*-phenylenediaminesulfonic acid.

The active compounds are applied in an aqueous soap or detergent solution containing ammonia to make the solution basic. The dye material is then oxidized by hydrogen peroxide to develop the desired color. The amines are oxidized to nitro compounds.

$$-NH_2 + 3H_2O_2 \xrightarrow{\text{Oxidation}} -NO_2 + 4H_2O$$

AMINE

NITRO COMPOUND

Hair can be bleached by a more concentrated solution of hydrogen peroxide, which destroys the hair pigments by oxidation. The solutions are made basic with ammonia to enhance the oxidizing power of the peroxide. Parts of the chemical process are given in Figure 16–18. This

Figure 16–18 Bleaching of the hair by hydrogen peroxide. There are several chemical intermediates between the amino acid—tyrosine—and the hair pigment—melanin, which is partly protein. Hydrogen peroxide oxidizes melanin back to colorless compounds, which are stable in the absence of tyrosinase (found only in the hair roots). Melanin is a high molecular weight polymeric material of unknown structure. The structure shown here is only a segment of the total structure.

Catalyzed by enzyme tyrosinase (a copper-containing protein)

Tyrosine

5,6-Dihydroxyindole (colorless)

H_2O_2

Melanin fragment (dark brown-black pigment)

Figure 16-19 Film of hair spray. Hair spray was allowed to dry on white surface and was then pulled up to reveal film.

drastic treatment of hair does more than just change the color. It may destroy sufficient structure to render the hair brittle and coarse.

Hair Sprays

Hair spray coats the hair with a plastic film.

Hair sprays are essentially solutions of a resin in a very volatile solvent whose purpose, when sprayed on hair, is to furnish a film with sufficient strength to hold the hair in place after the solvent has evaporated. A very common resin in hair sprays is the addition polymer, polyvinylpyrrolidone (PVP).

POLYVINYLPYRROLIDONE (PVP)

There are several dangers in breathing the vapor of hair sprays, such as the danger of possible carcinogens acting on delicate lung tissue and the danger of asphyxiation by the plastic coating lining the lungs.

The resin is blended in hair spray formulations with a plasticizer, a water repellent, and a solvent-propellant mixture. The plasticizer makes the plastic more pliable. The resin concentration of hair sprays is of the order of 4 percent. Other additives, such as silicone oils, are often put into hair sprays to give the hair a sheen.

Since PVP tends to pick up moisture, other, less hygroscopic polymers are beginning to replace PVP in hair sprays. For example, significantly better moisture control is obtained with a copolymer made from a 60/40 ratio of vinylpyrrolidone and vinyl acetate.

Depilatories

The purpose of a depilatory is to remove hair chemically. Since skin is sensitive to the same kind of chemical attack as hair, such preparations should be used with caution and, even then, some attack on the skin is almost unavoidable. Because of this, the interval between applications of a depilatory should be of the order of a week or so. It should never be used on skin that is infected or that has a rash, and should not be followed by application of a deodorant with its *astringent* (contracting) action. If the sweat pores are closed by the deodorant, the caustic chemicals are retained and can do considerable harm. If the sweat pores are open, the body fluids will dilute and wash the caustic chemicals to the outside of the body.

The chemicals used as depilatories include sodium sulfide, calcium sulfide, strontium sulfide (water-soluble sulfides), and calcium thioglycolate [$Ca(HSCH_2COO)_2$], the calcium salt of the compound used to break S—S bonds between protein chains in permanent waving. A typical cream depilatory contains calcium thioglycolate (7.5 percent), calcium carbonate (filler, 20 percent), calcium hydroxide (provides basic solution, 1.5 percent), cetyl alcohol ([$CH_3(CH_2)_{15}OH$], skin conditioner, 6 percent), sodium lauryl sulfate (detergent, 0.5 percent), and water (64.5 percent).

The water-soluble sulfides are all strong bases in water, as indicated by the hydrolysis of the sulfide ion:

$$S^{2-} + H_2O \longrightarrow HS^- + OH^-$$
SULFIDE   HYDROXIDE

For example, a 0.1 M solution of Na_2S has a pH of about 13, a strongly basic solution. The compounds act chemically on the hair to disrupt bonds in the protein chains and cause it to disintegrate by hydrolyzing to soluble amino acids and small peptides, which may be removed.

0.1 M means 0.1 mole of Na_2S (7.81 g) dissolved in sufficient water to make one liter of solution.

The area on which a depilatory has been used should be washed with soap and water, dried, and then treated with small amounts of talcum powder.

Killing Germs

Harmful microorganisms (germs) are always with us in large numbers. Fortunately, there are so many nonpathogenic microorganisms among them that there is little room left for the pathogenic kinds. Depending on body location, healthy skin may host as many as a million microorganisms per square centimeter.

Things we do to our skin, such as shaving, can upset the balance between harmful and harmless bacteria and promote infection by pathogenic microorganisms. ***Disinfectants*** are chemicals used to kill these organisms before they can overcome the skin's defenses. Most commonly used are the alcohols. These are the only disinfectants used in many aftershave preparations. Maximum effectiveness of ethanol is reached at a concentration of 70 percent, while isopropyl (rubbing) alcohol is most effective at a concentration of 50 percent. These alcohols kill germs apparently by hydrogen bonding with water, which dehydrates the cellular structure of the germ.

Alcohols dehydrate microbes.

One widely used disinfectant is phenol; its aqueous solution is known as carbolic acid. An —OH group attached to the benzene ring is slightly acidic. The disinfectant action of phenol was discovered in 1867 by Sir Joseph Lister, who introduced it into surgery. It appears that phenol kills bacteria by denaturing their cellular proteins. Today, about one third of all toilet soaps sold contain some derivative of phenol.

PHENOL

To denature a protein is to break down its structure.

Deodorants

The 2,000,000 sweat glands on the body surface are primarily used to regulate body temperature via the cooling effect produced by the evaporation of the water they secrete. This evaporation of water leaves solid constituents, mostly sodium chloride, as well as smaller amounts of proteins and other organic compounds. Body odor results largely from amines and hydrolysis products of fatty oils (fatty acids, acrolein, etc.) emitted from the body, and from bacterial growth within the residue from sweat glands. Sweating is both normal and necessary for the proper functioning of the human body; sweat itself is quite odorless, but the bacterial decomposition products are not.

There are three kinds of deodorants: those that directly "dry up" perspiration or act as astringents, those that have an odor to mask the odor of sweat, and those that remove odorous compounds by chemical reaction. Among those that have astringent action are hydrated aluminum sulfate, hydrated aluminum chloride ($AlCl_3 \cdot 6H_2O$), aluminum chlorohydrate [actually aluminum hydroxychloride, $Al_2(OH)_5Cl \cdot 2H_2O$ or $Al(OH)_2Cl$ or $Al_6(OH)_{15}Cl_3$], and alcohols. Those compounds that act as deodorizing agents include zinc peroxide, essential oils and perfumes, and a variety of mild antiseptics. Zinc peroxide removes odorous compounds by oxidizing the amines and fatty acid compounds. The essential oils and perfumes absorb or otherwise mask the odors, and the antiseptics are generally oxidizing or reducing agents that kill the odor-causing bacteria.

Body odor is promoted by bacterial action.

An astringent closes the pores, thus stopping the flow of perspiration.

Skin Preparations for Health and Beauty

To remain healthy, the moisture content of skin must stay near 10 percent. If it is higher, microorganisms grow too easily; if lower, the corneal layer flakes off. Washing skin removes fats that help retain the right amount of moisture. If dry skin is treated with a fat after washing, it will be protected until enough natural fats have been regenerated.

Lanolin is an excellent skin softener (***emollient***) and is a component of many cosmetics. It is a complex mixture of esters from hydrated wool fat. The esters are derived from 33 different alcohols of large molecular weight and 37 fatty acids. Cholesterol, a common alcohol in lanolin, is found both free and in esters. Cholesterol appears to give fat mixtures

Skin with a low fat content tends to be dry.

Lanolin is grease from wool.

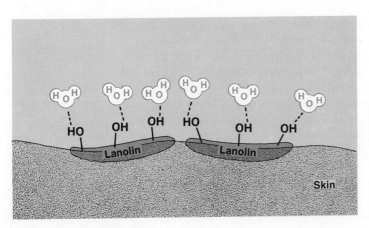

Figure 16—20 The hydroxyl groups of lanolin form hydrogen bonds with water and keep the skin moist. The fat parts of the molecule are "soluble" in the protein and fat layers of the skin.

the property of absorbing water. This is one factor that makes lanolin an excellent emollient. With its high proportion of free alcohols, particularly cholesterol, and hydroxyacid esters, lanolin has the structural groups ($-OH$) to hydrogen-bond water (to keep the skin moist) and to anchor within the skin (the fatty acid and ester hydrocarbon structures; see Figure 16–20).

Creams

Creams are generally emulsions of either an oil-in-water type or a water-in-oil type. An *emulsion* is simply a colloidal suspension of one liquid in another. The oil-in-water emulsion has tiny droplets of an oily or waxy nature dispersed throughout a water solution (homogenized milk is an example). The water-in-oil emulsion has tiny droplets of a water solution dispersed throughout an oil (natural petroleum and melted butter are examples). An oil-in-water emulsion can be washed off the hands with tap water, while a water-in-oil gives the hands a greasy, water-repellent surface.

Cold cream originally was a suspension of rose water in a mixture of almond oil and beeswax. Subsequently, other ingredients were added to get a more stable emulsion. An example of a modified cold cream composition is: almond oil, 35 percent; beeswax, 12 percent; lanolin, 15 percent; spermaceti (from whale oil), 8 percent; and strong rose water, 30 percent. Other oils can be substituted for some or all of the almond oil. Lanolin stabilizes the emulsion.

Vanishing cream is a suspension of stearic acid in water, to which a stabilizer has been added to prevent the ingredients from separating. The stabilizer may be a soap, such as potassium stearate. These creams do not actually vanish; they merely spread as a smooth, thin covering over the skin.

Creams of various sorts may be used as the base for other cosmetic preparations; other ingredients are added to give desired properties to the creams. As an example, hydrated aluminum chloride can be added to prepare a cream deodorant.

> Colloids are intermediate in size between small molecules and clumps of molecules sufficiently large to precipitate.

> Creams add oil or fat content to surface skin.

Lipstick

The skin on our lips is covered by a very thin corneal layer that is free of fat and consequently dries out easily. A normal moisture content is maintained from the mouth. In addition to being a beauty aid, lipstick can be helpful under harsh conditions that tend to dry lip tissue.

Lipstick consists of a solution or suspension of coloring agents in a mixture of high molecular weight hydrocarbons or their derivatives, or both. The material must be soft enough to produce an even application when pressed on the lips, yet the film must not be too easily removed, nor may the coloring matter run. Lipstick is perfumed to give it a pleasant odor. The color usually comes from a dye or "lake" from the eosin group of dyes. A *lake* is a precipitate of a metal ion (Fe^{3+}, Ni^{2+}, Co^{3+}) with an organic dye. The metal ion enhances the color or changes the color of the dye.

> A lake is a coloring agent made up of an organic dye adhering to an inorganic substance called a mordant. Some lakes are also approved as food colors.

Two suitable dyes, used in admixture and with their lakes, are dibromofluorescein (yellow-red) and tetrabromofluorescein (purple):

TETRABROMOFLUORESCEIN (Eosin)
(SODIUM SALT)

The ingredients in a typical formulation of lipstick include:

Dye	Furnishes color	4–8%
Castor oil, paraffins or fats	Dissolves dye	50%
Lanolin	Emollient	25%
Carnauba wax } Beeswax	{ Makes stick stiff by raising the melting point }	18%
Perfume	Imparts pleasant odor	small amount

Carnauba wax and beeswax are high molecular weight esters.

Suntan Lotions

One of the agents most harmful to skin is the short wavelength (ultraviolet) light from the sun. It is considered desirable to exclude the shorter, more harmful wavelengths, while transmitting enough less energetic, longer wavelength ultraviolet to permit gradual tanning.

Ultraviolet radiation tans skin.

The variety of suntan products ranges from lotions, which selectively filter out the higher energy ultraviolet rays of the sun, to preparations that essentially dye light-colored skin a tan color.

The lotions that filter out the ultraviolet rays are more accurately described as sunscreens, and their ingredients are often mixed with other materials, to give a lotion that both screens and tans. A common ingredient in preparations used to *prevent* sunburn is *p*-aminobenzoic acid.

Sunbathers refer to this compound as PABA.

p-AMINOBENZOIC ACID
(PABA)

Like most aromatic compounds, it absorbs strongly in the ultraviolet region of the spectrum (Figure 16–21).

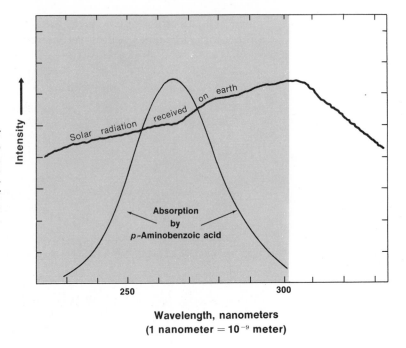

Figure 16–21 Absorption spectrum of *p*-aminobenzoic acid and its relationship to solar ultraviolet radiation received on earth. Maximum absorption occurs at 265 nanometers, although it absorbs at other wavelengths as shown. The maximum of the deep-burning ultraviolet radiation received on earth is at about 308 nanometers.

In tanning, the skin is stimulated to increase its production of the pigment **melanin.** At the same time, the skin thickens and becomes more resistant to deep burning.

Preparations for the relief of the pain of sunburn are solutions of local anesthetics such as benzocaine.

Increased melanin protects sensitive lower layers of skin.

Face Powder

Face powder is used to give the skin a smooth appearance by covering up any oil secretions, which would otherwise give it a shiny look. The powder must have some hiding ability, but if it is too opaque it will look too obvious. A powder that has the proper appearance, sticking properties, absorbance for oily skin secretions, and spreading ability usually requires several ingredients. A typical formula is:

Talc	Absorbent	65%
Precipitated chalk	Absorbent	10%
Zinc oxide	Astringent	20%
Zinc stearate	Binder	5%
Perfume, dye	Odor and color	trace

Perfume

A typical perfume has at least three components of somewhat different volatility and molecular weight. (Recall that lower molecular weight compounds are generally more volatile.) The first, called the **top note,** is the most volatile and is the obvious odor when the perfume is first applied. The second, called the **middle note,** is less volatile and is generally a flower extract (violet, lilac, etc.). The last, or **end note,** is least volatile, and is usually a resin or waxy polymer.

Perfumes are complex mixtures of odorous compounds.

Most perfumes contain many components, and chemically they are often complex mixtures. As the analysis of natural perfume materials has progressed, the use of pure synthetic organic compounds to duplicate specific odors has become very common. An example is the isolation of civetone (see structure, below), a cyclic ketone from civet, a secretion of the civet cat of Ethiopia. It is highly valued for perfumes.

Civatone is now available in a synthetic form. It is prepared by forming 8-hexadecene-1,16-dicarboxylic acid into a ring. The thorium ion (Th^{4+}) catalyzes the closure of the ring.

$$HC-(CH_2)_7COOH \atop HC-(CH_2)_7COOH \xrightarrow[\Delta]{Th^{4+}} \begin{matrix} HC-(CH_2)_7 \\ \| \\ HC-(CH_2)_7 \end{matrix}\bigg\rangle C=O + CO_2 + H_2O$$

CIVETONE

Other compounds used in perfumes include high molecular weight alcohols and esters. An example is geraniol (b.p. 230°C), a principal component of Turkish geranium oil.

$$\begin{matrix} H_3C \\ \\ H_3C \end{matrix}\bigg\rangle C=CH-CH_2-CH_2-\underset{\underset{HC-CH_2OH}{\|}}{C}-CH_3$$

GERANIOL

Esters of this alcohol are used to make synthetic rose aromas for perfumes. For example, the ester formed by reaction between geraniol and formic acid has a rose odor.

$$H-\overset{\overset{O}{\|}}{C}-OH + HOCH_2-\overset{\overset{H}{|}}{C}=R \longrightarrow$$

FORMIC ACID   GERANIOL

$$H-\overset{\overset{O}{\|}}{C}-O-CH_2CH=\overset{\overset{CH_3}{|}}{C}(CH_2)_2CH=C(CH_3)_2$$

GERANYL FORMATE

Typical perfumes are 10 to 25 percent perfume essence and 75 to 90 percent alcohol. Perfumes are added to most cosmetics to give the products desirable odors; they also mask the natural odors of other constituents. They are often mildly bactericidal and antiseptic.

Ethyl alcohol is a major constituent of most perfumes.

Eye Makeup

There are several types of eye makeup: eyebrow pencils, mascara for eyelashes, and shading, among others. Eyebrow pencils are very much like lipstick, but they contain a different coloring matter. The coloring matter is a pigment such as lampblack; the other ingredients include fats, oils, petrolatum, and lanolin, blended to give the desired melting point, which may be raised by the addition of beeswax or paraffin. Petrolatum is a semi-solid mixture of hydrocarbons (saturated, $C_{16}H_{34}$ to $C_{32}H_{66}$; and un-saturated, $C_{16}H_{32}$, etc.; melting point, 34 to 54°C). Brown pencils are made by using iron oxide pigments in place of lampblack.

Mascara is used to darken eyelashes and give them a longer appearance. The same colors as in eyebrow pencils are used, as well as other mineral coloring matters such as chromic oxide (dark green) and ultramarine (blue pigment of variable composition; probably a double silicate of sodium and aluminum silicate with some sodium sulfide). The coloring matter is suspended in a foundation that is a mixture of a soap, oils, fats, and waxes. The mascara may be water-soluble or water-resistant, depending upon the composition of the foundation. A typical formula consists of about 40 percent wax (beeswax, carnauba wax, and paraffin, adjusted for hardness), 50 percent soap, 5 percent lanolin and 5 percent coloring matter.

In the discussion of depilatories, the danger of the sulfide ion was emphasized. Through hydrolysis, the sulfide ion can produce a hydroxide ion, which functions as a corrosive poison. If very much mascara containing sodium sulfide gets into the eye, it can cause blindness.

$$S^{2-} + HOH \longrightarrow HS^- + OH^-$$

Eye shadow is a formulation of a coloring matter suspended in an oily-fatty-waxy base. A formula that has been used for this purpose is 60 percent petroleum jelly, 6 percent lanolin, 10 percent fats and waxes (approximately equal amounts of cocoa butter, beeswax, and spermaceti), and the balance zinc oxide (white) plus tinting or coloring dyes. Cocoa butter (melting point, 30 to 35°C) is composed of glycerides of stearic, palmitic, and lauric acids. It is obtained from natural products by compression of cacao seed. Spermaceti (melting point, 45°C) is chiefly cetyl palmitate, $CH_3(CH_2)_{14}COO(CH_2)_{15}CH_3$. It is taken from the solid fat from the head of the sperm whale.

Titanium dioxide, a white powder, is also used as a base for many eye makeup preparations.

Nail Polish

Nail polish is essentially a lacquer or varnish. It can be made of nitrocellulose, a plasticizer, a resin, a solvent, and perhaps a dye. The nitrocellulose can be replaced by another polymer molecule that possesses similar qualities. The evaporation of the solvent leaves a film of nitrocellulose, plasticizer, resin, and dye. The nitrocellulose furnishes the shiny film; the plasticizer is added to make the film less brittle; and the resin is added to make the film adhere to the nail better and prevent flaking. Perfumes are added to cover the odor of the other constituents. A typical formulation is:

Nail polish and hair sprays are formulated very much alike.

Nitrocellulose	15%
Acetone (solvent)	45%
Amyl acetate (solvent)	30%
Butyl stearate (plasticizer)	5%
Ester gum (resin)	5%

Perfumes and colors are added as needed.

Ester gum is a combination of esters—mainly glyceryl, methyl, and ethyl esters of rosin. Rosin is the resin remaining after distilling turpentine from pine exudate. It is 80 to 90 percent abietic acid. Rosin is slightly toxic to mucous membranes and slightly irritating to the skin. Its sticky

Abietic acid

nature is well known. The ester gum is prepared by heating rosin and the alcohol under pressure until the esterification occurs. The gums are soluble in nonpolar solvents.

Nail polish removers are simply solvents that dissolve the film left by the nail polish. They consist largely of acetone or ethyl acetate, or both, to which small amounts of butyl stearate and diethylene glycol monomethyl ether have been added to reduce the drying effect of the solvent. However, some formulations contain combinations of amyl acetate, butyl acetate, ethyl acetate, benzene, olive oil, lanolin, and alcohol. Both nail polish and nail polish removers are very flammable, and care should be taken never to use them in the presence of open flames or lighted cigarettes.

Cuticle softeners are primarily wetting agents and alkalies used to soften skin around the fingernails so it can be shaped as desired. The use of alkali to soften and swell protein is well known. A typical cuticle softener contains potassium hydroxide (3 percent), glycerol (12 percent), and water (85 percent). It may also contain sodium carbonate (an alkali), triethanol amine (a detergent, used as wetting agent), and trisodium phosphate (an alkali).

$$CH_2 - CH_2OCH_2 - CH_2$$
$$|\qquad\qquad\qquad |$$
$$OH\qquad\qquad\quad OCH_3$$

Diethylene glycol monomethyl ether

---

1. The surface pH of the human body is about _____.

2. Hair keratin contains considerable amounts of the sulfur-containing amino acid _____.

3. Two types of bonds that hold strands of protein in place include _____ and _____.

4. Wet hair swells and stretches because of the breaking of_____ bonds.

5. The peroxide oxidation of melanin produces a (colored/colorless) product.

6. A plasticizer causes a plastic material to be more _____.

7. What is the purpose of talc in powders? _____

8. Zinc peroxide acts in deodorants to oxidize amines and _____ _____.

9. Commercial lanolin comes from what animal? _____

10. Vanishing cream is completely absorbed through the pores of the skin. (True/False)

_____ 1. corneal layer   (a) bonds two protein chains

_____ 2. keratin   (b) reducing agent in waving lotion

_____ 3. sulfide   (c) hair bleach

_____ 4. isopropyl alcohol   (d) protein in skin

_____ 5. hydrogen peroxide   (e) resin in hair spray

_____ 6. hydrated aluminum chloride (f) alkaline ion in depilatory

_____ 7. disulfide linkage   (g) used in deodorants

_____ 8. lanolin   (h) surface skin

_____ 9. polyvinylpyrrolidone   (i) skin disinfectant

_____ 10. thioglycolic acid   (j) emollient

PART IV—AUTOMOTIVE PRODUCTS

A significant amount of chemistry is involved in the production and operation of the automobile. Before an automobile can exist, metals must be separated from their ores; plastics must be synthesized and fabricated; paints have to be formulated; sulfuric acid for the battery must be made; rubber must be synthesized, formed into shape, and vulcanized; and glass has to be mixed, fired, molded, and cut. Many chemical reactions are required to power the automobile. The two most familiar are the combustion of the fuel and the electrochemistry of the battery. In its wake the automobile leaves its chemical exhaust to undergo a variety of chemical reactions, some of which lead to smog formation. The chemistry of fuels and exhaust gases has been discussed in previous chapters. In this section, we shall discuss some of the chemicals that consumers buy for their automobiles.

Miscellaneous Gasoline Additives

In addition to tetraethyllead and/or aromatic compounds that reduce preignition or "knock" in an automobile engine, numerous other chemicals are added to gasoline to improve its properties.

Deposit modifiers are added to gasoline to prevent the ignition of new fuel by glowing particles from a previous ignition. Phosphorus compounds such as tricresyl phosphate (one trade name is TCP) and, more recently, boron compounds, have been used for this purpose. These reduce the melting point of the lead oxide residue (by forming a lead borate or phosphate glass), and the resulting hot liquid is blown out with the exhaust. The phosphorus compounds also prevent spark plug deposits from becoming so electrically conductive that the charge leaks away instead of firing the plug.

Tricresyl phosphate

Antioxidants, such as phenylenediamine, aminophenols, dibutyl-p-cresol, and ortho-alkylated phenols, are added to prevent the formation of peroxides that lead to knock and gum formation. About 2 or 3 pounds of these additives are added to every 1000 barrels of gasoline. The mechanism of antioxidation of automotive oils is very similar to the mechanism described for BHA and BHT in food additives. Because copper ions

Water molecule

Nonpolar end of molecule

Polar end of molecule

Steel or iron

Figure 16—22 The action of surface active agents, such as rust inhibitors, mild antiwear agents, and some deicing agents.

catalyze gum formation, metal ion scavengers such as ethylenediamine are added to chelate the trace amounts of copper ions and render them ineffective. The copper gets into the gasoline from the copper tubing used for fuel lines and from brass parts of the engine.

Brass is an alloy of copper and zinc.

To inhibit water from corroding and rusting storage tanks, pipelines, tankers, and fuel systems of engines, ***antirust agents*** are added. Four compounds used to prevent corrosion are trimethyl phosphate, sodium and calcium sulfonates, and N,N'-di-sec-butyl-*p*-phenylenediamine. All these compounds have a polar or ionic end and a nonpolar end in the molecule. These agents coat metal surfaces with a very thin protective film that keeps water from contacting the surfaces, as shown in Figure 16–22. This also helps prevent gummy deposits in the carburetor and combats carburetor icing during cold weather.

Antiicing agents coat metal surfaces, as do antirust agents, and thus prevent ice particles from accumulating on surfaces, and/or depress the freezing point. The small ice particles pass harmlessly through the carburetor and into the engine where the heat converts them to water vapor that eventually exits with the exhaust gases. The freezing point depressants, which include alcohols and glycols, act in the same manner as the antifreeze in the engine's cooling system.

OH OH
| |
H—C—C—H
| |
H   H
Ethylene glycol

Detergents, which include alkylammonium dialkyl phosphates, are added to prevent the accumulation of high-boiling components on the walls of the carburetor. These deposits interfere with the air flow into the carburetor and cause rough idling, frequent stalls and increased fuel consumption. The effectiveness of these detergents stems from their surface active properties, as shown in Figure 16–22. The film of detergent provides a thin, nonpolar coating on the metal surfaces that prevents high molecular weight, nonpolar gums from forming thick deposits on the surfaces.

H_2C—CH_2

H_2C CH_2

C
H_2
Cyclopentane

Upper cylinder lubricants are sometimes blended into gasoline. These are usually light mineral oils or low viscosity naphthenic distillates (such as cyclopentane) that lubricate the moving metal parts and dissolve away deposits in the intake system from the carburetor, cylinders, top piston rings, and valves. (Nonpolar deposits tend to dissolve in nonpolar solvents.)

At one time ***dyes*** were added to distinguish antiknock "ethyl" gasolines from nonantiknock gasolines. Always present in very small amounts, they are now used to distinguish grades and brands.

Lubricants and Greases

Lubricating oils from petroleum consist essentially of complex mixtures of hydrocarbon molecules. These generally range from low viscosity oils, having molecular weights as low as 250 amu, to very viscous lubricants with molecular weights as high as about 1000 amu. The viscosity of an oil can often determine its use. For example, if the oil is too viscous, it offers too much resistance to the moving metal parts. On the other hand, if the oil is not viscous enough, it will be squeezed out from between the metal surfaces, and consequently offer insufficient lubricating action. For these and other reasons, motor oil is often a mixture of oils with varying viscosities. The common 10W-30 oil, for instance, combines the low-temperature viscosity of the Society of Automotive Engineers (SAE) 10W classification (Table 16–8) for easy low-temperature starting with SAE 30 high-temperature viscosity for better load capacity in bearings at the normal engine running temperature.

"Viscous" means resistant to flow, like molasses.

The usefulness of motor oil is improved by the addition of substances such as antiwear agents, oxidation inhibitors, rust inhibitors, detergents, viscosity improvers, and foam inhibitors.

A number of synthetic oil blends have shown superior performance in that the oil molecules are more stable to heat. Breakdown of oil is the result of the thermal decomposition of the oil molecules to smaller ones, and also the partial oxidation of hydrocarbons at high temperatures.

Greases are essentially lubricating oils thickened with a gelling agent such as fatty acid soaps. The soaps form a network of fibers that entrap the oil molecules within the interlacing fiber structures. Carbon black, silica gel, and clay are also used to thicken petroleum greases. Chemical additives similar to those used in lubricating oils and gasolines are added to greases to improve oxidation resistance, rust protection, and extreme pressure properties. Synthetic greases are being developed that deteriorate so slowly that longer intervals between grease jobs are now possible. Silicone greases have a useful life of up to 1000 hours at 450°F (232°C). Unfortunately, silicone greases provide relatively poor lubrication for gears and other sliding devices. Diester greases such as di(2-ethylhexyl)

$$
\begin{array}{c}
CH_3 \\
| \\
(CH_2)_3 \\
| \\
CH-C_2H_5 \\
| \\
CH_2 \\
| \\
O \\
| \\
C=O \\
| \\
(CH_2)_8 \quad \textit{Diester} \\
| \\
C=O \\
| \\
O \\
| \\
CH_2 \\
| \\
CH-C_2H_5 \\
| \\
(CH_2)_3 \\
| \\
CH_3
\end{array}
$$

Di(2-ethylhexyl) sebacate

TABLE 16—8 Viscosity Data at −18°C and 99°C for the SAE Method of Rating Motor Oils

Motor Oil	Viscosity (SUS)*			
	−18°C		99°C	
	Min.	Max.	Min.	Max.
5W		4,000	39	
10W	6,000	12,000	39	
20W	12,000	48,000	39	
20			45	58
30			58	70
40			70	85
50			85	110

*SUS is the Saybolt Universal Second, which is the time in seconds required for 60 ml of oil to empty out of the cup in a Saybolt viscometer through a carefully specified capillary opening. Note the extremely shortened time for the outflowing of the hotter oil.

Mo

S

Figure 16–23 The structure of molybdenum disulfide, MoS_2. The small spheres represent molybdenum atoms. Note how the sulfides are adjacent to sulfides in every other layer. This allows one layer to slide over another, accounting for the slippery nature of MoS_2.

sebacate have found extensive use among synthetic greases. Lithium soaps dissolve well in the diester oil and form a grease with equal or better lubrication characteristics and a considerably longer useful life than petroleum greases. Blends of silicone oil and diester oil provide greases with good low-resistance lubricating power even at low temperatures ($-73°C$).

Solid Lubricants

Perhaps the most tenacious and wear resistant solid lubricant is molybdenum disulfide, MoS_2. Like graphite, a very common solid lubricant, MoS_2 has a layered structure that enables one layer to float over another and provide lubrication (Figure 16–23). Strong bonds exist within each S—Mo—S layer, while weak S—S bonds between the layers allow easy sliding of one layer over another. MoS_2 is extensively used in greases for automotive chassis lubrication. When used alone, MoS_2 suffers like any other solid lubricant from having poor wear resistance and being unable to heal any breaks in its surface coating.

Antifreeze

An antifreeze is a substance that is added to a liquid, usually water, to lower its freezing point. Although various substances have been used as antifreezes in the past, nearly all of the current market is supplied by ethylene glycol and methyl alcohol.

Ethylene comes from cracking petroleum. Ethylene glycol is prepared from ethylene, $CH_2{=}CH_2$, by an oxidation reaction

$$2CH_2{=}CH_2 + O_2 \longrightarrow 2CH_2{-}CH_2$$
$$\underset{\text{Ethylene oxide}}{O}$$

followed by hydrolysis:

$$\underset{O}{CH_2{-}CH_2} + H_2O \longrightarrow \underset{OH \quad OH}{CH_2{-}CH_2}$$

$$\begin{array}{ccc} & OH & OH \\ & | & | \\ H - & C - & C - H \\ & | & | \\ & H & H \end{array}$$
ETHYLENE GLYCOL

$$\begin{array}{c} OH \\ | \\ H - C - H \\ | \\ H \end{array}$$
METHYL ALCOHOL

More than 95 percent of the antifreeze on the market is "permanent" antifreeze, having ethylene glycol as the major constituent. The largest use of antifreeze is in protection of water-cooled engines. Water has two serious disadvantages as an engine coolant. First, it has a relatively high freezing point and second, under normal operating conditions, it is corrosive. Modern antifreeze mixtures effectively counteract these problems.

Antifreeze protection charts always show the temperatures at which the first ice crystals form. Below this temperature the antifreeze solution

a   b

Figure 16–24 How foreign molecules prevent water from forming its normal crystal structure. For example, by forming hydrogen bonds to water molecules, ethylene glycol (b) prevents water molecules from assuming their places in the ice structure (a).

turns to slush. If the slush is unable to circulate through the radiator, overheating, boiling and engine damage can result. For this reason it is best to add sufficient antifreeze to prevent the formation of the first ice crystals even at the lowest anticipated temperature (Figure 16–24).

A service station attendant measures the effectiveness of the antifreeze in your car by reading the position at which the hydrometer floats in a sample of a radiator solution. He is really measuring the solution's density, which varies with the amount of antifreeze present, since ethylene glycol has a density greater than that of water.

The density of an antifreeze solution can be used to determine its freezing temperature.

The prevention of corrosion is the second most important job of antifreeze. Metals in the cooling system that are subject to corrosion are copper, steel, cast iron, aluminum, solder (lead and tin), and brass (copper and zinc). The presence of oxygen, along with high temperatures, pressures, and flow rates, increases the possibility of general corrosion. Although corrosion can "eat" through the walls of the cooling system, the most general trouble is overheating caused by flakes of metal oxide clogging the radiator. Some additives inhibit corrosion by acting as reducing agents (e.g., nitrites), some as ion scavengers (e.g., phosphates), and some as surface active agents (detergents; e.g., triethanolamine). Most antifreezes contain two or more inhibitors for the different metals. All inhibitors are depleted with use.

Good auto antifreeze formulations also contain rust inhibitors.

Radiator sealants have been on the market for years. Modern sealants in antifreeze include asbestos fiber and polystyrene spheres. When a radiator springs a leak, the coolant penetrates the crack because the pressure inside the cooling system is greater than atmospheric pressure. Asbestos fibers are often too big to squeeze through and thus get caught in the crack, plugging the hole in the radiator. On the other hand, the larger polystyrene particles initially plug up most of the leak while the smaller ones build up behind them. The spheres fuse together under the pressure and temperature conditions that exist and form a solid plug in the crack (Figure 16–25). This is effective in stopping up holes or cracks up to 0.5 millimeter width, which includes about 90 percent of all radiator leaks.

Foaming is caused by one or more of several factors: air leaks in hoses,

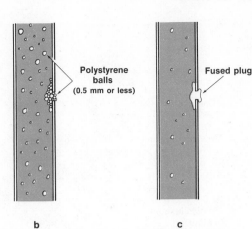

Figure 16–25 How polystyrene works to plug radiator leaks. Details are given in the text.

water pump, or radiator; exhaust gas leaking into the cooling system; failure to drain out cooling system cleansers; or extended use of antifreeze. Foaming can be corrected by tightening the system or by use of antifoam additives, such as silicones, polyglycols, mineral oils, high molecular weight alcohols, organic phosphates, alkyl lactates, castor oil soaps, and calcium acetate. These substances reduce the surface tension so bubbles cannot hold together.

Corrosion begins in the cooling system when the pH of the coolant drops below 7. Most commercial antifreeze contains an extra amount of alkali to maintain the pH above 7. As time passes this alkali is exhausted and corrosion begins. A check of your antifreeze's corrosion-fighting capability can be made with a piece of litmus paper. Pink means acid and trouble through corrosion. Adding a can of rust inhibitor (which contains an alkali) will adjust the pH back to a value above 7. A retest with litmus paper shows blue (basic). Your car is now protected as long as the mixture composition doesn't change due to boil-away or leaks.

Deicing fluids are a type of antifreeze, but they also will melt ice and frost. They are chiefly employed in removing ice and frost from parked aircraft and car windows. The glycol-alcohol type used on cars came on the market in 1959. The better formulations contain the following ingredients: ethylene or propylene glycol (for protection against refogging); one or more of the lower alcohols, such as 2-propanol, denatured alcohol, etc. (for low viscosity and good spray pattern); water (to minimize inside fogging by diminishing the evaporative-cooling effect of the formulation); nitrite or other inhibitor (to prevent corrosion of the container); and wetting agent, such as ethanolamine (to break surface tension so fluid will attack the ice readily).

A simple alcohol-water mixture works satisfactorily for quick frost removal and to defog the inside of glass, but it evaporates quickly and refogging occurs unless glycol is present. Before a deicer formulation is applied, the snow should be removed and the ice should be scored to permit faster penetration.

Pink litmus—acidic. Blue litmus—basic.

Deicers lower the freezing point of water.

Miscellaneous Automotive Products

Consumers spend millions of dollars each year on a large variety of specialized automotive products. The formulations of some typical examples of these products are given in Table 16–9.

TABLE 16–9   Typical Composition of Some
Automotive Products

Engine and Motor Cleaners

Ethylene dichloride	63%
p-Cresol	25%
Oleic acid	7.2%
Potassium (or sodium) hydroxide	1.4%
Water	3.0%

Radiator Cleaners

(1)	Oxalic acid	40%
	Boric acid	60%
(2)	Sodium carbonate	85%
	Potassium dichromate	15%

May contain also:
sulfamic acid
1-butanol (n-butyl alcohol)
2-propanol (isopropyl alcohol

Radiator Stop Leak

Dextrin	5–10%
Cellulose gum	0.5%
Asbestos	5%
Sodium carbonate	0.8%
Isopropyl alcohol	10–15%
Water	to 100%

Shock Absorber Fluids

(1)	Petroleum ether	97%
	Kerosene	3%
(2)	Mineral oils	90–100%
	Fatty oils	0–5%

May contain also:
viscosity improvers (polymethacrylate esters)	0–5%
dyes	0–150 ppm

Automatic Transmission Fluids

Mineral oils	75–100%
Oxidation inhibitors and detergents	0–20%
Viscosity improvers Polymethacrylate esters Polyisobutylenes	0–5%
Anti-wear agents Organic borates	0–2%
Antifoam agents Polysiloxanes	0–15%
Sealant Triaryl phosphate	to 100%
Dyes	0–200 ppm

Brake Fluids

Lubricant Castor oil Butyl or glyceryl ether of polyoxyethylene propylene glycol Prolypropylene glycol	20–25%
Solvent Methyl, ethyl and butyl ethers of ethylene glycol and related glycols	80–85%

May contain also:
Inhibitors
Amine soaps
Potassium soaps
Borax
Antioxidants
Hydroquinone
Dyes

---

1. Antirust agents such as trimethyl phosphate in gasoline prevent the formation of rust by: (a) reacting with oxygen; (b) coating metal surfaces; (c) converting rust (Fe_2O_3) into iron metal; (d) absorbing water. Select one answer.

2. Gum in the automobile fuel system is formed by _____ of hydrocarbons which must have one or more _____ bonds per molecule.

3. Since gums are nonpolar, a _____ solvent such as cyclopentane is used to dissolve the gum in an engine.

4. Antioxidants such as aminophenols in gasoline prevent oxygen from reacting with gasoline prior to ignition by destroying _____, which have unpaired electrons.

5. Lubricating oils with a rating of 10W-30: (a) have one type of molecule; (b) have the same viscosity at high and low temperatures; (c) are a mixture of oils with varying viscosities. Select one answer.

6. Molybdenum disulfide (MoS_2) is an excellent solid lubricant because it consists of _____, which slide over each other.

7. The viscosity of oils and greases is increased by adding linear polymers because the chains of the polymers become _____.

8. Automotive greases are essentially thickened (or gelled) motor oils. (True/False)

9. The development of what two kinds of greases makes it possible to go longer between automotive lubrication? _____ and _____

10. Ethylene glycol makes it more difficult for water to boil because the —OH groups on an ethylene glycol molecule form hydrogen _____ with water molecules and detain them from escaping into the atmosphere.

_____ 1. tricresyl phosphate   (a) phenylenediamine

_____ 2. antioxidant   (b) oil gelled in soap

_____ 3. antifreeze   (c) in some antifreeze mixtures to stop leaks

_____ 4. grease   (d) permanent antifreeze mixture

_____ 5. MoS_2 (e) lightweight motor oil

_____ 6. polystyrene   (f) ethylene glycol

_____ 7. greater density than water   (g) deposit modifier

_____ 8. 5W   (h) added to gasoline to keep carburetor clean

_____ 9. dyes   (i) used to distinguish grades of gasoline

_____ 10. detergent   (j) solid lubricant

MATCHING SET

PART V—THE CHEMISTRY OF PHOTOGRAPHY

Black and White Photography

Many chemical substances are known to be photosensitive. The chloride, bromide, and iodide salts of silver can illustrate the useful ability of storing an image of the absorbed light. The various processes known as *photography* are based upon this property.

J. H. Schulze observed in 1727 that a mixture of silver nitrate and chalk darkened on exposure to light. The first permanent images were obtained in 1824 by Nicéphore Niepce, a French physicist, by means of glass plates coated with a coal derivative (called bitumen) containing silver salts. In the early 1830's, Niepce's partner, Louis Daguerre, discovered by accident that mercury vapor was capable of developing an image from a silver-plated copper sheet that had been sensitized by iodine vapor. The *daguerreotype* image was rendered permanent by washing the plate with hot concentrated salt solution. In 1839 Daguerre demonstrated his photographic process to the Academy of Sciences in Paris. The process was improved by using sodium thiosulfate to wash off the unexposed silver salts.

In 1841, an Englishman, William Henry Fox Talbot, announced the calotype process. The Talbot process (Figure 16–26) involved a paper made sensitive to light by silver iodide. The light-sensitive paper could be developed into a negative image with gallic acid in a development process essentially the same as is used today. When made with semitransparent paper, Talbot's negatives could be laid over another piece of photographic paper which, when exposed and developed, yielded a "positive," or direct copy of the original. Although the Talbot process required less time than the Daguerre process, the Talbot images were not sharp.

Photochemistry deals with the chemical changes produced by absorbed radiant energy.

Pronunciation of daguerreotype: (dä-ger'-ō-tīp)

Figure 16–26   Talbot process.

Emulsion
(gelatin with grains of silver halide dispersed throughout)

Film support
(glass or flexible plastic)

Figure 16—27 A modern photosensitive gelatin emulsion.

It was obvious that some way of holding the silver halides on a transparent material would have to be devised.

At first, the silver salts were held on glass with egg white as binder. This provided sharp, though easily damaged, pictures. By 1871, the problem had been solved by an amateur photographer and physician, Dr. R. L. Maddox. He discovered a way to make a gelatin emulsion of silver salts and apply it to glass. In 1887, George Eastman introduced the Kodak, a camera using film made by attaching a gelatin emulsion to a plastic (cellulose nitrate) base (Figure 16–27). The camera could take 100 pictures and then camera and film had to be sent to Rochester, N.Y. for processing. The age of modern photography had arrived.

Cellulose acetate replaced easily combustible cellulose nitrate as the film support in 1951.

Photochemistry of Silver Salts

To understand the chemistry of photography, we must first look at the photochemistry of silver salts. A typical photographic film contains tiny crystallites called *grains* (Figure 16–28), which are composed of a very slightly soluble silver salt, such as silver bromide, AgBr. The grains

Figure 16—28 (*a*) Photomicrograph of the grains of a slow positive emulsion. (*b*) The grains of a high speed negative emulsion at the same magnification.

Figure 16–29 Grain in a photographic image is illustrated by five degrees of magnification: (*a*) original size; (*b*) $\times$ 25; (*c*) $\times$ 250; (*d*) $\times$ 2500; and (*e*) $\times$ 25,000 (by electron microscope).

are suspended in gelatin, and the resulting gelatin emulsion is melted and applied as a coating on glass plates or plastic film.

When light of an appropriate wavelength strikes one of the grains, a series of reactions begins that leaves a small amount of free silver in the grain. Initially, a free bromine atom is produced when the bromide ion absorbs the photon of light:

$$Ag^+Br^- \xrightarrow{\text{Light absorption}} Ag^+ + Br^0 + e^-$$

The silver ion can combine with the electron to produce a silver atom.

$$Ag^+ + e^- \longrightarrow Ag^0$$

Association within the grains produces species such as Ag_2^+, Ag_2^0, Ag_3^+, Ag_3^0, Ag_4^+, and Ag_4^0. The presence of this free silver in the exposed silver bromide grains provides the ***latent image,*** which is later brought out by the development process. The grains containing the free silver in the form of Ag_4^0 are readily reduced by the developer to form relatively massive amounts of free silver; hence a dark area appears at that point on the film. The unexposed grains are not reduced by the developer under the same critical conditions.

Film sensitivity is related to grain size and to the halide composition. As the grain size in the emulsion increases, the effective light sensitivity of the film increases (up to a point; Figure 16–28). The reason for this is that the same number of silver atoms are needed to initiate reduction of the entire grain by the developer despite the grain size.

In order for an exposed AgBr grain to be developable, it will need a minimum of four silver atoms as Ag_4^0.

The latent image is the "invisible developable image" stored in the silver halide grains.

Film sensitivity is rated on the American Standards Association (ASA) scale. The larger the number, the more sensitive the film is to light.

Amplification of the Latent Image—Development

Silver halides are not the most photosensitive materials known. Why, then, are they effective image producers? The answer lies in the fact that the impact of a single photon on a silver halide grain produces a nucleus of at least four silver atoms, and this effect is amplified as much as a billion times by the action of a proper reducing agent *(developer)*.

When an exposed film is placed in developer, the grains that contain silver atom nuclei are reduced faster than those grains which do not. The more nuclei present in a given grain, the faster the reaction. The reduction reaction is

$$Ag^+ + e^- \longrightarrow Ag^0$$

IN GRAINS
CONTAINING Ag_4^0

The blackness on the negative is due to free silver atoms, Ag^0.

Factors such as temperature, concentration of the developer, pH, and the total number of nuclei in each grain determine the extent of development and the intensity of free silver (blackness) deposited in the film emulsion.

Not only must the developer be capable of reducing silver ions to free silver, but it must be selective enough not to reduce the unexposed grains, a process known as "fogging." Table 16–10 lists some substances that are used as developing agents.

Most developers used for black and white photography are composed of hydroquinone and metol or hydroquinone and Phenidone. A typical developer consists of a developing agent (or two), a preservative to prevent air oxidation, and an alkaline buffer to prevent the actual reduction reaction from being retarded (Table 16–11). Other chemicals might be added but they are not absolutely necessary.

TABLE 16–10   Some Compounds Used as Photographic Developers

Name	Formula
Gallic acid	
o-Aminophenol	
Hydroquinone	
p-Methylaminophenol (metol)	
1-Phenyl-3-pyrazolidone (Phenidone)	

TABLE 16–11   Formula for a Typical
Developer for Black and White Films

750 ml water at 50°C; dissolve in this water:

Metol	2.0 g
Hydroquinone	5.0
Sodium sulfite	100.0
Borax, $Na_2B_4O_7 \cdot 10H_2O$	2.0

Add cold water to make 1000 ml of solution.

When hydroquinone acts as a developer, quinone is formed. Two protons are also produced for every two silver atoms:

HYDROQUINONE   QUINONE

Since this reaction is reversible, a buildup of either protons or quinone would impede the development process. The sodium sulfite reacts with quinone and destroys its ability to revert back to hydroquinone.

The protons are neutralized effectively by the hydroxide ions.

$$H^+ + OH^- \longrightarrow H_2O$$

If development proceeds either too long or at a higher temperature than recommended, sufficient fogging occurs to render a negative useless. Since the rates of the development reactions increase with increasing temperature, the photographer usually controls the temperature of the development bath very carefully.

The development process is terminated by a **stop bath.** The stop bath usually contains a weak acid such as acetic acid, which decreases the pH. The action of a stop bath is to build up the amount of hydrogen ions, which effectively stops the hydroquinone $\longrightarrow$ quinone reaction.

Fixing

One of the principal problems in the early days of photography was the lack of permanence of the image. If development only produces free silver where the light intensity was greatest and nothing further is done to the negative, the undeveloped silver halide will be exposed the instant

Figure 16–30 Black and white enlargement printing involves choosing a negative and placing it in an enlarger (*a*). The print paper is exposed for a length of time necessary to produce the desired print (usually by trial and error), and then placed in developer (*b*). After development, the print is fixed with hypo and washed thoroughly to produce the finished product (*c*), which is a direct opposite of the negative.

it is taken into the light. After that, almost any reducing agent will completely fog the negative. In order to overcome this problem, a suitable substance had to be found to remove the unreduced silver halides. The most commonly used *fixing agent* in black and white photography is the thiosulfate ion ($S_2O_3^{2-}$). Thiosulfate ions form stable complexes with silver ions in aqueous solution:

Solutions of sodium thiosulfate, $Na_2S_2O_3$, known as "hypo," were first used by Sir J. W. F. Herschel to "fix" negatives.

$$AgBr(s) \quad + \quad 2S_2O_3^{2-} \quad \longrightarrow \quad Ag(S_2O_3)_2^{3-} \quad + \quad Br^-$$

INSOLUBLE	FROM	WATER-SOLUBLE
SALT	"HYPO"	COMPLEX
(UNDEVELOPED)	SOLUTION	

Instant Black and White Pictures

In 1947, Dr. Edwin H. Land introduced his invention of a process that produced a finished picture in one minute. Since their introduction,

Figure 16–31 A schematic diagram of the chemistry of the Polaroid process.

the Polaroid and similar processes have been popular for just this unique feature.

After exposure, a Polaroid film is brought into contact with a piece of receiver paper; at the same time a pod containing both a developer and a silver solvent is broken and the pasty mixture is spread over the film. As the developer reduces the exposed silver halide grains in the film emulsion, the silver solvent picks up the unexposed silver ions, which then diffuse across the boundary onto the receiver paper. There, in contact with minute grains of silver already in the paper, the developer reduces the silver in the hypo complex to free silver and a *positive* image. This positive image forms since black areas in the original object photographed exposed no silver grains in the emulsion, and it was this silver, as Ag+ ions, which was then carried onto the receiver paper and reduced (Figure 16–31).

Spectral Sensitivity

Probably the most important ingredients in a black and white photographic emulsion, other than the silver halide salts themselves, are the spectral sensitizing dyes. Silver halides are most sensitive to blue light or higher energy electromagnetic radiation such as uv light (Figure 16–32). A film manufactured with only silver halides as the photosensitive agents will be only blue-sensitive and will not "see" reds, yellows, greens, and so on, as ordinary colors.

In 1873, while trying to eliminate light scattering problems in photographing the solar spectrum, W. H. Vogel, a German chemist, added a yellow dye to his emulsion. To his surprise he discovered that he could now record images in the green region of the visible spectrum. Later, in 1904, another German, B. Homolka, discovered a dye, pinacyanol, which when added to a silver halide emulsion rendered it sensitive to the entire visible spectrum (Figure 16–33). Films of this type are called *panchromatic* or "pan" films.

The mechanism by which a dye molecule can impart spectral sensitivity to silver halide grains seems to involve initially the absorption of a photon of light by the dye molecule. Next, the excited molecule ejects

By adding spectral sensitizing dyes, photographic emulsions can be made that are sensitive to selected regions of the spectrum with wavelengths from 100 nanometers (ultraviolet) to 1300 nanometers (infrared).

Figure 16–32 Spectral sensitivity of a typical AgBr emulsion.

Wavelength, nm

(nm = nanometer = 10^{-9} meter)

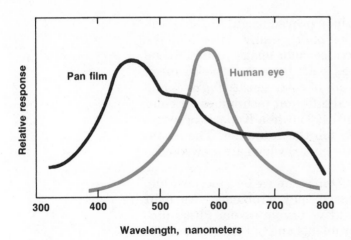

Figure 16–33 Spectral sensitivity of a panchromatic film compared to that of the human eye.

A panchromatic film is sensitive to the entire range of visible wavelengths.

an electron into the silver halide grain, where a free silver atom is formed. The electron-deficient dye molecule then oxidizes the bromide ion, producing a bromine atom:

$$Dye \xrightarrow{\text{Light}} Dye^+ + e^-$$

$$Ag^+ + e^- \longrightarrow Ag^0 \longrightarrow \text{Silver nuclei}$$

$$Dye^+ + Br^- \longrightarrow Dye + Br^0$$

Thus, the process is effectively the same as a photon striking the bromide in the grain itself, the dye serving as a catalyst.

Color Photography

The chemistry of color photography, which dates back to 1861, is more complicated than that of black and white photography. James Clerk Maxwell, the famous English scientist, was the first to photograph an object in color. He used three exposures through three primary color filters and then resynthesized the color of the object by projecting the image through the same filters. This experiment actually predated panchromatic film, but because a peculiarity of his particular emulsion, it was essentially panchromatic nonetheless. The important point is that the results were consistent with the then emerging theory of color vision and thus led to other more significant results.

Additive and Subtractive Primary Colors

As early as 1611, De Dominis showed that the visible spectrum is composed of three fundamental colors: red, green, and blue (known as

Figure 16–34 Pinacyanol, a cyanine dye. The conjugated group (double bond, single bond, etc.—one sequence in the white block) is the chromophore (Chapter 8). Other cyanine dyes have more CH groups, absorb longer wavelengths of light, and shift the film sensitivity toward the red.

additive primaries). This concept has since proved useful in the development of color vision theory and in color photography. After 1861 the idea slowly evolved that in order to reproduce color images, a film would have to be made with three different layers, each layer sensitive to one of the three primary colors. After the discovery of color-sensitizing dyes and panchromatic black and white film, several different techniques for color photography were developed; but not until 1935, when Kodachrome was placed on the market, did the products reach the consumer. The Kodachrome process produces transparencies (or slides) which are viewed with transmitted light.

If additive primary colors are used to form an image by superposition, such as we might expect in a color transparency, problems with light transmittance arise. Combinations of additive primary-color filters produce black. In order to overcome these problems and obtain a color slide (or color negative, for that matter), another system of primary-color filters was developed, known as *subtractive primary* colors. These colors are produced by dyes that absorb the additive primary colors. Thus a dye that absorbs red light transmits or reflects the remainder of the spectrum and appears greenish-blue (*cyan*). Absorption of blue light renders a dye yellow, and absorption of green light makes the dye appear bluish-red (*magenta*).

Blue light +
Green light } Cyan

Green light +
Red light } Yellow

Blue light +
Red light } Magenta

When the proper mixture of subtractive primary dyes is formed in a photographic emulsion during the development process, an image is produced with the desired color. For example, a mixture of magenta and cyan dyes would appear blue, since the magenta dye absorbs green light and the cyan dye absorbs red light, leaving only blue to be transmitted out of the three components of white light.

The use of subtractive primaries in color photography was suggested as early as 1869, but it was much later before the chemistry was worked out in enough detail to yield good results. The problem is to get the right amount of the correct subtractive primary dye in the right place to reproduce the correct true-to-life color. White is produced by the absence of all three subtractive primaries, and black is produced by an equal balance of all three.

Color Film

Generally, a color film consists of a support and three color-sensitive emulsion layers. The blue-sensitive layer is usually on the top since silver halides are inherently blue-sensitive. Next, a yellow colored filter layer is added. This layer absorbs blue light and serves to protect the lower emulsion layers from blue light. A green-sensitive layer is added and followed by a red-sensitive layer and the support (Figure 16–35). These layers are rendered color-sensitive by dyes similar to those in the cyanine class, which render black and white film panchromatic. It should be realized, however, that the color-sensitizing dyes are *not* generally involved in producing the final primary colors responsible for the color of the image. It is the final processing of the color film that yields the color image.

The thickness of the entire emulsion of color film is only about 0.0254 mm (0.001 inch) thick.

Color Development

Most color films are developed with the aid of a dye-forming color process first introduced by a German chemist, R. Fischer, in 1912. The

Figure 16–35 A typical arrangement of color-sensitive emulsion layers in a color film.

basis for this process is the oxidation of the developer to a dye-forming substance, which is then allowed to react with a molecule called a *coupler* to form the dye.

Color developers are generally substituted amines and as such are reducing agents. An example is N,N-diethyl-*p*-phenylenediamine:

N,N-DIETHYL-*p*-PHENYLENEDIAMINE
(A COLOR DEVELOPER)

To form a cyan dye during the development process, a phenol compound such as α-naphthol acts as a coupler:

COUPLER
(α-NAPHTHOL)

DEVELOPER

A CYAN DYE (ABSORBS AT 630 nm)

Light with a wavelength of 630 nm is red.

Thus, in the development of an exposed silver halide grain in the red-sensitive emulsion layer, a small amount of cyan dye is produced. The free silver must be bleached out prior to finishing.

Figure 16—36 Simplified color image-forming process. B—blue-sensitive layer; Y—yellow filters; G—green-sensitive layer; R—red-sensitive layer.

The Kodachrome Process

An interesting example of a widely used color photography system is the Kodachrome process of Eastman Kodak Company. The Kodachrome process is a *reversal* process; this means colors are reproduced in terms of their correct values and not their negative or complementary colors. The first developer in the Kodachrome process is a black and white developer. By careful temperature control, development of the exposed silver halide is made essentially complete.

The remaining unexposed silver halide in the three color-sensitive emulsions is a positive record of the original exposure. For example, red light striking the film would, upon black and white development, leave free silver in the red-sensitive layer (Figure 16–36). Since no other color-sensitive layers were exposed by the original image, they contain no information. Now, selective reexposure and color development will produce free silver throughout the emulsion layers, along with the colored dyes, *except* where the red light originally struck the film. No dye forms there since the silver was previously reduced with a black and white developer.

Next, all the silver in the three emulsion layers, as well as the yellow-colored protective layer, is bleached with an oxidant such as cyanoferrate ion, $Fe(CN)_6^{3-}$:

All the silver is bleached out of color negative and reversal films during processing.

$$Ag^0 + Fe(CN)_6^{3-} \longrightarrow Ag^+ + Fe(CN)_6^{4-}$$

Figure 16—37 White light passes through a three-layer transparency. Since blue light and green light are absorbed, the transmitted light is red.

Once oxidized, the silver is treated with hypo and washed from the emulsion. The resulting emulsion is colored, but transparent. Considering that red light originally exposed the film, we see that the transmitted light will appear red (Figure 16–37).

"Instant" color pictures involve the same principles described above, but require a delicate balance of light exposure, photochemical reagents, dyes, developers and couplers.

---

1. The branch of chemistry that deals with the changes in matter produced by absorbed radiant energy is _____.   SELF-TEST 16-F

2. What chemical composes the dark regions on a negative? _____

3. Name a chemical that can be used as a black-and-white developer.

   _____.

4. In the development process, silver is (oxidized/reduced).

5. Complete the following equation for the fixing process.

   $$Ag^+ + 2S_2O_3^{2-} \longrightarrow \text{\underline{\hspace{4cm}}}$$

6. Which are more sensitive to light, very small or somewhat larger grains of AgBr? _____

7. Photographic developers are (oxidizing/reducing) agents.

8. Hypo is another name for _____ _____.

9. (a) cyan + magenta = _____

   (b) cyan + yellow = _____

   (c) magenta + yellow = _____

10. The primary additive colors are _____, _____,

    and _____.

11. A Kodachrome color transparency has no silver in it. (True/False)

12. In the formation of a dye in color film, silver is (oxidized/reduced).

13. Film sensitive to all light in the visible range is known as _____ film.

_____ 1. negative	(a) introduced Kodak	**MATCHING SET**
_____ 2. larger grain size	(b) developer	
_____ 3. daguerreotype	(c) stop bath	

_____ 4. George Eastman   (d)   more light sensitive

_____ 5. hydroquinone   (e)   additive primary colors

_____ 6. acid solution   (f)   early photographic form

_____ 7. sodium thiosulfate   (g)   a cyanine dye

_____ 8. panchromatic film   (h)   introduced instant photography

_____ 9. red, green and blue   (i)   exposed areas are dark

_____ 10. pinacyanol   (j)   sensitive to broad band of wavelengths

_____ 11. Edwin H. Land   (k)   fixer

QUESTIONS ON PART I: FOODS

1. What are the primary reasons that foods spoil?

2. Which of the following food additives should be avoided? Give your reasons.

   Butter yellow   Glycerin
   Propionic acid   Sodium cyclamate

3. Why does it take less time to cook food in a pressure cooker than in an open pot of boiling water?

4. What would happen to your ability to digest protein if you kept the acid in your stomach neutralized all the time? Is acid bad for your stomach?

5. Why is saccharin preferable to chloroform as a sweetener?

6. A label on a brand of breakfast pastries lists the following additives: dextrose, glycerin, citric acid, potassium sorbate, Vitamin C, sodium iron pyrophosphate, and BHA. What is the purpose of each substance?

7. Choose a label from a food item and try to identify the purpose of each additive.

8. Describe some of the chemical changes that occur during the cooking of

   (a) a carbohydrate,
   (b) a protein,
   (c) a fat.

9. What causes fat in foods to become rancid? How can this be avoided?

10. What causes bread to rise?

11. What are the pros and cons of eating "natural" foods as opposed to foods containing chemical additives?

12. Why were cyclamates taken off the market?

13. Do you think it is wise to use animals in safety tests for drugs and food additives? Should mental patients and prisoners be used for this purpose?

14. Many consumer products are almost identical in chemical composition but are sold at widely different prices under different trade names. Do you think the products should be identified by their chemical names or their trade names? Why?

15. See what you can find out about correlation between taste and smell. Are they the same sensation? Are they independent of each other?

QUESTIONS ON PART II: MEDICINES

1. Explain the need for and the chemical action of an antacid.

2. What is an analgesic? Give three examples, two of which are habit-forming.

3. What are enkephalins? How are they related to drug use?

4. What is the common name for procaine? In what area is it often used?

5. What is the chemical formula for aspirin?

6. Name three common antiseptics.

7. What chemical has taken the "sting" out of iodine?

8. How does penicillin kill bacteria?

9. What is the common molecular feature of steroid drugs?

10. What are the active chemicals in birth control pills?

11. What is the purpose of pyribenzamine in medical use?

12. Write a paragraph to give your evaluation of diet pills.

QUESTIONS ON PART III: BEAUTY AIDS

1. You read in the newspaper about a new compound that will break disulfide bonds in proteins. What potential use might it have?

2. (a) What is the purpose of an emulsifier?
   (b) In which of the following cosmetics is an emulsifier important: suntan lotion, hair spray, cold cream?

3. What is the structure of the monomer unit in polyvinylpyrrolidone?

4. What is the purpose of each of the following:

   (a) Polyvinylpyrrolidone in hair sprays?
   (b) Aluminum chloride in deodorants?
   (c) p-Aminobenzoic acid in suntan lotion?

5. What specific substance is broken down during the bleaching of hair?

6. Use the structures of the constituents of lanolin to justify its ability to emulsify face creams.

7. Hydrogen bonding is a very handy theoretical tool. Name three applications of hydrogen bonding in consumer products.

8. If you were going to formulate a suntan lotion, what particular spectral property would you look for in choosing the active compound?

9. What is the purpose of talc in face powder?

10. Name an astringent widely used in deodorants.

11. What is the major ingredient in lipstick?

12. What is the chemical action of hydrogen peroxide on hair?

13. Commercial lanolin comes from what animal?

QUESTIONS ON PART IV: AUTOMOTIVE PRODUCTS

1. Write a balanced equation for the complete combustion of 2,2,4-trimethylpentane.

2. What is meant by high detergency (HD) gasoline? Specifically, what do the detergents do for a gasoline engine?

3. What is the purpose of each of these substances in gasoline?

   (a) ethylenediamine   (c) ethylene glycol
   (b) TCP   (d) dyes

4. What is the meaning of 10W-30 oil?

5. How do the antiwear agents in motor oils work?

6. What is the difference in the composition of motor oils and vegetable oils, such as corn oil or palm oil?

7. What is the basic difference in the composition of motor oils and automotive greases?

8. Referring to the structure of ethylene glycol, explain why this compound is so soluble in water.

9. Why would it be undesirable to have a methanol-type antifreeze in your radiator during the summer?

10. Explain how a deicer melts ice.

11. Why is the methanol-type antifreeze called temporary and the ethylene glycol-type called permanent antifreeze?

12. Write a chemical equation for the rusting of iron in a radiator.

13. List three typical additives in antifreeze formulations and describe the action of each.

QUESTIONS ON PART V: PHOTOGRAPHY

1. What would be the effect of low pH on a typical developer?

2. Explain the term fixing. What is a fixer, chemically, and why is it important in photography?

3. What are the subtractive primary colors?

4. Explain how a red dot would be photographed with a color reversal film such as Kodachrome.

5. What color is produced by the superposition of the following subtractive primary colors?

   (a) magenta and yellow
   (b) cyan and yellow

    (c)   magenta and cyan

    (d)   magenta, cyan and yellow

6.   Is all of the silver reduced in the development of a black and white film? Explain.

7.   What different support materials have been used for the light sensitive emulsion in photography?

8.   What is the purpose of sodium sulfite in black and white film development?

9.   Where are the developer and fixer in the black and white Polaroid film?

10.   What was done to increase the spectral range of black and white film?

11.   Explain the action of a coupler in color photography.

ENERGY AND OUR SOCIETY

It is fitting as we draw this text to a close to consider one of the greatest unsolved problems of mankind: How will we meet our energy needs? The problem will have to be solved satisfactorily before our other major problem—food supply—can be solved. In the United States the average yearly consumption of energy is now in excess of 1.8×10^{16} kilocalories (energy units are given in Table 17–1) and this consumption has been increasing at the rate of about 3.5 percent per year over the last few years. The average consumption is therefore about 240,000 kilocalories per person per day for all forms of energy. This energy is currently being supplied by petroleum, coal, natural gas, water power, and nuclear sources (Figure 17–1). The predicted energy shortage, however, is now a reality. The power brownouts in the eastern United States in the late 1960's were a signal that something was amiss. In the early 1970's gasoline shortages were caused by inadequate refining capacity, low return on investment capital, and a Mid-East war. In 1973, the Arab nations declared an embargo on oil shipments to all nations that supported Israel, demanding and getting political concessions. The price of oil was quadrupled, and has been raised a number of times since. Meanwhile, the United States pro-

3412 British Thermal Units (BTU) = 1 kilowatt-hour

TABLE 17–1   A Handy Chart of Energy Units[a]

Cubic Feet of Natural Gas	Barrels of Oil	Tons of Bituminous Coal	British Thermal Units (BTU)	Kilowatt Hours of Electricity	Joules	Kilo-calories[b]
1	0.00018	0.00004	1000	0.293	1.055×10^6	252
1000	0.18	0.04	1×10^6	293	1.055×10^9	0.25×10^6
5556	1	0.22	5.6×10^6	1628	5.9×10^9	1.40×10^6
25,000	4.50	1	25×10^6	7326	26.4×10^9	6.30×10^6
1×10^6	180	40	1×10^9	293,000	1.055×10^{12}	0.25×10^9
3.41×10^6	614	137	3.41×10^9	1×10^6	3.6×10^{12}	0.86×10^9
1×10^9	180,000	40,000	1×10^{12}	293×10^6	1.055×10^{15}	0.25×10^{12}
1×10^{12}	180×10^6	40×10^6	1×10^{15}	293×10^9	1.055×10^{18}	0.25×10^{15}

[a]Based on normal fuel heating values. 10^6 = 1 million, 10^9 = 1 billion, 10^{12} = 1 trillion, 10^{15} = quadrillion (quad).
[b]A food Calorie = 1000 calories = 1.000 kilocalorie.

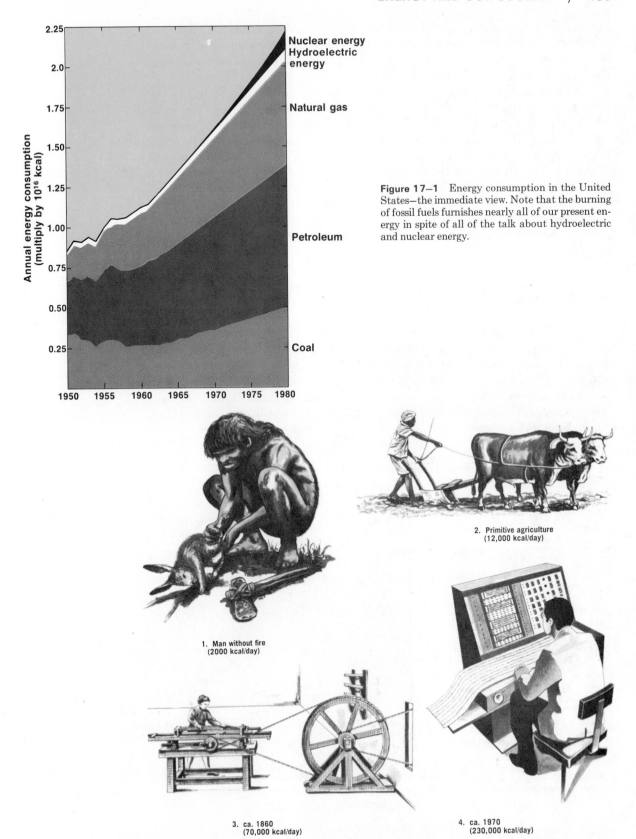

Figure 17–1   Energy consumption in the United States—the immediate view. Note that the burning of fossil fuels furnishes nearly all of our present energy in spite of all of the talk about hydroelectric and nuclear energy.

1. Man without fire
(2000 kcal/day)

2. Primitive agriculture
(12,000 kcal/day)

3. ca. 1860
(70,000 kcal/day)

4. ca. 1970
(230,000 kcal/day)

Figure 17–2   Energy use per capita based on various types of societies. (From Turk, A., et al.: *Environmental Science.* Philadelphia, W. B. Saunders Co., 1974.)

duction has decreased, and we are more dependent than ever on imported oil. Problems of this type were certain to occur because of two factors: (1) the world is using more energy while (2) the supplies of the type being used are becoming depleted.

As late as 1940, 80 percent of the world's population was living at the primitive or advanced agricultural level. Consequently, relatively little energy was being used. Today, the situation is in a rapid state of change. About 25 to 30 percent of the present world population lives in a very highly industrialized society and uses 80 percent of the world's energy (Figure 17–2). In the United States, 6 percent of the world's population uses 31 percent of the world's energy production in any given year. However, the relative use of energy in the United States is on the decline as compared with the rest of the world. This means that there is now taking place a serious worldwide competition for the types of energy sources currently available.

FOSSIL FUELS

As indicated in Figures 17–1 and 17–3, nearly all of the present energy needs in the U.S. are being supplied by fossil fuels: petroleum, natural gas, and coal. This is also true on a worldwide basis. Although the chemistry and geology of fossil fuel formation are not thoroughly understood, it is generally agreed that buried plant material formed these fuels over millions of years. Perhaps the most significant point about fossil fuels is that modern society appears capable of using up in a few hundred years what it took nature millions of years to produce (Figure 17–3). So far as we know, no coal and oil are being formed underground today.

Fossil fuels store the energy of the sun.

Figure 17–3 Energy consumption in the United States—the longer view. The future estimates are based on 400,000 kilocalories per day per person after 2000 A.D., a zero population growth by 2100 A.D., and a total population of 400 million. (From *Chemical and Engineering News*, January 10, 1972.)

TABLE 17–2   Energy Content of Fossil Fuels

Fuel	Reaction Equation	Heat Energy Evolved
Coal	$C + O_2 \longrightarrow CO_2$	94 kcal/mole; 7.8 kcal/g fuel
Natural Gas*	$CH_4 + 2O_2 \longrightarrow CO_2 + 2H_2O$	211 kcal/mole; 13.2 kcal/g fuel
Petroleum†	$2C_8H_{18} + 25O_2 \longrightarrow 16CO_2 + 18H_2O$	1,303 kcal/mole; 11.4 kcal/g fuel

*CH_4, methane, is the principal constituent in natural gas (up to 97%).
†C_8H_{18}, octane, is only one of many hydrocarbons present in petroleum.

When fossil fuels are burned, chemical energy stored long ago from sunlight via photosynthesis is released. The amount of energy varies with the type of fuel (Table 17–2). In each case the complete combustion products are carbon dioxide and water.

Petroleum

Vast deposits of petroleum were first discovered in the United States (Pennsylvania) in 1859 and in the Middle East (Iran) in 1908. Since that time petroleum has found wide use as an energy source, first as kerosene for lighting, then as gasoline and aircraft fuel for transportation, and more recently as fuel oil to produce electricity. All of this has resulted in the use of over 5.4 billion barrels of oil per year in the U.S. in recent years.

One barrel of petroleum contains 42 gallons.

Figure 17–4 illustrates the overall U.S. oil supply. In 1960, less than 20% of the U.S. oil consumption was imported oil. The drop in late 1973 represented an oil embargo imposed by the Organization of Petroleum

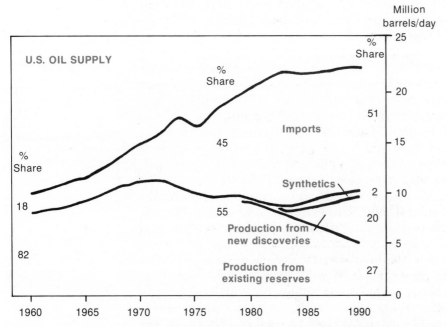

Figure 17–4   The U.S. oil supply, since 1970, has come increasingly from imported oil. Our demand for petroleum has continued to increase despite greater gasoline mileage from automobiles and somewhat lessened driving.

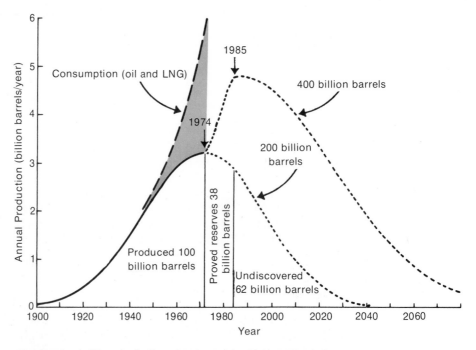

Figure 17–5 History of oil production in the United States (including Alaska). The dashed curve represents oil imports. The lower curve is based on known oil reserves, and the higher dotted curve is based on estimates of what is likely to be found with large deposits expected under the continental shelf. Regardless of how much is actually present, it is a finite amount and apparently will be consumed in a relatively short period of time, (NGL = natural gas liquefied). (From *Science,* April 19, 1974.)

Exporting Countries (OPEC). Since the embargo was lifted, late that year, the U.S. has continued to import much of its oil, until by 1978 approximately 45% of our oil was imported.

It has been estimated that there were about 200 billion barrels of oil to be recovered in the United States (including Alaska) when oil production first began. Half of this amount has now been produced (see Figure 17–5). Numbers of this type are often quoted, and a few qualifications should be noted. ***"Recoverable oil"*** is based on 30 percent recovery, i.e., pumping 30 percent from the ground and leaving 70 percent behind, since to recover more would cost more and hence raise the cost of the average barrel of oil. It stands to reason that as supplies decrease and costs go up it will prove economically feasible to "recover" more of the oil.

The definition of recoverable oil is based on economics.

A second estimate of U.S. reserves of oil can be based on as yet undiscovered oil under the continental shelf. This estimate is 400 billion barrels, thus doubling the amount of U.S. oil. Either way, the consumption of petroleum by the United States is increasing at too fast a rate; obviously, this cannot continue. Figure 17–6 illustrates the situation worldwide. The assumption underlying the projection in Figure 17–6 is that the rest of the world will continue in its lower rate of energy consumption. This is not a feasible assumption. Based on the present trends, a reasonable projection is that 80 percent of the world's petroleum will be used in a 60- to 70-year period, ending at about the year 2025. Unless drastic changes are made at some point in the relatively near future, perhaps as early as the year 2000, petroleum will become an increasingly rare commodity.

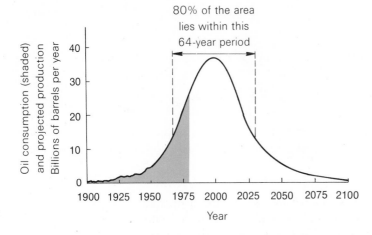

80% of the area lies within this 64-year period

Figure 17–6 World oil use: past and projected.

As indicated earlier in this text, petroleum is the base for a very large petrochemical industry, including plastics, fibers, paints and medicines. A real chemical consideration that must come into focus in the future as the petroleum supply decreases is whether we want to burn petroleum for its energy content or save it as a starting material for petrochemical products.

Oil may become more valuable as a chemical raw material than as an energy source.

The petrochemical industry may very well be required to produce basic foodstuffs in the future. Edible fats were produced in Germany during World War II. Glycerol is presently made from petroleum on a commercial scale and we know how to make sugar from oil.

Oil Shale

Extracting oil from oil shale has not been economical owing to the estimated $15 to $25 per barrel cost. With the recent increases in the costs of crude oil, at least a dozen major oil companies are getting involved. It has been estimated that the United States has reserves of shale oil that total 1,800 billion barrels and that 600 billion barrels could be economically recovered at current prices. Occidental Petroleum Corporation expects to produce shale oil near Grand Junction, Colo. by 1986, and the production of this mine could reach as much as 200,000 barrels per day.

The oil comes out of the rock on heating in the absence of air. The technique used to get the oil to the surface involves: (1) Blasting the rock to break it up, (2) burning a part of the oil underground to force the oil out of the rock, and (3) pumping the collected oil to the surface.

Coal

Unlike petroleum, where discoveries are being made almost daily, geologists believe that all the world's coal supplies have now been discovered. ***Minable coal*** is defined as 50 percent of all coal which is in a seam at least 12 inches thick and within 4000 feet of the surface. Estimates have been made as to the length of time this minable coal will supply our energy needs. Figure 17–7 shows that coal will be available as a fuel for a much longer period of time than oil. If new mining techniques are developed, then more of the deposited coal might be termed minable.

The largest supply of fossil fuel is in the form of coal.

Two of coal's major drawbacks are that it is a relatively dirty fuel and dificult to handle. Coupled with the atmospheric pollution caused by

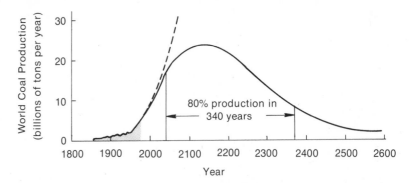

Figure 17–7 The coal mined to date (*shaded area*) represents only a small fraction of the minable coal. The rate of increase in coal consumption (*dashed line*) is 3.56 percent per year. It is obvious that such an exponential rise cannot continue long after the year 2000. At the present usage (held constant) coal would last for many hundreds of years.

sulfur-containing coal, these drawbacks were prime reasons for a major shift from coal to petroleum in electricl generating plants, particularly in the industrialized nations, late in the 1960's. The dangerous and unhealthful character of deep coal mining and the environmental disruption caused by strip mining contributed to the shift.

Coal Gasification

Coal can be converted into a relatively clean-burning fuel by a process known as ***gasification*** (Figure 17–8). In this process, coal is made to react with either a limited supply of hot air or steam. In the first reaction the product is a gaseous mixture known as *"power gas."*

Gasification can make coal cleaner to burn and easier to handle from supplier to user.

$$\text{Coal} + \text{Air} \longrightarrow CO(g) + H_2(g) + N_2(g)$$
POWER GAS

Power gas contains up to 50 percent by volume nitrogen, which does not burn and lowers the heating value of the gas considerably. In fact,

Figure 17–8 Schematic drawing of coal gasifier. A relatively cool combustion of powdered coal in a limited supply of oxygen produces a mixture of carbon monoxide and hydrogen along with other gases. The mineral content in the coal collects in the slag.

power gas of this composition has only one sixth the heat content of methane.

If the coal is allowed to react with high-temperature steam, a mixture of carbon monoxide and hydrogen (which contains no nitrogen) known as **synthesis gas** or **coal gas** is obtained.

Coal gas can be used like natural gas.

$$C + H_2O(g) \longrightarrow CO(g) + H_2(g)$$

COAL   SYNTHESIS GAS
OR COAL GAS

In either mixture, the CO and H_2 are burned in the air to produce heat. The heat produced is about one third that of an equal volume of methane (natural gas).

$$2CO + O_2 \longrightarrow 2CO_2 + Heat$$
$$2H_2 + O_2 \longrightarrow 2H_2O + Heat$$

Other opportunities exist in the area of coal modification, since liquid fuels can also be obtained from coal. Knowledgeable estimates indicate that between six and eight years will be needed to make such changes on an extensive scale.

Natural Gas

Natural gas approaches an ideal fuel. It burns with a high heat output (Table 17–2), with little or no residue, and is easily transported. The problem with natural gas is that its production is decreasing in this country. Federal price regulations begun in 1954 have held down prices, stimulated demand, and decreased incentives for exploration of new gas deposits. These price regulations are presently being lifted. It remains to be seen whether or not new gas will become available. Certainly the present gas will become higher priced.

In 1973, U.S. gas production peaked at 22 trillion (10^{12}) cubic feet

Figure 17–9 The U.S. gas supply has been mostly from existing reserves within the U.S., unlike the oil situation in recent years. Natural gas consumption may never again equal that of the mid-1970's. [Source: Exxon Corporation.]

*Trillion Cubic Feet of Gas/Year

per year. Even if significant new deposits of natural gas are discovered in such locations as the outer continental shelf, the North American natural gas deposits are about 50 percent depleted. Figure 17–9 illustrates the U.S. gas supply. Importation of natural gas is complicated by difficulties encountered in condensing natural gas to a liquid for ocean transit and the danger of transporting and storing such concentrated and very volatile fuel.

Electricity Production

When electricity is discussed, electrical power is usually mentioned. While energy is the ability to do work, power is the *rate* of doing work (Table 17–3).

Most consumers purchase electrical energy (kWhr) and not power, although appliances and other electrical devices are rated according to power. A 100-watt light bulb operating at 100 volts would draw 1 ampere of current; in one hour it would use 100 watt-hours or 0.1 kWhr of energy.

Watts = Volts × Amperes

Ultimately, all electricity costs (except hydro-electricity) must be related to the costs of the fuel used to produce the electricity. In 1978, about 30 percent of all the energy consumed in the United States was used to produce electricity. This accounted for 22 percent of all the fossil fuels consumed in this country. However, only 30 percent of the energy contained in those fuels is available at the point where the electricity is used (Figure 17–10).

Part of the energy loss in electricity production is illustrated in a schematic diagram of a large modern fossil-fuel generating plant (Figure 17–11). For a 1,000-megawatt coal-burning plant, one hour of operation might look like this:

Coal consumed	696 tons producing 2.270 billion kilocalories
Smokestack heat loss	0.227 billion kilocalories
Heat loss in plant	0.106 billion kilocalories
Heat loss in evaporator to cool condenser	1.080 billion kilocalories
Electrical energy delivered to power lines	0.857 billion kilocalories
Percent of energy delivered as electricity after transmission losses	$\frac{0.857}{2.27} \times 100\% = 37.8\%$

Electric generating plants yield about one-third of the fuel energy in the form of electrical energy.

TABLE 17–3 Power Units

1 watt (W) = 1 joule per second*
1 kilowatt (kW) = 10^3 W
1 megawatt (MW) = 10^3 kW = 10^6 W
1 watt = 1 volt × 1 ampere

*watts × time = energy units, i.e.,

$$1 \text{ kilowatt-hour} = 10^3 \text{ W} \times 1 \text{ hr} = \frac{10^3 \text{ joules}}{\text{second}} \times \frac{3600 \text{ sec}}{\text{hr}} \times 1 \text{ hr}$$

$$= 3.6 \times 10^6 \text{ joules.}$$

Figure 17—10 The annual fuel "pie" in the United States. Of the fuel sources left, this energy is more than half (59 percent) wasted. Only 41 percent of the fuel we pay for is actually used. This situation is due partly to the laws of thermodynamics and partly to wasteful practices, sloppy planning, and poor design of fuel-using devices.

Figure 17—11 The heat balance of a 1,000-megawatt coal burning electrical generating plant. Note that the 969 tons of coal burned per hour furnish 2.27×10^9 kcal of heat energy, but only 0.857×10^9 kcal of energy, or 38 percent, is converted to electricity. Note also the large amounts of heat energy lost to the cooling water and atmosphere.

There is a further energy loss in the power lines and the transformers, which lowers the useful output of the plant to 30 percent of the energy consumed. This is the *efficiency* figure for the overall operation. It is important to note that we pay for 300 kcal of heat energy in the form of coal or fuel oil but receive less than 100 kcal of energy in the form of electricity. Obviously it requires much less fuel to heat homes with the fuel itself, rather than by electricity made from the fuel.

SELF-TEST 17-A   1.   In 1980, approximately what percentage of the total U.S. energy consumption will be furnished by coal? See Figure 17–1. _____

2.   Which furnishes the most heat energy per gram—coal, petroleum, or natural gas? _____

3.   How many gallons of oil are there in one barrel? _____

4.   Using the recovery rate of 30 percent for oil in place, approximately how large a percentage of the oil in the United States has been thus far pumped from the earth (Figure 17–5)? _____

5.   Is the composition of "power gas" obtained from coal gasification CO, H_2, N_2 or CO, H_2?

6.   The typical efficiency of an electrical generating plant is about (100, 50, 33, 10) percent.

NUCLEAR ENERGY

Few things have captured the awe and imagination of mankind to quite the extent that nuclear energy has in the past half-century. Vast amounts of energy are released in splitting heavy atomic nuclei, the *fission process,* and the same is also true in combining small atomic nuclei to make heavier ones, the *fusion process.* Consider the energy contrast between an oxidation of a fossil fuel and a nuclear fusion reaction. When one mole (6.02×10^{23} molecules, or 16 grams) of methane from natural gas is burned, over 200 kilocalories of heat are liberated:

$$CH_4 + 2O_2 \longrightarrow CO_2 + 2H_2O + 211 \text{ kilocalories (kcal)}$$

In contrast, a lithium nucleus can be made to react with a hydrogen nucleus to form two helium nuclei in a nuclear reaction. The energy released per mole of lithium in this reaction is 23,000,000 kilocalories. This means that 7 grams of lithium and 1 gram of hydrogen produce 100,000 times more energy through fusion of nuclei than 16 grams of methane and 64 grams of oxygen produce by exchanging electrons.

Energy changes associated with nuclear events may be many thousands of times larger than those associated with chemical events.

$$^{7}_{3}\text{Li} + ^{1}_{1}\text{H} \longrightarrow 2^{4}_{2}\text{He} + 23,000,000 \text{ kcal}$$

Realizing that nuclear changes could involve giant amounts of energy relative to chemical changes for a given amount of matter, Otto Hahn, Fritz Strassman, Lise Meitner, and Otto Frisch discovered in 1938 that $^{235}_{92}U$ is fissionable. Subsequently the dream of controlled nuclear energy became a reality, followed by the bomb and nuclear power plants. In the 1950's it was hoped that nuclear energy would soon relieve the shortage of fossil fuels. To date this has not been accomplished, although the production of nuclear energy has grown very rapidly in recent years.

Fission Reactions

Fission is the break-up of heavy nuclei.

Fission can occur when a thermal neutron (with a kinetic energy about the same as that of gaseous molecules at ordinary temperatures) enters certain heavy nuclei with an odd number of neutrons ($^{235}_{92}U$, $^{233}_{92}U$, $^{239}_{94}Pu$). The splitting of the heavy nucleus produces two smaller nuclei, two or more neutrons (an average of 2.5 neutrons for $^{235}_{92}U$), and much energy. Typical nuclear fission reactions are written as follows:

^1_0n is a neutron.

$$^{235}_{92}U + ^1_0n \longrightarrow ^{141}_{56}Ba + ^{92}_{36}Kr + 3^1_0n + Energy$$

$$^{235}_{92}U + ^1_0n \longrightarrow ^{103}_{42}Mo + ^{131}_{50}Sn + 2^1_0n + Energy$$

A low-energy neutron will disrupt some large nuclei.

Note that the same nucleus may split in more than one way. The fission products, such as $^{141}_{56}Ba$ and $^{92}_{36}Kr$, emit beta particles and gamma rays until stable isotopes are reached.

The neutrons emitted can cause the fission of other heavy atoms if they are slowed down by a moderator, such as graphite. For example, the three neutrons emitted in the first reaction above could produce fission in three more uranium atoms, the nine neutrons emitted by those nuclei could produce nine more fissions, the 27 neutrons from these fissions could produce 81 neutrons, the 81 neutrons could produce 243, the 243 neutrons could produce 729, and so on. This process is called a *"chain reaction"* (Figure 17–12), and it occurs at a maximum rate when the uranium sample is large enough for most of the neutrons emitted to be captured by other nuclei before passing out of the sample. Sufficient sample in a certain volume to sustain a chain reaction is termed the *critical mass*.

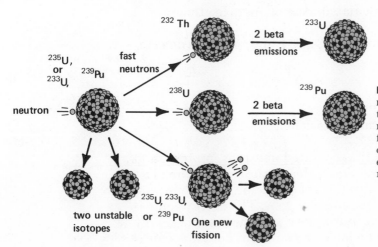

Figure 17–12 A chain reaction. A thermal neutron collides with a fissionable nucleus and the resulting reaction produces three additional neutrons. These neutrons can either convert non-fissionable nuclei, such as $^{232}_{90}Th$, to fissionable ones or cause additional fission reactions. If enough fissionable nuclei are present, a chain reaction will be sustained.

In the atomic bomb the critical mass is kept separated into several smaller subcritical masses until detonation, at which time the masses are driven together by an implosive device. It is then that the tremendous energy is liberated and everything in the immediate vicinity is heated to temperatures of 5 to 10 million degrees. The sudden expansion of hot gases literally explodes everything nearby and scatters the radioactive fission fragments over a wide area. In addition to the movement of gases, there is the tremendous vaporizing heat that makes the atomic bomb so very devastating.

There is no danger of an atomic explosion in the uranium mineral deposits in the earth for two reasons. First, uranium is not found pure in nature—it is found only in compounds, which in turn are mixed with other compounds. Second, less than 1 percent of the uranium found in nature is fissionable $^{235}_{92}$U. The other 99 percent is $^{238}_{92}$U, which is not fissionable by thermal neutrons. In order to make nuclear bombs or nuclear fuel for electrical generation, a purification enrichment process must be carried out on the uranium isotopes, thus increasing the relative amount of $^{235}_{92}$U atoms in a sample. Ordinary isotopic uranium such as that found in ores is only 0.711 percent $^{235}_{92}$U.

Separation of uranium isotopes had to precede the control of atomic energy.

It is interesting to note that fission products can be found in the Gabon Republic of West Africa, which indicate that a uranium ore deposit "went critical" about 150,000 years ago. At that time the natural uranium-235 content would have been higher than it is now.

Mass Defect—The Ultimate Nuclear Energy Source

What is the source of the tremendous energy of the fission process? It ultimately comes from mass being converted into energy, according to Einstein's famous equation, $E = mc^2$, where E is energy that results from the loss of an amount of mass, m, and c^2 is the speed of light squared. If separate neutrons, electrons, and protons are combined to form any particular atom, there is a loss of mass called the ***mass defect.*** For example, the calculated mass of one $^4_2$He atom from the masses of the constituent particles is 4.032982 amu:

$2 \times 1.007826 = 2.015652$ amu, mass of two protons and two electrons

$2 \times 1.008665 = \underline{2.017330}$ amu, mass of two neutrons

Total $= 4.032982$ amu, calculated mass of one $^4_2$He atom

Since the measured mass of a $^4_2$He atom is 4.002604 amu, the mass defect is 0.030378 amu:

The mass that is lost leaves in the form of energy; $E = mc^2$.

$$\begin{array}{r} 4.032982 \text{ amu} \\ - \ \underline{4.002604} \text{ amu} \\ 0.030378 \text{ amu, mass defect} \end{array}$$

Since the atom is more stable than the separated neutrons, protons, and electrons, the atom is in a lower energy state. Hence, the 0.030378 amu lost per atom would be released in the form of energy if the $^4_2$He atom were made from separate protons, electrons, and neutrons. The energy equivalent of the mass defect is called the ***binding energy.*** The binding energy is analogous to the earlier concept of bond energy, in that both are a mea-

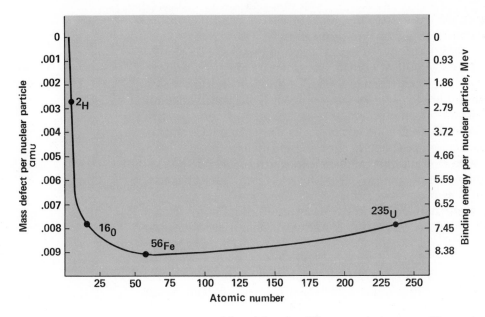

Figure 17–13 Mass defect for different nuclear masses. The most stable nuclei center around $^{56}_{26}$Fe, which has the largest mass defect per nuclear particle.

sure of the energy necessary to separate the package (nucleus or molecule) into its parts.

Atoms with atomic numbers between 30 and 63 have a greater mass defect per nuclear particle than very light elements or very heavy ones, as shown in Figure 17–13. This means the most stable nuclei are found in the atomic number range from 30 to 63.

Because of the relative stabilities, it is in the intermediate range of atomic numbers that most of the products of nuclear fission are found. Therefore, when fission occurs and smaller, more stable nuclei result, these nuclei will contain less mass per nuclear particle. In the process, mass must be changed into energy. This energy gives the fission process its tremendous energy. It takes only about 1 kg of $^{235}_{92}$U or $^{239}_{94}$Pu undergoing fission to be equivalent to the energy released by 20,000 tons (20 kilotons) of ordinary explosives like TNT. The energy content in matter is further dramatized when it is realized that the atomic fragments from the 1 kg of nuclear fuel weigh 999 g, so only one tenth of one percent of the mass is actually converted to energy. The fission bombs dropped on Japan during World War II contained approximately this much fissionable material.

Separated nuclear particles have more mass than when combined in a nucleus.

Intermediate-sized nuclei tend to have the greatest nuclear stability.

Controlled Nuclear Energy

The fission of a $^{235}_{92}$U nucleus by a slow-moving neutron to produce smaller nuclei, extra neutrons, and large amounts of energy suggested to Enrico Fermi and others that the reaction could proceed at a moderate rate if the number of neutrons could be controlled. If a neutron control could be found, the concentration of neutrons could be maintained at a level sufficient to keep the fission process going but not high enough to allow an uncontrolled explosion. It would then be possible to drain the heat away from such a reactor on a continuing basis to do useful work. In

1942, Fermi, working at the University of Chicago, was successful in building the first atomic reactor, called an *"atomic pile."*

An atomic reactor has a number of essential components. The charge material (fuel) must be fissionable or contain significant concentrations of a fissionable isotope such as $^{235}_{92}U$, $^{239}_{94}Pu$, or $^{233}_{92}U$. Ordinary uranium, which is mostly the nonfissionable $^{238}_{92}U$, can be used since it has a small concentration of the $^{235}_{92}U$ isotope. A moderator is required to slow the speed of the neutrons produced in the reactions without absorbing them. Graphite, water, and other substances have been used successfully as moderators. A substance that will absorb neutrons, such as cadmium or boron steel, is present in order to have a fine control over the neutron concentration. Shielding, to protect the workers from dangerous radiation, is an absolute necessity. Shielding tends to make reactors heavy and bulky installations. A heat-transfer fluid provides a large and even flow of heat away from the reaction center.

A schematic diagram of the X-10 graphite pile in Oak Ridge, Tennessee, is illustrated in Figure 17–14. Here the fuel is packed in openings in the graphite pile. Unlike the atomic bomb, the reaction cannot lead to an atomic explosion because heat expansion would dilute the fuel or the heat would drive out a moderator such as water. Even though the material of which the pile is made may absorb neutrons, the control rods offer a final and sensitive control to stop the fission reactions. With the control rods properly placed, the reaction can be maintained at various safe levels of activity.

Once the heat is produced in a nuclear reactor and safety measures are employed to protect against radiation, conventional technology

Atomic pile:
(1) carefully diluted fissionable material;
(2) moderator to control fission reaction;
(3) coolant to control heat;
(4) shielding to limit radiation.
The first one was piled together at the University of Chicago in 1942.

Figure 17—14 Schematic diagram of X-10 reactor at Oak Ridge, Tennessee.

Figure 17—15 Schematic illustration of a nuclear power plant.

allows this energy to be used to generate electricity, to power ships, or to operate any device that uses heat energy. A system for the nuclear production of electricity is illustrated in Figure 17–15.

What are the fuel requirements in nuclear fission energy production? In a typical fission event such as

$$\ _0^1\text{n} + \ _{92}^{235}\text{U} \longrightarrow \ _{37}^{93}\text{Rb} + \ _{55}^{141}\text{Cs} + 2\,_0^1\text{n} + 200 \text{ Mev},$$

the energy release, 200 Mev, is equivalent to 7.7×10^{-12} calorie per atom of $_{92}^{235}$U, or 4.64×10^9 kcal/mole. Since 1 gram of pure $_{92}^{235}$U contains 2.56×10^{21} atoms, the total energy release for 1 gram of uranium-235 undergoing fission would be

$$1\,g \times 2.56 \times 10^{21}\,\frac{\text{atoms}}{1\,g} \times 7.7 \times 10^{-12}\,\frac{\text{cal}}{\text{atom}} = 2.0 \times 10^{10} \text{ cal}$$

When one gram of ^{235}U undergoes fission, it provides the same energy as burning about six tons of coal.

This is the amount of energy that would be released if 5.95 tons of coal were burned, or if 13.7 barrels of oil were burned to produce heat to power a boiler. This means that about 3 kilograms of $_{92}^{235}$U fuel per day would be required for a 1,000-megawatt electric generator. The fuel used, however, is not pure $_{92}^{235}$U, but **enriched** uranium containing up to 3 percent $_{92}^{235}$U.

Nuclear energy is mostly used for electricity production. In 1965, when nuclear energy usage began, and until about 1974, more energy was produced in the U.S. from burning firewood than from nuclear energy. In 1977, 3 percent of the U.S. energy supply came from nuclear sources; by 1990, almost 10 percent of our energy needs is expected to be met by nuclear sources. There appears to be enough uranium ore available to meet the electric generating needs until the last decade of this century; by then, however, the cheap ore will have run out and dramatic increases in electric rates will be necessary, brought on by fuel costs alone.

$_{92}^{235}$U is in very short supply.

It is possible to convert the nonfissionable $_{92}^{238}$U and $_{90}^{232}$Th into fissionable fuels by using a **breeder reactor**. In such a reactor, a blanket of non-

Figure 17–16 Schematic diagram of a fast breeder reactor and steam-turbine power generator.

fissionable material is placed outside the fissioning $^{235}_{92}U$ fuel (Figure 17–16), which serves as the source of neutrons in the breeder reactions. The two breeder reaction sequences are the following:

$$^{238}_{92}U + {}^{1}_{0}n \longrightarrow {}^{239}_{92}U \xrightarrow{\beta} {}^{239}_{93}Np \xrightarrow{\beta} {}^{239}_{94}Pu$$

$$^{232}_{90}Th + {}^{1}_{0}n \longrightarrow {}^{233}_{90}Th \xrightarrow{\beta} {}^{233}_{91}Pa \xrightarrow{\beta} {}^{233}_{92}U$$

The products of the breeder reactions, $^{233}_{92}U$ and $^{239}_{94}Pu$, are both fissionable, and neither is found in the earth's crust.

Breeder reactors present many technological problems, not the least of which is the potential of a disaster caused by mishandling of the $^{239}_{94}Pu$ isotope, which is extremely toxic and can also be fabricated into a fission bomb. Nevertheless, the expected benefit from the breeder program is massive amounts of energy. For example, if all the uranium used for electrical generation were used in breeder reactors, instead of running out of uranium fuel in several decades, the breeder fuels would supply the United States' electrical requirements for about 2600 years, assuming 1970 electricity-use levels were maintained, something euphemistically called **_Zero Energy Growth!_** If a breeder program cannot start in earnest soon enough, then the $^{235}_{92}U$ isotope will be used up only to generate electricity and another, as yet undiscovered, source of neutrons will be needed to convert the $^{238}_{92}U$ and $^{232}_{90}Th$ into fissionable fuels.

Radioactive isotopes are produced in all nuclear fission reactors. The radiation levels produced by these isotopes are dangerous to life, and some of the half-lives involved extend into millions of years. We are, then, in the production of nuclear energy, building up large amounts of very dangerous materials. Serious objections have been raised to every suggested method of radioactive waste disposal. The radioactive waste is presently being held in underground depots and surface tanks. It is argued by some that the exploitation of fission energy should await adequate methods for the disposal of these radioactive wastes.

$^{239}_{94}Pu$ is toxic from a radiation as well as from a chemical point of view.

Radiation Damage

We are constantly bombarded by radiation from a number of sources; this radiation includes cosmic rays, medical X rays, radioactive fallout from countries that do nuclear testing, and naturally occurring radioisotopes, which are widespread. Fortunately, most radiation damage is too slight to be noticed immediately, although its very presence should be regarded as one of the hazards of everyday life.

A radioisotope is an atom that will disintegrate, producing one of three types of radiation, alpha (α), beta (β), or gamma (γ) (Figure 17–17). The resulting atom may be stable or another radioisotope. In a sample of radioactive matter large enough to measure, there will be many disintegrations over a given time if the half-life is short, or few disintegrations over the same time interval if the half-life is long. A half-life represents the period of time required for half of the radioactive material originally present to undergo transmutation. Normal background radiation on the human body is 2 or 3 disintegrations per second.

There are three principal factors that render a radioactive substance dangerous: (1) the number of disintegrations per second, (2) the half-life of the isotope, and (3) the type or energy of the radiation produced. In addition, radiation can be very damaging if the radioactive substance is of a chemical nature such that it can be incorporated into a food chain or otherwise enter a living organism.

Radioactive disintegrations are measured in **curies** (Ci; one Ci is 37 billion disintegrations per second). A more suitable unit is the microcurie (μCi), which is 37,000 disintegrations per second. One curie of a radioisotope is a potent sample if the energy per quantum is large enough to cause a biochemical change.

Wilhelm Roentgen discovered X rays in 1895 and was awarded the Nobel Prize for this work in 1901.

The unit **roentgen** is used to measure the intensity of X rays or gamma rays. One roentgen is the quantity of X or gamma radiation delivered to 0.001293 gram of air such that the ions produced in the air carry 3.34×10^{-10} coulomb of charge. A single dental x-ray represents about one roentgen.

The three types of natural radioactive emissions differ in their penetrating ability (Figure 17–17), with gamma rays being by far the most penetrating.

Damage by radiation is due to ionization caused by the fast-moving particles colliding with matter, and by the excitation of matter by gamma and X rays that in turn produces ionization. Neutrons are produced in nuclear explosions, in nuclear reactions, and by background cosmic radiation. A neutron does not itself produce ionization but instead imparts its

Paper   1-mm Aluminum foil   8-mm Lead sheet

Figure 17–17 Penetrating ability of alpha (α), beta (β), and gamma (γ) radiation. Gamma rays even penetrate an 8-mm lead sheet. Skin will stop alpha rays but not beta rays.

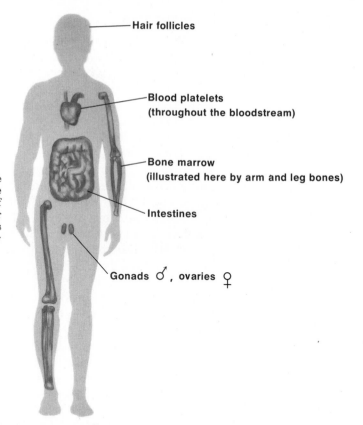

Hair follicles

Blood platelets
(throughout the bloodstream)

Bone marrow
(illustrated here by arm and leg bones)

Intestines

Gonads ♂, ovaries ♀

Figure 17–18 The fast-dividing cells within the body are the ones most harmed by radiation. These include cells in bone marrow, white cells, platelets of the blood, those lining the gastrointestinal tract, hair follicles, and gonads. In addition, the lymphocytes (cells producing the immune responses) are easily killed by radiation.

kinetic energy to atoms, which in turn may ionize or break away from the atom to which they are bonded. Neutrons render many engineering materials such as plastics and metals structurally weak over long periods of time because of the decay caused by breaking chemical bonds.

Biological tissue is easily harmed by radiation. A flow of high-energy particles may cause destruction of a vital enzyme, hormone, or chromosome needed for life of a cell. The radiation may also produce free radicals, which poison the cell. In general, those cells that divide most rapidly are those most easily harmed by radiation (Figure 17–18).

Whole body radiation effects are divided into ***somatic effects,*** which are confined to the population exposed, and ***genetic effects,*** which are passed on to subsequent generations. A unit of measurement of radiation density is helpful in measuring the effect of radiation on tissue. The ***rad*** is defined as 100 ergs of energy imposed on a gram of tissue. Whole body doses of radiation of up to 150 rads produce scarcely any symptoms, whereas doses of 700 rads produce death. Intermediate doses produce vomiting, diarrhea, fatigue, and loss of hair. Often the somatic effects are delayed. Perhaps the best studied of the delayed effects are the incidences of cancer related to exposure of radiation. It has been estimated that 11% of all leukemia cases and about 10% of all forms of cancer are attributable to background radiations. Certainly an individual who is exposed to a higher than normal level of radiation over a considerable length of time increases the chances of cancer. The alteration of normal cells to cancerous cells caused by radiation is undoubtedly a series of changes, since in almost all cases the onset of cancer lags behind the exposure to radiation by an induction period of 5 to 20 years.

Neutrons can damage metals, causing structural failure. This is a severe problem in nuclear reactors.

One *rad* is roughly the energy absorbed by tissue exposed to one roentgen of gamma rays.

1 joule = 10^7 ergs
1 calorie = 4.184 × 10^7 ergs

The genetic effects of radiation are the result of radiation damage to the germ cells of the testes (sperm) or the ovary (egg cells). Ionization caused by radiation passing through a germ cell may break a DNA strand or cause it to be altered in some other way. Replication of this altered DNA means transmission of a new message to successive generations, a *mutation.* Every type of laboratory animal upon which radiation damage experiments have been performed has responded by increased incidence of mutation. Therefore, the necessity of protecting the population of child-bearing age from radiation should be apparent. Theoretically, at least, one photon or one high-energy particle can ionize a chromosomal DNA structure and produce a genetic effect that will be carried for generations.

Radiation Leak at Three Mile Island

On March 28, 1979 the nuclear reactor cooling system failed at the Three Mile Island plant near Middletown, Pennsylvania. In the confusion that followed, operators did not get a clear picture of what was happening in the nuclear reactor from the readout gages, and did not follow a course of action that would have corrected the bad situation. As a result, the reactor core overheated, pressure-relief valves allowed radioactive material to escape to the atmosphere, the reactor core was exposed to superheated steam, the control rods shut down the nuclear reactor, and a bubble of hydrogen gas blocked efforts to turn on the cooling water

Cooling towers at Three Mile Island nuclear plant. Planned releases from these towers include only water vapor and heat. Solid and liquid radioactive wastes are transported from the associated nuclear reactor building under normal operation. (Photo by Laurie Usher.)

again. Officials, not knowing that nuclear fission had stopped, released radioactive gas in order to relieve the pressure. Although the human exposure to radiation was relatively minor, the question of a possible meltdown ("China Syndrome") remains. Would it have been possible for the reactor core to get so hot that the meltdown could have started prior to the automatic insertion of the control rods into the reactor core? The debate continues. However, there is general agreement on this point: Even though a nuclear reactor is in no danger of a nuclear explosion (the fissionable atoms are not close enough together), it can and will melt without a cooling system and/or control rods to stop the nuclear fission. If a meltdown does occur at a nuclear power plant, radiation exposure to humans is likely to be massive.

Fusion Reactions

When very light nuclei, such as H, He, and Li, are combined or *fused* to form an element of higher atomic number, energy must be given off consistent with the greater stability of the elements in this intermediate atomic number range (Figure 17–13). This energy, which comes from a decrease in mass, is the source of the energy released by the sun and by hydrogen bombs. Typical examples of fusion reactions are the following:

Fusion is the combination of very light nuclei.

$$4_1^1H \longrightarrow {}_2^4He + 2_{+1}^0e + 26.7 \text{ Mev}$$

$$_1^2H + {}_1^2H \longrightarrow {}_2^3He + {}_0^1n + 3.2 \text{ Mev}$$

$$_1^2H + {}_1^2H \longrightarrow {}_1^3H + {}_1^1H + 4.0 \text{ Mev}$$

$$_1^3H + {}_1^2H \longrightarrow {}_2^4He + {}_0^1n + 17.6 \text{ Mev}$$

$_1^2H$ = Deuterium
$_1^3H$ = Tritium
$_{+1}^0e$ = Positron

The net reaction for the last three reactions given here is:

$$5_1^2H \longrightarrow {}_2^4He + {}_2^3He + {}_1^1H + 2_0^1n + 24.8 \text{ Mev}$$

or, for every five $_1^2H$ fused, 24.8 Mev of energy are released.

Deuterium is a relatively abundant isotope—out of 6500 atoms of hydrogen in sea water, for example, one is a deuterium atom. What this means is that the oceans are a potential source of fantastic amounts of deuterium. There are 1.03×10^{22} atoms of deuterium in a single liter of seawater. In a single cubic kilometer of sea water, therefore, there would be enough deuterium atoms with enough potential energy to equal the burning of 1360 billion barrels of crude oil, and this is approximately the total amount of oil originally present in this planet.

Materials for fusion reactions are available in enormous amounts.

Fusion reactions take place rapidly only when the temperature is of the order of 100 million degrees or more. At these high temperatures atoms do not exist as such; instead, there is a *plasma* consisting of unbound nuclei and electrons. In this plasma nuclei merge or combine. In order to achieve the high temperatures required for the fusion reaction of the hydrogen bomb, a fission bomb (atomic bomb) is first set off.

High temperature and pressure are required for a fusion reaction to occur.

One type of hydrogen bomb depends on the production of tritium ($_1^3H$) in the bomb. In this type lithium deuteride ($_3^6Li_1^2H$, a solid salt) is placed around an ordinary $_{92}^{235}U$ or $_{94}^{239}Pu$ fission bomb. The fission is set off in the usual way. A $^{6}_{3}Li$ nucleus absorbs one of the neutrons produced and splits into tritium, $^{3}_{1}H$, and helium, $^{4}_{2}He$.

$$^{6}_{3}Li + ^{1}_{0}n \longrightarrow ^{3}_{1}H + ^{2}_{1}H$$

The temperature reached by the fission of $^{235}_{92}U$ or $^{239}_{94}Pu$ is sufficiently high to bring about the fusion of tritium and deuterium:

$$^{3}_{1}H + ^{2}_{1}H \longrightarrow ^{4}_{2}He + ^{1}_{0}n + 17.6 \text{ Mev}$$

A 20-megaton bomb usually contains about 300 lbs of lithium deuteride, as well as a considerable amount of plutonium and uranium.

Controlled Fusion

There are three critical requirements for controlled fusion. First, the temperature must be high enough for ignition to occur. For the deuterium-tritium combination given earlier, a temperature of about 10^8 to 10^9 degrees is needed. Second, the plasma must be confined for a long enough time to release a net output of energy. Third, the energy must be recoverable in some usable form.

As yet the fusion reactions have not been "controlled." No physical container can contain the plasma without cooling it below the critical fusion temperature. Magnetic "bottles," enclosures in space bounded by a magnetic field, have confined the plasma, but not for long enough periods of time (Figure 17–19). Recent developments suggest that these "bottles" may soon hold the plasma long enough so that the fusion reaction can occur.

Thermal energy conversion (Figure 17–20) could be used to take the power from a deuterium-tritium fueled fusion reaction. Liquid lithium would be used to absorb kinetic energy of fast-moving neutrons and then exchange this heat with water to drive a steam turbine, thus producing electricity. This system is actually a breeder reactor as some of the lithium is converted into fuel—tritium, $^{3}_{1}H$—by neutron absorption.

A newer confinement method is based on a laser system that simultaneously strikes tiny hollow glass spheres called **microballoons** which

Containment is one of the biggest of the problems in developing controlled fusion.

Figure 17–19 A controlled fusion experiment. (From *Chemical and Engineering News,* August 30, 1971, p. 5.)

Figure 17–20   Thermal energy conversion from nuclear fusion. Such a scheme, based on the deuterium-tritium fuel cycle, relies on the energy of highly energetic neutrons. A liquid lithium shield absorbs these neutrons and is heated. The lithium then exchanges this heat with water to generate steam.

enclose the fuel, consisting of equal parts of deuterium and tritium gas, at high pressures (Figure 17–21).

There is hope that controlled fusion will be demonstrated during the next decade, but it appears that it will only be after the turn of the century that fusion will furnish any significant fraction of the world's energy needs.

Controlled fusion might end many of the world's energy problems.

Figure 17–21 Focused laser light strikes the microballoon filled with deuterium and tritium, causing a plastic outer layer to burn off (or *ablate*). The outwardly expanding gases from the plastic material drive the glass sphere and its fuel contents inward. The high density and high temperatures produced might result in fusion.

Controlled fusion energy should result in a rather limited production of dangerous radioactivity. The lighter elements involved can be radioactive enough to be a serious hazard, but only for a short period of time; the half-lives of these isotopes are short. Storage and then return to the environment would be quite satisfactory.

SOLAR ENERGY

Solar energy is transmitted nuclear energy.

Earth's ultimate source of energy is the sun. The sun provides energy for photosynthesis, which gives us food to power our bodies, wood to burn, and the starting materials for coal and petroleum. In addition, the sun provides the energy for the water cycle, which is the cause of our weather.

Although Earth receives only about three ten-millionths (0.0000003) of the total energy emitted from the sun, this amount of energy is enormous—about 2×10^{15} kcal/min, or 2.0 cal/cm^2/min. Owing to reradiation from the atmosphere, and the absorption and scattering of radiant energy by molecules in the lower portions of the atmosphere, the amount of radiation actually reaching the surface of this planet is about 1 cal/cm^2/min. The actual value depends on location, season, and weather conditions. Even this is a large amount of energy. For example, the roof of an average-sized house will receive about 10^8 calories/day when the radiation level is 1 cal/cm^2/min. This is equivalent to the heat energy derived from burning about 32 pounds of coal per day. This is also equal to 120 kilowatt hours per day of electrical energy—more than enough to heat an average American home on a winter day.

Only about 1 percent of the solar radiation used by plants in photosynthesis ends up as stored chemical energy such as foodstuffs. After fossil fuels are depleted, could photosynthesis be considered as a viable source of energy outside of food materials? Much of the energy used in photosynthesis is for "fixing" carbon in the form of cellulose and carbohydrates. Recently it has been shown that certain blue-green algae, *Anabeana cylindrica,* can convert sunlight and water into hydrogen and oxygen. A

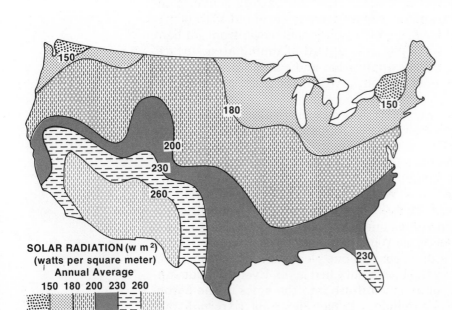

Figure 17–22 Annual rate of delivery of solar radiation in the United States.

SOLAR RADIATION (w m²)
(watts per square meter)
Annual Average
150 180 200 230 260

Figure 17–23 Schematic diagram of an electricity-producing photosynthesis process. H_2 and O_2 produced by the *Anabeana* are separated by palladium metal which is permeable toward H_2 but not O_2. The H_2 and O_2 are then combined in the fuel cell to produce electricity.

colony of such algae, coupled with a fuel cell utilizing hydrogen and oxygen, could be a source of electricity during sunlight hours (Figure 17–23).

The attractiveness of solar energy is very great. For example, the world's present energy requirements per year are about 10^{18} kcal. An area of desert of about 28,000 square miles with little cloud cover or dust, near the equator, would receive about 10^{17} kcal/year of solar radiation. Such deserts exist in northern Chile, and could, if the need were great enough to justify the costs, supply a large portion of the world's energy needs.

The solar energy could heat water to steam, which in turn could generate electricity (Figure 17–24), which could electrolyze water to hydrogen and oxygen. The hydrogen could be piped to where the energy is needed and then converted to electricity. Such an arrangement would give rise to a ***hydrogen economy,*** one that has many advantages over present energy sources such as fossil fuels.

Another approach to the direct utilization of solar energy is the ***solar battery,*** known as a photovoltaic device. The solar battery converts energy from the sun into electron flow. Solar batteries are about 13 to 14 percent efficient and are capable of generating electrical power from sunlight at the rate of at least 90 watts per square yard of illuminated surface. They are now used in space flight application and communication satellites, and in Israel, India, Pakistan, South Africa, and Azerbaijan SSR to obtain electric power.

One type of solar battery consists of two layers of almost pure silicon (Figure 17–25). The lower, thicker layer contains a trace of arsenic and the upper, thinner layer a trace of boron (Figure 17–26). Silicon has four valence electrons and forms a tetrahedral, diamond-like, crystalline structure. Each silicon atom is covalently bonded to four other silicon atoms. Arsenic has five valence electrons. When arsenic atoms are included in the silicon structure, only four of the five valence electrons of arsenic are used for bonding with four silicon atoms; one electron is relatively free to roam. Boron has three valence electrons. When boron atoms are included in the silicon structure, there is a deficiency of one electron around the boron atom; this creates "holes" in the boron-enriched layer. Even without sunlight, the 'extra" electrons of arsenic diffuse into the holes in the boron. The driving force is the strong tendency to pair electrons. Externally ap-

Hydrogen can be burned in most devices that now burn natural gas.

The goal is to capture and use the solar energy "on the run" without upsetting the energy flow into (light side) and away from (dark side) Mother Earth.

Figure 17–24 (*a*) Solar energy collector. (*b*) Solar energy conversion system.

Figure 17–25 Silicon photodiode (Bell solar battery).

Perfect crystal

Figure 17-26 Schematic drawing of semi-conductor crystal layers derived from silicon (after Masterton and Slowinski).

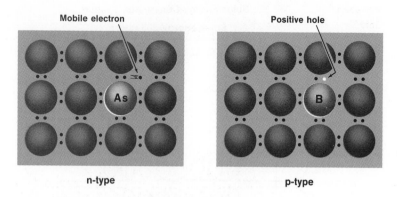

n-type · p-type

plied potentials as high as 1000 volts are incapable of reversing the flow. The negative charge built up in the boron layer would hinder the flow of electrons into that layer and eventually stop the flow. The opposing factors—repulsion between free electrons and the drive to pair electrons—finally bring about an equilibrium.

When sunlight strikes the boron layer, the equilibrium is disturbed. If the wafer is connected as an ordinary battery would be, electrons flow from what appears to be the negative arsenic layer through the circuit and back into the boron layer (Figure 17–25). The fact that electrons enter the circuit from the arsenic layer can be explained if sunlight unpairs electrons in the boron layer, and the freed electrons are repelled to the arsenic layer and into the circuit. This pulls electrons into the boron layer to maintain the new balance between electron pairing and electron repulsion.

The advantage of the solar battery is that it has no moving parts, no liquids, and no corrosive chemicals—it just keeps on generating electricity indefinitely while exposed to sunlight. The drawbacks of the solar battery are the very large area required for large amounts of power, the high costs of the very pure materials, and the fact that they work only when the sun is shining. Since the first practical use of solar batteries in 1955 to power eight rural telephones in Georgia, they have undergone a great deal of development, and much more is expected because of their great potential use.

Solar batteries are still very expensive.

In theory, a number of huge energy sources are potentially available for our use. As yet we have not learned to control them sufficiently well to have cheap energy for all. However, progress is being made. It is evident that both chemical sources of energy and the fission of heavy nuclei are limited in their ability to supply effectively our growing energy needs, and we must develop a wide variety of other energy sources.

1. The splitting of an unstable nucleus to produce energy is termed (fission/ fusion).

2. A sufficiently large sample to sustain a chain reaction in a fission process is called the _____ _____.

3. Of the two major isotopes of uranium, which is fissionable, uranium-238 or uranium-235? _____

4. When light nuclei combine to form heavy nuclei and energy, the process is (fission/fusion).

5. The major problem with obtaining fusion energy is (containment of reactants at high temperature/enough fuel/costs).

6. When and if fusion is used to produce energy in a breeder reaction, the fuel produced will be _____.

7. When fission is used to produce energy in a breeder reaction, the fuels produced will be _____ and _____.

8. Which is the more penetrating type of radiation? (alpha/beta) _____

9. The curie is a unit related to (numbers of disintegrations/intensity). _____

_____ 1. user of 35 percent of world's energy   (a) 1859

_____ 2. fossil fuels   (b) CO_2 and H_2O

_____ 3. combustion products of fossil fuels   (c) uranium-235 ($^{235}_{92}U$)

_____ 4. minable coal   (d) mass defect

_____ 5. synthesis gas   (e) plutonium-239 ($^{239}_{94}Pu$)

_____ 6. fissionable isotope   (f) United States

_____ 7. basis of nuclear energy   (g) ^2_1H

_____ 8. product of a fission breeder reactor   (h) sea water

_____ 9. deuterium   (i) 700 rads

_____ 10. date of petroleum discovery in U.S.   (j) within 4000 feet of surface

_____ 11. tritium   (k) microballoons

_____ 12.	source of deuterium	(l)	coal, petroleum, natural gas
_____ 13.	used to confine fusion fuel	(m)	photosynthesis
_____ 14.	one use of solar radiation	(n)	CO and H_2
_____ 15.	approximate efficiency of a solar battery	(o)	10 to 14 percent
_____ 16.	nuclear radiation death	(p)	$_1^3H$

QUESTIONS

1. What is your attitude toward using up the fossil fuels within a few decades? Do we owe future generations a supply of these resources? Would you agree to give up air-conditioning, private cars, and power tools, to mention a few examples, and to limit heating and cooking if necessary to share these fuels with your grandchildren?

2. Which theoretically yields the greatest energy per mole?

   (a) the burning of gasoline
   (b) the fission of uranium-235
   (c) the fusion of hydrogen

3. Which is the more efficient use of energy: burning coal in a house to heat it or heating the house electrically with energy produced in a coal-burning power plant?

4. Give three examples of systems that contain chemical energy that can be used as a source of heat energy.

5. Is the electrical energy where you live produced by burning fossil fuels? If not, what is the energy source? Are there pollution problems associated with the generation of the electrical power?

6. What produces the tremendous energy of a fission reaction?

7. What was the original source of energy that is tied up in fossil fuels?

8. Why is it difficult to fuse two $_{24}^{52}Cr$ nuclei? Explain.

9. What is meant by a chain reaction?

10. What major problem is associated with harnessing the energy from a fusion reaction?

11. Suggest several ways solar energy might be harnessed.

12. Name two sources of energy not specifically mentioned in this chapter.

13. Explain how useful energy might be obtained from garbage.

14. Which is more fundamental—a supply of energy or a supply of food? Explain.

15. The energy consumption of the United States in 1970 was 2×10^{13} kilowatt-hours. What is this amount of energy expressed in kilocalories? In BTU's?

16. Assume the world population to be 4.5 billion and calculate the earth's energy needs if everyone used as much energy as is used in the United States.

17. What is meant by "delayed somatic effect"? Given an example.

THE INTERNATIONAL SYSTEM OF UNITS (SI)

Since 1960, a coherent system of units known as the Système International (SI system), bearing the authority of the International Bureau of Weights and Measures, has been in effect and is gaining acceptance among scientists. It is an extension of the metric system that began in 1790, with each physical quantity assigned an SI unit uniquely. An essential feature of both the older metric system and now the newer SI is a series of prefixes that indicate a power of ten multiple or submultiple of the unit.

I. Units of Length

The standard unit of length is the *meter.* It was originally meant to be one ten-millionth of the distance along a meridian from the North Pole to the equator. However, the lack of precise geographical information necessitated a better definition. For a number of years the meter was defined as the distance between two etched lines on a platinum-iridium bar kept at 0°C (32°F) in the International Bureau of Weights and

14 feet
4.26 meters

250 pounds
113.5 kılograms

Figure A–1 The pole-vaulter is easily recognized as hefty when described by two hundred and fifty pounds and his jump something less than a record at fourteen feet. As Americans move closer to the use of the system of international measurements, the 113.5 kilograms and the 4.26 meters will produce similar conceptualizations related to previous experience.

Figure A–2   A meter equals 1.094 yards.

Measures, at Sèvres, France. The inability to measure this distance as accurately as desired prompted a recent redefinition of the meter as being a length equal to 1,650,763.73 times the wavelength of the orange-red spectrographic line of $^{86}_{36}Kr$.

The meter (39.37 inches) is a convenient unit with which to measure the height of a basketball goal (3.05 meters), but it is unwieldy for measuring the parts of a watch or the distance between continents. For this reason, prefixes are defined in such a way that, when placed before the meter, they define distances convenient for our particular purposes. Some of the prefixes with their meanings are:

> nano—1/1,000,000,000 or 0.000000001
> micro—1/1,000,000 or 0.000001
> milli—1/1000 or 0.001
> centi—1/100 or 0.01
> deci—1/10 or 0.1
> deka—10
> hecto—100
> kilo—1000
> mega—1,000,000

The corresponding units of length with their abbreviations are the following:

> nanometer (nm)—0.000000001 meter
> micrometer (μm)—0.000001 meter
> millimeter (mm)—0.001 meter
> centimeter (cm)—0.01 meter
> decimeter (dm)—0.1 meter
> meter (m)—1 meter
> dekameter (dam)—10 meters
> hectometer (hm)—100 meters
> kilometer (km)—1000 meters
> megameter (Mm)—1,000,000 meters

Since the prefixes are defined in terms of the decimal system, the conversion from one metric length to another involves only shifting the decimal point. Mental calculations are quickly accomplished.

TABLE A—1   Conversion Factors*

Length:	1 inch (in)	=	2.54 centimeter (cm)
	1 yard (yd)	=	0.914 meter (m)
	1 mile (mi)	=	1.609 kilometers (km)
Volume	1 ounce (oz)	=	29.57 milliliters (ml)
	1 quart (qt)	=	0.946 liter (l)
	1.06 quart (qt)	=	1 liter (l)
	1 gallon (gal)	=	3.78 liters (l)
Mass (weight)†:	1 ounce (oz)	=	28.35 grams (g)
	1 pound (lb)	=	453.6 grams (g)
	1 ton (tn)	=	907.2 kilograms (kg)

*Common English units are used.

†Mass is a measure of the amount of matter, whereas weight is a measure of the attraction of the earth for an object at the earth's surface. The mass of a sample of matter is constant, but its weight varies with position and velocity. For example, the space traveler, having lost no mass, becomes weightless in earth orbit. Although mass and weight are basically different in meaning, they are often used interchangeably in the environment of the earth's surface.

How many centimeters are in a meter? Think: Since a centimeter is the one-hundredth part of a meter, there would be 100 centimeters in a meter.

Conversion of measurements from one system to the other is a common problem. Some commonly used English-SI equivalents (conversion factors) are given in Table A-1.

II.   Units of Mass

The primary unit of mass is the **kilogram** (1000 grams). This unit is the mass of a platinum-iridium alloy sample deposited at the International Bureau of Weights and Measures. One pound contains a mass of 453.6 grams (a five-cent nickel coin contains about 5 grams).

Conveniently enough, the same prefixes defined in the discussion of length are used in units of mass, as well as in other units of measure.

III.   Units of Volume

The SI unit of volume is the **cubic meter** (m^3). However, the volume capacity used most frequently in chemistry is the liter, which is defined as 1 cubic decimeter (1 dm^3). Since a decimeter is equal to 10 centimeters (cm), the cubic decimeter is equal to $(10 \text{ cm})^3$ or 1000 cubic centimeters (cc). One cc, then, is equal to one milliliter (the thousandth part of a liter). The ml (or cc) is a common unit that is often used in the measurement of medicinal and laboratory quantities. There are then 1000 liters in a kiloliter or cubic meter.

IV.   Units of Energy

The SI unit for energy is the **joule** (J), which is defined as the work performed by a force of one newton acting through a distance of one meter. A newton is defined as that force which produces an acceleration of one meter per second per second when applied to a mass of one kilogram. Conversion units for energy are:

$$1 \text{ calorie} = 4.184 \text{ joules}$$
$$1 \text{ kilowatt hour} = 3.5 \times 10^6 \text{ joules}$$

V. *Other SI Units*

Other SI units are listed below.

Time	second (s)
Temperature	Kelvin (K)
Electric current	ampere (A) = 1 coulomb per second
Amount of molecular substance	mole (mol) = 6.023×10^{23} molecules
Pressure	pascal (Pa) = 1 newton per square meter
Power	watt (W) = 1 joule per second
Electric charge	coulomb (C)
	= 6.24196×10^{18} electronic charges
	= 1.036086×10^5 faradays

Further information on SI units can be obtained from "SI Metric Units—An Introduction," by H. F. R. Adams, McGraw-Hill Ryerson Ltd., Toronto, 1974.

TEMPERATURE SCALES

The system of measuring temperature that is used in scientific work is based on the *centigrade* or *Celsius* temperature scale. This temperature scale was defined by Anders Celsius, a Swedish astronomer, in 1742. The Celsius scale is based upon the expansion of a column of mercury that occurs when it is transferred from a cold standard temperature to a hot standard temperature. The cold standard temperature is the temperature of melting ice and is defined as 0°C. The hot standard temperature is the temperature at which water boils under standard conditions of pressure; it is defined as 100°C. The expansion of the mercury is assumed to be linear over this range, and the distance through which the mercury column expands between 0°C and 100°C is divided into 100 equal parts, each corresponding to one degree.

On the scale commonly used in the United States, the Fahrenheit scale (developed by and named after Gabriel Daniel Fahrenheit, a German-Dutch physicist of the early 18th century), the freezing mark is 32° and the boiling mark is 212°. Obviously, the ice is no hotter when a Fahrenheit thermometer is stuck into the system than when a Celsius thermometer is used. The marks of the tubes are simply different names for the same thing.

The ideal gas law predicts another temperature scale based on the temperature-volume behavior of a gas (Charles' law). Lord Kelvin reasoned that if an ideal gas lost $\frac{1}{273}$ of its volume during a temperature decrease from 0°C to −1°C, then at −273°C, theoretically at least, there should be no volume at all. All gases liquefy before this temperature is reached. However, this temperature, −273°C, appears to be a zero temperature defined by nature. Thus the Kelvin temperature scale was developed. Minus 273°C becomes 0 Kelvin and 0°C becomes 273 Kelvin. (More accurately, 273.15 K.) Note that current SI usage does not write °K, but K.

Because all these systems are now in common use, it is necessary at times to convert from one system to another.

Note that 100 Celsius degrees = 180 Fahrenheit degrees or 1 C degree = 1.8 (or $\frac{9}{5}$) F degrees.

But if we wish to convert temperature from one scale to another, we must remember that 0°C is the same as 32°F. Therefore, 32 must be added to the calculated number of degrees Fahrenheit in order to revert to the start of the counting in the Fahrenheit system.

Therefore,

$$°F = \frac{9}{5}(°C) + 32$$

which can be arranged to

$$°C = \frac{5}{9}(°F - 32).$$

EXAMPLE 1
Convert 50°F to the corresponding Celsius temperature:

$$
\begin{aligned}
°C &= \frac{5}{9}(°F - 32) \\
&= \frac{5}{9}(50 - 32) \\
&= \frac{5}{9}(18) \\
&= 10°C
\end{aligned}
$$

EXAMPLE 2
Convert 25°C to the corresponding Fahrenheit temperature:

$$
\begin{aligned}
°F &= \frac{9}{5}(°C) + 32 \\
&= \frac{9}{5}(25) + 32 \\
&= 45 + 32 \\
&= 77°F
\end{aligned}
$$

To convert Celsius to Kelvin:

$$K = °C + 273$$

EXAMPLE 3
Convert 23°C to Kelvin:

$$
\begin{aligned}
K &= 23° + 273 \\
&= 296\ K
\end{aligned}
$$

Rather than memorizing formulas, many students prefer to figure out temperature conversions using common sense. A common-sense approach is shown in Table B–1.

TABLE B–1 Common-Sense Method for Converting Temperatures

Starting point: $\begin{cases} \text{freezing point (f.p.) of water is } 0°C = 32°F \\ \text{boiling point (b.p.) of water is } 100°C = 212°F \end{cases}$

Other information needed: 100 C degrees = 180 F degrees

General Procedure	Example 1 (50°F = ?°C)	Example 2 (25°C = ?°F)	Example 3 (98.6°F = ?°C)
1. Select a reference point.	1. f.p. of water	1. f.p. of water	1. b.p. of water
2. Determine relationship of temperature of interest to reference point.	2. 50°F is 18 F degrees above f.p. of water	2. 25°C is 25 C degrees above f.p. of water	2. 98.6°F is 113.4 F degrees below b.p. of water
3. Convert the number of degrees.	3. 18 F degrees is $18\,F° \times \dfrac{100\,C°}{180\,F°} =$ 10 C degrees	3. 25 C degrees is $25\,C° \times \dfrac{180\,F°}{100\,C°} =$ 45 F degrees	3. 113.4 F degrees is $113.4\,F° \times \dfrac{100\,C°}{180\,F°} =$ 63.0 C degrees
4. Express temperature, taking into account the selected reference point.	4. The temperature is 10 C degrees above the f.p. of water (which is 0°C), so the temperature is *10°C*	4. The temperature is 45 F degrees above the f.p. of water (which is 32°F), so the temperature is $45 + 32 = 77°F$	4. The temperature is 63.0 C degrees below the b.p. of water (which is 100.0°C), so the temperature is $100.0 - 63.0 = 37.0°C$

FACTOR-LABEL APPROACH TO CONVERSION PROBLEMS

For converting a measurement from one system of units to another, the following method is straightforward and does not require the decision of whether to divide or multiply to obtain the proper answer. This method makes the decision for you, a decision that is sometimes difficult when dealing with new units with which you have had little experience.

EXAMPLE
How many liters are in 6 quarts?

Factor-Label Method of Solution:

1. Write down the unit to which you are converting. The question mark indicates the number of liters to be determined.

   1. ? liters =

2. On the right hand side of the equal sign, write down the quantity given. Write both the number and the name or label.

   2. ? liters = 6 quarts

3. Now look at the two units. Recall a conversion between these two units. These conversions must be learned or looked up.

   3. 1 liter = 1.06 quarts,
      1 liter per 1.06 quarts

      $$\text{or } \frac{1 \text{ liter}}{1.06 \text{ quarts}}$$

      $$\text{or } \frac{1.06 \text{ quarts}}{\text{liter}}$$

4. Write the conversion factor on the right hand side so that the unwanted units will cancel; that is, a unit in the numerator will cancel the same unit in the denominator, and only the unit you want will remain. Do not, of course, cancel the numbers, just the units.

   4. ? liters =

   $$6 \text{ quarts} \times \frac{1 \text{ liter}}{1.06 \text{ quarts}}$$

5. Do the indicated multiplication and division. The line, of course, means divided by. Check the units on both sides of the equation. The units should be the same and in the same position (numerator or denominator).

5. $? \text{ liters} = \dfrac{6}{1.06} \text{ liters}$

$= 5.66 \text{ liters}$

Now, suppose you needed to convert 3 pints to ml, and you do not know a conversion factor that will make the conversion in one step.

Proceed as before:

$$? \text{ ml} = 3 \text{ pints}$$

Note this is a volume conversion from the English system to the metric system as before. Recall the volume conversion factor that you know between these systems.

$$1 \text{ liter} = 1.06 \text{ quarts}$$

Convert in the system of the given quantity until you reach the unit in the conversion factor for this system. Then write down the conversion factor between systems so that like units in separate factors will cancel.

$$? \text{ ml} = 3 \text{ pints} \times \frac{1 \text{ quart}}{2 \text{ pints}} \times \frac{1 \text{ liter}}{1.06 \text{ quarts}}$$

Cancel units and recall a conversion factor that converts the units you have left to the unit you want. Do the indicated multiplication and division.

$$? \text{ ml} = \frac{3 \times 1 \times 1 \text{ liter}}{2 \times 1.06} \times \frac{1000 \text{ ml}}{1 \text{ liter}}$$

$$= \frac{3 \times 1 \times 1 \times 1000}{2 \times 1.06 \times 1} \text{ ml}$$

$$= 1420 \text{ ml}$$

In brief, the method is very simple:

1. Write the units you want, an equals sign, and the quantity you you have given.
2. Write conversion factors so the unwanted factors will cancel.
3. Keep in one system until you come to units in a familiar intersystem conversion factor.
4. Do the indicated arithmetic and check the units.
5. Examine the size of the answer for reasonableness.

Note: This approach to problem solving has wide applicability to many other types of problems. It is especially useful in problems pertaining to weight relationships in chemical reactions. See Appendix D.

STOICHIOMETRIC CALCULATIONS

The bases for stoichiometric calculations were presented in Chapter 4. Problems of a more complex nature and a systematic approach to their solution are presented in the following examples. Finally, a list of exercise problems is given for further study.

EXAMPLE 1

Balanced Equations Express Number Ratios for Particles

In the reaction of hydrogen with oxygen to form water, how many molecules of hydrogen are required to combine with 19 oxygen molecules?

Solution: A chemical equation can be written for a reaction only if the reactants and products are identified and the respective formulas determined. In this problem, the formulas are known and the unbalanced equation is:

$$H_2 + O_2 \longrightarrow H_2O$$

It is evident that one molecule of oxygen contains enough oxygen for two water molecules and the equation, as written, does not account for what happens to the second oxygen atom. As it is, the equation is in conflict with the conservation of atoms in chemical changes. This conflict is easily corrected by balancing the equation:

$$2H_2 + O_2 \longrightarrow 2H_2O$$

Now, all atoms are accounted for in the equation and it is obvious that two hydrogen molecules are required for each oxygen molecule. In other words, two hydrogen molecules are equivalent to one oxygen molecule in their usage. This can be expressed as follows:

2 hydrogen molecules {are equivalent to} 1 oxygen molecule;

or,

$$2H_2 \text{ molecules} \sim O_2 \text{ molecule};$$

or,

$$\frac{2H_2 \text{ molecules}}{O_2 \text{ molecule}},$$

which can be read as two hydrogen molecules per one oxygen molecule.

Using now the factor-label approch developed in Appendix C, the solution is readily achieved.

$$? \, H_2 \text{ molecules} = 19O_2 \text{ molecules}$$

$$? \, H_2 \text{ molecules} = 19O_2 \, \cancel{\text{molecules}} \times \frac{2H_2 \text{ molecules}}{O_2 \, \cancel{\text{molecule}}}$$

$$= 38H_2 \text{ molecules}$$

Note: The reader is likely to say at this point that the method is cumbersome and that he can quickly see the answer to be $38H_2$ molecules without "the method." However, problems to follow are made much easier if a systematic method of approach is used.

EXAMPLE 2

Laboratory Mole Ratios Identical With Particle Number Ratios

How many moles of hydrogen molecules must be burned in oxygen (the reaction of example 1) to produce 15 moles of water molecules (about a glassful)?

Solution: The balanced equation,

$$2H_2 + O_2 \longrightarrow 2H_2O,$$

tells us that two molecules of hydrogen produce two molecules of water;

or

$$2 \text{ molecules hydrogen} \sim 2 \text{ molecules water,}$$

and therefore,

$$1 \text{ molecule hydrogen} \sim 1 \text{ molecule water.}$$

It is obvious then that the number of water molecules produced will be equal to the number of hydrogen molecules consumed regardless of the actual number involved. Therefore,

$$\left. \begin{array}{c} 6.02 \times 10^{23} \text{ molecules} \\ \text{of hydrogen} \end{array} \right\} \sim \begin{array}{c} 6.02 \times 10^{23} \text{ molecules} \\ \text{of water} \end{array}$$

Since 6.02×10^{23} is a number called the mole, it follows that one mole of hydrogen molecules will produce one mole of water molecules. The general conclusion, then, is the following: the ratio of particles in the balanced equation is the same as the ratio of moles in the laboratory. The solution to the problem logically follows:

$$\left. \begin{array}{c} ? \text{ moles of hydrogen} \\ \text{molecules} \end{array} \right\} = \left\{ \begin{array}{c} 15 \, \cancel{\text{moles}} \\ \cancel{\text{water molecules}} \end{array} \right\} \times \frac{1 \text{ mole hydrogen molecules}}{1 \, \cancel{\text{mole water molecules}}}$$

$$= 15 \text{ moles hydrogen molecules}$$

Note: Again the solution to the problem looks simple enough without resorting to the factor-label method. However in examples 3 and 4, the numbers become such that a quick mental solution is not readily achieved by most students.

EXAMPLE 3

Mole Weights Yield Weight Relationships

How many grams of oxygen are necessary to react with an excess of hydrogen to produce 270 grams of water?
Solution: From the balanced equation,

$$2H_2 + O_2 \longrightarrow 2H_2O,$$

the mole ratio between oxygen and water is immediately evident and is one mole of oxygen molecules per two moles of water molecules, or

$$\frac{1 \text{ mole oxygen molecules}}{2 \text{ moles water molecules}},$$

This mole ratio can be changed into a weight ratio since the mole weight can be easily calculated from the atomic weights involved. One molecule of oxygen (O_2) weighs 32 amu (16 amu for each oxygen atom). Therefore, a mole of oxygen molecules weighs 32 grams. Similarly, two moles of water weigh 36 grams [2(16 + 2 × 1)]. Therefore, the weight ratio is:

$$\frac{1 \text{ mole oxygen molecules} \times \dfrac{32 \text{ grams oxygen}}{\text{mole oxygen molecules}}}{2 \text{ moles water molecules} \times \dfrac{18 \text{ grams water}}{\text{mole water molecules}}}$$

or,

$$\frac{32 \text{ grams oxygen}}{36 \text{ grams water}}.$$

This weight relationship is exactly the conversion factor needed to answer the original question:

$$? \text{ grams oxygen} = 270 \text{ grams water} \times \frac{32 \text{ grams oxygen}}{36 \text{ grams water}}$$
$$= 240 \text{ grams oxygen}$$

Note: It should be observed that a weight relationship could be established between any two of the three pure substances involved in the reaction, regardless of whether they are reactants or products.

EXAMPLE 4

How many molecules of water are produced in the decomposition of 8 molecules of table sugar? The unbalanced equation is as follows:

$$C_{12}H_{22}O_{11} \longrightarrow C + H_2O$$

Solution: Balance the equation,

$$C_{12}H_{22}O_{11} \longrightarrow 12C + 11H_2O$$

$$? \text{ molecules of water} = 8 \text{ molecules sugar} \times \frac{11 \text{ molecules water}}{1 \text{ molecule sugar}}$$

$$= 88 \text{ molecules of water}$$

EXAMPLE 5

How many grams of mercuric oxide are necessary to produce 50 grams of oxygen? Mercuric oxide decomposes as follows:

$$2HgO \longrightarrow 2Hg + O_2$$

Solution:

(a) Weight of two moles of HgO = 2(201 + 16) = 2(217) = 434 g
(b) Weight of 1 mole of O_2 = 2(16) = 32 g
(c) $? \text{ g HgO} = 50 \text{ g oxygen} \times \dfrac{434 \text{ g mercuric oxide}}{32 \text{ g oxygen}}$

$$= 678 \text{ g mercuric oxide}$$

EXAMPLE 6

How many pounds of mercuric oxide are necessary to produce 50 pounds of oxygen by the reaction:

$$2HgO \longrightarrow 2Hg + O_2$$

Solution: Note that the problem is the same as example 5 except for the units of chemicals. Also note that the conversion factor of example 5,

$$\frac{434 \text{ g mercuric oxide}}{32 \text{ g oxygen}}$$

can be converted to any other units desired,

$$\frac{434 \text{ g mercuric oxide} \times \dfrac{1 \text{ pound}}{454 \text{ g}}}{32 \text{ g oxygen} \times \dfrac{1 \text{ pound}}{454 \text{ g}}} = \frac{434 \text{ pounds mercuric oxide}}{32 \text{ pounds oxygen}}$$

It is evident that the ratio, $\dfrac{434}{32}$, expresses the ratio between weights of mercuric oxide and oxygen in this reaction regardless of the units employed.

$$? \text{ pounds mercuric oxide} = 50 \text{ pounds oxygen}$$

$$\times \frac{434 \text{ pounds mercuric oxide}}{32 \text{ pounds oxygen}}$$

$$= 678 \text{ pounds of mercuric oxide}$$

1. What weight of oxygen is necessary to burn 28 g of methane, CH_4? The equation is:

    $$CH_4 + 2O_2 \longrightarrow CO_2 + 2H_2O$$

2. Potassium chlorate, $KClO_3$, releases oxygen when heated according to the equation:

    $$2KClO_3 \longrightarrow 2KCl + 3O_2.$$

    What weight of potassium chlorate is necessary to produce 1.43 grams of oxygen?

3. Fe_2O_4 is a magnetic oxide of iron. What weight of this oxide can be produced from 150 grams of iron?

4. Steam reacts with hot carbon to produce a fuel called water gas; it is a mixture of carbon monoxide and hydrogen. The equation is:

    $$H_2O + C \longrightarrow CO + H_2$$

    What weight of carbon is necessary to produce 10 grams of hydrogen by this reaction?

5. Iron oxide, Fe_2O_3, can be reduced to metallic iron by heating it with carbon.

    $$2Fe_2O_3 + 3C \longrightarrow 4Fe + 3CO_2$$

    How many tons of carbon would be necessary to reduce 5 tons of the iron oxide in this reaction?

6. How many grams of hydrogen are necessary to reduce 1 pound (454 grams) of lead oxide (PbO) by the reaction:

    $$PbO + H_2 \longrightarrow Pb + H_2O$$

7. Hydrogen can be produced by the reaction of iron with steam.

    $$4H_2O + 3Fe \longrightarrow 4H_2 + Fe_3O_4$$

    What weight of iron would be needed to produce one-half pound of hydrogen?

8. Tin ore, containing SnO_2, can be reduced to tin by heating with carbon.

    $$SnO_2 + C \longrightarrow Sn + CO_2$$

    How many tons of tin can be produced from 100 tons of SnO_2?

PROBLEMS

Ans. 112 g oxygen

Ans. 3.65 g $KClO_3$

Ans. 207 g oxide

Ans. 60 g carbon

Ans. 0.56 ton carbon

Ans. 3.91 g hydrogen

Ans. 10.5 lb iron

Ans. 79 tons tin

ANSWERS TO SELF-TESTS

CHAPTER 1

SELF-TEST 1-A

1. (a) air, (b) concrete, (c) vegetables, (d) paint
2. operational definition
3. (a) table sugar, (b) distilled water, (c) diamond, (d) copper wire
4. false

SELF-TEST 1-B

1. (a) metals and (b) non-metals
   metals: (a) iron, (b) copper, (c) gold, (d) silver, (e) chromium, (f) magnesium
   non-metals: (a) oxygen, (b) silicon, (c) carbon, (d) nitrogen, (e) chlorine, (f) fluorine
2. 106
3. false
4. (a) burning of coal, (b) dissolving of iron ore in an acid, (c) making steel from iron ore, (d) making of aspirin
5. (a) cutting diamond, (b) blowing glass, (c) molding plastic, (d) slicing bread
6. chemical substance
7. submicroscopic
8. (a) macroscopic, (b) microscopic, (c) molecular
9. (a) theories, (b) laws, (c) facts
10. (a) a sodium atom, or a mole of sodium; (b) two sodium atoms, or two moles of sodium; (c) a hydrogen chloride molecule, or a mole of hydrogen chloride; (d) reacts to form; (e) a hydrogen molecule, or a mole of hydrogen; (f) 2 "molecules" of sodium chloride, or 2 moles of sodium chloride

MATCHING SET

1. b	5. c	9. g
2. e	6. f	10. h
3. d	7. i	11. j
4. a	8. k	

CHAPTER 2

SELF-TEST 2-A

1. Leucippus and Democritus
2. (b) philosophy
3. gained in a chemical reaction
4. CO and CO_2
5. (a) a new compound, (b) 2 : 4 : 1, (c) Law of Multiple Proportions
6. (d) atoms are recombined in different arrangements
7. (b) the same, (b) atoms
8. repel; attract
9. alpha (α), beta (β), gamma (γ), gamma (γ)

SELF-TEST 2-B

1. protons, neutrons
2. small
3. nucleus
4. electrons, protons, neutrons
5. electrons and protons
6. atomic
7. about 1836
8. 33, 33, and 42
9. different, identical
10. electrons
11. electrons and protons
12. false

SELF-TEST 2-C

1. particles, waves
2. spectrum
3. farther from, closer to
4. (d) wave nature of the electron
5. wavelength
6. 18, 2
7. Group IA, Group IVA, Group IIA
8. (a) no, not the paths
   (b) in a way, yes; they are representations of the space in which we can expect to find the electron with 90% certainty.
   (c) of a given type, yes, with 90% certainty

MATCHING SET

1.	e	6.	n	11.	h
2.	a	7.	b	12.	c
3.	j	8.	f	13.	k
4.	i	9.	m	14.	l
5.	d	10.	g		

CHAPTER 3

SELF-TEST 3-A

1. ions
2. ionic lattice
3. one
4. helium, cesium
5. CaI_2
6. Cl^-
7. valence
8. losing
9. gaining
10. Rb; 1 electron lost
    S; 2 electrons gained
    Ca; 2 electrons lost
    Mg; 2 electrons lost
    Al; 3 electrons lost
    Br; 1 electron gained

SELF-TEST 3-B

1. (a) hydrogen or fluorine, (b) water or hydrogen chloride
2. six
3. three
4. fluorine
5. (a) eight, (b) octet, (c) most of the time
6. fluorine
7. fluorine, oxygen, or nitrogen

SELF-TEST 3-C

1. σ (sigma) bond, π (pi) bond
2. σ bond, π bonds
3. s, p
4. nonbonding more than bonding
5. single
6. three
7. AX_2, linear; BZ_3, triangular (trigonal) planar; DQ_6, octahedral

MATCHING SET I

1. l	6. g	10. c
2. d	7. b	11. h
3. j	8. e	12. a
4. f	9. k	13. m
5. i		

MATCHING SET II

1. d	5. a
2. g	6. e
3. h	7. b
4. f	8. c

MATCHING SET III

1. a	4. e
2. c	5. b
3. d	

CHAPTER 4

SELF-TEST 4-A

1. temperature, because rate of bacterial growth slows down, as does the rate of any chemical reaction, with decreasing temperature
2. hydrogen and oxygen
3. hemoglobin and oxygen, and calcium oxide (CaO) and water
4. absorbs
5. no, this is not true
6. 136 kcal
7. (2) flour in dust form
8. hemoglobin oxygen uptake and release
9. formation of salt from sodium and chlorine, and formation of SO_2 from S and O_2

SELF-TEST 4-B

1. (a) $2Mg + O_2 \longrightarrow 2MgO$
   (b) $Si + 2Cl_2 \longrightarrow SiCl_4$
   (c) $4Al + 3O_2 \longrightarrow 2Al_2O_3$
2. (a) SeO_2 and SeO_3, (b) SeF_4 and SeF_6, (c) H_2SeO_3 and H_2SeO_4
3. (a) $2H_2O \longrightarrow 2H_2 + O_2$
   (b) 18 grams of water, (c) 16 grams of oxygen, (d) 18 tons of water

MATCHING SET I

1. h   5. c
2. f   6. d
3. e   7. a
4. b   8. g

MATCHING SET II

1. c   4. f
2. e   5. b
3. a   6. d

CHAPTER 5

SELF-TEST 5-A

1. an electrolyte
2. a base, an acid
3. a base, an acid
4. amphiprotic
5. neutral
6. water
7. an acid reacts with a base

SELF-TEST 5-B

1. 3(a) acidic, 10(b) basic, 7(c) neutral
2. 1400 grams or 1.4 kilograms
3. $NH_4^+ + OH^-$
4. $H_3O^+ + Cl^-$
5. pH = 2 more acidic than pH = 6
6. yes
7. insoluble
8. CO_2
9. fertilizer manufacture, explosives manufacture

SELF-TEST 5-C

1. gains, gets reduced; loses, gets oxidized
2. oxidizing. reducing
3. reduced. cathode
4. $2Ag + Cu^{2+} \longrightarrow$ No
   $F_2 + Zn \longrightarrow$ Yes
   $Fe^{2+} + Ag^+ \longrightarrow$ Yes
   $Mg^{2+} + 2Fe^{2+} \longrightarrow$ No
5. a reduction
6. an oxidation

MATCHING SET

1. h	5. f	9. c
2. d	6. l	10. e
3. k	7. j	11. g
4. a	8. b	12. i

CHAPTER 6

SELF-TEST 6-A

1. oxygen
2. aluminum
3. Minnesota
4. limestone
5. copper
6. copper, aluminum, magnesium
7. slag
8. reduced
9. cathode
10. magnesium
11. positive ions

SELF-TEST 6-B

1. oxygen
2. lead
3. ammonia
4. potassium, nitrogen, and phosphorus
5. petroleum hydrocarbons
6. sulfuric acid
7. KCl
8. iron compounds

MATCHING SET

1. h	6. d
2. i	7. e
3. b	8. a
4. f	9. c
5. g	

CHAPTER 7

SELF-TEST 7-A

1. organic
2. tetrahedral
3. 13, no
4. false
5. two (a or b or d, and e or f)
6. double bond
7. 2,4-dimethylhexane
8. $-CH_3$

SELF-TEST 7-B

1. four
2. rotation of plane polarized light
3. 32
4. delocalized
5. 12 (6 C, 6 H)
6. 1,2,4-trimethylbenzene
7. false

MATCHING SET

1. e	4. c	7. b

2. g   5. f   8. h
3. a   6. i   9. d

CHAPTER 8

SELF-TEST 8-A

1. c
2. (a) methanol   (e) ethanoic acid or acetic acid
    (b) ethanol   (f) ethylene glycol
    (c) methanoic acid or formic acid (g) acetaldehyde
    (d) 2-butanol   (h) 2-bromopropane
3. (a) fermentation of starch, (b) hydration of ethylene
4. stearic acid, palmitic acid, or oleic acid, among others
5. (a) alcohol, (b) carboxylic acid, (c) aldehyde, (d) ketone

SELF-TEST 8-B

1. $CH_3CH_2CH_2OCCH_2CH_3$ (an ester) $+ H_2O$

    with $\overset{\parallel}{O}$

2. a salt of a long chain fatty acid
3. (a) A fat is solid at room temperature, while an oil is liquid. The fat molecule has
    fewer double bonds. (b) By hydrogenation (adding hydrogen to $C{=}C$ double bonds).
4. false
5. c
6. carbolic

MATCHING SET

1. g   4. i   7. b
2. e   5. a   8. c
3. h   6. d   9. f

CHAPTER 9

SELF-TEST 9-A

1. monomers
2. (a) $H_2C{=}CH$ (b) $HC{=}CH_2$ (c) $F_2C{=}CF_2$
        with CH_3

3. polyester
4. isoprene
5. copolymer
6. polyamide or condensation
7. water
8.
9. condensation
10. thinner, pigment
11. linseed

12. emulsions
13. alcohols, acids

SELF-TEST 9-B

1.

$$-O-\underset{\underset{\displaystyle CH_3}{|}}{\overset{\overset{\displaystyle CH_3}{|}}{Si}}-\left(O-\underset{\underset{\displaystyle CH_3}{|}}{\overset{\overset{\displaystyle CH_3}{|}}{Si}}-\right)_n O-$$

2. ultraviolet light
3. plasticizer
4. O (oxygen); it is a polymer held together by a network of Si—O bonds.
5. false, sometimes HCl
6. ultraviolet
7. plasticizers

MATCHING SET

1. i	5. b	8. f
2. g	6. c	9. e
3. h	7. d	10. j
4. a		

CHAPTER 10

SELF-TEST 10-A

1. carbon, hydrogen, and oxygen
2. monosaccharide
3. glucose, fructose
4. D-glucose
5. D-glucose
6. prostaglandins

SELF-TEST 10-B

1. amino acids
2. essential amino acids
3.
$$-\underset{}{\overset{\overset{\displaystyle H}{|}}{N}}-\underset{\underset{\displaystyle O}{\|}}{C}-$$

4. $R-\underset{\underset{\displaystyle NH_2}{|}}{C}HCOOH$

5.
$$H-\underset{\underset{\displaystyle NH_2}{|}}{\overset{\overset{\displaystyle H}{|}}{C}}-\underset{\underset{\displaystyle}{\|}}{\overset{\overset{\displaystyle O}{\|}}{C}}-\underset{}{\overset{\overset{\displaystyle H}{|}}{N}}-CH_2-C\overset{\displaystyle O}{\underset{\displaystyle OH}{\diagup}}$$

6. (a) sequence of amino acids
   (b) helical structure in which the amino acid chains are coiled
   (c) the way in which the helical sections are themselves folded
   (c) the aggregation of subunits
7. (a) 27, (b) 6

SELF-TEST 10-C

1. catalyst
2. key, lock
3. niacin
4. enzyme
5. enzyme
6. coenzyme
7. niacin
8. active site
9. (a) ribose, (b) deoxyribose
10. phosphoric acid, a sugar (ribose or deoxyribose), and a nitrogenous base
11. double helix
12. true, when it invades a host cell

MATCHING SET

1. d	6. b	10. l
2. g	7. c	11. m
3. e	8. a	12. j
4. i	9. h	13. f
5. k		

CHAPTER 11

SELF-TEST 11-A

1. the sun
2. ATP
3. ADP, phosphate or phosphoric acid, energy
4. free
5. CO_2, H_2O, energy
6. NADPH
7. 3-phosphoglyceric acid
8. glucose

SELF-TEST 11-B

1. hydrolysis
2. lipids
3. emulsifying agents
4. ATP, lactic acid
5. CO_2, H_2O
6. coupled

SELF-TEST 11-C

1. (a) DNA, (b) transfer RNA
2. hydrogen
3. ATP
4. adenine: thymine or uracil
   cytosine: guanine
   guanine: cytosine
   thymine: adenine
   uracil: adenine
5. false
6. false

MATCHING SET

1. i	5. g	9. j	
2. h	6. c	10. d	
3. a	7. e	11. k	
4. f	8. b		

CHAPTER 12

SELF-TEST 12-A

1. observed experimental facts
2. same
3. rate
4. change
5. radioactive
6. food

MATCHING SET

1. f	4. j	7. c	10. e
2. g	5. a	8. d	11. i
3. h	6. b	9. d	

CHAPTER 13

SELF-TEST 13-A

1. dehydration, hydrolysis
2. oxidizing
3. hemoglobin, oxygen
4. false
5. cyanide (CN^-), cytochrome oxidase, oxygen
6. fluoroacetic acid
7. heavy metal poisons, complex
8. true

SELF-TEST 13-B

1. neurotoxins
2. synapses
3. acetylcholine
4. genes, chromosomes
5. false
6. chimney sweeping
7. tetrahydrocannabinol (THC)
8. acetaldehyde

MATCHING SET

1. b	5. j	9. c	
2. f	6. d	10. k	
3. h	7. g	11. a	
4. i	8. e		

CHAPTER 14

SELF-TEST 14-A

1. water
2. chlorination
3. biological oxygen demand
4. oxygen, liter
5. limnology
6. sodium tripolyphosphate

SELF-TEST 14-B

1. chlordan
2. toxic
3. electric
4. calcium, magnesium
5. chlorine, fluorine
6. electric
7. two
8. more pure
9. settling, filtration
10. metal, organics

MATCHING SET

1. f
2. i
3. j
4. b
5. d
6. c
7. l
8. k
9. h
10. a
11. e
12. g

CHAPTER 15

SELF-TEST 15-A

1. parts per million, ten thousand, 920
2. carbon monoxide, hydrocarbons, peroxides, nitrogen oxides, ozone, sulfur dioxide, etc.
3. troposphere
4. aerosols
5. surface area
6. Cottrell precipitator
7. nitrogen dioxide
8. warm, cool
9. sulfur dioxide

SELF-TEST 15-B

1. sulfur dioxide
2. fuels
3. first
4. other elements
5. endothermic
6. oxides
7. false
8. false
9. limited
10. hemoglobin
11. false
12. benzo(α)pyrene (BaP)

SELF-TEST 15-C

1. eight
2. knocking
3. air
4. tetraethyllead
5. lead
6. 93
7. yes
8. false
9. hydrocarbons, carbon monoxide, nitrogen oxides
10. copper, and a number of other metals such as platinum or palladium

MATCHING SET

1. c
2. h
3. n
4. l
5. p
6. d
7. i
8. f
9. o
10. k
11. e
12. q
13. a
14. m
15. j
16. b

CHAPTER 16

SELF-TEST 16-A

1. synergism
2. salting, drying
3. false
4. volatile
5. more
6. sequestrant
7. false
8. flavor enhancer
9. dehydrating
10. free radicals

SELF-TEST 16-B

1. saccharin
2. double bonds and aromatic rings
3. citric and lactic acids
4. carbon dioxide
5. flavor enhancer
6. breaks
7. generally recognized as safe
8. active
9. acid

MATCHING SET (Food)

1. b
2. c
3. d
4. e
5. f
6. a

SELF-TEST 16-C

1. hydrochloric acid
2. pain killer
3. morphine
4. acid, ester
5. fever
6. Alexander Fleming
7. true
8. sulfa drugs, antihistamines

MATCHING SET (Medicine)

1. g
2. f
3. h
4. a
5. i
6. b
7. d
8. c
9. e

SELF-TEST 16-D

1. 4
2. cystine
3. disulfide, ionic
4. ionic
5. colorless
6. flexible
7. adsorbent
8. fatty acids
9. sheep
10. false

MATCHING SET (Beauty Aids)

1. h
2. d
3. f
4. i
5. c
6. g
7. a
8. j
9. e
10. b

SELF-TEST 16-E

1. b
2. polymerization, double

3. nonpolar
4. free radicals
5. c
6. layers
7. entangled
8. true
9. silicone and diester greases with lithium soaps
10. bonds

MATCHING SET (Automotive products)

1.	g	6.	c
2.	a	7.	d
3.	f	8.	e
4.	b	9.	i
5.	j	10.	h

SELF-TEST 16-F

1. photography
2. silver
3. hydroquinone
4. reduced
5. $Ag(S_2O_3)_2^{3-}$
6. somewhat larger
7. reducing
8. sodium thiosulfate
9. (a) blue, (b) green, (c) red
10. red, blue, and green
11. true
12. reduced
13. panchromatic

MATCHING SET (Photography)

1.	i	5.	b	9.	e
2.	d	6.	c	10.	g
3.	f	7.	k	11.	h
4.	a	8.	j		

CHAPTER 17

SELF-TEST 17-A

1. about 22 percent
2. natural gas
3. 42
4. over 50 percent
5. CO, H_2, N_2
6. 33 percent

SELF-TEST 17-B

1. fission
2. critical mass
3. uranium-235
4. fusion
5. containment of reactants at high temperature
6. tritium
7. plutonium-239, uranium-233

8. beta
9. intensity

MATCHING SET

1.	f	5.	n	9.	g	13.	k
2.	l	6.	c	10.	a	14.	m
3.	b	7.	d	11.	p	15.	o
4.	j	8.	e	12.	h	16.	i

INDEX

Note: Numbers in *italic* indicate a figure; d following a page number indicates a definition; s indicates a molecular structure; and t indicates a table.

A

Abietic acid, 456s
Absorption by particulates, 383
Acetamide, 174s
  use of, 174t
Acetaminophen (Tylenol), 432
Acetic acid, 114s, 174s, 194s, 215
  in photography, 470
  use of, 174t
Acetone, 174s
  in nail polish remover, 457
  use of, 174t
Acetylcholine, 324s, 326
  local anesthetics and, *431*
Acetylene, *84*s
  bonding in, *67*
Acetylsalicylic acid. See *Aspirin.*
Acid(s), 110d
  amino. See *Amino acids.*
  arachidonic, 241s
  cholic, 423
  fatty, 240–241
  in foods, 426t
  linoleic, 241s
  linolenic, 241s
  nucleic, 256–261
  organic, 193
  para-aminobenzoic, 435s
  rain, 392
  rebound, 429
  strength of, 114t
  thioglycolic, 445s
Acidulant, 422d
Acrilan, 224t
Acrylic latex paints, 214
Acrylic polymer, 224t
Acrylonitrile, 209t, 214
Activated sludge, 364
Activation energy, *251*d
Active sites, on enzymes, 252d
Activity series, 128
Addition polymers, 208, 212–215
Addition reactions, 165
Additive primary colors, 474
Additives, for polymers, 225–229
  gasoline, 458–459
Adenine, 257s
Adenosine diphosphate (ADP), 265s
Adenosine monophosphate (AMP), 265s
Adenosine triphosphate (ATP), 264, 265s, 425t
  Embden-Meyerhof pathway and, 276
  glucose metabolism and, 279
  Krebs cycle and, 277
  photosynthesis and, 268
  protein synthesis and, 285
Adipic acid, 218s
ADP. See *Adenosine diphosphate.*
Adrenalin. See *Epinephrine.*

Adsorption, by particulates, 383
Aeration, water purification by, 345
Aerobic, 275d
Aerosols, air pollution and, 382
African sleeping sickness, 434
Afterburner pollution control, 408
Aftershave lotions, 450
Agriculture, 149d
  land used by, *151*
Air, 298
  composition of, 92
Air-fuel ratios, in autos, 406
Air particulates, removal of, 385
Air pollution, 297, 375–411
  automobiles and, 404, 405t
  carbon dioxide, 401
  carbon monoxide, 396
  chlorinated hydrocarbons, 399
  diseases caused by, 377t
  dispersal by chimneys, 393
  nitrogen oxides, 375, 388, 389, 394
  ozone, 399
  smog, 386–389
  sulfur dioxide, 91, 375, 389, 390–394, 392t
  trends in, 378
Alanine, *170*
Alcohols, 186, 187t
  as disinfectants, 450
  boiling points of, 191
  distillation of, 6
  functional group, 174s
  hydrogen bonding in, 190, *192*
  in antiicing agents, 459
  in beverages, 189t
  in carbohydrates, 234
  polyhydric, 423
  propyl, 190s
  synergistic action of, 442
  synthesis with, 192
  toxicity of, *340*s
  types of, 187t
Aldehyde, functional group, 174s
  in carbohydrates, 234
Aldrin, 355s
Aldrin, Edwin E., Jr., 223
Algae, blooms, *304*
  solar conversion of water into H_2 and O_2, 505
Aliphatic hydrocarbon, 167d
Alkali metals, electron distribution of, 57
Alkalies, in foods, 426t
Alkaloid, 329d, 429d
Alkanes, 165d
Alkenes, 165d
Alkyd paint, 216
Alkynes, 165d
Allergens, 438
Allied Chemical Corp., 427
Allura Red AC, 427

TOP 25 CHEMICALS IN THE UNITED STATES, 1978

Rank	Name	Formula	Production (billions of pounds)	How Made	End Users
1.	Sulfuric acid	H_2SO_4	79.19	Burning sulfur to SO_2, oxidation of SO_2 to SO_3, reaction with water. Also recovered from metal smelting.	Fertilizers 65%, petroleum refining, manufacture of metals and chemicals.
2.	Calcium oxide (quicklime)	CaO	38.78	Heating limestone ($CaCO_3$).	Steel production 42%, water-treating, refractions, pulp and paper.
3.	Oxygen	O_2	35.45	Separated from liquid air.	Steel production 65%, metal fabricating and chemical processing.
4.	Ammonia	NH_3	33.89	Catalytic reaction of nitrogen, air, and hydrogen.	Fertilizers 75%, plastics, fibers and resins.
5.	Nitrogen	N_2	28.41	Separated from liquid air.	Blanketing atmospheres for metals, electronics, etc., freezing agent for foods.
6.	Ethylene	C_2H_4	28.13	Cracking hydrocarbons from oil and natural gas.	Plastics 65%, antifreeze production, fibers and solvents.
7.	Chlorine	Cl_2	21.92	Electrolysis of NaCl, recovery from HCl users.	Chemical production 40%, plastics 20%, solvents 15%, pulp and paper 15%.
8.	Sodium hydroxide	NaOH	21.42	Electrolysis of NaCl.	Chemicals 50%, pulp and paper 15%, aluminum 5%, textiles 5%, oil refining 5%.
9.	Phosphoric acid	H_3PO_4	19.13	Sulfuric acid reacted with phosphate rock; elemental phosphorus burned and dissolved in water.	Fertilizers 80%, detergents and water treating compounds 10%.
10.	Nitric acid	HNO_3	16.10	Oxidation of ammonia to nitrogen dioxide and dissolved in water.	Ammonium nitrate and phosphate fertilizers, nitro explosives, plastics, dyes and lacquers.
11.	Sodium carbonate	Na_2CO_3	15.16	Trona ore, made from NaCl and limestone with ammonia.	Glass 55%, chemicals 25%, paper and pulp.
12.	Ammonium nitrate	NH_4NO_3	14.41	Reaction of ammonia and nitric acid.	Explosives, fertilizers, source of laughing gas, matches.
13.	Propylene	C_3H_6	14.37	Cracking oil and oil products.	Plastics 50%, fibers, solvents.
14.	Benzene	C_6H_6	11.13	From oil and coal tar.	Polystyrene 25%, other resins 10%, nylon 20%, rubber.
15.	Urea	CN_2H_4O	10.76	React NH_3 and CO_2 under pressure.	Fertilizers 75%, animal feeds, adhesives and plastics.
16.	Ethylene dichloride	$C_2H_4Cl_2$	10.46	Chlorination of ethylene.	Production of vinyl chloride.
17.	Toluene	C_7H_x	9.25	Catalytic reforming of petroleum hydrocarbons.	High octane gasolines, urethane resins, solvent, intermediate to make TNT, benzaldehyde and benzoic acid.
18.	Ethylbenzene	C_xH_{10}	8.38	Alkylation of benzene.	Production of styrene.
19.	Vinyl chloride	C_2H_3Cl	6.96	Dehydrochlorination of ethylene dichloride.	Polymers 55%, films, coatings and moldings.
20.	Styrene	C_xH_x	6.88	Dehydrogenation of ethylbenzene.	Polymers 80%, rubber, polyesters.
21.	Formaldehyde (37%)	CH_2O	6.43	Oxidation of methyl alcohol.	Adhesives 60%, plastics.
22.	Methyl alcohol	CH_4O	6.36	From natural gas. Methane oxidized to CO and H_2; catalytic conversion to alcohol.	Polymers, adhesives 60%.
23.	Xylene	C_xH_{10}	6.16	From oil and oil cracking products.	Production of terephthalic acid.
24.	Terephthalic acid	$C_xH_6O_4$	5.97	Oxidation of xylene.	Polyester fibers and film 95%.
25.	Hydrochloric acid	HCl	5.55	React sodium chloride with sulfuric acid. Dissolve HCl gas in water.	Production of chlorides, cleaning metals and brick, neutralizing bases, organic synthesis, removing boiler scale, hydrolyzing starch and protein.

TOP 25 CHEMICAL PRODUCERS (U.S.), 1978
(Chemical Sales in Billions of Dollars)*

DuPont (8.08), Dow Chemical (4.98), Monsanto (4.32), Union Carbide (4.20), Exxon (4.03), Celanese (2.45), W. R. Grace (2.35), Shell Oil (1.90), Allied Chemical (1.85), Hercules (1.71), Gulf Oil (1.70), Occidental Petroleum (1.67), American Cyanamid (1.43), Eastman Kodak (1.34), Standard Oil (1.31), Stauffer (1.30), Borden (1.25), Ethyl Corp. (1.24), Rohm and Haas (1.22), Phillips Petroleum (1.11), Atlantic Richfield (1.09), Diamond Shamrock (1.00+), Mobil (1.00), FMC (1.00−), Air Products (0.91).

*Chemical sales are only a fraction of total sales. For example, total sales for Exxon were $60.33 billion, and total sales for DuPont were $10.58 billion.